WEST ORANGE PUBLIC LIBRARY
46 MT. PLEASANT AVENUE
WEST ORANGE, N. J. 07052

Grzimek's Encyclopedia of Mammals

Volume 2

THE GERMAN PUBLISHER
Dr. Bernhard Grzimek,
Frankfurt am Main

THE AUTHORS
Dr. Concepción L. Alados, Almería, Spain
Dr. Rudolf Altevogt, Münster
Dr. Raimund Apfelbach, Tübingen
Dr. Walter Arnold, Seewiesen, Upper Bavaria
Dr. Alison Badrian, Dublin, Ireland
Dr. Noel Badrian, Dublin, Ireland
Dr. Leonid Baskin, Moscow
Monica Borner, Arusha, Tanzania
Dr. Günter Bräuer, Hamburg
Dr. Johannes Bublitz, Kiel
Dr. Anton Bubenik, Thornhill, Canada
Dr. Christiane Buchholtz, Marburg
Dr. Wilfried Bützler, Göttingen
Dr. Fritz Dieterlen, Stuttgart
Dr. Lothar Dittrich, Hanover
Dr. Irenäus Eibl-Eibesfeldt, Seewiesen, Upper Bavaria
Dr. John F. Eisenberg, Gainesville, Florida
Dr. Wolfgang von Engelhardt, Hanover
Reinhild Etter-Ganslosser, Erlangen
Dr. Roger Fons, Banyuls-sur-Mer, France
Dr. Dian Fossey, Ruhengeri, Ruanda
Dr. William L. Franklin, Ames, Iowa
Dr. Udo Ganslosser, Erlangen
Dr. Valerius Geist, Calgary, Canada
Dr. Wolfgang Gewalt, Duisburg
Dr. Jane Goodall, Kigoma, Tanzania
Dr. David R. Gray, Ottawa, Canada
Dr. Colin P. Groves, Canberra, Australia
Dr. Bernhard Grzimek, Frankfurt am Main
Dr. Gerhard Haas, Wuppertal
Dr. Ursula Heckner, Heidelberg
Dr. Hubert Hendrichs, Bielefeld
Dr. Wolf Herre, Kiel
Dr. Hendrik N. Hoeck, Constanz
Ruedi Hess, Zurich
Dr. Dietrich von Holst, Bayreuth
Dr. Jan A. R. A. M. van Hooff, Utrecht, Netherlands
Dr. Klaus Immelmann, Bielefeld
Dr. Kosei Izawa, Sendai, Japan
Marvin L. Jones, San Diego, California
Dr. Milan Klima, Frankfurt am Main
Dr. Hans Klingel, Brunswick
Dr. Rainer Knussmann, Hamburg
Dr. Kurt Kolar, Vienna
Dr. Helmut Kraft, Munich
Dr. Richard Kraft, Munich
Dr. Franz Krapp, Bonn
Dr. Dieter Kruska, Kiel
Dr. Hans-Jürg Kuhn, Göttingen

Dr. Erwin Kulzer, Tübingen
Dr. Fred Kurt, Zurich
Dr. Ernst M. Lang, Mattweid, Switzerland
Dr. Paul Leyhausen, Windeck/Sieg
Dr. Kathy MacKinnon, Haddenham, England
Dr. Mark MacNamara, Ardley, New York
Dr. Wolfgang Maier, Frankfurt am Main
Dr. Patricia Major, Heidelberg
Dr. Heinrich Mendelssohn, Tel Aviv, Israel
Dr. Heinz F. Moeller, Heidelberg
Dr. Walburga Moeller, Heidelberg
Dr. Ewald Müller, Tübingen
Dr. Cornelis Naaktgeboren, Hoorn, Netherlands
Dr. Jochen Niethammer, Bonn
Dr. Bernhard Nievergelt, Zurich
Dr. Ivo Poglayen-Neuwall, Tucson, Arizona
Dr. A. George Pook, Cheltenham, England
Dr. Holger Preuschoft, Bochum
Hans Psenner, Innsbruck
Dr. Urs Rahm, Basle
Dr. Galen B. Rathbun, San Simeon, California
Dr. Josef Reichholf, Munich
Dr. Clifford G. Rice, New York, New York
Dr. Ingo Rieger, Zurich
Dr. Klaus Robin, Berne
Dr. Manfred Röhrs, Hanover
Dr. Hans Heinrich Sambraus,
  Freising-Weihenstephan, Upper Bavaria
Dr. Cornelia Schäfer-Witt, Cassel
Dr. George B. Schaller, New York, New York
Dr. Rudolf Schenkel, Basle
Dr. Harald Schliemann, Hamburg
Dr. Robert Schloeth, Zernez, Switzerland
Dr. Christian Schmidt, Zurich
Dr. Uwe Schmidt, Bonn
Dr. Eberhard Schneider, Göttingen
Ingrid Schneider, Leutershausen
Dr. Hiroaki Soma, Tokyo
Dr. Adelheid Stahnke, Tübingen
Dr. Gerhard Storch, Frankfurt am Main
Dr. Erich Thenius, Vienna
Eberhard Trumler, Birken-Honigsessen, Sieg
Dr. Raul Valdez, Las Cruces, New Mexico
Dr. Christian Vogel, Göttingen
Dr. Jiří Volf, Prague
Dr. Fritz Walther, Wienau, Westerwald
Dr. Christian Welker, Cassel
Dr. Paul Winkler, Göttingen
Jürgen Wolters, Bielefeld
Dr. Charles Woods, Gainesville, Florida
Dr. Victor Zhiwotschenko, Moscow
Dr. Erik Zimen, Dietersburg, Lower Bavaria
Dr. Waltraud Zimmermann, Cologne
Robert Zingg, Zurich

Grzimek's Encyclopedia of Mammals

VOLUME 2

McGRAW-HILL PUBLISHING COMPANY

VOLUME 1

Mammals: Introduction
Monotremata (Egg-laying mammals)
Marsupialia (Opossums, Marsupial mice, Bandicoots, Koalas, Wombats, Kangaroos)
Insectivora (Solenodons, Tenrecs, Hedgehogs, Golden moles, Shrews, Moles)
Macroscelidea (Elephant shrews)
Chiroptera (Bats)
Dermoptera (Flying lemurs)

VOLUME 2

Scandentia (Tree shrews)
Primates (Prosimians, Monkeys, Apes, Humans)
Xenarthra (Anteaters, Sloths, Armadillos)
Pholidota (Pangolins)

VOLUME 3

Rodentia (Squirrels, Beavers, Mice, Dormice, Porcupines)
Carnivora (Bears, Pandas, Viverrids, Hyenas, Cats)

VOLUME 4

Carnivora (Cats, Dogs, Seals, Sea lions)
Lagomorpha (Rabbits, Hares, Pikas)
Cetacea (Whales, Dolphins)
Tubulidentata (Aardvarks)
Proboscidea (Elephants)
Sirenia (Sea cows)
Hyracoidea (Hyraxes)
Perissodactyla (Horses, Tapirs, Rhinoceroses)

VOLUME 5

Artiodactyla (Pigs, Peccaries, Hippopotamuses, Camels, Deer, Giraffes, Cattle, Goats, Sheep, Gazelles, Antelopes, Reindeer)

Grzimek's Encyclopedia of Mammals

Volume 2

McGRAW-HILL PUBLISHING COMPANY

New York   St. Louis   San Francisco

Auckland   Hamburg   London   Mexico
Montreal   New Delhi   Oklahoma City   San Juan
Singapore   Sydney   Toronto

English Language Edition

Sybil P. Parker, Editor

The Language Center, Inc., South Orange, New Jersey
Translations of Volumes 1, 3, and 4.

The Language Service, Inc., Hastings-on-Hudson, New York
Volume 2 translated by Ellen and Ernst van Haagen.
Volume 5 translated by Sally E. Robertson and William P. Keasbey.

German Language Edition

Wolf Keienburg, Editor in chief
Dr. Dietrich Heinemann and Dr. Siegfried Schmitz, Editors
Dr. Siegfried Schmitz, Captions of illustrations
Dr. Dietrich Heinemann (Fossey, Jones),
Dr. Ingrid Horn (Badrian, Goodall, MacKinnon),
and Barbara Leyhausen (Izawa, Pook),
German translations
Dr. Siegfried Schmitz (zoological data)
and Prof. Dr. Milan Klima (pictorial synopses), Research
Silvestris Fotoservice, Kastl/Obb., Photograph library
Dr. Linde Lang, Final editing and index
Claus-J. Grube, Graphic design
Bernd Walser, Production

Library of Congress Cataloging-in-Publication Data

Grzimek's encyclopedia of mammals.
Translation of Grzimeks Enzyklopädie Säugetiere.
Bibliography: p.
Includes index.
1. Mammals. I. Grzimek, Bernhard. II. Title:
Encyclopedia of mammals.
QL201.G7913 1989   599   89-12542
ISBN 0-07-909508-9 (set)

Copyright © 1990 by McGraw-Hill, Inc. All rights reserved.
Except as permitted under the United States Copyright Act of 1976,
no part of this publication may be reproduced or distributed
in any form or by any means, or stored in a database or
retrieval system, without the prior written permission of the publisher.

The original German language edition of this book
was published as *Grzimeks Enzyklopädie*, copyright © 1988,
Kindler Verlag GmbH, München, West Germany.
Typesetting and Printing: Appl, Wemding

ISBN 0-07-909508-9

CONTENTS

Tree shrews
- General information — 2
- Phylogeny — 4 — Erich Thenius
- Modern tree shrews — 5 — Dietrich von Holst

Primates
- General information — 14
- Introduction — 16 — Hans Jürg Kuhn
- Phylogeny — 28 — Erich Thenius

Prosimians
- General information — 32
- Introduction — 34 — Kurt Kolar
- Phylogeny — 39 — Erich Thenius
- Lemurs — 42 — Kurt Kolar
- Lorises and galagos — 77 — Ewald Müller
- Tarsiers — 97 — Kurt Kolar

Simians
- General information — 106
- Introduction — 108 — Christian Welker
- Phylogeny — 112 — Erich Thenius

New World primates
- General information — 120
- New World monkeys — 122 — Christian Welker and Cornelia Schäfer-Witt
- Goeldi's monkey — 178 — A. George Pook
- Marmosets and tamarins — 183 — Jurgen Wolters and Klaus Immelmann

Old World primates
- General information — 206
- Macaques and allies — 208 — Jan A.R.A.M. van Hooff
- "Culture of the red-faced macaques" — 286 — Kosei Izawa
- Langurs and colobi — 296 — Christian Vogel and Paul Winkler

Gibbons
- General information — 326
- Modern species — 328 — Holger Preuschoft

Great apes
- General information — 358

Introduction	360	Holger Preuschoft
Orangutans	401	Kathy MacKinnon
Orangutan rehabilitation	419	Monica Borner
Breeding and life span of orangutans in captivity	422	Marvin Jones
The gorilla	424	Bernhard Grzimek
Our first field studies	441	George B. Schaller
New observations of gorillas in the wild	449	Dian Fossey
Chimpanzees	463	Jane Goodall
Chimpanzee rehabilitation	482	Bernhard Grzimek
The bonobo or pygmy chimpanzee	486	Alison and Noel Badrian

Humans

General information	488	
Human phylogeny	490	Gunter Brauer
Modern humans: physical form	520	Rainer Knussmann
Biological roots of human behavior	538	Irenäus Eibl-Eibesfeldt

New World edentates

General information	576	
Introduction	578	Walburga Moeller
Phylogeny	579	Erich Thenius
Modern xenarthrans	583	Walburga Moeller

Pangolins

General information	628	
Phylogeny	630	Erich Thenius
Modern pholidotes	630	Urs Rahm

Appendix

References	642	
Authors of this volume	642	
Illustration credits	643	
Index	645	

TREE SHREWS

Category
ORDER

Classification: An order of mammals comprising one family (Tupaiidae) that is divided into two subfamilies with six genera.

The classification of the tree shrews, or tupayas, was long disputed. Whereas formerly they were regarded first as insectivores and later as primitive lemurs, and therefore placed at the beginning of the Order PRIMATES, today they are accorded the rank of a distinct order, on a par with the Orders INSECTIVORA and PRIMATES. This long-standing uncertainty is accounted for primarily by the fact that too little was known about these small, inconspicuous creatures in their confined habitats. To this day, only a few species of tree shrews have been closely studied in the wild.

Family Tupaiidae
Subfamily Tupaiinae (bushy-tailed tree shrews)
Subfamily Ptilocercinae (pentail tree shrews)

BUSHY-TAILED TREE SHREWS (5 genera of probably 17 species)
Body length: 5–8.4 in.; 12.5–21 cm
Tail length: 5.2–7 in.; 13–17.5 cm
Weight: 1.7–8.9 oz; 50–250 g
Distinguishing features: Squirrel-like; snout more or less long and pointed in various species; primate-like external ears; tail is more or less bushy; makes characteristic tail movement when excited; brain is comparatively large; Jacobson's organ; caecum; predominant retinal cones; able to distinguish colors; according to species, females with one to three pairs of nipples.
Reproduction: Gestation, where known, 45–55 days; one to four, usually two young per birth; birth weight not known; in captivity, reproduction throughout the year; in nature, young also at any season, but climatically conditioned periods of reproduction possible.
Life cycle: Time of weaning not yet known; sexually mature at about two months; full growth at about three months; in captivity, life expectancy about 9–10 years; in nature, not yet known.
Food: Fruit, arthropods, small vertebrates.
Habit and habitat: All species active during the day; according to species, predominantly arboreal to almost exclusively terrestrial; in tropical forests at elevations up to 3200–4800 ft (1000–1500 m); pairs occupy territories, identified by marking with neck gland secretions or urine; Energetic defense of territory against conspecific intruders; extent of territory where known 5100–81,600 ft^2 (500–8000 m^2), according to habitat; at sexual maturity, young are driven off by parents.

PENTAIL TREE SHREWS (one genus, comprising one species)
Body length: about 5.8 in.; 14.5 cm
Tail length: about 7 in.; 17.5 cm

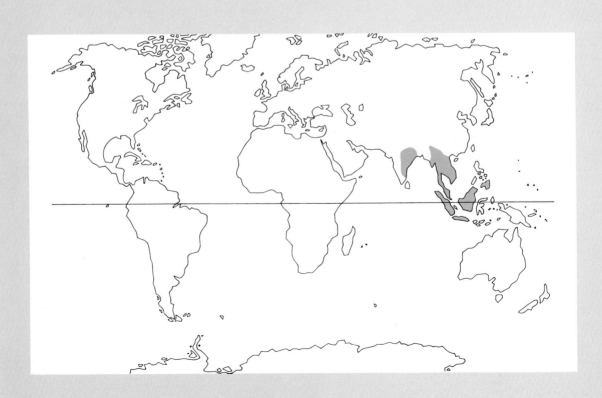

Scandentia
Tupayes, Tupaidés — FRENCH
Spitzhörnchen — GERMAN

Weight: About 1.7 oz; 50 g
Distinguishing features. Also squirrel-like, with pointed snout and long vibrissae; long, naked tail with whitish, fringe-like tuft of hair at tip; brain more primitive than in subfamily Tupaiinae; retinal rods only; females with two pairs of nipples.

Reproduction. Probably one to four young per birth; reproductive habits otherwise not known.
Life cycle. No reliable data as yet available.
Food. Fruit, arthropods, small vertebrates.
Habit and habitat. Sole nocturnal representative of the order; arboreal in tropical forests; lives in pairs; which possibly live with young until sexual maturity; presumably in territories which are aggressively defended against conspecific intruders and marked with glandular secretions or urine; habits otherwise still almost unknown.

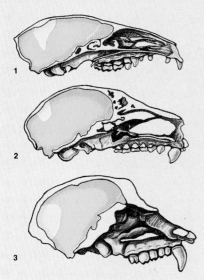

Sensory brain centers. Whereas in the tree shrews the sense of smell still plays an important part, among the primates the eye becomes increasingly dominant because of the transition to almost exclusively diurnal activity, as an aid to acrobatics in the treetops, and for optimum guidance of increasingly exacting manual tasks. The olfactory cranium is shortened, the optic parts enlarged.
The olfactory lobe of the brain (violet) recedes, the visual center (yellow) becomes more developed. 1 = tree shrew, 2 = lemur, 3 = macaque.

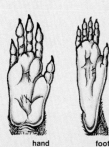

Tree shrews' hands and feet are reminiscent rather of the paws of some insectivores than of those of the primates. They have narrow digits with pointed claws, the thumbs or great toes not much set off from the others. They are adapted to fast running on the ground or in the trees but cannot grasp, and they lack any pronounced tactile sensitivity.

hand   foot

Comparison of teeth. Aside from some special adaptations, the evolution of primate teeth shows an increasing tendency to simplification of form, flattening of the crown, and reduction in the number of teeth. This is due in part to the relative shortening of the jaws and a transition from predominantly animal to predominantly vegetable food. The dentition of the tree shrews (*Tupaia*, 1), which resembles that of the insectivores, comprises 38 teeth. The dentition of the prosimians is likewise quite multiform, but as a rule the number of teeth is reduced to 36 (lemur, 2). As in the tree shrews, the lower incisors of the lemurs form a comb for grooming. In the New World primates (howler monkey, 3) and in primates of the Old World (macaque, 4), an increasing simplification of dental anatomy and a further reduction down to 32 teeth may be noted.

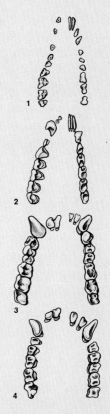

Tree Shrews

Phylogeny
by Erich Thenius

The phylogenetic position of the tree shrews (Tupaiidae) is by no means a matter of unanimous opinion. Are they insectivores, or primates, or neither? May they perhaps be regarded as latter-day survivors of the ancestors of the primates, or are they representatives of primitive placental mammals (having, that is, a placenta, the vascular organ uniting the fetus to the uterus)? Such are the questions to be answered in deciding the position of the tree shrews in phylogeny. At first regarded as insectivores, along with hedgehogs, shrewmice, and moles, they were classified with the elephant shrews in an insectivores suborder Menotyphla by the celebrated phylogenist Ernst Haeckel in 1866 upon grounds of the possession of a caecum, or blind gut, in contradistinction to the other insectivores (Lipotyphla).

Intimations that the tree shrews might be more closely akin to the primates go back to W.K. Gregory in the year 1910, and were confirmed by the British anatomist W. Le Gros Clark on the basis of anatomical studies in the sense that he regarded the tree shrews as an early collateral line of the primates. But it remained for the well-known American zoologist and paleontologist G.G. Simpson, in his classic work on systematics of the mammals (1945), to view the Tupaiidae as simians of lemuroid stock. They have since been usually classed as subprimates or as an independent group (Tupaiiformes) of simians, and this move was adopted in Grzimek's *Tierleben* (1968). Of course, there were those who hold that certain similarities and shared traits between tree shrews and primates, if not simply common primitive features, do not reflect relationships of actual kinship, but parallel evolution (that is, evolution in the same direction) associated with their arboreal habit as diurnal animals. This is confirmed in part by substantial differences in reproductive behav-

ior and early development. In 1969, therefore, I classed the tree shrews in a separate order of their own, Tupaioidea. But that designation, as was noted by the English zoologist P.M. Butler in 1972, had been anticipated by J.A. Wagner's use of the name Scandentia in 1855. The Scandentia split off from the other placental mammals at an early branching. This explains why the well-known American vertebrate paleontologist A.S. Romer in his *Vertebrate Paleontology* (1966) ranked them as "Proteutheria," thus assigning them a low (basal) position among the Placentalia (=Eutheria). For many scientists, however, they are the sister group of the Primates. That is to say, the two may be traced to a common stock. According to recent biochemical findings, however, the flying lemurs (Dermoptera) are the sister group of the primates, the two being contrasted as such with the Scandentia, a view that would seem to be borne out by W.K. Gregory's "archonta" theory, though in a narrower sense (that is, not including the Chiroptera).

(Right) The scaly tail, furred only at the tip, more than anything else distinguishes the pentail tupaya, sole representative of its subfamily, from the other tree shrews. Another distinguishing trait is its nocturnal habit. (Bottom) Tree shrews will not lie in a heap unless they are on good terms.

The fossil history of the Scandentia has so far been documented only by Paleotupaia of the Late Tertiary in southern Asia. Among present-day tree shrews, two subfamilies may be distinguished: the Ptilocercinae (one genus, *Ptilocercus*) and Tupaiinae (including remaining genera).

Modern Tree Shrews
by Dietrich von Holst

The tree shrews, or tupayas, are a fairly uniform-appearing group of small olive-gray to rust brown mammals that are widespread in the forests of South and Southeast Asia. They comprise a single family, Tupaiidae, having two subfamilies: the bushy-tailed tupayas (Tupaiinae), with 5 genera and some 17 species, active entirely in the daytime; and the nocturnal pentail tupayas (Ptilocercinae), with a single species, *Ptilocercus lowii*.

The first account of a tupaya is due to the English physician William Ellis, who accompanied Captain James Cook in 1780 on his third voyage, to the Malay Archipelago. In his journal we find a sketch of an animal that was previously which was unknown, collected in the vicinity of Saigon. Although the careful drawing of an insectivore-like jaw clearly shows that it cannot have been a squirrel or other rodent, Ellis called his find a "tree shrew," and in the English-speaking world that name has been retained to this day for all representatives of the order.

As a matter of fact, in appearance and behavior the tupayas do resemble squirrels with a long, pointed muzzle, as indeed the German name (Spitzhörnchen) suggests.

Although a few tupayas entered museum collections in ensuing years, it was not until 1821 that the English naturalist Sir Thomas Stamford Raffles classified the tupayas as insectivores, and coined the scientific generic name *Tupaia* from the Malay word for squirrel (tupai). In the next thirty years, the other genera were discovered and assigned to the insectivores. Scientific interest in these rather inconspicuous animals grew abruptly when, in 1872, the celebrated English zoologist Thomas Henry Huxley first pointed out certain astonishing agreements between the physical features of tree shrews and those of monkeys. A more highly developed brain than that of the insectivores, a ring of bone around the large, laterally placed eyes, an ape-like external ear, and many other physical traits prompted the noted American zoologist George Gaylord Simpson in 1945 to classify the tree shrews as prosimians.

Later on, however, this attribution came to be doubted on the basis of new findings concerning physical, physiological, biochemical, and behavioral traits of the tupayas, so that today are ranked

The remote tropical forests of South and Southeast Asia are the home of the tree shrews, or tupayas. No wonder that the biology and behavior of these animals long remained hidden from science and did not become accessible to close study until recent decades.

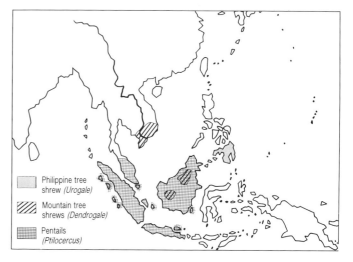

Philippine tree shrew (*Urogale*)

Mountain tree shrews (*Dendrogale*)

Pentails (*Ptilocercus*)

as a distinct order, the Scandentia, on an equal level with the Insectivora and the Primates. This systematization aside, however, there is no group of mammals living today that more closely resembles the form from which the monkeys arose over 60 million years ago. Doubtless this resemblance is due in part to parallel (convergent) adaptations to the arboreal habit of the two groups. Most of all, however, the reason for the resemblance may be that the tupayas as well as the monkeys have retained numerous primitive traits of their common ancestors. Hence the tupayas probably represent a form not so very different from the ancestors of all placental mammals, that is, those mammals whose offspring are nourished by way of a maternal organ, the placenta.

While the knowledge of their behavior and physiology would thus provide an essential foundation for an understanding of the more highly developed mammals, we as yet know surprisingly little about the group. Almost nothing is known about their behavior in nature and only a few species have been successfully kept and studied in captivity to date.

As previously mentioned, tupayas occur only in a fairly limited region of Asia, from India in the west to the Philippines in the east, and from South China in the north to Sumatra in the south. While the occurrence of the Indian tupayas (genus *Anathana*) is limited to India and that to the Philippine tupayas (genus *Urogale*) to the island of Mindanao in the Philippines, the other genera are distributed over wide regions of the Malay Archipelago. The greatest abundance of species prevails on Borneo, where no less than 10 of the 18 species of tree shrews are to be found.

In their regions of distribution, tupayas appear to be among the most abundant mammals; they colonize tropical rain forests as well as areas with tree plantings and park lands.

With the exception of the exclusively arboreal, nocturnal pentail tupayas and the diurnal dwarf tupaya *Tupaia minor*, most species spend their time chiefly on the ground where they find theifood of their main diet (insects, worms, and other invertebrates); but even warm-blooded animals the size of a mouse or a sparrow can be killed and eaten by tree shrews without difficulty. Besides animal matter, all species accept fruit and other vegetable fare as well.

To date, there have been field studies of only a few *Tupaia* species *(T. glis, T. belangeri, T. minor)* and of *Lyonogale tana*. Evidently, all these species live in pairs – with their young if any – in fixed territories that are vigorously defended against conspecifics. The size of these territories varies quite widely, even within a species, according to habitat – from less than 5100 ft^2 (500 m^2) on the rich fruit and coconut plantations, up to more than 81,600 ft^2 (8000 m^2) in virgin forest.

Territorial Demarcation and Defense

As we know in particular from the work of Japanese researchers, the Kawamichis, individual territories overlap very little, if at all. They are identified by scent markings that are placed by the males primarily at the boundary lines of neighbors

Tree shrews exhibit pronounced territorial behavior. They identify or "mark" the boundaries of their home territories diligently with odoriferous secretions, either from a neck gland (left) or from an abdominal gland anteroir to the scrotum (right).

and which are continually renewed. An especially important feature seems to be the secretion from a glandular area located on the neck of the males. In captivity they disperse their secretions into their surroundings in a highly characteristic behavior repeated some thousands of times a day. In such 'neck marking,' as it is called, the males press the ventral surface of the neck firmly against the ground or a tree branch and sweep the glandular area to and fro over the substrate once or several times, creating an oily scent mark smelling strongly of musk.

As studies by our group have shown, tree shrews are able to recognize the species, sex, sexual condition, and identity of the individual responsible for the mark.

Whereas scent marks of conspecific strangers will immediately trigger extreme alertness and 'overmarking' in response, the secretion of a known conspecific of higher rank always has a deterrent effect and may, therefore, serve for territorial demarcation in nature. Besides the neck gland secretion, tupayas in nature, as well as in the laboratory, will mark the same objects with urine, which contains the same information as to species, sex, and individual identity as the neck secretion. According to studies by my associate Falko von Stralendorff, it is even probable that the biologically important substances are not formed in the neck gland at all, but rather derive from the urine and become mingled with the oily secretion of the neck gland only contact with objects already marked with urine. Probably the high lipid content of the neck gland secretion (over 99%) is necessary for any biologically significant persistence of the scent marks under the climatic conditions in nature. According to our data, while a urine marking will have lost its effect on conspecifics after a matter of minutes, mingling with the oily neck gland secretion will cause it to last for hours or days.

Marking of boundaries and fighting over territories are among the most conspicuous behavior patterns observed in nature. Intruders are almost invariably driven off with success. But even in fairly large enclosures, conspecific strangers of the same sex are instantly attacked by males or females. The battles are extraordinarily violent, but anything more than superficial bites and scratches is very rare. Nevertheless, as a rule the intruder will have been conquered within seconds or minutes and will try to escape from the enclosure. Even where flight is not possible, the victor loses all interest in the vanquished within a very short time. The other will then withdraw to as sheltered a spot as possible, and will leave it only for hurried eating and drinking.

Despite adequate nourishment, the beaten individual will proceed to lose weight rapidly, becoming comatose within a few days and then dies unless removed from the victor's enclosure in time. Death is not caused by injuries of resulting infection, but rather by the constant presence of the victor. For if an individual is separated from the victor by an opaque partition, so that the victor is no longer seen, it will recover from the conflict almost as quickly as the victor. If the two are separated by a screen partition, so that the loser is safe from attack but constantly sees the threatening victor, death consistently results. Thus a defeated tupaya can die of its lasting fear.

All this likewise applies to the young. Tupayas leave the nest at about four weeks, and then form a close family group with their parents and sometimes also with older siblings. They all sleep in a nest togeth-

(Left) The dwarfs of the primeval forest, diminutive as they are, sometimes behave with much aggression. Its mouth opened menacingly, this tree shrew is showing its teeth. (Below) Perhaps the best-known representative of the bushy-tailed tree shrews is the species *Tupaia glis*.

er at night and also often rest together during the day. The young reach sexual maturity at about six to eight weeks and are full-grown at about three months. Even after sexual maturity, there is no strife within the family at first, but the parents increasingly detach themselves from their young; they are now seldom observed to rest close together. Without apparent cause, fights break out, after a few months between father and son or mother and daughter, the young usually being defeated. If emigration is not possible after such a defeat they will die within a few weeks, like defeated conspecific strangers. These laboratory findings are in accord with observations in the wild. Thus, all investigators have found tupayas in nature in pairs only, or in small groups consisting as a rule of one sexually mature pair and two or three young – family groups, in all probability.

This is true not only of various species of bushy-tailed tupayas, but also of pentail tupayas, according to observations by the Malayan zoologist Lim Boo Liat. In fact, the Kawamichis observed that young were driven out of the territory by their parents. The future fate of these young is not as yet known. They would certainly have to find unclaimed territories of their own, or failing this, gain one by conquest. Accordingly, in the densely settled fruit and coconut plantations, along with established territorial occupants, up to 80% of the juvenile tupayas are found wandering through the area and repeatedly being attacked and driven off by the territorials. These battles are noticeable enough to have been remarked upon in the very first behavioral description, by the naturalist T. Cantor (1846). Conceivably these struggles and the resulting stress eliminate those individuals who fail to find an unoccupied territory in nature. This is particularly significant since the tree shrews are not known to have any special enemies. This is made all the more likely by the fact that tupayas, even in nature, attain an age of several years (as much as ten years in captivity) and are able to reproduce throughout life. The number of possible descendants is, therefore, greater by a multiple than required to maintain the population.

As has been mentioned, tree shrews live in pairs in nature. Surprisingly enough, however, the formation of such a pair in captivity is no simple matter. If, for example, a grown male and female are put together in an enclosure, in about 20% of all cases this will lead immediately to such violent combat that one of the animals will die unless removed in time. In about 60% of all cases – especially in larger enclosures – a more or less amicable pairing will take place. The animals avoid confrontations

Tree shrews play an important part in present-day stress research. The bristling tail is a sure sign of intense excitement or severe stress (Below). Definite kidney changes are also observable: under conditions of stress - after two days confrontation with a high-ranking conspecific - the capillaries are dilated and contain few red blood corpuscles or none at all (Upper right). A kidney under normal conditions is shown for comparison (Lower right).

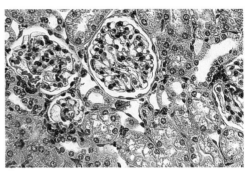
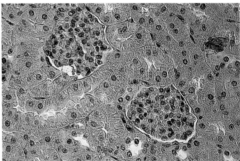

by giving each other a wide berth; as a result, quarrels are few or non-existent.

Nevertheless, physiological data give evidence of constant stress. According to our studies, levels of the stress hormones cortisol and corticosterone are constantly more or less elevated in both individuals, depending on degree of social stress, and both have heart rates accelerated by as much as 50% day and night. In the long term, such uncongenial couples develop high blood pressure and may actually die of cardiovascular failure. Successful matings may sometimes occur, but the young are never brought up; instead, the mother or the father will enter the nest with the young after the birth and devour them like any other prey. The reason for this, according to our findings, is that a female thus stressed by the presence of her partner will fail to provide the young after birth with the scents that ordinarily protect them from conspecifics. This difficulty of finding compatible pairs is a principal cause of failure of nearly all attempts to breed tupayas in any numbers.

Only in about 20% of all cases, according to our findings, the meeting of a male and a female will result from the first moment in a definite pairing bond. It would seem to be a matter of "love at first sight", even though the smell may be more of a factor than the appearance. The two will nuzzle each other all over, keep marking each other, and lick each other affectionately. One will present the muzzle to the other, with saliva dripping from it, to be lapped up by the partner. This affectionate licking, or "kissing," as the English anthropologist Robert Martin terms it, is a characteristic of all harmonious pairs, and recurs daily for up to an hour at a time. Furthermore, the members of such a pair nearly always rest together the day, and always spend the night in the same sleeping box. Matings may occur on the first day but are not a prerequisite for successful bonding; successful matings may not occur until after some days. About 45 days after mating, the females will have one to four young (usually two) which are always raised successfully. Immediately after the birth of the young, the females are once more ready to conceive and are covered by the male; consequently, a harmonious pair will produce young every 45 days for years.

Surprisingly enough, such a harmonious pair relationship has striking physiological repercussions, precisely opposite to those of members of incompatible pairs. Production of the sex hormones is increased by up to 100%, the heart rate of the partners is lowered by as much as 20% compared to conditions before pairing, and the animals respond to stresses of all sorts with substantially less release of stress hormones, so that their state of health is much improved over all. Curiously enough, such a harmonious pairing relationship is based on purely individual inclination; thus, for example, a female rejected and very violently combatted by one male may be accepted immediately as a partner by another male, and vice versa.

(Left) Fighting tree shrews charge each other at lightning speed, with bristling tails. (Right) The young of pairs that are not compatible are not raised, but devoured by their parents like any ordinary prey.

Tree shrews (Scandentia)

Nomenclature English common name Scientific name French German	Approximate Size Body length Tail length Weight	Distinguishing Freatures	Reproduction Gestation period Young per birth Weight at birth
Tree shrews *Tupaia glis, T. Belangeri, T. longipes,* *T. montana, T. nicobaria, T. picta,* *T. palawanensis, T. splendidula, T. javanica,* *T. gracilis, T. minor* Toupaïes Tupajas; Spitzhörnchen	5–7.4 in; 12.5–18.5 cm 5.8–7 in; 14.5–17.5 cm 1.7–6.4 oz; 50–180 g	Bushy tail; dark olive-green to gray-brown upper parts with beige shoulder stripes; beige under parts; depending on species, one to three pairs of nipples	45–55 days 1–4 (usually 2) 0.2–0.4 oz; 6–10 g
Malayan tree shrews *Lyonogale tana, L. dorsalis* Toupaïes de la Malaisie Malaiische Tupajas	8.4 in; 21 cm 6.6 in; 16.5 cm 7.8 oz; 220 g	Bushy tail; dark rust-brown upper parts with black lengthwise stripes on back and beige shoulder stripes; two pairs of nipples	45–55 days 1–4 (usually 2) About 0.4 oz; 10 g
Philippine tree shrew *Urogale everetti* Toupaïes des Philippines Philippinentupaja	8 in; 20 cm 6 in; 15 cm 7.8 oz; 220 g (to 12.5 oz or 350 g in captivity)	Largest species of tree shrews; snout guite elongated; bushy tail; dark brown upper parts with beige shoulder stripes; beige under parts; two pairs of nipples	About 55 days 1–4 (usually 2) About 0.4 oz; 10 g
Indian tree shrew *Anathana ellioti* Toupaïe des Indes Indisches Tupaja	7.6 in; 19 cm 7 in; 17.5 cm 6.4 oz; 180 g	Bushy tail; gray-brown upper parts with beige shoulder stripes; beige under parts; three pairs of nipples	Probably much like *Tupaia*
Smooth-tailed tree shrews *Dendrogale, melanoura, D. murina* Toupaïes des montagnes Bergtupajas	5.2 in; 13 cm 5.2 in; 13 cm 2.1 oz; 60 g	Tail thin and short-haired; rusty red to black-brown upper parts; no shoulder stripes; beige under parts; one pair of nipples	Probably much like *Tupaia*
Pentail tree shrew *Ptilocercus lowii* Ptilocerque Federschwanztupaja	5.8 in; 14.5 cm 7 in; 17.5 cm 1.7 oz; 50 g	Long, naked tail with whitish, plume-like tuft of hair at tip; dark gray-brown upper parts; beige under parts; no shoulder stripes; brain more primitive than in subfamily Tupaiinae, two pairs of nipples	Not known Probably 1–4 (usually 2) About 0.4 oz; 10 g

Thus at the very root of the placental family tree, we have a pair bond based on "personal" inclinations of the individuals to each other – a pattern of behavior that we tend to attribute to ourselves alone.

Rearing of the Young

Surely one of the most unusual peculiarities of the tupayas is the rearing of their young. This was first observed in *Tupaia belangeri* in seaside meadows by the English researcher Robert Martin of the Max Planck Institute of Behavioral Physiology, but since verified in other species such as *Tupaia glis, T. minor, Lyonogale tana,* and *Urogale everetti*. It is likely that this peculiarity is common to all bushytailed tupayas.

A few days to some hours before the birth, the female will make a large nest of leaves, wood chips, or other available materials in a nesting box in which the young will be brought forth. The young arrive naked, blind, and with closed ears. Immediately after birth, the young, each weighing about 0.4 oz (10 g), will take on about 0.2 oz (5 g) of milk from the mother within a few minutes, causing the milk-filled belly to bulge and become tautly distended from the body. The young then crawl to the mother's mouth and take up her saliva in prolonged lickings of greeting. No other contact takes place between the mother and her little ones. The mother does not bite off the umbilical cord, nor does she wash the young; instead, she leaves them alone and merely returns every 48 hours for 5–10 minutes of further suckling. Consistent with this infrequent feeding, tupaya milk has a peculiar composition: it has an uncom-

By lickings of affectionate greeting, harmonious pairs keep announcing their commitment repeatedly.

COMPARISON OF SPECIES

Life Cycle Weaning Sexual maturity Life span	Food	Enemies	Habit and Habitat	Occurrence
About 30 days About 2 months In captivity, 9–10 years (*T. belangeri*); over 12 years (*T. glis*)	Fruits, arthropods, small vertebrates	None known	Tropical forests to elevations of 3200 ft (1000 m) above sea level; more or less terrestrial to arboreal; pairs occupy territories, territory size 5100–81,600 ft² (500–8000 m²), according to biotope; young driven off by parents after sexual maturity	Abundant
About 30 days About 2 months Probably much like *Tupaia*	Fruits, arthropods, small vertebrates	None known	Predominantly terrestrial; otherwise much like *Tupaia*	Abundant
About 30 days Probably about 2 months Probably much like *Tupaia*	Fruits, arthropods, small vertebrates	None known	Terretrial; otherwise probably like *Tupaia*	Abundant
Probably much like *Tupaia*	Fruits, arthropods, small vertebrates	None known	Much like *Tupaia*	Abundant
Probably much like *Tupaia*	Fruits, arthropods, small vertebrates	None known	Predominantly arboreal; to be found at elevations up to 4800 ft (1500 m); otherwise probably like *Tupaia*	Apparently abundant
Not known	Fruits, arthropods, small vertebrates	None known	Nocturnal tree dwellers in tropical forests; live in pairs (together with any young up to sexual maturity); behavior for the most part not known	Abundant

monly high fat content of 26%, enabling the young to maintain a body temperature of about 98.6°F (37 °C) from birth without maternal contact. Besides, the milk contains a great deal of protein (about 10%) ensuring rapid growth of the young. Their ears open at about 10 days and their eyes at about 20; the young leave the nest and are largely independent after only about four weeks. They now weigh over 3.5 oz (100 g), although by this time the mother has nursed them only 14 times for a total time of about 90 minutes.

Strangely enough, as a rule neither the father nor other conspecifics present in the enclosure will enter the nest containing the young, even if they formerly used it for sleeping quarters. This apparently is because of scents given off by the mother at the time of birth, protecting the young from conspecifics. The origin of these scents is not known, but they affect all conspecifics alike. Thus, it is possible to interchange such scent-marked young of different mothers, even with an age difference of more than 10 days between them. Even delicacies ordinarily devoured forthwith, such as meal worms, grasshoppers, of young mice, will not be touched by tupayas after these protective scents have been artificially applied.

It is an interesting feature that any disturbance to the mother before the young are born (such as conflict with the "spouse" or sexually mature daughters, as well as noises or strange persons) always results in failure to provide the young with the protective scents, so that they are regarded as prey and devoured as previously described. By contrast, such disturbances after birth of the young lead to irregularities in the rhythm of feeding; the mother will seek out the young for nurs-

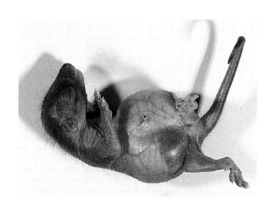

(Above) Two tree shrews engaged in close 'infighting'.
(Left) Immediately after birth, the baby tree shrew sucks itself so full of mother's milk that its belly is tautly distended. This reservoir is necessary, for tupaya mothers nurse their little ones only every 48 hours.

(Opposite page) The forest is the primeval home of all the primates. Most species - in the photograph, a gibbon, as a typical tree dweller - have remained faithful to that habitat, but some have become ground dwellers in greater or less degree; most decidedly so, humans.

Tree shrews often use their hands for face-washing.

ing once or several times a day. Insufficient notice of the individual inclinations of the members of a pair, as well as disturbance by other tupayas or by persons, may help account for the fact that tupayas have hitherto been regularly bred successfully at only a few research stations and that the 48-hour feeding rhythm has not been observed by most investigators.

Although this astonishing mode of rearing the young has so far been verified in only three genera of tupayas, it is very likely characteristic of all bushy-tailed species. Whether the 48-hour feeding rhythm also applies to the pentail tupayas cannot be decided because this species has not yet been bred in captivity.

Comparably reduced maternal care of the young is not known among other mammals except for the Lagomorpha which visit their young to nurse them only once in 24 hours. Certainly this is not a primitive characteristic. Rather, the infrequency of the mother's visits to the helpless young in the nest must tend to diminish the likelihood that the young will be found in their hiding place by predators.

Whereas we have at least some field observations and a number of comprehensive laboratory studies of the bushy-tailed tupayas, virtually nothing is known of the ecology or behavior of the pentail tupayas, even though this species also appears to be abundant in the primeval forests and plantations of Malaysia and Borneo. The zoologist Lim Boo Liat, who has captured pentails in nature on various occasions, found them sleeping in hollow trees during the day in small groups, doubtless family groups. Trapping data suggest that the animals are strictly arboreal. Their diet is much the same as that of the bushy-tailed tupayas, ranging from insects and small vertebrates to fruit. Occasionally, pentails have been kept under human care for a time, but the results are so scanty that one can hardly generalize. The animals seem to be active only at night, and very aggressive against strangers of the same species. Whether pentails are territorial like the other tree shrews, living in private, defended territory, is not known, but the pronounced marking of laboratory animals with secretions from glands in the posterior area and with urine suggest that possibility. The American investigator Edwin Gould, who put a male together with various females, in some cases found a reciprocal marking of the animals, such as had already been described for the bushy-tailed tree shrews. Whether there ultimately are points of contact between the ecologically otherwise adapted, diurnally active bushy-tailed tupayas and the nocturnally active pentailed tupayas remains to be shown by future field and laboratory studies.

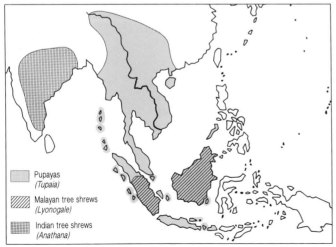

- Pupayas (Tupaia)
- Malayan tree shrews (Lyonogale)
- Indian tree shrews (Anathana)

THE PRIMATES

Category
ORDER

Classification: Order of mammals comprising two suborders (Prosimiae, Simiae), which, however, are not phylogenetically homogeneous groups, but rather represent different levels of evolution. Accordingly, a more recent major subdivision of the primates, to be adopted in the present volume as well, distinguishes two categories only: Prosimians having a nose leather (Strepsirhini; all the prosimians except the tarsiers); and primates without nose leather (Haplorhini; the tarsiers and all monkeys and apes). This highly multiform order comprises 15 families of roughly 50 genera and about 500 species and subspecies.

Strepsirhini (Prosimians with nose leather)
Family Cheirogaleidae (mouse lemurs)
Family Lemuridae (lemurs)
Family Lepilemuridae (sportive lemurs)
Family Indriidae (short-tailed lemurs)
Family Daubentoniidae (aye-ayes)
Family Lorisidae (lorises)
Family Galagidae (galagos)

Haplorhini (Prosimians without nose leather)
Family Tarsiidae (tarsiers)

Platyrrhini (New-World or broad-nosed primates)
Family Cebidae (cebids)
Family Callimiconidae (callimicos)
Family Callithricidae (marmosets and tamarins)

Catarrhini (Old-World or narrow-nosed primates)
Family Cercopithecidae (guenons)
Family Hylobatidae (gibbons)
Family Pongidae (large anthropoid apes)
Family Hominidae (hominids)
Body length: 4.4 in. (11 cm; mouse lemurs) to about 6 ft (185 cm; gorilla, humans)
Tail length: 0 to about 3.6 ft (110 cm)
Weight: 1.8 oz (50 g; mouse lemur) to 605 lb (275 kg; male gorilla)
Distinguishing features: There are no conspicuous traits typical of primates; rather, the representatives of thir order are distinguished by combinations of various features, atypical in themselves. Optic cavities are encircled by closed rings of bone; eyes are in the anterior aspect of the head, and are capable of space perception and estimating distances; brain has posterior lobes and fissura calcarina; in higher forms, enlargement of cerebral areas is required for association of ideas; clavicle present; hands and feet, with few exceptions, having five digits; first finger and first toe are more or less opposable; fingers are capable of spreading and grasping; digits for the most part bear nails, less often claws; cecum; external, depending penis; testes in scrotum; one pair of thoracic nipples; tail originally present, but sometimes reduced, and wholly absent in the higher forms (gibbons, anthropoid apes, and humans).
Reproduction: Usually possible throughout the year; gestation or pregnancy 60 days (mouse lemur) to about 266 days (human); as a rule, only one young per birth; birth weight ranges from 0.4 oz (10 g) to about 4.5 lb (2000 g) according to species, more in humans; invariably close mother-child bond in early life.

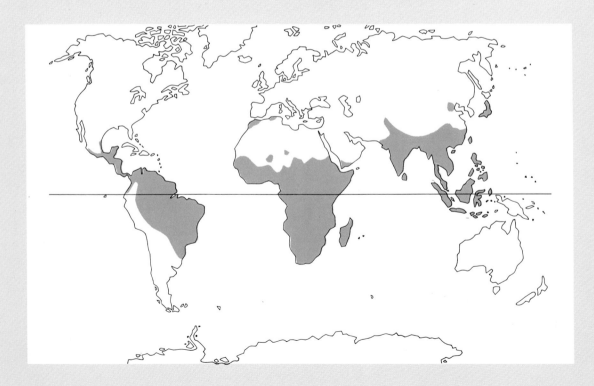

Primates
Primates FRENCH
Herrentiere, Primaten GERMAN

Life cycle. Weaning at some time between about 2 months and 4 years; sexual maturity between about 9 months and 9 years, rather later in humans only; life expectancy in nature evidently at most about 40 years, nearly twice as long in modern humans.

Food. Predominantly vegetable fare; in some species, chiefly or additionally animal food as well, occasionally even smaller mammals; humans are the only true omnivore.

Habit and habitat: Predominantly active by day; only the South American nocturnal monkeys and most prosimians are active at night; except for humans and some Asian species, exclusively in tropical and subtropical latitudes between about 25°N and 30°S; principal habitat forests, especially tropical rainforests, but also arid forests, fringing forests, coastal forests, tree and bush savannas, rocky or mountainous terrain; some species even in environs of human settlements, including some large cities; only humans are uniformly represented in virtually all environments on Earth; originally fully adapted to arboreal life, but various species become partly or entirely terrestrial; mostly excellent climbers, leapers, or acrobats; as a rule, highly social with marked community behavior and strict ranking; live together in pairs, family groups, or more or less large bands; richly expressive behavior (utterance of sounds, scents, mime, and gesture) for intraspecific communication; occupancy of territories or tracts; demarcation and defense of territory by scent marking, cries, threatening, or intimidating behavior, sometimes physical aggression; owing to prolonged adolescence, capability of learning by experience; in higher forms, insightful behavior, certain intellectual faculties, ability to solve problems, occasional use of tools, and rudiments of cultural tradition ('simian' or 'pro'-culture); development of true culture, capability of abstract thought, speech articulation, production of tools by means of tools, and habitual erect posture, in humans only.

Forms of locomotion. The original mode of locomotion of the primates is a four-footed progress through the treetops. This may include slow climbing (potto, 3) as well as fast running (monkey, 4). The arms of these forms are of approximately the same length as the legs. Markes prolongation of the legs produces leapers (sifaka, 2); lengthening of the arms leads to swinging hand-over-hand (gibbon, 1). Evolved directly from the tree-dwelling quadrupeds are the terrestrial quadruped forms (baboon, 7); the hand is placed flat against the ground. It is otherwise with the anthropoid apes, with predominantly terrestrial locomotion (chimpanzee, 5; gorilla, 6); only the knuckles of the fingers touch the ground. These forms are presumably derived from the hand-over-hand swinging type.

Brain development. In primate evolution, behavior guided by instincts is increasingly replaced by conscious action. These changes are reflected notably in the development of the cerebrum, the seat of the higher faculties. The cerebral cortex gains size and capacity through the development of numerous furrows and convolutions (blue = brain stem with cerebellum; pink = olfactory lobe; yellow = cerebrum). 1. tree shrew; 2. slow loris; 3. macaque; 4. chimpanzee; and 5. human.

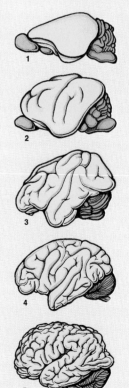

Primates

Introduction
by Hans-Jürg Kuhn

In the 10th edition (1758) of his *Systema Naturae* (see illustration below), the founder of modern zoological nomenclature, Carl von Linné (Linnaeus) placed the order Primates at the beginning, in the highest rank. On the basis of a quite simple definition, 'four upper parallel incisors, two thoracic mammary glands,' he assigned humans, the anthropoid apes, monkeys, and prosimians to this order, and indeed the bats and flying lemurs as well, which today are classed in orders of their own. Yet the results of very recent phylogenetic research indicate that the primates, the tree shrews (Scandentia), and the flying lemurs (Dermoptera), the latter two groups not known to Linnaeus, together might form a natural unit (Archonta), itself not so very remote from the bats (Chiroptera). Thus, after more than 200 years, the Linnaean system is far from outmoded. But how much has changed! Linnaeus listed 25 primate species, which, with the sure sense of an experienced systematist, he felt belonged together. Today, in South and Central America, Africa, and South and East Asia, more than 500 species and subspecies of primates are known, to which more are continually being added. Between 1977 and 1987, for example, a new, strikingly colored species of macaque in Zaïre and a new subspecies of the Preuss macaque in Gabon have been described.

The phylogeny of the primates, which extends back into the Cretaceous, has been documented today by thousands of fossils, and more thoroughly researched than that of any other order of mammals. A distinct branch of science, primatology, is devoted to the primates.

Why all this interest in Primates? Fascinating or indeed irritating as the fact may be to us as individuals, we are reminded by the non-human primates or our roots in the rest of creation. Substantially in the hundred years since Charles Darwin, primatology has thus made a major contribution to the unfolding of a new human self-perception.

But even if humans were not among them, the Primates would be a mammalian order of great phylogenetic interest, for the reason that, along with good documentation of extinct species, primates of quite diverse evolutionary levels still exist today and can be studied. On Madagascar, prosomians have been living in isolation with hardly any competition from other mammals ever since the Tertiary, and have adapted to every habitat on the Island. Of the abundance of Early Tertiary prosimian forms, only a few species (lorises, galagos) have survived elsewhere – in Africa and southern Asia – living a nocturnal life, as it were in the shadow of the higher primates and in competition with many other climbing mammals.

In the 18th century, Carl von Linné, the founder of modern systematics, placed the order Primates at the head of the animal kingdom, and included humans in that order.

INTRODUCTION

On the great islands of Southeast Asia, we find the tarsiers, the last, highly specialized representatives of a primate group likewise widely distributed in North America and Europe in the Early Tertiary, and the progenitors of the higher primates. They developed independently of each other in South America (Platyrrhini) and in the Old World (Catarrhini: monkeys, apes and humans) to their present ascendancy.

The German terms "Halbaffen" (prosimians) and "Affen" (simians) do not designate natural kinship groups; they refer to primates of unlike evolutionary level. From prosimians, simians emerged not only in South America (Platyrrhini) and in continental Africa (Catarrhini); the prosimians of Madagascar also produced forms *(Archaeolemur, Hadropithecus)* – today, unfortunately, extinct – that we might call simian. The tarsiers of Southeast Asia (Tarsiidae) are "prosimians" in evolutionary level, but they form a natural kinship group (Haplorhini) together with the Platyrrhini and Catarrhini.

All we lack is a living representative of the paleoprimates (Paromomyoidea), dating back to the Cretaceous. This is regrettable, for we have no starting point for the reconstruction of many evolutionary processes that took place during the history of the primates. But here the tree shrews (Scandentia) come to our aid; of all the placental mammals (Eutheria, mammals whose offspring are nourished in the womb through a placenta), they have preserved the greatest number of the original mammalian traits. Although they cannot be classified as Primates, because no derived traits common to tree shrews and primates have yet been positively identified, in many areas they provide good models of the situation at the starting point of primate phylogeny. In the following, therefore, in the discussion of reproduction biology, locomotion, and body structure, a comparison of a tree shrews of the genus *Tupaia* with humans will illustrate the full range of evolutionary processes that have characterized this phylogeny.

All tree shrews are good climbers. Adaptation to locomotion in tree branches is certainly a key feature of the primates throughout. It was on this foundation alone that, for example, the evolution of the human grasping hand and of spatial vision was possible. Very few species of primates have completely given up their connection with that foundation, only the gelada (the ape *Theropithecus*) in adaptation to the high mountains of Ethiopia, and the first representatives of the genus *Homo* in the Late Tertiary savannas of Africa. All other species still climb regularly, at least to gather food, or else sleep in trees. The antithesis, at first glance seemingly irreconcilable, between a climbing four-footed tree shrew and a human being striding erect, is to a large extent resolved if we consider the many different modes of locomotion to be found among other primates, from the slow grasping climb of the loris, or potto, to the powerful leaps of the long-legged galagos and tarsiers, to the 'flying leaps' of the gibbons. In analyzing the

For most primates, life goes on almost exclusively in airy heights. (Left) The hanuman frequently comes down to the ground, but usually traverses the gap between two treetops by an acrobatic leap. (Right) The anubis baboon is one of the few species that are at home on solid ground.

Like most other primates, the ring-tailed lemur, a representative of the prosimians, has the habit of sitting up to eat, holding the food with deft fingers.

INTRODUCTION

process of humanization, it is worth noting that many higher primates, for example the macaques and baboons, spend much of the day with the trunk more or less upright, as when taking food, social grooming, suckling, and sleeping. The buttocks of the gibbons and many apes are protected by massive callosities. Many primates are excellent swimmers, in particular, for example, the proboscis monkey and the Java monkeys of Southeast Asia. Others, such as the baboons, the African anthropoid apes, and humans, lack the requisite innate coordination. It is a characteristic of the primates that their fingers and toes are armed with nails rather than the claws of most insectivores and of the tree shrews, and that the thumbs and the great toes of climbers grasp independantly. In detail, of course, there are many exceptions. The nails of the aye-aye and of the clawed lemurs are claw-shaped, except for the great toe, which enables these small monkeys to cling to the bark of large trees. Among prosimians, the second digit bears a nail used for grooming; in the tarsiers, so does the third.

Even though modern humans everywhere artificially creates their own savanna terrain – thereby

destroying the habitats of most other primates, especially tree dwellers – many human beings have preserved a sentimental attachment to the forest. Profiting from worldwide deforestation is rare among the other primates, except the savanna-dwelling baboons are green monkeys of Africa.

Endangerment by Humans

Where primates live in equilibrium with their natural habitat, for example a tropical rainforest in South America or West Africa, they often exceed all other mammals in biomass (total live weight). For this reason, they have always been hunted intensively, and were the chief source of flesh meat for the human population until very recent times. The traditional means, traps and snares, cruel though they may have been, were not efficient enough to destroy primate stocks. It was not until the coming of firearms and new facilities for the transport of smoked meat that the simian population of large parts of Africa and South America collapsed. Thus, according to conservative estimates, 60,000 monkeys were slaughtered in 1985

in the comparatively small territory of Sarawak, in northwestern Borneo, although some species there are already on the brink of extinction. Where agriculture advances into primate habitats, some species become "pests," and are then pursued inexorably. The "monkey drives" in Sierra Leone alone are supposed to have bagged about 30,000 animals annually into the 1960s. Only a short time ago, in preparation for a single agricultural project in Kenya, about the same number of baboons and green monkeys were destroyed. Exports of living primates and of hides for the fur trade (colobus monkeys) have had less impact

(Left) The Java monkey is one of the best swimmers among the primates. (Right) The monkey trade is a sinister chapter, unfortunately not yet concluded. From 1968 to 1972, 173,000 squirrel monkeys were imported into the United States - often, as here, in cages much too crowded - as pets, a purpose for which they are quite unsuited.

than biotope loss and hunting, even though a great many individuals have been affected. In the 1950s, for example, hundreds of thousands of rhesus monkeys were brought to Europe and North America from India for the development and production of vaccine against infantile paralysis; and, as late as the years 1968 to 1972, 173,000 live South American squirrel monkeys (Saimiri) were imported into the United States, to serve as household pets – a purpose for which they are quite unsuited. Of course, no zoo can be without primates; today, zoo primates are only occasionally imported from their countries of origin. For many species, now that their life requirements are quite well known and their diseases have been brought under control for the most part, the problem has rather become one of accommodating their progeny. Given the numbers of primate species, to be sure, only a small proportion can be kept in captivity. Some prosimians – the tarsiers – and many Colobini are still among the most difficult problem inmates of a zoological garden.

Because of destruction of their natural biotope, today all non-human primate species are listed as endangered in Appendix II of the Washington Convention on the protection of species, and some are among the most endangered species of all – for example the tamarins *(Leontopithecus)* of South America, the bearded monkey *(Macaca silenus)* of India, the aye-aye *(Daubentonia)* of Madagascar, and among anthropoid apes, the mountain gorilla. Usually, deforestation in an already limited range of distribution is responsible. However, there can be other contributing factors. Thus, the tamarins seem unable to hold their own with the marmoset *(Callithrix jacchus)* which has been released in parts of their historical range. The merciless persecution of the aye-aye in Madagascar, by a population that has traditionally spared the prosimians of the island on religious grounds, appears to have something to do with its reputation for bringing bad luck. The last of the mountain gorillas are threatened because, among many other reasons, poachers can sell the imposing skulls of the males to tourists as trophies.

Size, Food and Life Span

The earliest fossil primates (*Purgatorius*, Upper Cretaceous) may have been about the size of tree shrews, weighing 1.4–12.5 oz (40 to 350 grams) as adults. Size increased in many lines of descent, but rarely beyond 26–33 lb (12 to 15 kilograms) in exclusively arboreal species. An exception is the orangutan, which, however, has a slow, grasping climb and swing, but will not leap. The gain in absolute body weight had many consequences. Small primates, up to about 18 oz (500 g), can live largely by catching insects. But beginning at only about 18 oz (500 g) body weight, the balance of energy (energy expended against energy gained) in the capture of an individual insect becomes negative, and the animal must fall back on other food, such as fruits, obtainable in larger quantity in one location. Nearly all the larger primates, including chimpanzees, for example, will eat insects occasionally, even though they are hardly a net source of energy.

Independently of each other, various primates have acquired the ability to break down indigestible food, for example leaves, by means of enzymes in fermentation chambers, somewhat like the first stomach of a bovine or other ruminant, and to neutralize the toxins by which plants protect

Primate females, like this orangutan, make proverbially good mothers. They generally care for their young longer and more devotedly than other mammals.

No group of mammals exhibits such enormous differences in size as do the primates. (Left) The most massive species is the gorilla, attaining a height of 6 ft (185 cm) standing upright on two legs. (Above) The most minute is the mouse lemur, with a head-and-trunk length of only 4.4–5.2 in. (11 to 13 cm).

A sign of the high level of evolution of the primates is their wide range of forms of intraspecific communication, involving several sense organs. The howler monkey (below right) uses piercing sounds as acoustic signals; the mandrill (below left) has brilliant coloring for visual display; and the male anubis baboon (below center) uses the senses of sight, smell, and touch all at once to examine consorts.

themselves against herbivores. Such fermentation chambers are found in the large intestine of some prosimians and of the howler monkeys, and especially in the stomachs of all colobini. But since fermentation will take a considerable length of time, depending on the nature of the food, these chambers must be quite large to be commensurate with energy requirements. Given the law of metabolism that the energy requirement is proportional to body surface, not to body weight, this mode of digestion is available to large species only; for the smallest primates, a fermentation chamber would theoretically have to equal the volume of the body in order to supply its energy requirement.

Gain in body weight of primates was also an important requirement for the evolution of higher intellectual capabilities of the brain.

A tree shrew of the genus *Tupaia* may attain an age of 9 or 10 years in captivity. Humans are the longest-lived, among the placental mammals at least. The life spans of all other primates fall within this range. The smallest prosimians (dwarf galago) live not much longer than tree shrews, while the larger prosimians of Madagascar may reach 20 to 30 years in the zoo, and simians up to about 30 years. Anthropoid apes live to an age of from 40 to over 50 years. However, since it is not yet understood how to keep them successfully for a long time, there have been only scattered observations.

Newborn tree shrews are deposited by their mother in a nest, to be visited only every other day for nursing, and leave the nest at an age of 28 days. As a rule, the mother's next litter will be born when they are 43 days old. At the age of 8 to 10 weeks, the young become sexually mature and soon lose close contact with the mother. Hence there is very little time for social learning. By contrast, the simians and anthropoid apes characteristically have a comparatively long infancy and childhood in very intimate contact with the mother. The single offspring is usually transported on the mother's body for months, and in many species, the maternal family, consisting of a female and her sexually mature female descendants and their issue, is a basic unit of social organization. Social maturation is by no means completed at sexual maturity. From the researches of Hans Kummer in Ethiopia, we know, for example, that male hamadryads, which become sexually mature at about 5 years and may attain an age of 30 years in the zoo, can fully discharge their social duties as leaders and masters of a harem for only a comparatively short period – between the ages of 12 and 18. The long time spent in very close association with the mother and in the social group is necessary for normal behavioral development. It also creates prerequisites for social learning and hence eventually for the establishment of traditions and a culture.

A mother tree shrew usually produces two or three blind, naked young at a birth. Most primates, on the other hand, bear a single, open-eyed, and furred young, which is transported on the body of the mother from birth, usually clinging tightly to her coat. The grasping reflex required for this purpose ist still well developed in the human newborn. Twins usually cannot be raised in this way, unless in captivity. In the few species where twins or even triplets regularly occur (some galagos, varecias), they are deposited in a nest; among the South American clawed monkeys, very highly specialized modes of social care of the young have developed, whereby the mother is to a large extent relieved of the burden of twins, though at the expense of the reproductive success of her older descendants. But even a large family of clawed monkeys is unable to raise triplets.

Under favorable conditions, a female tree shrew may produce two or three young about every 43 days. A female chimpanzee in the wild, whose life span is four-times-longer, will hardly raise more than six, born at intervals of three years or so. Quite in general, the phylogeny of the primates shows a trend towards an ever greater social "investment" in an ever smaller number of progeny.

Sensory Capabilities and Communication

The eye is predominant among the sense organs of the primates. Climbing and leaping, as well as grasping minute objects under the guidance of the eye – as for example in the practice of so-called "like-picking" – require good stereoscopic (binocular) vision and accommodation (close-up focusing). Color vision is well developed in most of the higher primates, much as in humans. Accordingly, they include the only mammals with blue coloration as a social stimulus, for example on the nasal protuberances of the mandrill or the sexual parts of many long-tailed monkeys. The vivid markings of many tarsiers and long-tailed monkeys are effective signals for mutual recognition where several closely related species occur in the same habitat. In higher primates, conspecifics increasingly convey information concerning subjective mood by means of mime; lips, eyelids, and ears play a special role here.

Only a single genus of the higher primates, the South American night ape *(Aotus)*, with its large eyes and retina equipped with rods only, is adapted to a nocturnal life. On the other hand, with the exception of several large species in Madagascar, the modern prosimians are nocturnal animals, some with extreme development of the visual apparatus. In the tarsiers of the genus *Tarsius*, each eyeball alone is larger than the brain, and a very extensive area of the cerebral cortex is devoted to vision. It must here be said by way of reservation, however, that we as yet know little about the nocturnal behavior of many "diurnally active" primates, even in zoos. There are numerous individual observations to indicate that the division into diurnal and nocturnal species does not always do justice to the true state of affairs. Thus, the Malagasy brown lemur appears to be active largely at night, although the visual apparatus betrays hardly any such specialization.

Among the prosimians that hunt chiefly at night, the external ears have important functions. When one ear picks up a sound, the other sonic receiver will immediately zero in on the source, and a direction-finding operation is performed. During the ensuing comparatively slow rotation of the head, both ears continue to be aimed at the source, until the eyes take over from the organs of hearing. The prey is then attacked under the guidance of the eyes, while in some species the

One pair of milk glands located on the breast is the rule for primates, like this ring-tailed lemur mother. Two nipples suffice because as a rule there is only one infant to be nursed, though she happens to have twins. Smaller prosimians often have one or two additional pairs of nipples on the abdomen.

ears are folded up. Although sonic direction-finding is no longer achieved by means of the external ears in this way among the higher primates and in humans, reportedly, the corresponding nerve impulses to the muscles of the conchae can be detected even in humans.

The ears of higher primates, that are active by day are strikingly similar to humans. The tip, directed upward and towards the back in many Old World monkeys, can be identified in many human individuals as "Darwin's point." The sense of hearing is not only important in the recognition of prey or enemies; much intraspecific communication is acoustic as well. The rustling of leaves that accompanies tree climbing helps the members of the group to maintain contact; many anthropoid and other apes make a threatening display by vigorous shaking of branches and small trees; gorillas slap their rumps; and chimpanzees will beat anything that can somehow be set in vibration and make a noise. By far-carrying cries, neighbor males (orangutan), families (indri, gibbon), or groups (howler monkeys, guerezas) of many primates keep in touch with each other and mark off their territories without resorting to combat. Resonance chambers ("sound boxes") may consist of throat sacs (orangutan, many Old World primates) or of the lingual bones (howler monkeys). In the forest generally, acoustic communication is more important than visual.

The tree shrews as well as the prosimians possess an external nose leather (rhinarium), which is kept moist. It can trap scents and pass them between the upper incisors to the palate and then via the nasopalatal duct to a large chemical sense organ, like a cul-de-sac (Jacobson's or vomeronasal organ) at the base of the nasal cavity. Since we ourselves lack this sense organ, we cannot very well imagine its import. Still very much an enigma, it probably serves not so much for olfactory orientation at a distance as for short-range orientation in the context of reproduction and social relations generally. The nose leather and Jacobson's organ are found in the prosimians; the tarsiers and New World monkeys, though they lack the nose leather, retain a functional, if comparatively small, Jacobson's organ. The catarrhine primates (Old World monkeys, anthropoid apes, humans) have no trace of it at all. The olfactory mucous membrane in the nasal cavity is also at least comparatively less extensive in the higher primates. Certainly their olfactory orientation is a subordinate factor in the search for prey, in tracking, and in the avoidance of enemies. But in the social sphere, olfactory communication has remained important to all primates. Tree shrews have large areas of skin glands between the throat region and the rump, with secretions that rub off on branches. Many prosimians and New World monkeys mark the territories claimed by a family or group with scents that may consists of mixtures of glandular and other secretions, to identify them for neighboring groups. Gisela Epple has shown in the laboratory that among callitrichid monkeys, such scent marks convey information as to species, sex, and social rank of the individual responsible for them. In the higher Old World primates, the skin glands become less important – we know very little even about the functions of the scent glands that develop in human beings at sexual maturity. In many Old World primates, for example, rhesus monkeys, scents that do not derive from skin glands, convey information about the female cycle.

The sense of touch in primates is especially well developed in the area of the ventral skin of the hands and feet, as well as of the "fifth hand" at the end of the tail of some large New World monkeys. The ability of many apes and monkeys to precisely grasp minute objects between the

Primates have evolved an extraordinarily multiform social life, ranging from the solitary, by way of pairing and the family group, to organized communities of many individuals. The hamadryas baboons are pictured here, for example, in "single-male harem groups", in which several females and their offspring are attended by one strong male.

thumb and the side of the index finger is best appreciated by watching them at social grooming – "nit picking" – as they remove flakes of skin, foreign matter, and small parasites from each other's fur. Correspondingly, the importance of sensory hairs (vibrissae) is diminished. Tree shrews and the Malagasy prosimians retain on the head and forearms nearly all of the groups of sensory hairs characteristic of the earliest mammals. The New World monkeys have them at the eyebrows only, and the marmosets on the forearm as well. In adult Old World primates they are much reduced, yet humans appears to be the only species that show no trace of sensory hairs, or whiskers, even in the embryo.

Seemingly the coat of hair is nearly absent in humans. The "naked ape", the term used by Desmond Morris, has passed into the language. Upon closer inspection, as the late Zurich master of primatology A.H. Schultz noted, although the hairs on the human trunk are much shorter, finer, and lighter, there are about the same number per square centimeter (1 cm^2 = 0.2 in.^{2}) as on a chimpanzee's. Other primates, such as galagos, gibbons, and some platyrrhines of South America, are densely furred, a feature that established the woolly monkey's name. Remarkable hair configurations occur as secondary sex characteristics of male primates, such as the hamadryas baboon's mane, and the beard. Especially where several species occur together, striking patterns of line and color are found in their coats, doubtless favoring species recognition, and rendering the many forest monkeys of Africa, marmosets of South America, and langurs of southeast Asia attractive as zoo mammals. The newborn of many long-tailed monkeys have especially conspicuous coloration; while they wear this infant dress, they are accorded an indulgent tolerance in the community.

STREPSIRHINI – prosimians with nose leather

Lemuriformes – prosimians of Madagascar
　Cheirogaleidae – mouse lemurs
　Lemuridae – prolemurs, true lemurs
　Leipilemuridae – weasel lemurs
　Indriidae – indris
　Daubentoniidae – aye-ayes
Lorisiformes – prosimians of Africa and Asia
　Lorisidae – lorises and pottos
　Galagidae – galagos

HAPLORHINI – primates without nose leather

Tarsiiformes – prosimians without nose leather
　Tarsiidae – tarsiers
Platyrrhini – Ceboidea – New World or broad-nose monkeys
　Cebidae – owl monkeys, squirrel monkeys, titis, sakis and short-tailed monkeys, howler monkeys, capuchins, spider monkeys
　Callimiconidae – callimicos
　Callithricidae – marmosets, tamarins
Catarrhini – higher Old World or narrow-nose primates
　Cercopithecoidea – tailed (dog-like) Old World monkeys
　　Cercopithecidae – long-tailed monkeys
　　　Cercopithecinae – macaques, baboons, mangabeys, guenons
　　　Colobinae – langurs and colobi, proboscis monkeys, guerezas
　Hominoidea – human-like apes
　　Hylobatidae – lesser anthropoid apes, or gibbons
　　Pongidae – great anthropoid apes: orangutan, gorilla, chimpanzee
　　Hominidae – human species

Reproduction

The mammary glands do not by any means meet Linné's definition for all primates. Besides the thoracic pair of mammary glands, many smaller prosimians and the tarsiers have one or two additional pairs in the abdominal wall. The aye-aye has only one pair, in the inguinal region.

Reproductive periods are adapted to the external conditions of life. In the north of the main island of Japan, where winter reigns for six months, all the female red-faced macaques habe their young in spring. Only then are there good prospects for survival. Many tropical species have their offspring more or less evenly distributed throughout

(Above) The sexual swelling at the rump of this female chacma baboon seems grotesquely enlarged. In many species, such conspicuous swelling indicate readiness for mating. (Left) The current systematic classification of the Order Primates. Two major groups are distinguished, those with and those without a rhinarium, or nose leather.

the year, although there may be time coordination within the group. Ovulation is induced by mating in the tree shrews; in the higher primates, much as in humans, it is probably spontaneous as a rule, within the confines of a cycle. In most of the higher Old World primates, the length of the cycle is about a month. Matings of tree shrews occur almost exclusively before an ovulation. In many higher Old World primates, sexuality has taken on important social functions apart from impregnation, as may be seen from the fact that matings will occur even during pregnancy, and that the mounting that was originally preliminary to mating in many cases occurs in other social contexts.

Different evolutionary levels may be identified in the structure of the womb. From left, a tree shrew uterus, a prosimian uterus, and a simian, anthropoid, or human uterus.

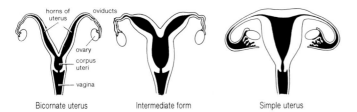

Bicornate uterus · Intermediate form · Simple uterus

In the large multi-male communities of the red colobus monkey *(Procolobus badius)*, the pig-tailes macaque *(Macaca nemestrina)*, the South African chacma baboon *(Papio ursinus)*, and the chimpanzees, at about the time of ovulation the females exhibit remarkably pronounced sexual swellings of the external genitalia, with vivid red coloration. These swellings indicate estrus – that is, readiness for mating – to the males. In these species, matings are to be observed very frequently, often with several males. In closely related species that live in well-established, single-male groups or families, where females do not compete for the favor of the male, such signals originally proclaiming the estrus do not occur, and matings are seldom observed (guereza, long-tailed monkeys – except talapoins – gibbons, gorilla). The fact that the sexual swellings sometimes take on grotesque dimensions – in young pig-tailed monkeys and red colobi, up to more than one-fourth the body weight – and that they outlast the time of ovulation, having become dissociated from their original function as a sign of estrus, is an indication that these signals procure social advantages to the bearer over other females in the group. They are a result to intra-specific selection. If that is so, it might be asked why these signals are not displayed at all times. There is a simple explanation. True, in many species the sexual swellings are a necessary condition for mating, but on the other hand they would render birth impossible. Remarkably enough, young males of the West African red colobus appear to have "stolen" the same social advantages by developing red cushions in the sexual area, on an entirely different physiological basis from the females, but imitating their sexual swellings in fine detail.

The primate penis contains a bone (the baculum); it is absent secondarily, that is, it has been "lost" in the course of the phylogeny, only in the group of the South American woolly and spider monkeys, the tarsiers, and humans. The paired embryonic Müllerian ducts, from which the oviducts, the uterus, and parts of the vagina arise, do not unite as far in the prosimians and tarsiers as in the higher primates. In the former, a bicornuate uterus is therefore found, and a simple uterus, as in humans, in the latter. In the prosimians, with few exceptions, the fertilized ovum is implanted superficially in the mucous membrane of the maternal womb (epitheliochorial placentation); in the tarsiers and the higher primates, it penetrates the mucosa to a varying extent and comes into contact with the maternal bloodstream (hemochorial placentation). Gravidity (pregnany) lasts 60 days in the smallest primate, the mouse lemur of Madagascar, and 266 days on the average in humans. The gestation periods of other primates life between these extremes; as a rule they are shorter in the smaller species than in the larger ones, but there are many exceptions – 168 days for the macaques and 177 for the larger baboons, as contrasted with 196 days for the diminutive talapoin. Monkeys are almost invariably born at night, whereas the anthropoid apes and human are born either during the day or at night.

Dentition

The first primates had three incisors, one canine, four premolars, and three molars in each half-jaw, making a total of 44 teeth. This complete dental formula is not to be found in any primate living today. The incisors are generally reduced to two, as Linné's definition postulates. The aye-aye has only one incisor in each half-jaw, with the root perpetually growing as in the rodents. In modern prosimians, the lower incisors and canines form a comb with which the fur is groomed. This comb is cleaned with the under side of the tongue. The upper incisors are vestigial in these prosimians.

Only the aye-aye lacks canine teeth. In many primates, on the other hand, the canines of the male are especially pronounced; in the baboons and their nearest relatives, they are formidable weapons, and serve visually as social signals as well. Their posterior edge grinds against the first lower premolars. In all primates, the upper canines serve to guide jaw movements. Prosimians and the South American primates have three premolars. Even when they are absent in adult prosimians, they may often be detected in the embryo. The higher primates of the Old World (Catarrhini) have only two premolars. The least modified element is the number of molars. Only a few South American clawed monkeys lack our "wisdom tooth", while all other primates have three molars in each half-jaw.

Social Behavior and Brain Development

The key to an understanding of the primates certainly lies in their social behavior. Among three shrews of the genus *Tupaia*, a male has a territory of his own, defended against other males and possibly overlapping the territories of several females. Similar, comparatively simple social systems are found among various nocturnal prosimians, and an apparently similar one – remarkably enough – is represented by the orangutan. Then both New World and Old World primates have evolved perhaps the most multiform social structures to be found among mammals – with the possible exception of some social carnivora or the toothed whales. Large bands of chimpanzees with several adult males and numerous "clans" of females involve such an intricate fabric of relationships among individuals that it took Jane Goodall many years to decipher it to some degree for a single group. The behavior of the individual is formed by social learning and play in a long juvenile phase. As experiments with mirrors have shown, chimpanzees are able to see themselves in their environment; no such physical self-awareness has yet been demonstrated elsewhere in the animal kingdom. This may be a prerequisite for the abili-

In the highly developed social behavior of primates, juvenile play assumes very great significance. It is primarily "learning" play; the individual's behavior is shaped by playful learning throughout a long juvenile phase.

ty to reproduce certain body movements after they have been observed on the part of a conspecific. A secure life, satisfying all requirements, is possible only within the group, but that same group almost daily involves its members in confrontations, occasionally extremely severe or to the death. Life-threatening combat with a neighboring group may result inadvertently. In such a social fabric, the individual's Darwinian "fitness" is enhanced by anything that sharpens the senses, facilitates learning, promotes the power of association, adapts behavior to the requirements of a given situation, and increases the ability to prevail in social relations by physical strength or other appropriate means. At the same time, however, the social setting must remain capable of satisfying the individual's needs. The organ that balances all of these partly contradictory considerations is the brain, or more precisely the "new" brain (neencephalon) that was rapidly developed in primate history, as casts of fossil skulls indicate. According to H. Stephan, the neocortex of a Malagasy weasel lemur is ten times as large as it would be in a primitive insectivore of the same body weight; the factor is between about 25 and 75 for New and Old World monkeys, 85 for the chimpanzee, and 214 for humans. There has been no other comparable development in the animal kingdom. The phylogenetic roots of human social behavior and the evolutionary trends enforced by it have given humans their paramount position in nature – but humans themselves may eventually fall victim to this bizarre product of intraspecific selection.

Phylogeny
by Erich Thenius

By kinship classification (taxonomy), *Homo sapiens* belongs to the order Primates. Hence the phylogeny of this order of mammals is of special interest to us as human beings. It is true that phylogenetic evolution cannot be documented directly, but, by using various methods, the development of traits, behavior patterns, life processes, and life activity can be reconstructed to a certain degree of probability. Such reconstructions are expressed in genealogical form. Of course we must bear in mind that these "pedigrees" are based on hypotheses, and may be subject to amendment in the light of new findings. Hence, there is constant revision of family trees in the course of time. It is an easier matter to appraise the evolution of particular traits or conditions, and in this connection the scientist must inquire whether a particular idea of their evolution corresponds to the actual historical development or is to be regarded merely as an approximate model.

Some examples will serve to make this clear. In accordance with the various evolutionary conditions among living primates (for example, prosimians, simians, anthropoid apes), the species may be arranged in a series morphologically (according to form), or – where only particular organs are concerned – anatomically, to provide a model of past evolution in which the several species included do not stand in a relationship of descent. A further step is taken by paleontology, the science of living creatures of past ages, with its fossil finds which – if they occur in a time sequence – may represent successive stages of development. In addition to these models, based on traits of present-day (recent) or extinct (fossil) species, there are other clues to primate evolution. With the growth of interest in primatology, which in the past few years or decades has become a scientific discipline

Skillfull hands are indispensable to apes and monkeys. This spider monkey also uses its prehensile tail as a "fifth hand."

in its own right, with relevance to the study of humans in particular, there are now biochemical data. These are concerned with the proteins (amino acid sequences, albumins, globulins) and blood pigments (hemoglobins), as well as the study of the cell nucleus (karyology) and chromosomes (vehicles of the genes). Some of the data are in accord with the results of studies in morphology and anatomy, while some are at variance with them. It turns out that evolution by type of cell nucleus, that is, by chromosomes (number and configuration) does not necessarily agree with evolution according to blood serum findings (reactivity to extra-specific proteins) or morphological and anatomical features. Thus according to blood serum findings, the African anthropoid apes (gorilla and chimpanzee) are closer to humans than to the orangutan of Asia. This result has led some to join chimpanzee, gorilla, and human in a single family, the Hominidae, but this move may not do justice to the special position of humans.

Nevertheless, such serological findings, supported in part by similarity or agreement in chromosomes, support the statement that the anrhopoid apes and humans are to be traced to common ancestors. This again is in agreement with the fossil evidence.

But first we must answer the question: Just what are the primates? This issue is not as abstract as it seems. The early exponents of evolutionary research, in the time of Charles Darwin and Ernst Haeckel, were aware that the primates arose from the insectivores. To be sure, shrews, moles, and hedgehogs were not what they had in mind; they were thinking of the tree shrews (Tupaiidae), which are reminiscent of squirrels in appearance. Some zoologists classified these in the Insectivora, others in the Primates (or as subprimates). Understandably enough, fossil finds, often consisting of tooth and jaw relics only, present still more difficult questions. Thus, the position of various extinct families of mammals (for example, Carpolestidae, Amphilemuridae, Apatemyedae, Picrodontidae) has long been disputed. Do they "rank" as primates, or "merely" as insectivores? So far as the Carpolestidae and the Picrodontidae are concerned, primate membership seems secure.

Modern tree shrews, which in many respects indeed represent a model of archaic primates ("eoprimates"), and agree with primates in a number of

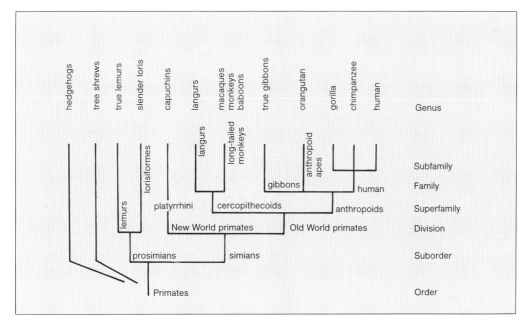

Similarity diagram based on comparison of blood fluid composition (serodiagnosis). Note the placement of the African anthropoid apes and human in one family. The tarsiers were not included in the study.

(Opposite page) The term "prosimian" subsumes three divisions of the primates. This is a mother sifaka with her little one. Sifakas belong to the indri family of lemurs.

(Below) Diagram of the evolution of the Old World primates according to chromosomes, the genetic vehicle. The numbers of chromosomes as well as their configurations suggest that the African human-like apes (chimpanzee and gorilla) are closest to humans and may be traced to common ancestors.

traits, but differ from them, for example, in mode of reproduction and rearing, brain structure, and the auditory region, are therefore almost universally regarded today as belonging to an order of their own (Scandentia). Ernst Haeckel had already separated them – together with the elephant shrews (Macrocelidea) – from the other insectivores in a group of Menotyphla (cecum present, well-developed organs of vision, and corresponding brain development). However, they, together with the primates, may be traceable to a common group among the insectivores, namely the Lepticoidea. The primate remains of the greatest geological age data from the Late Upper Cretaceous of North America. They have been described under the name *Purgatorius*, and belong, as is documented by more complete remains from the earliest Tertiary (Old Paleocene), to an independent, entirely extinct group of prosimians (Plesiadapiformes) that still lived in North America and Europe at that time. They were prosimians with a highly specialized dentition, reminiscent of some modern marsupials (for example, *Dactylopsila*). For a time, because of this dentition, relations to the aye-aye of Madagascar (*Daubentonia* = "Chiromys") were suspected, as is indicated by various older names (for example, *Chiromyoides*). *Plesiadapis*, the best-known genus, indicates that they were tree dwellers with clawed digits, not unlike present-day squirrels. The first digit was not opposable; a well-developed olfactory brain and the placement of the eyes shows that these animals were olfactory (macrosmatic) rather than visual, so that they lacked such key features as grasping hands and depth perception, typical of modern primates. In view of the marked specialization (of the jaw), they cannot be regarded as ancestral forms of the "euprimates." Through enlargement of the eyes and corresponding parts of the cerebrum, these became distinctly visual, with an ever diminishing sense of smell with reduction of the olfactory brain, so that the macrosmats ultimately became "microsmats." The facial skull was shortened and the cranium became more spacious. The evolution of grasping hands and feet, with true opposability of the first finger and toe, permitted claws to be transformed into nails, and clawed climbers became grasping climbers. Thus, the prosimian type arose among the primates.

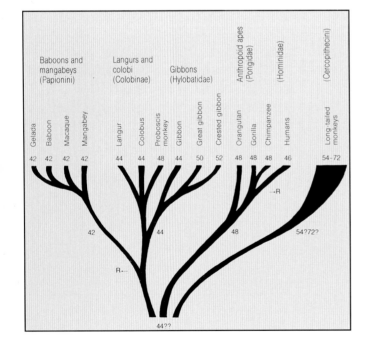

PROSIMIANS

Category
SUBORDER

Classification: The traditional category of Prosimiae subsumes three infraorders of the order Primates: the Strepsirhini (primates with nose leather) are represented by two, the lemurs and the lorises, and the Haplorhini (primates without nose leather) by one, the tarsiers.

Infraorder Lemuriformes (lemurs)
Family Cheirogaleidae (mouse lemurs)
Family Lemuridae (true lemurs)
Family Lepilemuridae (sportive lemurs)
Family Indriidae (indris)
Family Daubentoniidae (aye-ayes)

Infraorder Lorisiformes (lorises and galagos)
Family Lorisidae (lorises)
Family Galagidae (galagos)

Infraorder Tarsiiformes
Family Tarsiidae (tarsiers)

LEMURS
(12 genera with 21 species)
Body length: 4.4 in. (11 cm; least mouse lemur) to 32 in. (80 cm; indri)
Tail length: 2–14 in. (5–60 cm)
Weight: 1.8 oz (50 g; least mouse lemur) to 15 lb (7 kg; indri)
Distinguishing features: Tail commonly long and bushy (except indri); flat nails; broad fingertips; second toe with grooming claw; first finger and first toe opposable; generally long muzzle; moist rhinarium; nocturnally active forms with large eyes, commonly without cones, reflective tapetum lucidum; lower incisors and canines nearly horizontal (grooming comb); some species with fat storage at base of tail, incomplete temperature regulation, and resting period, depending on temperature and humidity.
Reproduction: Gestation 60–150 days; 1–4 young per birth, depending on species.
Life cycle: Time of weaning, where known, about 45 days (least mouse lemur) to 9 months (indri); sexual maturity generally at 7–24 months; lifespan in the wild not known, up to 30 years in captivity (black lemur).
Food: Predominantly vegetable, but occasionally insects and small vertebrates as well.
Habit and habitat: Predominantly nocturnal, but some species active by day; arboreal, almost exclusively in forest land; good climbers and leapers; often living in family groups.

LORISES AND GALAGOS
(5 genera with 10 species)
Body length: 4.8–14.8 in.; 12–37 cm
Tail length: galagos 6.4–18.4 in. 16–46 cm, others 0–2.8 in. (0–7 cm)
Weight: 2.2 oz–3.5 lb; 60–1600 g
Distinguishing features. Large eyes, directed forward, with tapetum lucidum; comparatively short muzzle; anterior part of lower jaw forms a tooth "comb"; left and right upper incisors separated by a gap; in the galagos, hind legs much prolonged; thumb and great toe spread wide and opposable; in lorises, second finger and second toe reduced, sometimes almost absent; external ears of galagos and slender loris very large and membranous, folded up by muscles.
Reproduction: Gestation 110–193 days; usually only 1 young per birth; birth weight 0.4–1.8 oz (10–50 g).
Life cycle: Weaning usually at 45–70 days; sex-

Prosimiae
Prosimiens FRENCH
Halbaffen GERMAN

ual maturity at 8–12 months; life span in captivity 12–15 years or more.
Food: More or less vegetable or animal according to species.
Habit and habitat: Exclusively nocturnal tree dwellers; rainforests, dry forests, and tree savannas; roam about singly or in groups; territorial behavior.

TARSIERS
(1 genus with 3 species)
Body length: 4.8–5.2 in.; 12–13 cm
Tail length: 22–42 cm
Weight: 4.3–5.7 oz; 120–160 g
Distinguishing features: Conspicuously large eyes; fairly large and independently mobile external ears; head rounded and rotatable 180 degrees to either side; hind legs much prolonged; grooming claws at second and third toes.

Reproduction: Gestation 180 days; 1 young per birth; birth weight 0.7–1.1 oz (20–30 g).
Life cycle: Weaning age not yet known; sexual maturity probably about 12 months; life span probably over 8 years.
Food: Evidently animal only, predominantly insects.
Habit and habitat: Arboreal, active in twilight and at night; in forest, occasionally in bush or thicket; good leapers (leaps in perpendicular posture); average extent of territory 1 hectare.

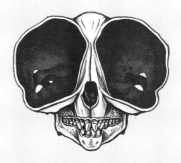

Skull of a nocturnal prosimian. Note enormous eye sockets. In most prosimians, the eyes are set laterally in the head, but the tarsiers' gaze is forward, much as in the higher primates. This makes possible some depth perception and ability to estimate distances.

	🦗	🍒	🌿
Tarsier	●		
Galago	●		
Slender loris	●		
Potto	●	○	○
Squirrel monkey	○	●	
Capuchin	○	●	
Spider monkey		●	
Howler monkey			●
Langur			●
Macaque	○	●	○
Baboon	○	●	○
Gibbon		●	○
Chimpanzee	○	●	○
Orangutan		●	○
Gorilla		○	●

Comparison of diets (above). Most primates are omnivorous, using worms, snails, insects, and other small vertebrates (first column of table above); or flowers, buds, soft and hard fruits, seeds, and grasses (second column); or leaves, stalks, and roots (third column). Among small forms, animal sources of nourishment predominate. The diet of terrestrial species is generally more varied than that of the arboreal species. Solid circles indicate predominant food sources.

Prosimians' hands and feet (below). Despite differences in form, both hands and feet exhibit some common features: broad palm and sole of foot; flattened fingers and toes increasingly equipped with nails; sensitive pads and balls of fingers as seat of a refined sense of touch; and detachment of the thumb or the great toe from the other digits for grasping branches firmly. In the slow loris and the potto, the extremities form tongs, with supernumerary fingers much reduced. The long tarsals of the bush babies are fitted for long leaps. Suction cups at the tips of the fingers and toes of tarsiers provide a firm hold, even on a smooth surface. The aye-aye's middle finger serves to dig food out of narrow crevices. Hand (top) and foot (bottom) of (1) bush baby, (2) tarsier, (3) lemur, (4) aye-aye, (5) slender loris, (6) slow loris, and (7) potto.

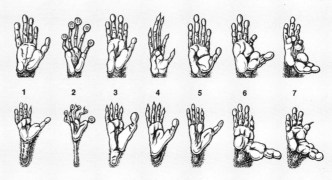

Prosimians

Introduction
by Kurt Kolar

As the name suggests, this group of the Primates is comprised of forms that share only certain characteristics with the true simians. Thus a trait of mammals that depend predominantly on their sense of smell, namely the hairless nose leather, is well developed in nearly all prosimians and absent in the simians. Only the tarsiers, the most simian-like forms of this entire group, lack the rhinarium. Thus, the prosimians qualify as predominantly visual animals, with a substantially smaller olfactory center in the brain than in the more macrosmatic mammals. Whereas the eyes and ears are employed primarily for orientation and searching for food, most prosimian species use scents secreted by their own glands to identify their individual territories. Other traits of the prosimians are the bicornuate uterus, opposability of the first finger and first toe, and the claw at the second toe. All other fingers and toes are equipped with nails.

When the founder of systematic zoology, Carl von Linné (Carolus Linnaeus), first had the opportunity to examine prosimians – probably sifakas – he at once drew the correct conclusion, and coined the term Prosimiae.

In their ranges of distribution, the prosimians – the lemurs of Madagascar and on the Comoro Islands, the galagos and lorises of Africa and Southeast Asia, and the tarsiers of Indonesia and the Philippines – have evolved all sorts of specializations to occupy a wide variety of ecological niches. We find diurnal and nocturnal species (active, respectively, by day or at night), herbivores and predators, swift leapers and slow climbers, and solitary and gregarious forms. The diversity of habits of the several genera of prosimians is matched by their diversity in outward appearance and certain physical traits. There are prosimians of all statures, from the smallest of all primates, the least mouse lemur – which resembles a dormouse – to forms of medium dog size. Some might be mistaken for arboreal squirrel types; the indri, for size and vociferous occupation of its home territory, is comparable to the gibbons; the aye-aye has powerful, rootless incisors like a rodent, and most lemurs have a fox-like countenance.

This is an abundance of forms that science has not yet been able to classify satisfactorily in every detail. Especially in the finer systematics of the prosimians, there will certainly be many changes as our knowledge of these animals increases.

The position of the "problem group," the tree shrews, long regarded as an infraorder of Primates, may now have been established. By reason of their long muzzle, tree shrews immediately remind one of insectivores, and many researchers originally placed them in that order. The bony orbits and other cranial traits, however, are characteristic of primates, as is the flexible hypoglottis. Although the face, with eyes set in the side of the head, is not really primate-like, and moreover an essential faculty of the primates, namely opposability of the thumb, is lacking, they nevertheless were formerly classed as prosimians on the basis of the traits mentioned above.

That conception has recently come into question, for in mode of reproduction and in structure of the brain and auditory region, the tree shrews are fundamentally different from the primates. The tree shrews have now been assigned their own order Scandentia, separating them from the Insectivora also. Tree shrews and primates are related to each other only through common ancestors, surmised to have been insectivores. Today, the tree shrews may at least be taken as models of what earlier forms of primates, or links between them and the insectivores, may have looked like and how they may have behaved.

INTRODUCTION

All lemurs, galagos, lorises, and tarsiers are entered today in the lists of the International Convention for the Protection of Species; that is, they are among the endangered species of animals, and some are already threatened with impending extinction.

One of the earliest Malagasy species, the giant lemur *(Megaladapis edwardsi)*, was exterminated about three thousand years ago, probably by the early settlers of this, the Earth's fourth-largest, island. As we know, the four species of giant ostrich-like flightless birds also vanished after the arrival of humans.

Madagascar is in any event a tragic example of unrestrained annihilation of tropical forest. A mere tenth of the original vegetable cover has been preserved; vanilla and cocoa plantations now stretch far and wide where once there was uninterrupted primeval forest. Clearcutting for timber as well as overgrazing have transformed large areas of the island into steppeland. Even the semideserts in the southwest have been partly destroyed by slash-and-burn agriculture. The prohibition of animal exportation enacted decades ago has not sufficed to preserve species that have become rare. In the country itself, the need for better protective measures is very urgent. Unfortunately, the twelve existing preserves are inadequately policed. Hence we do not know whether the lemurs' last habitats, already much constricted, can be preserved in Madagascar, and we must fear that before long certain species will have ceased to exist except in zoological gardens and special institutions.

But not all species thrive in captivity. Two members of the indri family, the avahi and the indri, have so far failed to survive for any length of time even in the Jardin Zoologique de Tsimbazaza, the zoo at the Malagasy capital of Tananarive. Aye-ayes have occasionally lived in zoos for more than twenty years, but they did not reproduce, and here again, it is hardly to be expected that this prosimian species, reduced to a total stock of perhaps only fifty individuals, can survive in human keeping. We can but hope that the last habitats of the remarkable animal forms of Madagascar may perhaps yet be rescued, with international aid.

Prosimians live on vegetable and animal food. The potto (top left), likes insects as well as fruits; the black lemur (top right; this one is a female) lives almost exclusively on fruits and other vegetable matter. The food is brought to the mouth from the side. Outside of Madagascar, all prosimians are nocturnal, like the slender loris (next page). On the island, some species such as the ring-tailed lemur (bottom) are active during the day; without daylight, the conspicuously marked tail could hardly perform its function as a "semaphor."

In the zoos, only the sturdy and easily bred ring-tailed lemur is regularly found. Other lemurs, if only because of the aforementioned export embargo, rank as rare zoo inmates. In West Germany, the Cologne zoo has specialized in the keeping of prosimians, and achieved some notable breeding successes. In some other zoos also, in primate centers, and in zoological departments of universities, some prosimian species have not only been successfully kept for years but also bred for generations. The value of such breeding successes cannot be overestimated in the efforts being made towards the preservation of these species.

Of the nocturnal prosimians living in Africa and Asia, the galagos, the slow loris, and the slender loris are especially charming denizens of "darkness exhibits" in zoos. They may be kept without major difficulty, will reproduce, and present themselves in constant motion under artificial nighttime conditions.

(Above) Two prominent enemies of the Malagasy prosimians: the dogheaded python of Madagascar (top) and the civet cat or fossa (bottom). (Right) Like all lorises, the slender loris moves slowly and deliberately through the branches, often twisting its body in bizarre fashion. (Far right) The galagos, on the other hand, are remarkably swift climbers and leapers.

Because of the difficulties of observing arboreal prosimians, some active only at night, in the wild, or – in the case of the rarer species – finding them at all, available data on their behavior derives largely from observation of zoo animals. The Malagasy sifakas were probably the first prosimians ever known in Europe. The then French governor of the island, Gaston Étienne de Flacourt, in 1658 reported "white, dogheaded beasts with manlike bodies." Live prosimians, too, first came to Europe about that time. They were chiefly African galagos, at first taken for squirrels because of their size, their bushy tails, and their leaping skill. Pottos and golden pottos, supposedly close relatives of the sloths, were also discovered towards the close of the seventeenth century. Their Asian counterpart, a loris, was seen by a French traveler in 1684 at the court of an Indian Grand Mogul.

In 1748, the first living ring-tailed lemur was brought to Europe. Thereafter, other species of lemurs came into zoos, where they survived for longer or shorter periods of time. Housing conditions at that time of course by no means met the animals' requirements, but dealers and sailors steadily provided replacements into the present century. These practices came to an end about two decades ago. Madagascar enacted an export embargo, and the animals still to be found in zoos were assembled to form breeding groups. The private zoo of Georges Basilewsky at Cros-de-Cagne near Nice, which made a name for itself as a breeding place for lemurs, is especially notable.

The striking figure of the big Malagasy lemurs, accompanied by widespread misinterpretation by the aborigines of their behavior, and their cries – audible over long distances – resulted in their being regarded with awe, indeed as spirits of the departed. That was the best protection the lemurs

ever enjoyed, as taboo animals not to be killed on any account. Unfortunately, the taboo has long since ceased to exist.

The big eyes of the tarsiers and their ability to rotate the head 180 degrees to either side rendered them harbingers of ill fortune on Borneo, to be avoided or killed. Thus, these animals likewise played a major role in the folk beliefs of their homeland.

Phylogeny
by Erich Thenius

The multiformity of the prosimians raises the question as to whether this group of animals really represents a natural unity, that is, a "monophyletic" (Greek *mono* "single," *phylon* "tribe") entity, proceeding from a single stock or root, or merely a developmental level in evolution. The answer to this question is crucial, not only in terms of phylogeny but also more generally for the purpose of classifying the animals by as natural a system as possible, in accord with actual relationships of kinship.

Among the living prosimians of today, there are highly diverse adaptive forms, ranging from the lemurs through the galagos and lorises to the tarsier at one extreme, and to the Malagasy aye-aye *Daubentonia* at the other extreme. Detailed investigations of external and internal anatomy (for example, structure of fetal membrane) and of the blood serum confirm the classification of the Primates (due to the work of English zoologist R.I.Pocock) into two groups according to the presence or absence of a rhinarium: Strepsirhini (lemurs, lorises, and galagos) and Haplorhini (tarsiers and simians). These names are derived from the Greek *streptos* "twisted, convoluted"; *haplos* "simple"; and *rhino-* "pertaining to the nose." Pocock's classification is no longer fully accepted today, but there are other traits to document his division. This means that the prosimians do not constitute a natural entity.

The "strepsirhine" prosimians are confined to the Old World today, where they are represented by the infraorder Lemuriformes on Madagascar and by the infraorder Lorisiformes in Africa and Southern Asia. In the Early Tertiary, strepsirhine prosimians were distributed chiefly over the northern hemisphere as adapids (Adapiformes, for example, *Adapis, Notharctus, Pelycodus*), which is understandable given the then subtropical to tropical climates in much of Eurasia and North America. These adapids lacked the "comb" dentition so characteristic of modern lemurs, but the postcranial skeleton (the bones of the body and limbs "posterior to the skull") differed only insignificantly from those of the genus *Lemur*. The adapids vanished at the close of the Eocene. Whereas there are only Quaternary fossil finds of the Malagasy prosimians, galagos and lorises are to be found in the Miocene of Africa (for example, *Progalago, Komba*) and Southern Asia (for example, *Nycticeboides*). The few fossil finds bear witness not only to the completion of separation of galagos (Galagidae) and lorises (Lorisidae) but also – in postcranial skeletal remains of *Nycticeboides* – to the occurrence of slow climbers, similar to the slow lorises and pottos of today.

Because of the absence of Tertiary fossil finds of other lemur and loris relatives, any phylogenetic statements must refer to forms living today or in the Ice Age. Thus, the mouse lemurs (Cheiroga-

The moist nose leather is accepted as the common trait of all prosimians but the tarsiers.

Evolution of prosimians and New World simians. *Lepilemur* (sportive lemurs) and *Megaladapis* (giant lemur) are often regarded as belonging to separate families.

leidae), usually accounted to be true lemurs, and in the opinion of M. Cartmill, J. Tattersall, and H. Schwartz constitute a "sister" group of the Lorisidae. Thus they represent loris relatives (Lorisiformes). According to the serum composition of their blood, however, the mouse lemurs are, after all, closer to the Lemuridae.

Were one to assign the mouse lemurs to the Lorisiformes, a question of distributional history would arise: Where, then, did the Lorisiformes first appear – Africa, Southern Asia, or Madagascar? Therefore, while the question of the phylogenetic provenance of the Lorisiformes remains unanswered, the exclusively Malagasy "true" lemurs and Indriidae may certainly be derived from Early Tertiary *Adapis* relatives (Adapiformes). When the ancestors of the lemurs and indris (Lemuriformes) reached Madagascar however, remains to be determined from additional fossil finds. In view of the abundance of forms and the diversity of the Malagasy prosimians, which was much greater still in the Early Holocene – before the human settlement of Madagascar – a very early "immigration" in the Tertiary is to be assumed. Besides the actual lemurs (for example, *Lemur, Varecia*), indris (for example, *Indri, Propithecus*) and aye-aye (*Daubentonia madagascariensis*; sole species of the family Daubentoniidae) still living today, in the Quaternary there were giant lemurs (*Megaladapis*) of anthropoid size and numerous other prosimians that are now extinct (for example, *Archaeolemur, Hadropithecus, Palaeopropithecus, Mesopropithecus*). Some of these had nearly reached the evolutionary level of true simians, as indicated by the size and fissures of the cerebrum, the eye sockets directed forward, and the two-ridged (bilophodont) molar teeth. In *Palaeopropithecus*, the postcranial skeleton is distantly reminiscent of that of the sloths, to which, incidentally, the remains were originally ascribed. The classification of the genera last mentioned is still in dispute (members of the Indriidae, or representatives of their own families Archaeolemuridae and Palaeopropithecidae?). The same is true of *Megaladapis*, who was a pronounced leaf-eater like the sportive lemur *Lepilemur*, and a grasping climber like the marsupial koala of today. Detachment from the Lemuridae in a separate family Megaladapidae seems unnecessary. The skull of the giant lemur *(Megaladapis edwardsi)* varied much in size and shape but attained 12 in. (30 cm) in length. *Megaladapis* was known to the first settlers of Madagascar, and may have been exterminated by humans.

The phylogenetic origin of the aye-aye *(Daubentonia madagascariensis)*, at first regarded as a rodent because of its rootless incisors, is not settled, but there might be rather close connections with the Indriidae. A high degree of specialization (not only in dentition) is evidence of a long separate line of descent. In terms of brain development also, the aye-aye is the most highly evolved prosimian. Because of its peculiar manner of feeding, it is regarded as an ecological equivalent of the woodpeckers, which are absent in Madagascar; it seeks out insect larvae in hollow branches.

Remarkably, the prosimians of today, where they occur together with simians, have for the most part become nocturnal, thus escaping direct competition with the higher primates.

The exception to the rule: The tarsier, by contrast with all other prosimians, has no moist rhinarium. Thus, like all the simian species, it belongs to the group of the Haplorhini.

Lemurs

by Kurt Kolar

Among the ancient Romans, the word *lemures* referred to the spirits of the departed, who appeared especially in the month of May to haunt their posterity. The prosimians of Madagascar were probably given this name because, with their eyes shining in the dark and their howls, they seemed rather uncanny to the first Europeans to encounter them.

All the prosimians of Madagascar, about 40 percent of the mammalian species living on the island, belong to this group. Because of an early separation from the African mainland, the lemurs were able to occupy a wide variety of habitats of Madagascar and the Comoros – their farther range of distribution – without competition from more highly evolved primates. Through that separation, Madagascar became much the same sort of zoological "curiosity shop" as Australia.

The lemurs evolved forms of adaptation that remind one of certain African prosimians, or of higher primates generally. Thus the mouse lemur – except that the ears will not fold up – looks like a galago; sifakas and indris howl like gibbons, and have, like them, an erect posture; the ring-tailed lemurs have been compared in their behavior with the baboons and macaques, which similarly live in communities. All possible habitats and modes of life have been utilized by lemurs. They occur in the tropical rainforest as well as in dry thornbush terrain, in trees and on the ground, feeding on insects and other small prey, mixed diets, or special plants and plant parts; they may be active by day or by night, travel by leaping and by climbing, and use either two or all four limbs for locomotion.

Yet despite their various adaptations and their differences in size and form, there are essential common traits. Thus, the sense of smell is still rather more important than in the true simians, as the longer facial skull suggests, and it plays a part in territorial and social behavior as well. All species mark their home territories with scent, using urine in the simplest case, but usually excretions from specialized scent glands. The claim to a particular territory is thereby asserted, or – as among the ring-tailed lemurs – regular "stink battles" are waged between males.

Another common feature is a horizontal comb, formed by the four lower incisors and two canines and used as a grooming tool. Matting and any foreign objects caught in the fur are easily eliminated with this six-toothed comb – four-toothed in the case of the indris. Also, the tree resins in the diet of some species are readily scraped from the bark by this means. There is a "toothbrush," namely a

(Above) In the dark of the forest night, the lemurs' big eyes shine with a ghostly luminescence; they are equipped with a reflective layer, better utilizing the scanty rays of light. (Right) With increasing clearing of the forests of Madagascar, humans are destroying the habitat of the lemurs and many other arboreal species.

hypoglottis, for cleaning the comb. Remarkably enough, the first premolars have taken over the function of the modified canines, having become long and pointed.

Another grooming tool is the claw of the second toe, with which the lemurs can scratch themselves, dog-fashion. Then there is mutual grooming, in which, however, by contrast with the grooming behavior of simians, the hands are employed only to grasp the partner.

Reproduction is generally confined to certain months, depending on the rainy season. These animals belong to an archaic reproductive type in which menstruation is absent.

Family Cheirogaleidae (mouse lemurs)

All species in this group are smaller than the other lemurs. They are active at night and in twilight, have large eyes with a reflective layer, the tapetum lucidum, and large external ears. Despite their thick and woolly fur, they cannot completely regulate their body temperature, which ranges between 11 and 95°F (25° and 32°C). As soon as

the weather grows cool, but also during the dry season, these lemurs fall into a state of torpor, such as we commonly find only among insectivores, bats, and a few rodents. This "winter sleep" may last for days or even months. Fat stored in the hind legs and in the tail region is consumed during this time.

The mouse lemur *(Microcebus murinus)*, the smallest living primate, and furthermore a nocturnal animal, is not very conspicuous even to the natives. Fortunately, these tiny creatures have never been considered particularly good eating, and they are still comparatively abundant; in the east and west of Madagascar, they live in dry forests and rain forests, in swampy and scrub terrain.

Like a small rodent, this lemur progresses in a series of short sprints. However, it is also a good leaper, the tail being useful as a balancing pole. The mouse lemur sits up to eat, and curls up to sleep.

During the dry season, fat stored in the body is an aid to survival. There is no prolonged hibernation, but the body temperature may fluctuate between 77 and 95°F (25° and 35°C) within only a few days. The animals kept by the French zoologist Jean Jacques Petter became torpid at a surrounding temperature as high as 64°F (18°C). This was not observable in the case of the mouse lemurs kept at the University of Vienna department of zoology.

The diet consists of nectar from flowers and other vegetable juices, insects, lizards such as chameleons, and probably also small birds that are surprised during sleep. These lemurs themselves may fall prey to owls and occasionally diurnal birds of prey.

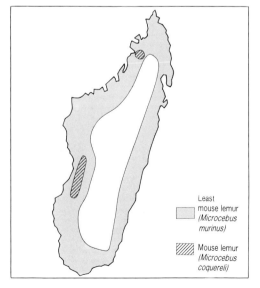

The pygmy among the prosimians, most of which are rather small, and the smallest of all primates, is the mouse lemur, which is still widely distributed on the island of Madagascar (see map). Its very small size and nocturnal habit protect it quite well from enemies, especially humans.

▷ Lemurs are definitely crepuscular and nocturnal animals, but occasionally they are active by day.

Since the mouse lemurs are found singly when seeking food at night, they were long thought to have a solitary habit. It is now known, however, whether they form local reproductive communities. The females spend the day with their young in social groups of as many as 15 individuals in their sleeping nests or in hollow trees, while the males occupy hiding places by ones and twos. Sleeping places in hollow trees are lined with dry leaves. In eastern Madagascar, hollows in the "traveler's tree," a member of the banana family, are preferred. Free-standing nests made of twigs are found in underbrush. They are some 8 in. (20 cm) in diameter. Individual territories extend for a radius of about 160 ft (50 m) around the sleeping place; a male territory may overlap those of the females. The animals mark their territories by sprinkling some urine on their hands and rubbing it on the soles of the feet. Traces of scent are thereby left on all twigs and branches visited inside the territory.

High-pitched cries, extending into the ultrasonic range, maintain communication between group members. Other known vocalization includes chattering in anger; a high peeping when a male pursues a willing female; three to four screeches in succession as a warning cry; and the contented purring of the babies. Mouse lemurs reproduce at least twice a year. At that time, the males are exceptionally busy laying scent with excretions from their ischial and genital regions.

We have some information about other reproductive traits from the researches of Jean Jacques Petter. During pregnancy, the females gain 0.21–0.25 oz (6–7 g) in weight, and bring their two or three young forth in the nest or in a hollow tree. The births occur in the months between October and March. After the mother has bitten off the umbilical cord, she consumes the afterbirth. The newborn have gray-brown fur, and have their incisor and canine teeth, but their eyes are not open until after four days.

Immediately after birth, the babies are thoroughly licked by the mother, while being held with the hands. They do not cling to the mother's belly, nor are they afterwards carried on the back like the young of larger lemurs. The mother grasps them at the flanks and so carries them. At two weeks, the little ones are leaping about, and a may be seen climbing eagerly. typical of hunting week later they siblings, try They play chasing games each mammals; they tussle with their to catch tails, and pursue and topple other. The mother gets into the game and allows the young to climb about on her back and tug her ears. Nursing continues only to the seventh week of life, the first solid food having already been taken at three-and-a-half weeks. At two months, a young mouse lemur is independent and behaves like an adult.

These dwarf prosimians may be admired in the "darkness houses" of many zoos, where they reproduce regularly.

Coquerel's lemur *(Microcebus coquereli)*, a larger edition of the mouse lemur, is still fairly abundant in some isolated forests. In the northwestern coastal woods, a substantial population density as high as about 250 individuals per square kilometer (1 km^2 = 0.36 mi^2) has been estimated. To be sure, these areas of occurrence comprise one private and several State preserves. Elsewhere, Coquerel's

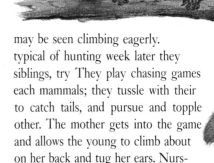

(Right) Bushy-eared or lesser mouse lemur. This extremely rare species has hardly been studied at all, and only four specimens have been reported to date. (Below) Coquerel's mouse, or dwarf, lemur.

lemur, like all kindred species, is endangered by the steadily advancing destruction of habitat.

No marking behavior has been observed in the case of this species, and special scent glands are absent. Its habits probably resemble those of the mouse lemurs. One to four young are born in the second half of November and cared for in the nest by the mother. These nests are somewhat larger, about 12 in. (30 cm) in diameter.

These mouse lemurs have quite often been kept in zoological gardens. In fact, a pair at the London Zoo has bred several times. An extreme age of 15 years and 5 months is reported by that source. These animals are hardly to be seen at all today, outside their homeland.

In the medium-sized fat-tailed dwarf lemur *(Cheirogaleus medius)*, the reservoir of stored fat is especially noticeable, the deposit in the tail being up to 0.8 in. (2 cm) thick.

The animals occur in most forests of the west and south of Madagascar, but they are seen very little, because during the dry season – about a quarter of the year – they are dormant, with much reduced body temperature. At a surrounding temperature of 60°F (16°C, a body temperature of 63.5°F (17.5°C) was measured in one case. Since the animals merely withdraw into the rather damp hollows of living trees to sleep in the hot, dry woods, considerable assemblages of mouse lemurs result. Several will lie side-by-side and one on top of another in the hollows, covering themselves with leaves. Otherwise, however, they are not very gregarious, and tend to be solitary foragers. Not only the adverse climatic conditions, but more particularly the shortage of food during the dry season, probably accounts for the long periods of dormancy of these lemurs. Fruits, flowers, nectar, and pollen are not available until October, when the rainy season has begun. This is also the mating season.

Territorial behavior and marking have not been observed in the dwarf lemur.

The greater dwarf lemur *(Cheirogaleus major)* likewise has no special scent glands, but distinct markings are made in the form of feces and urine left on branches. During the day, these lemurs occasionally even sleep in company, either in hollow trees or in the large nests, 12 in. (30 cm) in diameter, that they build in the treetops out of grass and leaves. During the dry season, they retreat into

The drawings show (left) the fat-tailed dwarf lemur and (right) the greater dwarf lemur, which is shown in the photograph resting on a human arm.

hollowed-out tree trunks or burrows between the roots. However, it may be that they remain in a state of lethargy for a few days longer.

Large mouse lemurs generally move slowly and stealthily, without much leaping. Besides fruits and flowers, insects are eaten, as well as small vertebrates. Nocturnal birds of prey and the civet cat *(Cryptoprocta ferox)*, Madagascar's most formidable predator, can be a threat to them.

These lemurs are not noticeably clamorous, only the juveniles being heard to chirp when distressed. Their life is presumably as amiable and peaceable as their looks; at least one gains that impression when observing two greater dwarf lemurs licking and nibbling each other's coats.

In December or January, the two or three young are born in a hollow tree lined with leaves. Their eyes open on the second day, at three weeks they are climbing about, and a week later they will be

The fork-crowned dwarf lemur's black dorsal stripe parts at the top of the head, giving this unmistakable facial marking. Note the grooming claw at the second toe, a characteristic of all prosimians.

following their mother. At the age of a month and a half, they are independent.

Only four specimens of the bushy-eared dwarf lemur *(Allocebus trichotis)* have so far been found, one in a wood near Mananara on the east coast in 1966. This is the least-known of the prosimians. Outwardly, the species resembles the somewhat larger fat-tailed dwarf lemur, and both sleep by day in hollow trees.

We know a bit more about the fork-crowned dwarf lemur *(Phaner furcifer)*. The upper incisors are much prolonged in this species, and each upper premolar resembles a canine. Thus, it looks as though the animal had two canines on each side in the upper jaw. These variants in dentition have prompted taxonomists to place the forkcrowned dwarf lemur in a genus of its own.

The hairless glandular area under the throat of the male clearly indicates that this is another territorial form. The home territory, about 2.5 acres (1 hectare) in extent, is marked out by glandular secretions and loud cries. Sleeping hollows and feeding trees are found within the territory.

The animals are liveliest after sundown. They leave their sleeping places, and at some distance they begin to call, so loudly that the woods seem full of them. Where environmental conditions are still favorable, actually as many as 8 individuals can be supported on 2.5 acres (1 hectare). They are on the move until first light. They move on all fours with raised tail, and will leap a distance of up to 16 ft (5 m). Sharp fingernails and toenails are useful for landing on tree trunks.

In mid-November, at the beginning of the rainy season, the single offspring is born. First it is cared for in the hollow tree, then its clings to the mother's furry underside and is carried about. Later, like most baby lemurs, it will crouch on the mother's back.

In the Berlin Zoo, there were dwarf lemurs before the First World War. Later, others were kept in the primate section of the National Institute for Natural History in Paris.

Family Lemuridae (True lemurs)

In markings, the gray gentle lemur *(Hapalemur griseus)* is very similar to the ring-tailed lemur. A milky-white secretion from the glands of the upper arm, having an odor of beeswax, is transferred to a horny area on the forearms. Thus, objects in the environment of the lemur come to be scented. The acoustic territorial claim takes the form of a loud yodel. In confrontation with a conspecific, use is made of olfactory threats or intimidations: The tail is passed several times over the inside of the forearms in applying scent gland secretions; then the tail is waved, fanning a cloud of scent towards the adversary. Whistling sounds are made as well, probably very impressive to lemur's ears, and the ears are laid back, producing a hostile facial expression to the human view.

Bamboo thickets are the chief abode aof gray gentle lemurs. There, even the human nose readily detects the animals' scent, which clings to the tree trunks. These lemurs are to be found at Lake Alaotra. Probably, the animals living there in the rushes, traveling over the mats of aquatic plants on all fours, represent a subspecies. Unfortunately, its days may be numbered, for the lake area is increasingly being used for planting rice. In the dry season, the fishermen set fire to the heaps of reeds

and skill the animals when they try to escape the fire. The stock of gray gentle lemurs elsewhere is actually increasing somewhat. Extensive clearing of forest – generally a catastrophe for wildlife – has extended the habitat of the gentle lemurs. They especially prefer the new growths of bamboo thickets. They are adapted to climbing and leaping among vertical stalks. Even at rest, they tend to lean against vertical trunks or branches.

During their daytime sleep, the animals are quite lethargic and easily captured. They are often hunted for food. In the evening and early morning twilight, groups of three to five individuals, roaming about and maintaining communication with each other by expressive grunts and cries, are very lively.

Sugarcane and grasses form an important part of the diet. Enemies include snakes, kites, buzzards, and mongooses.

The single young fully furred and with its eyes open, is born in the nest hollow between October and January. At first it is carried with the mouth, then it clings to the fur of the mother's abdomen, and later it rides on her back. If the baby is deposited in the hollow or among the branches, it sits silent and motionless until the mother returns. The relationship between mother and offspring is very close; the little one is licked repeatedly. The father also takes an interest in his offspring. The lemur's trick of suddenly falling in flight from a dangerous enemy and disappearing into a hiding place on the ground is mastered by the young at two weeks of age. They are nursed for six months, and become sexually mature at two years.

In captivity, the dwarf lemurs are quite lovable. They will adapt to the life rhythm of their keep-

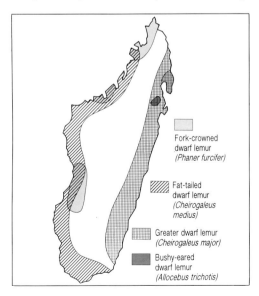

The gray gentle lemur is aptly so called in English because of its mild appearance. These attractive animals are chiefly active in the early morning and late evening, and they are excellent swimmers.

ers, staying awake all day. The English designation "gentle" is very appropriate. As in the wild, gray gentle lemurs prefer a fibrous diet in the zoo. This preference goes so far that when bananas are offered, they may eat only the skins. But it is not very difficult to switch them to a different diet. Breeding has been successful on some occasions.

By contrast with the previous species, the broad-nosed gentle lemur *(Hapalemur simus)* has no upper or lower arm glands. But in habit and habitat, there are only slight differences between the two. This lemur also is at home in bamboo thickets, rushes, and stands of sugarcane, but is already so rare that it has been thought extinct. Fortunately the report was premature.

A pair observed by the zoologist Gustav Peters in the Tananarive Zoo became active in the afternoon, when their outcry would be noticeable. The food, consisting of sugarcane and bamboo shoots, was held in the hand, the rolled-up leaves being peeled off layer by layer with the teeth. Usually only the fleshy base of the blade would be eaten. In the Saarbrücken Zoo, a broad-nosed lemur thrived on a diet of fruit, carrots, chicory, and willow and acacia leaves. By the way, this individual lived on excellent terms together with a crowned mongoose lemur *(Lemur mongos coronatus)*.

The best-known representative of the group is certainly the ring-tailed lemur *(Lemur catta)*. Like the other five species of the genus *Lemur* proper, the ring-tails are characterized by a fox-like face. Because of sounds reminiscent of the purring of cats, they are sometimes called cat lemurs. The term is not well chosen, and invites confusion with the dwarf lemurs, called *Katzenmakis* in German.

Ring-tails live not only in wooded areas but also on stony ground with little vegetation in the southwest of Madagascar. But even here, their habitat is being increasingly restricted, because human settlements are expanding steadily. Predomi-

Two species of gentle lemurs: (drawing and close-up photograph at right) the gray or lesser gentle lemur, and (drawing at left) the broad-nosed or greater gentle lemur, which is now rare.

nantly fructivorous, in the wild these lemurs live chiefly on wild figs, the fruit of the fig cactus *Opuntia*, and bananas. Flowers, leaves, grasses, herbs, and bark are also on the menu. Hard husks like those of the fig thistles are opened with the powerful incisors. Pieces are then bitten off with the molars. To avoid soiling the fur at such a juicy meal, the ring-tails hold their heads high when eating fruit. It need hardly be mentioned that these lemurs use their hands to grasp and hold their food.

Ring-tailed lemurs like company; being kept in pairs at the zoo does not suit them or their near relatives. As has been noted by Georges Basilewsky, long experienced in the keeping of lemurs, the ring-tails are happy only in groups of at least five or six. Such a population is necessary for reproductive behavior to get started. According to J.R. Napier, a leading primatologist, a typical ring-tail group consists of six males, eight females, four half-grown juveniles, and four babies. In 1960, an observation area of 25 acres (10 hectares) was home to a group of 20 ring-tailed lemurs, increased to 24 by 1964. Such groups pass the night in a common sleeping tree, which, however, is changed frequently. Adult females are highly respected by the other members of the group. They are life members, whereas males may change from one group to another.

After the Sun rises, the ring-tails wake up and gradually set themselves in motion towards a feeding place. The younger females usually begin, followed by the males, and finally the mothers and infants. Without interruption, they will cover a distance of about 640 ft (200 m); a day's march is said to be 1900 ft (600 m). After eating, the animals sit astraddle for a Sun bath, stretching their arms wide also.

In cooler weather, group life has the advantage of mutual warmth. Tame ring-tails are also affectionate towards their human keeper, and like to be stroked. They then make the purring and mewing sounds reminiscent of cats. Excited individuals utter bark-like cries to warn their fellows. The cry of fear and the infant signal of abandonment are a shrill scream. Before going to sleep, there is always a hooting cry, audible over a long distance, that presumably supports group cohesion, but also notifies other groups of the ring-tail's presence.

The behavior of such social animals is of course always of special interest to us. Groups of ring-tailed lemurs have been closely observed not only in zoological gardens but also in the wild, and the importance of the well-developed marking behavior of the species has always been noted.

The immediate living area is identified by means of secretions from the ischial and brachial glands. Often the animals do a regular headstand, pressing the posterior against a tree trunk or, in the zoo, against walls or cage fittings. The black-and-white ringed tail serves as a visual means of intimidation and as a scent organ. After the the male ring-tail has briefly rubbed the forearm glands over an upper arm gland located near the shoulder, he will alternately draw his tail past the insides of the left and right forearm, thus supplying them with the secretions of both glands. Thus impregnated, the tail is turned upright and the last third is waved in the direction of a rival, thus wafting the scent towards the latter. Such "stink battles"

Ring-tailed lemurs inhabit all levels of the primeval forest. They frequently occupy the ground, but they also climb to the highest treetops with astonishing agility.

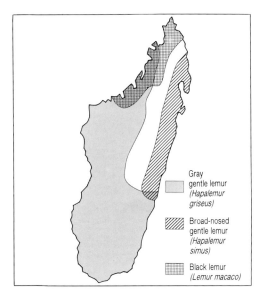

Gray gentle lemur (*Hapalemur griseus*)

Broad-nosed gentle lemur (*Hapalemur simus*)

Black lemur (*Lemur macaco*)

determine rank within the group, but they also occur at the boundaries of territories. Although in this way much is expressed primarily by olfactory means, we may nevertheless observe a facial expression characteristic of many more highly developed mammals in anger or in combat, namely, baring the threatening canine teeth.

May is the ring-tails' breeding month as well. At this season, pairs are seen crowded close together on the sleeping trees. Since the females are fertile for only a very short interval at a time, all the babies in a group come into the world almost simultaneously.

The newborn have blue eyes; later, the color of the iris changes to luminous yellow. All are usually single offspring; twins occur only now and then, and triplets are highly exceptional. The babies are carried over the belly, not crosswise in the usual lemurian manner, but lengthwise, as among the simians. After two weeks, the little ones move to the mother's back; at three weeks, they first make trial of their climbing abilities, and by one week later they have become quite enterprising, returning to the mother only for sleeping and drinking.

As in a housing community, the young are raised by the entire group. The little ring-tails climb about on other females unconcernedly, and let them search their fur. The mothers themselves will exchange babies, and even adopt orphans from other groups.

The infant ring-tails are nursed for five months. Weaning is similarly timed in related species. The young then require no further maternal care. However, it is not the age of two years that the males are big and strong enough to get a mate.

In most zoos, ring-tailed lemurs are a familiar sight; there has been enough reproduction to keep all zoos supplied with these prosimians. Unlike most other lemurs, they adapt quite well to our climatic conditions, and can spend much of the year in the open. At the Cologne Zoo, ring-tails live on two connecting islands. The keeper would bring food in a boat, into which the tame animals would jump without hesitation. At the Copenhagen Zoo, a large group of ring-tails was kept in an outdoor facility formerly used for baboons. There, the animals may choose at will between the fresh air and the heated interior.

In naming the black lemur *(Lemur macaco)*, zoologists made an unfortunate choice. Strictly speaking, only the males should be so called, for only they are black, while the females are fox-red. At birth, both sexes are similarly dark, but the females change color after six months. Thought to belong to a different species, the females were once described as white-bearded lemurs *(Lemur leucomystax)*. That situation has been cleared up, but the black and brown lemurs still pose some riddles for taxonomists. Sometimes the two species are cited as one, and besides there are differing opinions about the division into subspecies. Here – with inclusion of the brown lemur and the mongoose lemur, of very similar behavior and habits – we shall refer to them all as black lemurs.

In the dense forest where they live, these lemurs are adapted to dwelling in the trees. They move through the branches with great speed, leaping over distances up to 25 ft (8 m), and have developed a special technique for throwing off pursu-

The pointed "fox" face, along with the strikingly ringed tail, is highly characteristic of this lemur.

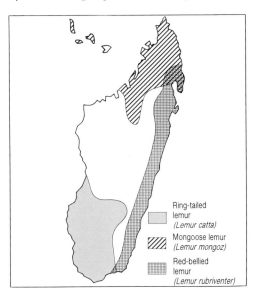

With outspread arms and tail hanging down, the ringtail can easily balance on a slender swaying branch.

▷ In an acrobatic leap from tree to tree, the long bushy tail serves as a balancing rod.

ers; they will suddenly drop, and then run some distance on the forest floor under the protection of the underbrush, until they once again climb a tree in "monkey" mode. This is indeed effective protection against birds of prey, which might be dangerous to them in the treetops. To rest, they lie on horizontal branches, legs hanging down to left and right.

Rarely, a prosimian species may not only adapt to changes made by humans in the natural terrain, but actually draw benefits from them. black lemurs find good feeding grounds in some plantations, and in a few places, they have actually enlarged their former range. Unfortunately, however, such cases are exceptional, for despite protection by law, these lemurs are not only shot but are also poisoned and trapped. Jean Jacques Petter, the French lemurologist, has therefore proposed that in areas where they are too troublesome to the farmers, the lemurs should be captured and turned over to zoos. As almost everywhere else in the world, some protective measures in Madagascar are only on paper, and inadequately enforced. The lemurs are not safe today even in government preserves.

Secretions from posterior and forearm glands are used for marking. Urine also serves for olfactory identification of the home territory. Besides, the black lemurs – much like the gibbons and many New World monkeys – make their presence in an area known by sound. Especially animals gathered at sleeping places will utter these "terrifying" cries. Once upon a time, the Malagasy people feared these "akumbas," and would not harm them.

According to Petter's researches, groups consist of six to fifteen individuals, with males in the majority but females calling the shots. Group members maintain communication by deep grunting sounds. More intimate relations, such as mutual grooming, occur chiefly between members of a pair. This reciprocal grooming has developed into a peculiar habit; conspecifics, or familiar keepers in a zoo, are greeted by protruding the lower jaw, with symbolic grooming and licking gestures. A quite intimate relationship is suggested in this way, as in the symbolic feeding known as a friendly gesture among parrots.

In the Lokobé preserve on the island of Nosy-Komba, according to Petter's observations, it seems that each group has its own living area, but in the evening two or three neighboring groups

(Top right) After a meal, the ring-tail likes to take a Sun bath, usually seated legs apart and arms outspread. (Bottom left) The tail serves not only as a threatening visual signal but also as an olfactory weapon. The lemur draws it past the inside of the forearms to treat it with secretions from glands in that location. (Bottom right) Two ring-tails in playful combat.

will gather in a common sleeping place. The ten groups on the island would always assemble at four sleeping places. Before retirement for the night, to be sure, there would be a long interval of noise and restlessness.

In zoos, black lemurs are agreeable inmates, much gentler than the sometimes rather rough ring-tails. Breeding is not infrequent. When it occurs, the thinly furred, gray-black infant becomes the center of the entire group; all group members pay a visit, to lick it and play with it.

Similar to the black lemur, but larger, is the ruffed lemur *(Varecia variegata).* This species was long assigned to the genus Lemur, but now that more is known about it, it is clear that the ruffed lemurs differ in such important features, including behavior, from the other lemurs proper, that they merit a genus to themselves. Thus, the gestation period of the ruffed lemur is shorter, the young are cared for in a nest, and the mother carries them only with the mouth. Furthermore, ruffed lemurs are mostly active at night.

They are appreciably larger than the lemurs we have previously considered. The right and left halves of the body may differ in appearance. The coloration, so striking in the zoo environment, is not conspicuous in the foliage of treetops. Like camouflage, the colors dissolve in the flickering surroundings. The loud calls, however, are clearer signals, especially when the series of roars swells and swells, followed by chucking sounds. In the quiet of the nocturnal rainforest, the sudden eruption of these cries is like a horror film effect.

During their nightly excursions, violent downpours of rain are no problem at all, for the dense fur provides excellent protection against the wet. Then, if enough fruit and leaves have been eaten by morning, it will be time for sunbathing. The legs are far outstretched and the face is turned to the Sun. For the human natives, this was formerly a reason to regard the "varikandanas" as sacred animals, worshipping the Sun. This belief, which preserved the ruffed lemurs from persecution by humans for centuries, has died out in the days of

A peculiarity of the black lemurs is the strong contrast in coloration of the sexes. Whereas the males (bottom right) are black, the females (top and bottom left) are less conspicuously colored. Few people have ever seen representatives of the rare subspecies *Lemur macaco flavifrons* (bottom left) alive. All black lemur babies - regardless of sex - are dark in color during their first six months.

so-called civilization. Now that there are firearms, the lemurs command none of the respect inspired by a benevolent superstition. The ruffed lemurs are being eliminated, not only by the continual diminution of their habitat but also due to persecution by humans.

Marking behavior is not very pronounced. The males have a throat gland, and tree trunks in the territory are sprayed with urine, or the animals will rub their posteriors against them.

Before the two or three young are born, the mother prepares a nest of twigs and leaves. The little ones have blue eyes at first, which change color after twelve days. The female spends a great deal of time in the nest the first few days to warm the babies and nurse them. After each meal, they are thoroughly licked over. At six days they are scrambling about in the nest, and five days later they are in motion outside of the nest. The first solid nourishment is taken at 24 days. At two months of age, the juveniles will even play with their father, rolling about and trying their little teeth.

Ruffed lemurs were formerly often kept, and also bred, as special exhibits in zoos. In the old Berlin Zoo, there was a cross between the black-and-white and the black-and-red-brown subspecies; the offspring were black and golden brown and white. Today, these lemurs have become rare in zoos, but with care they can be kept quite well, and reproduce regularly. Examples include the Thuringian Zoopark in Erfurt with its repeated breeding successes, and the Cologne Zoo, which harbors a goodly number of ruffed lemurs bred in captivity, in its Lemur House and extensive outdoor enclosures. So we may hope that at least the stocks in human care can continue to be preserved and increased.

Family Lepilemuridae (Sportive lemurs)

The several forms of the greater sportive lemur *(Lepilemur mustelinus)* are outwardly quite similar to each other, but because of different chromosome patterns, 7 species are counted today. Absence of the upper incisors, which fall out at a very early age, and some organic adaptations to the processing of leaves, the main component of the diet, clearly distinguish the sportive lemurs from the lemurs proper.

The name *Mustelinus* ("weasel-like") seems somewhat inappropriate, for there is nothing in appearance or habit to suggest a weasel. The English word "sportive" is more evocative, for, when threatened, this lemur will put up his hands like a boxer fending off an attack.

The sportive lemurs' habitat is forest land, from the evergreen forests on the east coast to the hot dry forests of southwestern Madagascar. Leaps executed in erect posture – like an indri – between vertical trunks and boughs are characteris-

(Above) Grasping hand of the black lemur, with the broad fingertips and opposable thumb typical of prosimians. (Right) Brown lemurs as "Sun worshippers"; after the cool of the night, they fill up on heat in this way. The female is carrying her young across the belly like a girdle. (Bottom left) The ruffed lemur, active in twilight and at night, is the largest representative of the lemurs. Note the moist nose leather.

In appearance and behavior, the brown lemur resembles the black lemur so closely that the two are sometimes regarded as a single species. The brown lemur is by no means always so brown as the one in this photograph; coloration is highly variable not only between the sexes, but also from one subspecies to another.

▷ Two lemurs, a brown lemur and a ring-tail, both active by day, in peaceful association.

tic. The tail is not important as a means for balancing. These lemurs are also able to run on all fours, or hop on two legs, along branches or on the ground.

Being creatures of twilight and darkness, they are rather inconspicuous, and hardly anyone has seen them in the evening emerging from their hiding places, gathering in rather large companies, and then making for their feeding places. Ordinarily, they spend the day curled up asleep in a hollow tree. On the island of Nossi-Bé, however, where there are no animals that might endanger them, they will sleep out on a branch in the open. Even in the Lokobé preserve, however, timbering is inadequately controlled, so that the continued survival of these lemurs is not assured.

Breeding in zoos is hardly to be reckoned with at all, because the sportive lemurs are such specialized leaf eaters that to date not even the Tananarive Zoo (it does not indeed meet European standards) has been able to keep them over periods of years. Besides flowers and bark, they will accept the hard leaves of the cactus-like plant *Alluaudia* – not a very rich fare, for after preliminary digestion in the cecum, they are eliminated and re-ingested. It is said that the flesh of the sportive lemur is particularly tasty because of this vegetable diet.

Hence no protective measures can prevent woodsmen from plundering hollow trees during the day.

Sportive lemurs live singly in their territories, which is marked with urine and the secretion of posterior glands. At night, boundaries are defined by cries as well. Each male inhabits a territory of about 30,000 ft^2 (3000 m^2) in immediate proximity to two or three females, who in turn will hold 20,000 ft^2 (2000 m^2) each. The female territories are overlapped by those of the males.

During the mating season, from May to July, the calls of the animals chasing each other are often heard. One young is born in September or October; it is quite well developed and fairly lively immediately after birth. If it must be transported, the mother will pick it up with her mouth. After a month, the little one will strike out on its own. At an age of two and a half months, it is no longer directly dependent on the mother, but will remain with her until the birth of the next infant.

As has been mentioned, there has been little suc-

cess with raising sportive lemurs in zoos: one lived for twelve months at the Jardin des Plantes in Paris, and one for three months at the Philadelphia Zoo. A breeding has been reported at the Tananarive Zoo, where the animals can be offered their accustomed diet.

(Upper right) Grasping foot of a sifaka, with grooming claw on the second digit. (Bottom right) In the forests of Madagascar, the sportive lemurs lead a life of concealment. They are active at night, and highly specialized as to diet, consuming chiefly leaves. The drawing shows the lesser sportive lemur.

Family Indriidae (Indris and their allies)

Peculiar external features of all four indri species are an erect posture, both in locomotion and at rest, making for a more ape-like appearance than that of other lemurs, and the development of the feet as powerful grasping members. Much like the sportive lemurs, they are specially adapted to a diet of leaves, flowers, bark, and fruits. Large salivary glands and the digestive organs are adapted to the processing of leaves.

The four species living today are a remnant from the time when Madagascar was the home of many large prosimian forms. The giant lemurs of the extinct genus *Megaladapis* must have been impressive, for fossil skulls as large as that of an ass have been found. Other such giant forms resembled the langurs of today in the structure of the skull, jaw, and brain. Of all these large forms, only the indri type survives. Some that probably died out only after the human immigration from the Malayan and Polynesian regions were as large as anthropoid apes of today.

Verreaux's sifaka *(Propithecus verreauxi)* and the crowned sifaka *(P. diadema)* are forest dwellers. Verreaux's sifaka lives in the western *Euphorbia* woods and in dry forests, while the crowned sifaka is at home in the rainforests on the east coast. Their fur is dense and silky-soft, and a fold of skin with long fringes of hair extends from the upper arm to the trunk. When the animals leap and spread out this fold, it gives the impression of a parachute. The name sifaka derives from a native language and probably refers to a sneeze-like call, the warning cry when enemies are sighted on the ground.

The males have a gland at the throat, and branches are marked with the secretion. The territories, however, with adequate food supply, are not very large; sifakas are not eager to move about.

These social, durnal animals live in groups of four to six, all members of one family. They are liveliest about the middle of the day. Then one sees their astonishingly long leaps, in which the body can be projected a distance of up to 32 ft (10 m) by the powerful legs. While the body flies through the air horizontally during the first part of the trajectory, the landing is always at a tree trunk with the body in vertical posture. The tail does not seem to perform any function in these leaps. But except for these leaps, the sifakas'

The sifakas are gentle, diurenal forest dwellers; they are social plant-eaters, most of the time climbing about in their lofty abode, where there are ample supplies of food. Here, two vividly colored sifakas are seen at their common meal.

movements are far more deliberate than those of the lemurs. When they do come down from the trees, they do it in an almost human style, feet first and very carefully. On the ground, they can run upright or hop like kangaroos, with the arms held at head height. They can manage up to 13 ft (4 m) at a bound.

When sitting at rest in the trees, they lay their hands on their knees with the tail curled up like a watch spring between the legs. Sleeping animals hide the head between the knees, with hands and feet clasping a vertical branch and the tail laid over the body. When it is hot, the sifakas come down to lower boughs. They will either lean against the trunk or lie stretched out on a strong horizontal bough with arms and legs dangling.

The subspecies Coquereli of Verreaux's sifaka ex-

▷ A picture of grace and elegance: Verreaux's sifaka in midleap from treetop to treetop.

Verreaux's sifakas sunbathing together. Every morning, they climb a tall tree and turn their chilled body with uplifted arms towards the rising Sun, to absorb the warming rays.

periences nocturnal temperatures near the freezing point in the months from June to August on the high plains. The animals sleep huddled close together: The first will sit facing the trunk of the tree and hold on to it, and the others will cling to their fellows.

They do not of course like such low temperatures. This may be seen in the morning, when they simply cannot get enough Sun. Like some of their relatives, they will then climb a tall tree, turn towards the Sun, and lift their arms, to collect as much of its heat as possible. The morning dew will be dried up from their fur at the same time. Because of this morning sun bath, the sifakas, like the ruffed lemurs, used to be considered sacred Sun worshipers, to whom special abilities were ascribed. Thus it was said that injured sifakas would place certain leaves on their wounds, making them heal quickly.

Animals of social habit naturally have modes of behavior that serve to secure the bonds between individual group members. This includes greeting behavior, which in the case of the sifakas consists of touching noses, as well as expressive cries that sound like a soft "coo." The warning cry for enemies on the ground, sounding like a sneeze, has been mentioned, and there is also a special piping sound, uttered with the muzzle held high, as a warning when predators are sighted.

From sleeping place to place of sun bathing will be only a few tens of feet. A similar distance will then be traversed in the search for food. A whole day's traveling is said to be only about 1600 ft (500 m), territory size of course depending on the available food supply. In the dense forest, with abundant vegetation, there is no need to travel far to find enough nourishment. Leaves, flowers and bark of *Tamarindus indicus*, *Mangifera indica* and *Lemuropsium edule* (all plant species confined to Madagascar), are eaten, as are acacia buds, berries, and other fruits. Hard husks are filed open with the teeth of the lower jaw, and then the pulp is dug out morsel by morsel. When necessary in the search for food, they will hang head down by their hind legs, and they can also travel hand-over-hand on horizontal branches. Usually the food is seized directly with the mouth, but leaves and fruits can also be plucked with the hands.

In the northwest of Madagascar, the young sifakas are chiefly born in June, but in the east, about the first of August. The newborn have their eyes open, and with their big heads and short arms, they are somewhat like human infants. In the first days after birth, the little one is continually licked by the mother. She carries it across her stomach, its head hidden in her fur, and moves with great care through the branches. No nest is needed. The old story told by the natives that the sifakas will pull out fur from the breast and forearms shortly before giving birth, to make a nest, is a

The sifakas can float through the air a distance of 32 ft (10 m). In the air, the body is at first held horizontally, then swung erect (left), so the sifaka lands with outspread arms and extended feet on a tree trunk (above).

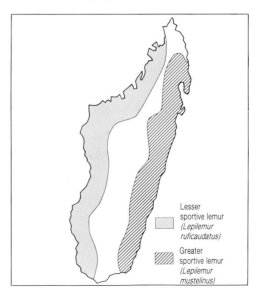

Lesser sportive lemur (*Lepilemur ruficaudatus*)

Greater sportive lemur (*Lepilemur mustelinus*)

myth invented at one time when these animals were held sacred.

In the Tananarive Zoo, the zoologist Gustav Peters observed a baby Verreaux's sifaka born there in the second half of July. The under parts, tail, and limbs were nearly hairless for the first few days, while the head and back were already heavily furred. The infant would actively hold tight to its mother's fur, and at two weeks it was climbing about in her coat, the mother supporting it with her hand.

By the time the little sifaka is a month old, it is riding on its mother's back. The slender tail is coiled like a spring. If the mother is in motion, the baby puts its arms around her back. It can now leap a little, and run a few steps with arms dangling. A soft mewing is heard. The first solid food is taken at three months. First contacts with conspecifics take place at the same age, and the juveniles will race about wildly together. The close relationship between mother and offspring lasts six to seven months, and all that time the young one will be carried around by its mother. It will have reached about two-thirds of full size. Sifakas are full grown at 21 months, and males can reproduce at two and a half years.

Ludwig Koch-Isenburg once raised a four-week-old sifaka in Madagascar on cow's milk, tea, and sugar, with surprising success. As a substitute for its mother's fur, the baby prosimian had a rabbit to cling to. Soon it adopted its human keeper and made a playmate of him as well.

Sifakas are gentle and not inclined to bite, but they must be allowed ample freedom of movement. The few people who have known tame sifakas in Madagascar have been quite charmed by their personalities.

Nearly all attempts to keep the animals in zoos outside their homeland have failed because of dif-

The sifaka seems almost weightless at take-off and landing.

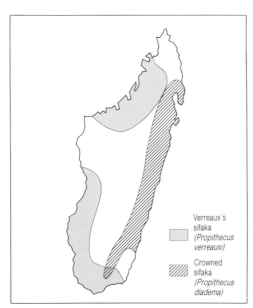

Verreaux's sifaka *(Propithecus verreauxi)*

Crowned sifaka *(Propithecus diadema)*

ficulties in procuring the right diet. Sifakas who reached zoos in Berlin and London before the First World War survived there for only a short time. Outside of Madagascar, only a single subspecies, Coquerel's sifaka *(Propithecus verreauxi coquereli)*, has been bred. That extraordinary event occurred in a department of Duke University in Durham, North Carolina.

At the Tananarive Zoo, the sifakas, represented there by both species and five of the nine subspecies, are fed buds and leaves of the white mulberry *(Morus alba)*, the guajava *(Psidium guajava)*, and the tall grasses *Panicum maximum* and *Neyraudia madagascariensis*. Bamboo, which lemurs find so delicious, was ignored.

A round, almost spherical head, small ears concealed in the fur, staring eyes, and an extremely soft coat are the unmistakable characteristics of the sifakas' smaller relative, the woolly lemur, or avahi *(Avahi laniger)*.

The nocturnal avahi is hardly known to the inhabitants of Madagascar. We know much less about its habits than about those of the sifakas. In the trees, it moves in the same fashion, if anything still more deliberately. On the ground, which is seldom visited, it walks erect.

During the day, the avahi will either sleep in the fork of a branch or cling to a vertical trunk. Two or three individuals crowded close together form a sleeping community. Presumably this comprises a pair and their latest offspring. Avahis are believed to mate for life. Given the limits of knowledge of their habits, such information is open to some doubt. However, it has been observed that the male will also care for the baby and carry it about.

No marking behavior has yet been observed, either in the wild or on the part of zoo animals.

The single young is born in September. It is at first carried crosswise over the mother's belly. After nursing for five months, it has attained about two-thirds of adult size, but is still carried on the back.

Avahis are even more difficult than sifakas to keep in the zoo. There was a woolly lemur living in the London Zoo in 1889. So far, it has not been possible even at the Tananarive Zoo to keep avahis alive longer than three months. It is questionable whether the difficulties of obtaining their accustomed food is the only reason for this.

The indri *(Indri indri)* is so named by misunderstanding on the part of this prosimian's discoverer. When the natives cried, "indri," that simply meant in their language, "There he is!" The natives of various regions call the animal "babakota" (father's son) or "amoanala" (bush dog).

Striking features are the large ears, exceptionally long hands, and almost complete taillessness, otherwise unknown among prosimians except for the lorises. A throat sac is connected with the vocal apparatus and serves as a resonance chamber.

A sifaka has made itself comfortable on a spiny *Euphorbia*. The *Euphorbia* woods and dry forests of Madagascar are home to these large prosimians.

Terpsichorean grace distinguishes Verreaux's sifakas on level ground also, where they usually run and hop as bipeds.

Today indris live only in a few forest regions, for, as animals unable to adapt to altered conditions, they have been steadily driven back. In the opinion of Petter, they even suffer from the aircraft overflying their preserve; the noise apparently interferes with reproduction.

These animals are now encountered, if at all, singly, in pairs, or in groups of up to five, comprising a pair with offspring of different ages. They are abroad by day, but they are heard rather than seen. About 5 o'clock in the morning, soon after sunrise, the first concert begins. In the afternoon, toward 4 p.m., their voices sound again, starting with a melodious bark and passing over into screams like those of human beings in pain, or howling dogs. These deafening performances mark the indris' rather large territories, much after the manner of gibbons. Other groups will take up the cry, filling the entire wood with their impressive voices. Each group occupies a territory between 31 and 15 acres (15 and 30 hectares) in area. It is marked with secretions from the throat gland as well as by sound.

A single young, born in May or June, is black at first, the white spots appearing later. It clings crosswise to the mother's ventral fur; after 4 months, it is carried on her back. The young indri undertakes its first little excursion away from the mother at an age of about 6 weeks. Two weeks later, solid food is tried, but nursing continues until the ninth month of life. The indri spends its first year with its mother, sleeping near her.

Today the indri is one of the species directly threatened with extinction, unable to survive even in the best-equipped zoos. Only thorough investigation of its mode of life might provide some hints to tell us how this largest of all present-day prosimians can be saved.

It has not been possible to keep indris even at the zoo in the capital of Madagascar itself. Their favorite food was provided, for example *Eugenia* species and *Uapaka thouarsii*, but the animals accepted none of it, suffered from diarrhea, and soon died. They were probably overstressed by capture, transport, and their new surroundings.

For a long time, the indri was deemed sacred among the Malagasies, for it was believed that human beings were transformed into indris after death. This prosimian is still a legendary beast, in the truest sense of the word, as evidenced by a very charming story. David Attenborough, who lived in Madagascar as a zoologist for some time, has recorded it:

"After a while, the woman bore a large number of children. When they were grown, some who were industrious by nature began to till the soil and plant rice. The others continued to live entirely on roots and leaves of plants growing wild. But in the course of time, the members of the first group fell to fighting among themselves. They were the ancestors of the human race. The others took flight into the treetops so that they might continue to live in peace. They were the first indris."

Family Daubentoniidae (Aye-ayes)

The much prolonged toes and fingers, especially the extremely long and withered-looking middle finger with its powerful claw, are the most

■ Woolly lemur *(Avahi laniger)*
▨ Indri *(Indri indri)*
▦ Aye-aye *(Daubentonia madagascariensis)*

(Right) An especially soft coat of fur, a rounded head with staring eyes, and small ears hidden in fur are the outward characteristics of the nocturnal woolly lemur, also known as the avahi.

conspicuous features of the aye-aye *(Daubentonia madagascariensis)*, so different from the other Malagasy prosimians. This characteristic accounts for the German name "Fingertier." The name "aye-aye", on the other hand, derives merely from a native interjection of surprise, mistaken by European naturalists for the name of the animal.

The aye-aye did indeed cause their discoverers some puzzlement, for this species, with its 18 teeth, including a large, rootless incisor in each half of each jaw, and an appearance not resembling any other animals known at the time, was difficult to classify. Largely upon grounds of its gnawing dentition, it was thought to be a relative of the squirrels. Since its look is also somewhat reminiscent of an opposum, some observers thought it might be a climbing marsupial. The German zoologist Schreber, about 1775, first realized that the aye-aye was to be classed with the lemurs. Confirmation did not come until the middle of the following century when the English naturalist Richard Owen had an opportunity to examine the milk dentition of aye-ayes. And behold, the situation was clearly the characteristic one for prosimians.

A form one-third larger than the aye-aye, the greater aye-aye *(Daubentonia robusta)*, died out a few centuries ago. The last surviving species of the family may well share that fate in our time. On Malagasy soil, the aye-aye today occurs only in two unconnected forest regions: first, in the rainforest on the east coast, where it has been found at elevations of 2200 ft (700 m), and perhaps also in the northern part of the Sambirano rainforest, in the northwest part of the island. Aye-ayes sometimes frequented the mango plantations laid out by the early settlers. Tall trees are indispensable to the aye-ayes, for it is only in such aged trunks that they can find enough beetle larvae under the bark. But just these trees are being felled everywhere. An alternative place of abode is the dense jungle of giant bamboo.

The island of Nossi-Mangabé is one of the few as yet undisturbed areas on the east coast of Madagascar, and has therefore been made a reserve for animals and plants. In 1966, eleven aye-ayes were brought there in the expectation that they would multiply. It is too early to say whether this species has found a last, safe refuge at least, on the island.

(Left) The indri, threatened with extinction, contrasts with its relatives in two characteristics: it is larger than any of the others, and it has only a little stump of a tail. (Below) One of the rarest and most remarkable of the world's animal forms is the aye-aye, which takes its German name "Fingertier" from its extremely long fingers and toes. The middle finger is expecially long and slender.

Lemurs (Lemuriformes)

Nomenclature English common name Scientific name French German	Approximate Size Body length Tail length Weight	Distinguishing Features	Reproduction Gestation period Young per birth Weight at birth
Lesser mouse lemur *Microcebus murinus*, 2 subspecies Chirogale mignon Mausmaki	4.4–5.2 in; 11–13 cm 5.2 in; 13 cm 1.8–2.1 oz (50–60 g), sometimes to 3.5 oz (100 g)	Smallest prosimian, and smallest primate; long, bushy tail; upper parts gray to red-brown, under parts white; imperfect homoiothermism; stores fat; perception of ultrasound	59–62 days 2–3 twice a year 0.1–0.2 oz; 3–5 g
Coquerel's mouse lemur *Microcebus coquereli* Microcèbe de Coquerel Rattenmaki; Coquerels Zwergmaki	9.6 in; 24 cm 12 in; 30 cm 13.9 oz; 390 g	Much like lesser mouse lemur, but double its size; general coloring olive-brownish; imperfect homoiothermism; stores fat	84–89 days 1–4; second half of November Not known
Fat-tailed dwarf lemur *Cheirogaleus medius*, 2 subspecies Chirogale à queue grasse Mittlerer Katzenmaki; Fettschwanzmaki	8.4 in; 21 cm 8 in; 20 cm 8.9 oz; 250 g	Much like greater dwarf lemur, but smaller; upper parts gray-red or gray-brown, under parts white to orange; thick layer of fat in tail region; imperfect homoiothermism (body temperature varies between 70–90° F [21° and 32°C])	65 days 2–3; December–January Not kown
Greater dwarf lemur *Cheirogaleus major*, 2 subspecies Chirogale grand Großer Katzenmaki	7.6–10.8 in; 19–27 cm 6.4–10 in; 16–25 cm 12.5–16 oz; 350–450 g	Upper parts gray-brown or brown-red, posterior parts usually somewhat darker, under parts lighter; imperfect homoiothermism; stores fat	70 days 2–3; December–January 0.6–0.7 oz; 17–19 g
Hairy-eared dwarf lemur *Allocebus trichotis* Chirogale aux oreilles velues Böschelohriger Katzenmaki; Kleiner Katzenmaki	5.6 in; 14 cm 6.4 in; 16 cm Not known	Upper parts brown-gray, under parts whitish-gray, tail red-brown; ears short with tufts of fur	Not known
Fork-crowned dwarf lemur *Phaner furcifer* Maki à fourche Gabelstreifiger Katzenmaki	10 in; 25 cm 13.6 in; 34 cm 14.2–16 oz; 400–450 g	Upper parts brown-gray to red-gray, under parts yellow; black mid-dorsal stripe forks into two black eye rings; finger nails and toe nails are pointed	Not known 1; November Not known
Gray gentle lemur *Hapalemur griseus*, 3 subspecies Hapalémur gris Grauer Halbmaki; Kleiner Halbmaki	13.6 in; 34 cm 13.6 in; 34 cm 2.2 lb; 1 kg	Thick, soft gray fur with brownish or greenish tinge, under parts lighter; lower-arm glands, males with upper-arm glands as well; body temperature varies between 90 and 97° F (32° and 36°C)	140 days 1; December–February about 1.4 oz; 40 g
Broad-nosed gentle lemur *Hapalemur simus* Hapalémur à nez large Breitschnauzen-Halbmaki; Großer Halbmaki	18 in; 45 cm 18 in; 45 cm 4.8 lb; 2.2 kg	Upper parts red-gray, under parts yellow-gray; tufts of hair at ears; no arm glands	140 days 1; December–January Not known
Ring-tailed lemur *Lemur catta* Lémur catta Katta	18 in; 45 cm 22 in; 55 cm 5.3–8.2 lb; 2.4–3.7 kg	General coloring pearl-gray; "fox-face" with white mask; black and white rings on tail; lower-arm glands in both sexes, upper-arm gland in male	132–134 days Usually 1 1.4–2.8 oz; 50–80 g
Black lemur *Lemur macaco*, 2 (?) subspecies Lémur macaco Mohrenmaki	14–16 in; 35–40 cm 18–20 in; 45–50 cm 4.4–6.4 lb; 2–2.9 kg	Conspicuous sex differences in fur coloring: males black, females red-brown; very good hearing and vision (color)	130 days 1–2; August–October 2.8 oz; 80 g
Brown lemur *Lemur fulvus*, 6 (?) subspecies Lémur brun Brauner Maki	16 in; 40 cm 20 in; 50 cm 6.6 lb; 3 kg	sex differences in fur coloring; coloring also highly variable among individuals; long ear fringes	130 days 1–2; August–October Not known
Mongoose lemur *Lemur mongoz*, 2 subspecies Lémur mongoz Mongozmaki	17.2 in; 43 cm 17.2 in; 43 cm 4.4 lb; 2 kg	Head to shoulders dark gray; back brown; sides of neck brown-red; bridge of nose, throat, and chest white-gray; male under parts red-yellow; tail red-brown (anterior half), and gray (posterior half); female, under parts white, black face and tail	120 days Usually 1 2.1 oz; 60 g
Red-bellied lemur *Lemur rubriventer* Lémur à ventre rouge Rotbauchmaki	16.8 in; 42 cm 18 in; 45 cm 4.4 lb; 2 kg	General coloring of male chestnut-brown, under parts red; female, umber with cream-colored under parts; brown-black tail	Probably much like mongoose lemur, but details not known
Ruffed lemur *Varecia variegata*, 2 subspecies Vari Vari	24 in; 60 cm 24 in; 60 cm 8.8–11 lb; 4–5 kg	Largest species of the family Lemuridae; long, soft, dense fur; highly contrasted and variable coloring (black and white, brown and white, reddish-white); long ruff at neck	100 days 1–3; November About 3.5 oz; 100 g

COMPARISON OF SPECIES

Life Cycle **Weaning** **Sexual maturity** **Life span**	Food	Enemies	Habit and Habitat	Occurrence
Age about 45 days Age 7–10 months In captivity, 14 years	Fruits, flowers, insects, lizards	Owls	Nocturnal, high-forest dweller among reeds and *Euphorbia* bushes; sleeps in hollow trees, arboreal nests; territory 1.8 acres (0.75 hectare); territories of males and females overlap	Comparatively abundant
Not known Not known In captivity, 15 years	Fruits, leaves, tree sap, insects	Owls, probably also mongoose	Nocturnal, dry-forest dweller; builds nests (10–12 in. or 25–30 cm diameter)	Endangered by progressive destruction of habitats
Not known Not known In captivity, to 18 years	Fruits, flowers, insects, small vertebrates	Owls, weasels	Nocturnal forest dweller; remains dormant for weeks; sleeps almost exclusively in hollow trees; no territorial behavior	No longer very abundant due to destruction of habitats; preserves inadequately protected
Age about 6 weeks Not known In captivity, to 15 years	Fruits, flowers, insects, small vertebrates, possibly young birds	Owls, ferret cat	Nocturnal, rain forest dweller; remains dormant for weeks in some regions; generally deliberate gait; days spent to community nests or hollow trees; engages in scent marking with urine and feces	Not yet endangered
Not known	Not known	None known	Nocturnal, forest dweller; days spent in hollow trees	Very rare; only 4 specimens found to date
Not known	Fruits, flowers (especially, nectar), tree sap, leaf-louse secretion, insects	Not known	Nocturnally active forest dweller; generally quadruped locomotion, but also long leaps (to 32 ft; 10 m); territory size about 2.5 acres (1 hectare)	Not accurately known; 550–870 counted per km^2 (1 km^2 = 0.36 mi^2) species believed endangered
Age about 6 months Age 2 years In captivity, at least 12 years 9 months	Bamboo shoots and leaves preferred, reed and papyrus sprouts	Snakes, birds of prey, mongoose	Predominantly active at twilight in bamboo thickets and forests; subspecies at Lake Alaotra travels over carpet of aquatic plants, good swimmer; engages in scent marking (urine); usually in groups (probably family groups) of 3–5 individuals	Declining due to habitat destruction but populations sometimes increase locally
Not known Not known In captivity, 12 years	Fibrous plants (grasses, bamboo, reed, sugar cane)	Not known	Predominantly active at twilight, reed dweller; eats throughout the day; found in small groups	Rare; limited range of distribution; endangered by hunting and destruction of habitat
Age 6 months Age 15 months In captivity, over 20 years	Fruits, flowers, leaves, grasses, greens, bark	Birds of prey, ferret cat	Diurnal forest dweller; pronounced scent marking behavior; engages in sun bathing with outstretched arms; usually in groups of 20–25; territory size 13.5–50 acres (5–20 hectares) per group; easy to keep and breed	Despite high local population densities, endangered by growth of human settlements and by hunting
Age about 5–6 months Age 18 months In captivity, 30 years	Vegetable fare, will accept meat in captivity	Ferret cat, civet, birds of prey	Diurnal forest dweller; family groups of up to 15, under female leadership; territory size 13.5–100 acres (5–40 hectares), depending on forest type	Endangered; subspecies *L. m. flavifrons* possibly already extinct; declining steadily, with infrequent gains
Age about 5–6 months Age 18 months Probably much like black lemur	Vegetable fare	Ferret cat, civet, birds of prey	Diurnal forest dweller; found in family groups	Endangered, especially subspecies red-faced lemur (*L. f. rufus*) and Sanford's lemur (*L. f. sandordi*)
Age 5 months Age 18 months In captivity, 26 years	Vegetable fare	Ferret cat, civet, birds of prey	Diurnal in forests and savannas; found in family groups of 6–8	Species endangered by habitat destruction and hunting
Probably much like mongoose lemur	Probably much like mongoose lemur	Probably same as mongoose lemur	Probably diurnal in rain forests; found in groups of 4–5	Endangered; rare
Age 135 days Not known In captivity, 19 years	Leaves and fruits	Not known	Predominantly active at twilight and nocturnally in rain forests; engages in early morning sun bath with outstretched arms and legs; little scent marking behavior; builds nests for raising young; found in family groups of 3–5	Endangered; steady decline due to deforestation

Nomenclature **English common name** Scientific name French German	Approximate Size Body length Tail length Weight	Distinguishing Features	Reproduction Gestation period Young per birth Weight at birth
Sportive lemur; weasel lemur *Lepilemur mustelinus*, 5 subspecies (often ranked as separate species) Lépilémur mustélin Großer Wieselmaki	14 in; 35 cm 12 in; 30 cm 2.8 oz; 800 g	General coloring usually red-brown, under parts lighter; tail darker towards the tip; salivary gland large as adaptation to diet of leaves; very good vision	140 days 1; September–October About 1.8 oz; 50 g
Lesser sportive lemur *Lepilemur ruficandatus* Lépilémur à queue rouge Kleiner Wieselmaki	10 in; 25 cm 10 in; 20 cm 1.3 lbs; 600 g	General coloring red-gray; red-brown tail; salivary gland large as adaptation to diet of leaves	135 days 1 Not known
Verreaux's sifaka *Propithecus verreauxi*, 4 subspecies Propithèque de Verreaux Larvensifaka	20 in; 50 cm 22 in; 55 cm 11 lbs; 5 kg	General coloring white; brown, gray, black, or rust-red areas with distinct boundaries; fold of skin with long fringes along side of trunk	150 days 1; June–August 1.4 oz; 40 g
Crowned sifaka *Propithecus diadema*, 5 subspecies Propithèque diadème Diademsifaka	21.2 in; 53 cm 20 in; 50 cm 12 lbs; 5.5 kg	Vividly colored; soft, silky coat; extensive color variations within subspecies	150 days 1 Not known
Woolly lemur; avahi *Avahi laniger*, 2 subspecies Avahi (laineux) Wollmaki; Fliesmaki; Avahi	14 in; 35 cm 16 in; 40 cm 1.3–2.2 lbs; 600–1000 g	Short, woolly fur; ears small, hidden in fur; round head; coloring highly variable	150 days 1; September Not known
Indri *Indri indri* Indri Indri	2.5–2.6 ft; 75–80 cm 2 in; 5 cm 15.5 lbs; 7 kg	Largest living prosimian species; nearly tailless; dense, long, silky coat; coloring variable; black, hairless face; throat sac functions as resonance structure	150 days 1; May–June Not known
Aye-aye *Daubentonia madagascariensis* Aye-aye Fingertier; Aye-Aye	18 in; 45 cm 22 in; 55 cm 4.4 lbs; 2 kg	Long, coarse coat: general coloring dark brown to black; large eyes; large membranous ears; long, slender middle finger with long claw; one pair of nipples in inguinal region (exceptional location for primates)	Not known 1 every 2–3 years; October–November Not known

During the day, the aye-ayes stay in the trees or in dense brush were they build their nests. Such nest building occurs among some lemurs, as we have seen, but is otherwise exceptional for mammals. The structures, installed in forked branches, are about 2 ft (60 cm) in diameter. The main building material is rolled leaves of the traveler's tree, *Ravenala madagascariensis*. The side entrance hole, 6 in. (15 cm) across, is closed when the owner is at home. Some five dozen leaves and twigs are interwoven, and a few fresh ones are added daily. Each individual will have two to five such nests, at heights from 30–45 ft (10–15 m). They spend the day rolled up in a ball, the long tail draped over the body, and sleep through the day.

At dusk, the aye-ayes awaken, and then leap about in the branches in a quite lemur-like manner. But unlike their relatives other than the lorises, they are also able to hang by their hind legs. The free hands are then used to hold food, or for grooming. Here the long middle finger is very useful, for with it the aye-aye can not only scratch and comb itself, but minutely cleans all parts of the face, including difficult places, such as the corners of the eyes, the ears, and the nose. The other fingers take no part in this operation; they are simply curled in. But in feeding, the long middle finger may be seen performing its most important function. One might truly say that the middle finger is to the aye-aye what the bill is to the woodpecker – even in its use as a drumstick. The bark of trees is tapped until a hollow sound indicates that a larva's tunnel might be underneath. To make quite sure, the large ears are held close to the trunk, lest the sound of some tasty morsel might be missed. When the prey has been located, the chisel-like incisors go into action. They bite a hole in the bark, and the middle finger pulls out the victim. The powerful teeth can also crack the hard outer layer of the giant bamboo, as well as coconuts, sugarcane stalks, and mango husks. After this use of the teeth, the middle finger again comes into play,

COMPARISON OF SPECIES

Life Cycle Weaning Sexual maturity Life span	Food	Enemies	Habit and Habitat	Occurrence
Age about 4 months Age 18 months In captivity, 12 years	Predominantly leaves	Owls, ferret cat	Nocturnal in forests; locomotion chiefly by leaping (from tree to tree), in erect posture; males' territories of males about 32,100 ft^2 (3000 m^2), territories of females 21,400 ft^2 (2000 m^2)	Threatened by habitat destruction, somewhat by hunting as well; subspecies *L. m. leucopus* most endangered
Age about 4 months Age 18 months Not known	Chiefly leaves and flowers	Owls, ferret cat	Nocturnal in rain forests and dry forests; occupies small territories	Endangered by destruction of habitat
Age about 6 months Age 2 years In captivity, over 18 years	Leaves, fruits, flowers, bark	Birds of prey, ferret cat	Diurnal in *Euphorbia* and dry forest; engages in sun bathing with outstretched arms and legs; found in family groups of 4–6; territory size 2.5–7.5 acres (1–3 hectares)	All subspecies threatened with extinction
Not known	Vegetable fare	Not known	Diurnal in forests; engages in sun bathing; habits not much studied; not yet successfully kept and bred in captivity	Rare
Age about 6 months Not known Not known	Leaves, buds, bark, fruits	Owls, weasels	Nocturnal in forests; mating possibly for life; habits little studied; hard to be kept in zoo	Rare and endangered; subspecies *A. l. occidentails* close to extinction by habitat destruction (burning)
Age 9 months Not known Not known	Leaves, flowers, fruits, buds	Birds of prey, ferret cat	Diurnal in coastal and mountain rain forests: found solitary or in groups of 3–5; excellent leaper; territory identified by howling; territory size 31.5–75 acres (15–30 hectares); cannot be kept in zoos	Threatened with extinction
Probably at age 1 year Age 3 years In captivity, over 23 years	Fruits, insect larvae	Ferret cat	Nocturnal in rain forests, mangrove and bamboo thickets; constructs elaborate sleeping nests (diameter about 20 in. or 50 cm); found solitary or in small family groups; territory size about 12.5 acres (5 hectares)	Most endangered of all mammals in Madagascar

as a multipurpose tool; it digs out the pith of the hard stalk, as well as the pulp of the hard-shelled fruit.

There have been only a very few observations in the wild. Aye-ayes have been seen climbing and leaping, and also running along on all fours, tail raised, on the forest floor. Known sounds include a scraping noise like two sheets of metal rubbing together, by way of communication, grunting sounds in the search for food, and something like "R-ront-sit!" when taken by surprise. If held in the searchlight beam, the animals will spit and snarl.

Hitherto, almost exclusively single aye-ayes have been encountered; meetings with two individuals, possibly a pair or a mother and her young, have been quite rare. The one baby is born in February or March, or in October according to some reports. The nest is the nursery, and remains so for about a year. Presumably the offspring is nursed throughout that time. At one year, it is about two-thirds adult size. Probably births occur only in alternate years. The gestation period is not known. Incidentally, the aye-aye is the only species of the order Primates in which the pair of nipples is located in the abdominal region.

Despite their pronounced preference for beetle larvae and bamboo pith, aye-ayes can be kept in captivity without difficulty on sugarcane, coconuts, mangoes, lichees, bananas, dates, eggs, and cooked rice. Eggs are gnawed open and the con-

With its chisel-like incisors, the aye-aye bites holes in the bark, and then pulls out insect larvae with its long middle finger. Vegetable fare is also regularly on the menu.

tents spooned out with the middle finger. At least in the zoo, even water has been taken with the aid of this finger: it is dipped into the water and then drawn through the mouth. An aye-aye living in the Berlin Zoo many years ago would mark with urine, was very lively at night, and would leap about on the branches in its cage. It liked to crack nuts. Although keeping has actually presented no difficulties, and an aye-aye lived no less than 23 years in the Amsterdam Zoo, there has been no successful breeding outside of Madagascar. The Tananarive Zoo reports only one single birth.

So remarkable a creature has played a large part in popular beliefs among the native population in Madagascar. Happily for the aye-aye, it was firmly beliefed that anybody who killed one would himself die within the year. So everyone avoided doing an aye-aye any harm. If a person falls asleep in the forest, according to another legend, an aye-aye will come and make a grass cushion for him. If one finds this under one's head upon waking, wealth in the near future is a certainty. If the cushion is under one's feet, one will fall under the evil spell of a sorcerer.

Only in such remote rainforest areas as this, with tall trees, have the strange aye-ayes been able to survive to the present day. But the future of the small remaining populations is uncertain.

Lorises and Galagos

by Ewald Müller

Two quite different types of prosimians, in appearance and in gait, constitute the infraorder Lorisiformes. At first glance, one would hardly suppose that the quick, agile bush babies are close relatives of the lorises and pottos, which generally move through the branches slowly and cautiously. The name "loris" is probably derived from Dutch *loeres* "sluggish," or Flemish *lorrias* "lazy, tardy;" it has also been traced to the world *loeris* by which the old Holland seafarers meant a clown. Because of their "slow motion" ways, the lorises were in fact at first taken for sloths; it remained for the great French naturalist Buffon to recognize, in 1766, that they are prosimians. Today, scientists place the lorises together with the bush babies in a common infraorder. The two groups are very similar in many respects, for example the number and shape of the teeth and the structure of the skull, the digestive tract, and the reproductive organs. The Lorisiformes are divided into the two families; Galagidae (galagos or bush babies) and Lorisidae (lorises and pottos), chiefly on the basis of their different modes of movement and certain related peculiarities of physical structure. The bush babies have much prolonged hind limbs, and a long tail that serves as a balancing organ in leaping. In the Lorisidae, the front and hind limbs are of nearly the same length, and at most a stump of a tail is present.

Because of the absence of all but a very short tail and their dense woolly fur, the lorises and pottos resemble little bears, so that one would hardly take them for relatives of the monkeys. The slow loris *(Nycticebus coucang)*, with its round head, small ears almost hidden in the fur, and short, study arms and legs, looks very much like a teddy bear. The head is usually marked with a dark dorsal midline stripe making a Y at the forehead, the two branches forming a dark ring around each of the large close-set eyes. The dark rings are emphasized by the snow-white bridge of the nose and lower part of the forehead between them.

The prosimians, almost without exception, are at home in Africa and on its offshore island of Madagascar. The exceptions are the two loris species, the slow loris (left) and the slender loris (below), living in Southeast Asia. The lorises are solitary nocturnal animals, moving through the branches of the tropical rainforest with great deliberateness.

During sleep, the slow loris draws the head and limbs close to the breast and belly, forming an almost perfect sphere; the thick fur affords excellent protection against loss of heat.

Very similar to the slow loris in fur coloration and facial marking is its smaller cousin, the slender loris *(Loris tardigradus)*. To be sure, as the name suggests, these are far less robust in build. Their limbs are long and slender and seem fragile. They have thus been described as resembling a "banana on stilts."

The ears of the

slender loris are much larger than those of the slow loris, and often yellowish in color.

The striking facial marking of these two Asian species is not found in the African representatives of the family Lorisidae, namely the potto *(Perodicticus potto)* and the angwantibo, or golden potto *(Arctocebus calabarensis)*. The latter's German name, *Bärenmaki*, again alludes to a resemblance to little bears. However, it happens that the angwantibo has a comparatively long and pointed muzzle, which blurs the bear-like image. The other name, "golden potto," refers to the rust-brown coloring of the fur of the back. In general appearance, the angwantibo rather resembles the slender loris.

The other African species, the potto, is of about the same size as the slow loris, and has a comparably robust and compact build. Pottos are born with a silvery-white coat, which takes on the adult color, a dark yellow-brown, after about six months. Of all the Lorisidae, the pottos have the longest tail, but even so, it is no more than 2.4 in. (6 cm) in length. Remarkably long sensory hairs, or vibrissae, are seen to protrude far out of the thick fur in the neck region. Beneath, usually hidden in the fur, are spines covered with thick horny skin. They represent especially long processes of certain cervical and thoracic vertebrae, and serve the potto as an effective weapon of defense against enemies.

In all the Lorisidae, the second finger of the hand is much reduced, especially so in the potto and the angwantibo. As the thumbs will also spread very wide, the hand forms a forceps with an exceptionally large grip.

Despite their close kinship, the bush babies or galagos, which today occur only in Africa, show little outward similarity to the slow and dignified lorises and pottos. Bush babies move swiftly and adroitly through the branches, and are distinguished by their great leaping ability. The basis for this is the unusual length of two bones at the base of the foot. As a result, the hind legs are nearly twice the length of the forelimbs.

In their acrobatics among the branches and especially in their leaps, often 3 ft (1 m) long, the bush babies use their long, usually heavily furred tails, as a balancing means. The neck is very supple, so that the head can be rotated almost 180 degrees; such head mobility is very helpful in locating prey. Another important factor is the large membranous ear, moved by muscles and capable of being folded up at fine creases in the cartilage.

The largest species of bush baby, the giant galago *(Galago crassicaudatus)*, attains about the same body weight as the slow loris and the pottos. Of all the Lorisiformes, these kombas, as they are also called, have perhaps the densest fur. The tail, especially is long and bushy, so that they have also been named thick-tailed galagos. However, this has sometimes been misunderstood to suggest that the giant bush babies use their tails for fat storage to survive the famine of the dry season, much like a hibernating animal. Probably because of their size, the kombas do not leap as actively as the smaller species.

Among these is the Senegal bush baby *(Galago senegalensis)*, especially the East African and South African subspecies, the moholi *(G. senegalensis mo-*

With its large rounded head, very large eyes, and cuddly fur, the Senegal, or lesser, bush baby fits the Lorenz "infantile" pattern that spontaneously triggers our nurturing and sheltering urges. These cute little creatures were not named bush babies for that reason, but because of their cries, reminiscent of those of human babies.

The Senegal galago, like all bush babies, is nocturnal. Its eyes glow in the dark like little lamps when the beam of a searchlight strikes the reflecting layer of the retina.

holi), perhaps the best-known of all the Lorisiformes, being often kept in zoos. Its build resembles that of the giant bush baby, but it is less than half as large. The dense, silky fur on its back is generally gray to gray-brown; the ventral fur is sparser and often snow-white. In some subspecies, patches of fur on the flanks and thighs have an orange to yellow-brown cast. The Senegal bush babies inhabit chiefly the dry tree and bush savannas.

Next to the least mouse lemur of Madagascar, the dwarf, or Demidoff's, bush baby *(Galago demidovii)* is the smallest primate; this inhabitant of the rainforests weighs only 1.8 to 2.8 oz (50 to 80 g). It prefers the so-called liana "curtain." Occasionally, the dwarf bush babies, usually of brown coloration, form sleeping communities of up to 30 individuals.

Another rainforest dweller is the black-tailed, or Allen's bush baby *(Galago alleni)*. It is distinguished by exceptionally large ears and excellent leaping ability. This species is light brown to red-brown in the upper parts and more yellowish underneath; the tip of the tail is often brilliant white.

Very little is known about the needle-clawed bush babies that inhabit the upper levels of the African rainforest. There are two species, the western needle-clawed bush baby *(Galago elegantulus)* and the eastern *(Galago inustus)*. These bush babies are so called because of a special configuration of the toe nails and finger nails, which have a central lengthwise keel terminating in a sharp point. They enable the needle-clawed bush babies to cling to smooth tree trunks, where, sometimes climbing head downward, they seek out the running sap that is their main source of nourishment.

Physical Peculiarities

Despite the scarcity of fossil finds, it is established that the Lorisiformes had separated from the other prosimians towards the close of the Oligocene, at least 30 million years ago, and have gone through their own evolution since. It may be thought surprising that the earliest representatives of the Lorisiformes to have been discovered to date resembled today's bush babies, and probably had a leaping locomotion like theirs. The anthro-

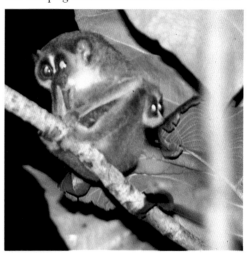

pologist Ernest Walker therefore concludes that the pottos and lorises are the phylogenetically younger, derived branch, although to us their slow movements seem more primitive than the liveliness of the bush babies.

The lorises and galagos still have many features in common with the other large groups of living prosimians, the lemurs of Madagascar, even though their paths of development have been sep-

Slender loris mother and baby. This unusual photograph was taken in a woodland district of southern India.

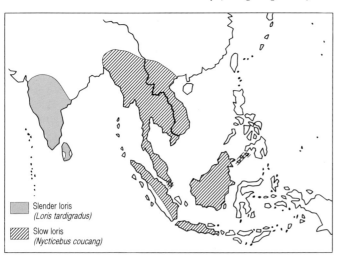

Slender loris *(Loris tardigradus)*

Slow loris *(Nycticebus coucang)*

arate for so long. The bush babies, particularly, show a great resemblance to the dwarf lemur family Cheirogaleidae; in fact, some investigators surmise a closer relationship between the two groups. To be sure, there is some evidence that the similarity is rather to be accounted for by parallel (convergent) evolution under comparable ecological (environmental) conditions, and by retention of conservative original features.

Like the lemurs, the lorises and galagos have a highly developed sense of smell, Accordingly, the olfactory lobe of the brain is prominent compared to those of the higher primates, to whom smell is less important. The moist rhinarium (nose leather), absent in all the simians, is to be seen in the same context.

The dentition of the lorises and galagos, like that of most lemurs, exhibits a quite archaic state. The dental formulas are much like those of the earliest prosimian-like mammals that lived about

50 million years ago: in the lower and the upper jaw, there are two incisors, one canine, three premolars, and three molars on each side. Like most lemurs, the lorises and galagos have a peculiar feature of the jaw; the anterior part of the mandible is modified to provide a comb-like structure. This so-called "tooth comb" is formed by the four incisors and the two canines, all of which are slender and pointed, and directed forward. The first premolar, however, has taken on a canine-like shape; in the upper jaw, there is a wide gap (diastema) between the left and right incisors.

The tooth comb is used to clean encrusted or soiled fur areas. Particles are thus caught between the teeth, and the lorises and galagos, in addition to a tongue, have developed a so-called "undertongue" (sublingua). It is located underneath the tongue proper, and is armed in front with hardened points of horn. This second fleshy comb is used to clean the tooth comb. The tooth comb and sublingua, not the hands as in the higher primates, are also employed in the social practice of mutual fur care (grooming).

Another important function of the tooth comb was discovered only in the last few years. Observations in the wild have shown that for many species of Lorisiformes, tree saps are an important part of the diet. Like some South American clawed monkeys, they scrape the resin from the trees or out of recesses in the bark, and they use the tooth comb for this purpose.

The lorises and galagos, like the lemurs, have a claw at the second toe, a "grooming claw." It is

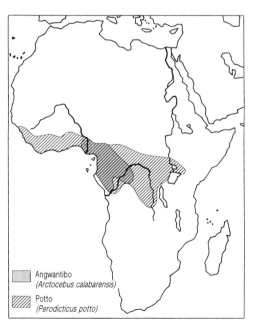

Angwantibo
(Arctocebus calabarensis)

Potto
(Perodicticus potto)

In slow motion, the rotund slow loris stalks about the nocturnal forest on its comparatively short, stumpy limbs.

used especially to clean parts of the body that cannot be reached with the tooth comb.

Determining the sex of lorises or galagos is no simple matter for the inexperienced. The external sex characteristics are visible enough, but the female clitoris is deceptively similar to a penis on superficial examination. Thus, it is traversed by the urethra, opening at the tip. The vaginal opening is at the base and visible only during estrus; the rest of the time it is overgrown with tissue. The most reliable method of sexing is therefore to check whether a scrotum is present.

Nocturnal Habit

While some of the prosimians living on Madagascar are active by day, all of the lorises and galagos are more or less dormant until nightfall. Reports of daytime activity of bush babies are based primarily on observations of animals in captivity; in the wild, activity in the light of day is exceptional. The reason for the nocturnal habit of the lorises and galagos presumably is that they share their habitat with higher primates. These are abroad exclusively in the daytime, and represent overwhelming competition which the small lorises and bush babies must avoid. On Madagascar, however, where there are no higher primates, some prosimians have adopted a diurnal habit.

The eyes of the Lorisiformes, especially those of the slender loris and the bush babies, are very large. To enhance light sensitivity, all species have a reflective layer, the tapetum lucidum, in the retina. It consists of riboflavin, and like a mirror it reflects all light rays that have traversed the retina without stimulating the sensory cells. So when they pass through the layer of sensitive cells again, there is a chance that they will take effect after all. The reflective properties of the tapetum are utilized by scientists in field research; they sweep the trees with long-range searchlights, and when the beam catches a prosimian, its eyes shine bright. Otherwise it would be almost impossible to locate the animals at night in the dense growth of the tropical rainforest.

On Sri Lanka, the large eyes of the *thevangu* or *una happolava* (slender loris) are a liability as well. The Singhalese used the animal's tears as a love potion. To collect them, they would hold red-hot

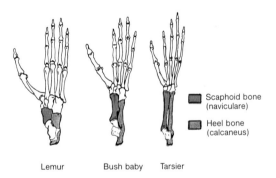

Lemur   Bush baby   Tarsier

■ Scaphoid bone (naviculare)
■ Heel bone (calcaneus)

iron rods near its eyes. It is to be hoped that this cruel practice is a thing of the past.

Like many other animals that are active only in darkness, the lorises and bush babies are fairly color blind. In the retina there are but few if any of the cones required for color vision. Experiments too have indicated that they will not respond to differences of hue, and probably cannot distinguish them.

(Far right) The conspicuous lengthening of the scaphoid and heel bones in the bush baby and tarsier adds much length to the hind limbs, with a great gain in leaping capability.

(Below left) The African potto appears, and is, quite relaxed. Unlike their hyperactive relatives, the bush babies, they climb about slowly, usually solitary, in the treetops of the tropical rainforest. The potto is characterized by three peculiar features: the short stubby tail, the marked reduction of the second finger, and the prominent spinal processes at the nape, which probably serve for defense.

Habitat and Adaptation

Among the lorisiformes living today, scientists distinguish ten species: six of them are bush babies, family Galagidae, and four are lorises or pottos, family Lorisidae. All bush babies, as well as the potto and the angwantibo, occur only in Africa south of the Sahara. The slender loris inhabits Sri Lanka and the southern part of India, while the slow loris is at home in Southeast Asia and on Sumatra, Java and Borneo. The typical loris and galago habitat is the evergreen rainforest. Only the giant galago, and especially the Senegal galago, have succeeded in advancing into drier climates (dry forest, gallery forest, treed savanna). Some slender lorises also live in regions where there are long dry seasons and the trees shed their foliage.

Whereas fairly uniform climatic conditions prevail in the tropical rainforest, the Senegal and giant bush babies are exposed to wide fluctuations of ambient temperature of air and water in much of their range of distribution. The primatologists Bearder and Doyle, during field research in South Africa, measured temperatures between 21 and 106°F (−6°−+41°C); annual precipitation was only 18.5 in. (464 mm), of which 17.9 in. (449 mm) fell between October and April. In the rainforest of Gabon, where five species of Lorisiformes occur side by side, temperatures fluctuate only between 57 and 92°F (14°−33.5°C) through the year, and the average annual rainfall is 60 in. (1500 mm). Dry seasons of two to three months may indeed occur, but the relative humidity is always high.

Notwithstanding the sometimes severe stress of heat or cold, the Senegal and giant bush babies manage to keep their body temperature at a level between 97 and 104°F (36°−40°C). There is no quasi-hibernation, with lowered body temperature, such as with the dwarf lemurs of Madagascar. In studies at the Tübingen University department of zoophysiology, both of these Galago species were subjected to an artificially simulated dry season with reduced food supply. But even after several weeks, they showed no changes in body temperature, even though they had lost as much as 20 percent of their baseline weight. Suppositions that the giant bush babies use the tail as a fat reservoir to tide them over a long sleep in the dry season, therefore, seem implausible. Very likely, observations of the Malagasy fat-tailed dwarf lemur *(Cheirogaleus medius)*, which actually does store fat in the tail and sleep in dry seasons, were simply extended to the giant bush baby, whose specific name *crassicaudatus* happens to mean "thick-tailed." But in their case, this name merely refers to the dense and bushy fur of the tail.

On the trunk also, the bush babies, like the lorises and pottos, wear a thick, woolly coat of hair. This may seem surprising, since most of the species, as inhabitants of the tropical rainforest, are exposed to only moderate cold. The explanation presumably lies in the comparatively low endogenous generation of heat in the prosimians. Measurements by this writer as well as studies by other authors have shown that when physically at rest, the bush babies generate about 20–40 percent less heat than most mammals of comparable size. In the case of the lorises and pottos, the heat generation is only about half the norm. Their body temperature is also somewhat lower than that of the bush babies. Because of their lesser generation of heat, the Lorisiformes need to lose as little of the heat as possible, and this is

The leaping form of the Senegal bush baby: While still in mid-air, the upper body, at first extended, is drawn far back, so that the feet land first on the branch of destination. This perhaps tends to protect the vulnerable head region from injury.

achieved primarily through the insulating effect of the thick coat of fur.

The Lorisidae have evolved a further adaptation in this connection. In their arms and legs, blood vessels are arranged in bundles that have been termed the *rete mirabile* (miracle net). In entering the limbs, arteries and veins branch into many smaller vessels closely adjacent to each other. Thus the warm blood arriving from the heart can transfer much of its heat to the cooler blood on the way from the hands and feet to the heart in the neighboring veins. Thus the bundle functions like an industrial heat exchanger, so that the blood reaching the hands and feet has been precooled. In this way, loss of heat through the comparatively large areas of these body parts can be kept very low. The bush babies have so such vascular bundles. Possibly too abrupt a cooling of the musculature is not compatible with their constant readiness for long leaps.

Locomotion

In older accounts, the leaping prowess of the bush babies was occasionally much exaggerated. But the results of reliable measurements are impressive enough. In one experiment, a Senegal galago leaped almost vertically from the ground to a platform at a height of 7.2 ft (2.25 m) above him. This is more than ten times the body length, not counting the tail. A human being can jump not even one-and-a-half times his height. Horizontally, the bush baby can clear distances of at least 8 ft (2.5 m) from a standstill. Needle-clawed bush babies have been observed to drop a depth of 26 ft (8 m) to be sure of ending up in the foliage of a neighboring tree.

An important factor for these powerful leaps is the prolongation of the hind limbs. Whereas pottos and lorises have forelegs and hind legs of about the same length, the bush babies' hind limbs are longer by half. The increment in length is due almost entirely to a prolongation of two bones of the foot, the scaphoid and the heel bone. This lengthening does not interfere with the grasping function of the toes.

In the landing after the leap, the bush babies appear to have evolved two different procedures. The French primatologist Charles-Dominique observed that dwarf and black-tailed bush babies will first grasp the target branch or trunk with the hands. That is to say, it is chiefly their arms that cushion the impact. But photographs in our department have shown that Senegal bush babies usually land feet first, often drawing the upper body far back. This may be because in their habitat, trees and bushes often bear long, sharp thorns. The retraction of the anterior body might thus tend to abate the hazard of injury to the head region.

In complete contrast to the bush babies, whose locomotion consists largely of leaps and hops, the almost sloth-like, slow locomotion of the lorises and pottos has earned them the name "softly-softly" in the pidgin English of West African natives; in Swahili, the potto is called "pole-pole." Both expressions mean "easy does it." Of course, anyone familiar with these prosimians knows that they are perfectly capable of moving quickly. In warding off attackers, they can snap with lightning speed.

When a potto runs along a branch, at least one

(Right) This slender loris runs on its index digits alone; only the thumb and the last three fingers serve for grasping.
(Next page, photos) When danger threatens, the angwantibo (bottom) and the potto (top) will roll up in a ball. We can see both of the potto's hands; the absence of the index finger results in an even more effective forceps action.

hand and one foot are always in contact with the support. The strong need for a firm hold becomes especially clear when a loris or a potto is to be removed with gentle force from a branch. In this urge to cling, they will squirm violently and try to grasp any object they can. Sometimes it will even calm them if they can catch hold of their own arms with their feet. Of course this strong grasping reflex has much survival value for animals that normally live in trees, far above the ground.

To obtain a very firm grip, the Lorisidae are able to extend their thumbs and great toes very far. Besides, the second finger is residual, and nearly absent in the potto and the angwantibo. Thus these prosimians are able to grip branches as with a pair of tongs, for which reason they are referred to as grasping climbers. With the aid of the strong musculature of the hands and feet, they can hold on so tight that it is very difficult to detach a slow loris or a potto from a branch against its will. Both are also readily able to hang head downward from a branch, anchored by the feet only, so as to snatch fruit or prey with both hands, or play with their fellows. A great aid in this are the well-developed pads of the hands and feet, affording an excellent grip.

The slow as well as fast-flowing movements of the lorises and pottos are advantageous to them in two ways. Firstly, they can steal very close to their prey without attracting attention and then snatch it quickly; and secondly, they are themselves protected against discovery by enemies. In the Gabon rainforest, Charles-Dominique often observed pottos halting abruptly and freezing in place at the approach of danger (usually weasels). As if petrified, they would remain motionless for minutes until the coast was clear. Technically, this is known as "cryptic" behaviour. Bush babies, by contrast, when an enemy is discovered, perform great leaps of flight, uttering loud cries of alarm.

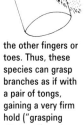

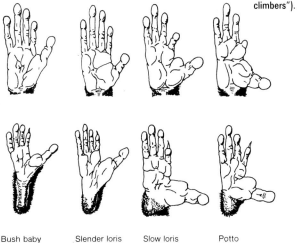

Bush baby   Slender loris   Slow loris   Potto

(Below) Hand of slow loris, clamped around a cylindrical branch represented as transparent. The highly developed pads provide a good grip.
(Bottom right) The extensibility of the thumb (top row) and great toe (bottom row) is comparatively limited in the bush baby. In the lorises and pottos, however, it is much greater; they can be spread very far and opposed to the other fingers or toes. Thus, these species can grasp branches as if with a pair of tongs, gaining a very firm hold ("grasping climbers").

When a potto or loris is, after all, discovered and attacked by an enemy, the small prosimian is by no means defenseless. Pottos have evolved a number of adaptations to ward off aggressors. They face the adversary, drawing the head in to the chest and sheltering it with the arms. Although he cannot see to fight in this posture, the nape region is foremost, with its peculiar features. The skin is much thickened and hardened, and the shoulder blades extend almost to the middle of the back. With this protective shield, the potto can fend off enemy bites. There are long hairs, as much as 4 in. (10 cm) in length, in the back of the neck, extending far beyond the coat of fur; they connect with sensory cells in the skin, monitoring any contact. In this way the potto can tell when the enemy is close. Then he will suddenly jerk his head up and bite vigorously. He also delivers powerful blows with his neck. They are especially effective because three neck vertebrae and the first three thoracic vertebrae have long spinal processes. These are covered with thick skin and protrude from the "neck shield" like horns. Since he himself is firmly anchored to a branch by his iron grip, the potto often succeeds in discouraging the attacker or tumbling him off the branch.

The smaller angwantibo, or golden potto, having no such neck shield, adopts a different strategy. When attacked by an enemy, it rolls up into a ball, head to chest and hidden under one arm. All that projects from this ball is the button of a tail, the hairs of which the angwantibo erects in the form of a ring. When the attacker, diverted by this maneuver, attempts to investigate the strange configuration more closely, the angwantibo darts out from under its arm to bite him.

The bush babies' resources for rapid flight and the cryptic behavior and effective defense measures of the lorises and pottos evidently provide good protection against being eaten. In Gabon, at any rate, examination of the stomach contents of weasels showed hardly any remains of prosimians.

A special difficulty arises from the distribution of the several species of the Lorisiformes. Whereas

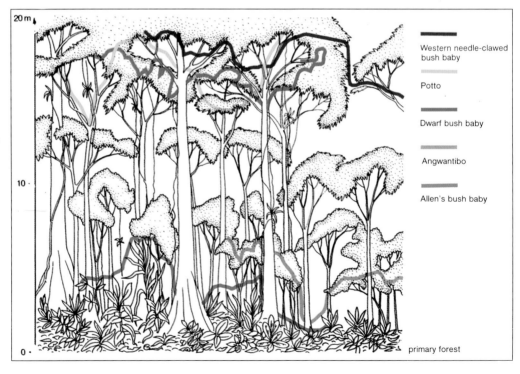

An impressive example of subdivision of a habitat and far-reaching avoidance of food competition is presented by the Lorisiformes in the rainforest of the Congo. They are distributed on different forest levels (right), and the species living in the same "storey" use different sources of nourishment (graph on opposite page: The figures ate increased up to 100 percent by the occasional use of other food, such as, leaves, mushrooms). 1 m = 3 ft the two Asian representatives, the slow loris and the slender loris, have non-overlapping ranges, a glance at the map shows that in Africa there are two habitats in which several Lorisiformes live side-by-side (in sympatry): the tropical rainforest of the Congo basin is inhabited by no less than five species, and in East and South Africa the coastal and gallery forests as well as the bush savanna are the abode of giant and Senegal bush babies. One naturally wonders whether this must not lead to keen competition for the available food sources.

Division of the Food Supply

Recent investigations have shown, however, that the sympatric Lorisiformes avoid competition for food to a large extent. In South Africa and Zimbabwe, Bearder and Doyle learned that the giant and Senegal bush baby (local subspecies *Galago senegalensis moholi*) again subdivide their common habitat. Whereas the giant bush babies attain their highest population density in comparatively moist forestland (over 100 individuals per square kilometer; $1 km^2 = 0.36 mi^2$), the small Moholi bush babies prefer the marginal forest areas, most especially open and dry treed savannas. In acacia woods, the primatologists found as many as 500 per square kilometer.

This subdivision of habitat is attended by differences in diet. Whereas the giant bush baby lives chiefly on fruits, the moholis have specialized in hunting small arthropods, predominantly insects. Tree sap is an important component of the diet of both species. At certain times of the year, when there are hardly any insects or fruits, they depend almost completely on this carbohydrate-rich fare. Moholis are thus found mostly where resinous tree species occur. They also profit by the activity of certain insect larvae (of the long-horned, jewel, and click beetle families, as well as the wood borers among butterflies), whose feeding activities stimulate the trees to produce resin. In the long run, to be sure, the tree saps cannot supply a complete diet, so that the bush babies tend to lose weight in the season of few insects.

It is reported that the giant bush babies are very fond of fermenting palm sap. They find it at the taps placed by the natives to make palm wine. Intoxicated on alcohol, the kombas, as they are called in Swahili, often fall from the tree and lie on the ground in a stupor.

In spite of the subdivision of habitat and differentiation of diet, some competition between giant and Senegal bush babies must remain. Where both species occur in close association, apparently the larger predominates; in Zimbabwe, for example, where giant bush babies occur even in more open and drier bush, the lowest population density of moholis was found.

Prepared to leap, a Senegal bush baby is on the lookout for prey. He is not interested in the gorgeous blossom.

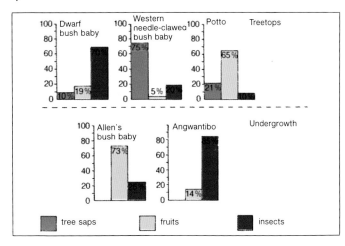

The Lorisiformes living in the African rainforest illustrate the avoidance of competition by subdivision of habitat and specialization of diet still more impressively. Simply because each species has taken on a different role in the ecological texture of the tropical rainforest (a different niche, to use the technical term), it becomes possible for five lorisiform species to exist together. The causes have been discovered by Charles-Dominique in the course of years of observations in Gabon. He learned that the lorises and galagos are distributed in different storeys of the rainforest. Of the two, the small angwantibo, or golden potto, occupies the lower level almost exclusively (up to about 16 ft or 5 m), while the pottos are met with predominantly in the treetop range between 32 and 96 ft (10 and 30 m). Among the bush babies, the black-tailed or Allen's galago keeps near the ground, while the dwarf and needle-clawed galagos dwell in the crowns.

Species living on the same storey are clearly distinguished in the composition of their diet. According to Charles-Dominique's observations, that of the dwarf bush babies consists about 70 percent of insects, the rest being fruits and tree sap. Their low weight specially qualifies the dwarf bush babies to climb out on the weak outer twigs and search them for insects. With their acute hearing, they can detect their prey at long distance. In an experiment, a dwarf bush baby was found able to track the movements of an insect accurately by sound through an opaque partition. Bush babies are aided in this by the mobility of their large, thin-skinned (membranous) ears, which have a number of rib-like structures about which they can be folded. The little beasts perform amazing acrobatics when they catch gnats in flight. They clamp themselves to a branch with their feet and dart out with trunk and arms to grasp the insect with both hands. As if retrieved by a spring, a fraction of a second later they are sitting in the same place as before, consuming their prize.

In the case of the larger needle-clawed bush babies and the pottos, however, insects account for only about 10–20 percent of total diet. The needle-clawed bush babies are very highly specialized, living primarily on flowing tree sap and resin. Only about a quarter of their diet is made up of fruits and insects.

This species too shows an adaptation to its special mode of foraging. The structure of their nails, for which they are named, in effect equips them with climbing irons. When the first joint of the finger is bent inward, the sharp, powerful nails project far forward and can be thrust into the bark of a tree. With this anchorage, its possessor can even run down a trunk head downward, reaching sources of nourishment inaccessible to other species.

The largest species, the potto, lives primarily on fruit. On the average, fruits provide about two-thirds of its diet. Resins, with a proportion of about 20 percent, are important as well. Besides, the potto does not disdain evil-smelling insects or other arthropods that the bush babies avoid.

Among the inhabitants of the lower storey, the angwantibos and the black-tailed bush babies, there are similar differences in feeding habits. Allen's black-tailed bush baby will pursue insects, but three-quarters of its nourishment consists of

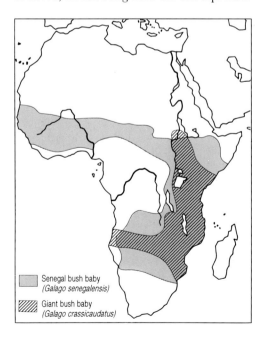
Senegal bush baby (*Galago senegalensis*)
Giant bush baby (*Galago crassicaudatus*)

fallen fruits. The angwantibos on the other hand have specialized in the hairiest butterfly caterpillars, scorned by other animals. They have evolved a special method of eliminating the long hairs, which are often severely irritating to mucous membranes. They take the head of the caterpillar between the teeth and roll its body between their hands until the hairs have been broken off. When they have chewed and swallowed the stripped prey, they rub their lips and snout on branches to remove adhering caterpillar hairs. They are assisted in hunting by their remarkable sense of smell, with which they detect a caterpillar at a range of 4.8 ft (1.5 m).

The lorises and galagos of Africa are thus a very

good example of the concept of ecological "niching." True, several species share the same "address," the tropical rainforest in this case, but each gains its livelihood in its own way, having in other words a particular "calling." It should not be overlooked that in principle probably all the Lorisiformes prefer animal fare. The chances of finding it are about the same for all. As a matter of fact, Charles-Dominique's researches have shown that both the small dwarf bush babies and the pottos, 10 to 20 times their size, daily consume one or two grams (28 g = 1 oz) of insects. This quantity is enough to meet nearly all of the dwarf bush baby's energy requirement, while the larger species need additional food. Thus, niching represents primarily the different ways that have been taken by different species in order to obtain that additional food.

Contrary to the widespread impression that the tropical rainforest is an opulent land of plenty, the lorises and bush babies living there are under considerably stronger pressure of competition than the giant and Senegal bush babies of the treed savannas and gallery forests. This is illustrated by the population densities of the several species. In his Gabon research field, Charles-Dominique found an average of 50 dwarf bush babies, 15 Allen's black-tailed bush babies, 15 needle-clawed bush babies, 8 pottos, and 2 angwantibos per square kilometer (1 km² = 0.36 mi²). In East and South Africa, however, Bearder and Doyle – despite comparatively adverse climatic conditions – generally found more than 100 giant and Senegal bush babies per square kilometer. The differences in frequency of occurrence are understand-

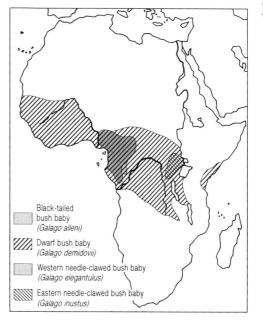

Senegal bush baby mother and offspring. In the first weeks of life, the little one is totally dependent on maternal care.

The giant bush baby looks like an enlarged edition of the Senegal bush baby (opposite page). However, its tail is bushy throughout its length.

able when it is recalled that the lorises and galagos in the rainforest must compete with about 100 other species of mammals and some 300 species of birds. The giant and Senegal bush babies, on the other hand, have far fewer food competitors to contend with.

Less is known about the feeding habits of the slender loris and the slow loris in the wild than about the African Lorisiformes. In captivity, both species eagerly accept animal fare, but they can be kept with good success on a high proportion of vegetable fare. In the wild, the slender loris presumably feeds primarily on insects, whereas the slow loris, much like the potto, may be more dependent on vegetable fare as well.

Social Behavior

The social structures of the lorises and galagos – their forms of social life – are comparatively simple. Again, some of the African species are best-known. Female giant and Senegal bush babies occupy individual territories which they mark and defend. Within these territories, they tolerate their own progeny only. In the long run, however, only the female young remain with the mother; the males leave the territory when sexually mature. Males may sometimes remain in the group longer, but in that case they exhibit no sexual activity. The system leads to the formation of small social units consisting of mothers, daughters and sisters, and their young.

As a rule, only a single adult male mates with the females of such a group. But the territories of high-ranking (dominant) males are much larger than those of the females, so that one male may easily control several groups of females. Accordingly, competition among male bush babies is intense. Whereas a female may occupy a territory for several years, the dominant males change almost annually. When the young male bush babies have left their maternal group, they at first wander about but then settle on the edge of a female territory. Sometimes they will form bachelor groups of two or three individuals. In that position, they bide their time until an opportunity of replacing the dominant male presents itself. Then much time will be spent in examining all females in the territory regularly to check whether they are in heat. If so, the male will remain with the female until breeding occurs.

Lorises and pottos have still simpler social structures than the galagos. Their dominant females and males likewise occupy territories, but no female groups are formed, because the females of sexual maturity do not remain with the mother either. The slow loris may be an exception. In captivity, monogamous pairing and the formation of family groups has been observed. When young males reach sexual maturity, they are expelled by their male parent. However, field studies of this species in the wild are as yet entirely lacking.

Identification of Individual Territory

Lorises and galagos identify their territories chiefly by scent. The most important method is marking with urine. The bush baby and the slender loris have a peculiar technique called "urine washing." They balance the trunk on the limbs of one side, while urinating on the palms and soles of the other side. Then the latter are rubbed together (hence "washing"). Going to and fro in their territory, the hand and foot surfaces thus moistened leave scent marks everywhere, notifying conspecifics that this is occupied territory. The slow loris and the pottos and angwantibo do not practice urine washing. They urinate directly on the substrate. They do this either spotwise in certain places (males especially will mark in this way), or they will hold the posterior against the substrate and advance some distance, leaving a track of urine behind them. The great significance of scent marking to lorises and galagos becomes clear when they are placed in a new cage, or when the old cage is thoroughly cleaned. They then exhibit intensified marking behavior for some days, and

Senegal bush baby.

do not rest content until the installation has again taken on their individual odor.

Besides marking with urine, some species also use the secretions of certain glands for olfactory intraspecific communication. For example, the giant and Senegal bush babies have glands over the posterior division of the breastbone, secretion from which is applied by rubbing against the substrate. Other scent glands are found in the skin of the scrotum and the external vagina. What information is transmitted by this scent marking has not yet been determined in detail. There is some evidence, however, that membership in certain subspecies, sex, or even the identity of an individual, may be read off from its scent marks. Thus the marking animal leaves, as it were, a print of its "personal" identification. Presumably the males can also tell by the scent marking whether the females are in heat.

Unique among the Lorisiformes is a special behavior pattern of the giant bush babies. Both sexes have a peculiarly structured surface at the posterior edge of the sole of the foot. It bears a large number of short horny combs, giving the whole somewhat the appearance of a file. Glands are not developed at this site. When excited, the giant bush babies rub this surface against the substrate, producing a loud scratching sound. It is not quite clear whether this serves to apply the urine more effectively or whether the noise itself is in the nature of a signal.

Communication by Utterance of Sounds

Communication by acoustic signals is in any event important among the lorises and galagos. It was first studied in detail by Elke Zimmermann in adult Senegal bush babies and slow lorises. The Senegal bush babies have at least eighteen different calls that can be correlated with definite modes of behavior. They belong to three functional groups: social contact, aggressive and defensive behavior, and annunciatory. The slow loris, by contrast, has only eight sounds, and none at all relating to annunciatory behavior. The greater prominence of acoustic communication among bush babies is part of their strategy upon recognition of an enemy. They are constantly prepared for instant flight, and have a series of alarm calls alerting conspecifics to threats of danger. They can also afford to identify their territory by loud

cries. Their frequently heard, somewhat mournful calls first earned them the name "bush baby."

The slow loris and other Lorisidae must follow a different strategy. Being unable to flee from danger quickly, they depend on remaining undiscovered as long as possible. Alarm calls would attract the attention of a possible predator. The lorises therefore use a different, silent method of warning conspecifics. On the inside of the upper

(Far right) All bush babies have large membranous ears, but those of Allen's black-tailed bush baby are gigantic. (Bottom left) The infant Senegal bush baby falls into a transportation trance when being carried about in its mother's mouth. It thus accompanies the mother on her nocturnal excursions. Only when she is searching for food will she set it down in the branches.

arm, they have a large, naked glandular area. Scents secreted by this brachial gland serve to heighten the alertness of conspecifics in the neighborhood. Such an olfactory warning system is of course not so effective at long range as the warning noise of the bush babies.

A form of intraspecific communication not recognizable by humans without technical aid was discovered by Elke Zimmermann only a few years ago. It had indeed been known for some time that slow lorises and Senegal bush babies can perceive sound in the ultrasonic range, above 20 kHz. But it was not known whether they themselves produced such high-frequency sounds. It now appears that the adults produce only sounds that are in the human audible range, but the young also produce sounds in the ultrasonic range. These sounds lead the mother back to the young. Assuming that a predator cannot hear these high frequencies, the young are thus enabled to call for help without disclosing themselves.

Reproduction and Raising of Young

Low infant mortality is especially important for the preservation of loris and galago species. Compared to other mammals of similar size, their birth rate is very low. With the exception of the small Galago species, which have twins often, generally there is only one young per year. This is an effect of the linkage of births to certain seasons and also to the unusually long gestation periods of prosimians. Even the little dwarf bush babies have a gestation period of about four months, and more than six months elapse between pairing and birth among the pottos and slow lorises. Despite the comparatively long periods of development in the maternal body, the young of prosimians weigh less at birth than the newborn of higher primates of similar size. Thus the young of the Senegal galago subspecies *Galago senegalensis senegalensis* are born weighing 0.6–0.8 oz (18–22 g), and those of the common marmoset *(Callithrix jacchus)*, which is about the same size, are born at 0.8–1 oz (22–28 g). Since, moreover, with about the same gestation period, the common marmosets usually have twins, their total litter weight is more than twice that of the Senegal bush baby.

These differences in birth weight are presumably attributable to more effective nutrition of the fetus in the higher primates through their differently formed placenta. For whereas in the prosimians the maternal uterine tissue is completely preserved, so that the epithelium of the uterus is adjacent to the chorion, in the higher primates the layers of tissue deriving from the mother are completely absorbed, so that the chorion is bathed directly with maternal blood. Hence, in the prosimian placenta, there are a total of six layers between the maternal bloodstream and that of the fetus, and in the higher primates, only three. Consequently only the respiratory gases can be exchanged between bloodstreams in the prosimians; the fetus cannot be nourished through the blood. This function is performed by disintegrating maternal uterus cells. A special position is occupied here by the dwarf bush babies, where, at least locally, the uterine connective tissue and the epithelium are disintegrated, so that the chorion is directly adjacent to the maternal vascular wall. This is thought to be related to the small body size of the dwarf galagos.

Newborn bush babies already have a thin coat of hair, and the eyes are open, but they are not yet able to go upright independently. The moholi mother leaves the young only briefly to feed during the first three or four days, returning immediately to the nest of twigs and leaves. Later, she takes the baby along on her nocturnal excursions; it is carried about hanging transversely from the mouth, inducing a dazed state. Now and then, she will deposit the baby among branches during her search for food. She will not return to the nest until dawn. After about two weeks, she will rejoin the female group. In the case of the larger giant and needle-clawed bush babies, the young are found clinging to the fur of the mother's back at the age of about a month. With increasing age, they be-

LORISES AND GALAGOS

Lorises and Galagos (Lorisiformes)

Nomenclature English common name Scientific name Frenc German	Approximate Size Body length Tail length Weight	Distinguishing Features	Reproduction Gestation period Young per birth Weight at birth
Slender loris *Loris tardigradus*, 6 subspecies Loris grêle Schlanklori	9.2–10.4 in.; 23–26 cm 8.9–11.4 oz; 250–320 g	Very long, slender arms and legs ("banana on stilts"); large eyes and ears	160–174 days Usually 1, occasionally twins 0.4–0.6 oz; 10–18 g
Slow loris *Nycticebus coucang*, 10 subspecies Loris paresseux Plumplori	10–14.8 in; 25–37 cm 0.4–0.8 in; 1–2 cm 0.5–3.5 lbs; 250–1600 g	Far more compact and stocky than slender loris; limbs shorter and sturdier; small, hairy ears	186–193 days Usually 1, rarely twins 1.3–1.6 oz; 35–45 g
Angwantibo; golden potto *Arctocebus calabarensis*, 2 subspecies Potto doré; potto de Calabar Bärenmaki; Angwantibo	9.2–12 in; 23–30 cm 0.4 in.; 1 cm 5.3–17.8 oz; 150–500 g	Pointier snout than other Lorisiformes; second finger and second toe almost totally absent	135 days Usually 1 0.8 oz; 24 g
Potto *Perodicticus potto*, 5 subspecies Potto de Bosman Potto	12.2–14.8 in.; 30.5–37 cm 1.6–2.8 in; 4–7 cm 1.9–3.5 lbs; 850–1600 g	Thickened skin (nuchal shield) on nape with and long hairs that communicate with sense organs in the skin; scapulae nearly to center of back; last 4 cervical and first 3 thoracic vertebrae with long spinous processes protruding from the nape as short pegs; newborn sometimes almost white	180–193 days Usually 1, twins very rare About 1.8 oz; 50 g
Thick-tailed bush baby; greater galago *Galago crasicaudatus*, 11 subspecies Galago à longue queue; galago à queue touffue Riesengalago	12.4–14 in; 31–35 cm 16–18.4 in; 40–46 cm 2–3.5 lbs; 900–1600 g	Long, bushy tail; horny papillae form file-like surface at posterior edge of the sole of the foot	130–136 days Usually 1, twin births not rare About 1.6 oz; 45 g
Lesser bush baby; Senegal bush baby *Galago senegalensis*, 9 subspecies Galago du Sénégal; galago commun Senegalgalago; Steppengalago	5.6–7.6 in; 14–19 cm 8.8–12 in; 22–30 cm 5.3–15 oz; 150–420 g	Similar to greater galago, but only about half its size; tail less thick and bushy, at least anterior half	120–144 days *G. s. moholi* usually 2; *G. s. senegalensis* Usually 1 0.3–0.1 oz; 8–22 g
Allen's bush baby; black-tailed bush baby *Galago alleni* Galago d'Allen Buschwaldgalago	7.4–8.2 in.; 18.5–20.5 cm 9.2–11.2 in.; 23–28 cm 6.7–12.1 oz; 190–340 g	Very large ears; tip of tail often white	About 133 days Usually 1 About 0.8 oz; 24 g
Dwarf bush baby; Demidoff's bush baby *Galago demidovii*, 7 subspecies Galago de Demidoff Zwerggalago; Urwaldgalago	4.4–5.2 in; 11–13 cm 5.6–7.2 in; 14–18 cm 1.7–3.0 oz; 50–85 g	With mouse lemur, the smallest primate species	110 days Usually 1 0.2–0.4 oz; 7–12 g
Western needle-clawed bush baby *Galago elegantulus*, 2 subspecies Galago mignon Westlicher Kielnagelgalago	7.2–8.4 in; 18–21 cm 11.2–12.4 in.; 28–31 cm 9.6–12.8 oz; 270–360 g	Nails long, centrally keeled, with sharp points, used like crampons for climbing; long comb teeth; intestinal tract remarkably long; very large golden eyes; tip of tail always white	Not known
Eastern needle-clawed bush baby *Galago inustus* Galago du Congo Östlicher Kielnagelgalago	6.4 in.; 16 cm 9.2 in.; 23 cm 8.9 oz; 250 g	Dark fur; nails long and with sharp keel	Not known

The needle-clawed bush babies owe their name to their long, sharp nails, with a central keel. These nails illustrate how a flat nail can become a claw. The Western needle-clawed bush baby is pictured.

come more independent and will follow the mother or other group members. The young of the moholi bush babies attain adult weight at about 20–25 weeks. Development takes a little longer for twins than for single young.

The lorises and pottos build no nests. Their young are somewhat more thickly furred, and can cling to the mother's coat. During the first days of life, they are usually seen hanging at the belly; they are carried about in that position. The young

COMPARISON OF SPECIES

Life Cycle Weaning Sexual maturity Life span	Food	Enemies	Habit and Habitat	Occurrence
Age about 10 weeks Age 11–12 months In captivity, to 15 years	Prefers small prey (insects, reptiles, birds); fruits	Not known	Nocturnal; during period of activity, usually solitary in rain forest and tree and bush jungles with dry summers	Numbers not known
Reports range between 10 weeks and several months Age at least 12 months In captivity, over 20 years	Fruits and small prey (insects, reptiles, birds, mammals)	Not known	Nocturnal in tropical rain forests; hardly any field observations	Subspecies *N.c. pygmaeus* (regarded by some as separate species) threatened
Age about 4 months Age 8–10 months In captivity, 13 years	Small prey, specializing in long-haired caterpillars; fruits	Not known	Nocturnal; mostly roaming solitary in lower levels of rain forest (0–16 ft; 0–5 m); prefers dense liana growth and bush vegetation of secondary forests	In primary forest, about 2 individuals per km^2 (1 km^2 = 0.36 mi^2)
Reports range between 5 weeks and 8 months Age about 12 months In captivity, over 22 years	Predominantly fruits; tree sap and small prey (insects, reptiles, birds, mammals)	Palm civets, genets, owls, snakes	Nocturnal; roams, usually solitary, in crowns of trees in tropical rain forests; size of female territory about 18.7 acres (7.5 hectares); male territory 30.7 acres (12.3 hectares)	In primary forest, about 8 individuals per km^2 (1 km^2 = 0.36 mi^2)
Age about 70 days Age about 12 months In captivity, to 15 years	Fruits, tree sap, insects	Owls, genets, snakes	Nocturnal in fairly dense and damp forests, occasionally also in treed savannas; roams solitary or in groups; usually sleeps in communities of several individuals; territory size about 17.5 acres (7 hectares).	Abundant, 70–125 individuals per km^2 (1 km^2 = 0.36 mi^2)
Age 45–70 days Females at 6–7 months, males later In captivity, to 14 years	Predominantly insects and tree sap	Same as greater galago	Nocturnal in margins of dry forest; abundant in treed savannas; territory size about 2.8 hectares; sleeping communities divide and territories usually are ranged by solitary individuals	Abundant, 80–500 individuals per km^2 (1 km^2 = 0.36 mi^2)
Not known Not known In captivity, to 12 years	Fruits and insects	Not known	Nocturnal (0–16 ft; in lower strata of tropical rain forests (0–16 ft; 0.5 m); days spent sleeping alone in hollow trees; territories of females about 10 hectares	Abundant; in virgin forest, about 15 individuals per km^2 (1 km^2 = 0.36 mi^2)
Age 45–50 days Age 8–10 months In captivity, to 12 years	Predominantly small insects, but also fruits and tree sap	Not known	Nocturnal in primary forest, almost exclusively in crown stratum; descends lower in secondary forest; favors liana "curtains"; abundant in thick bush beside paths and roads; size of female territory about 2.5 acres (1 hectare), male territory 4.5 acres (1.8 hectares)	In primary forest, about 50 individuals per km^2; roadsides, often over 110 per km^2 (1 km^2 = 0.36 mi^2)
Not known Not known In captivity, over 15 years	Predominantly tree sap (up to 80%), some insects	Not known	Nocturnal in tropical rain forests; does not build nests	Abundant in primary forest, about 15 individuals per km^2 (1 km^2 = 0.36 mi^2)
Not known	Not known	Not known	Nothing known concerning biology of this species; placed by some authors in genus *Euoticus* with western needle-clawed bush baby	Not known

of the slow loris, however, may be left alone from the first day. They will remain clinging to a branch while the mother goes in search of food. This behavior is technically termed "baby parking." Often the mother will not return until after quite a while, sometimes not until just before daybreak. Later on, the infant will be taken along on foraging expeditions. At first it clings to the back; after a time, it must go under its own power. It thus learns to recognize proper food. Weaning from milk to solid food takes place at an age of 40–60 days, but the young of the slow loris may be nursed into the fifth month. Young lorises and pottos become sexually mature after 8–12 months, but slow lorises possibly not until after two years.

Somewhat different observations have been made in the anatomy department of the University of Bochum, which at present harbors the largest breeding group of slender lorises outside of their natural habitat. Here, no baby parking by the mothers has yet been observed. The babies continued to cling to the mother's belly for a whole month. Only then will they leave the mother actively for a few minutes at a time. Beyond the 100th day of life, the young slender lorises slept through the day, clinging to the mother. At night, however, they would be carried about only until the 60th day of life. Independent feeding was begun at the age of about 75 days.

Until a few years ago, some species of lorises and galagos were occasionally kept as pets in Europe. The lively Senegal galagos were especially popular, becoming very affectionate if captured young. Being a savanna dweller, doubtless this species was hardy enough to endure the climatic conditions in European homes. Species from the rainforest are far more difficult to keep; their climatic and dietetic requirements can hardly be met in a private household.

A special difficulty continued to be encountered in breeding most of the Lorisiformes. Even in university departments and in zoos, there have been few successes so far in breeding lorises and galagos over successive generations. Yet for some species, breeding in groups under scientific care may be the only chance of survival in the near future. Of the ten species of lorises and galagos living today, at least seven are confined to the tropical rainforest. In view of the rapidity with which just this habitat is at present being destroyed by humans, there is reason to fear the worst. All lorises and galagos are therefore protected under the Washington Convention, and exportation from their home country is allowed only for good reasons in exceptional cases.

(Right) In captivity, bush babies become very trusting. One must, of course, be prepared to find that this lively guest will impregnate his entire living area with urine.
(Far right) The Eastern needle-clawed bush baby, distinguished from its western cousin by its smaller size and darker fur coloring.

Tarsiers

by Kurt Kolar

How to place the tarsiers, family Tarsiidae, in the order of Primates, is a question that is still not quite settled. Some regard the extinct forms in this family, of which about 25 different genera existed during Tertiary times, in Europe and in North America as well, as ancestors of the simians, or even of the anthropoid apes and humans in particular. According to studies of blood serum and structure of the auditory organs, tarsiers have more in common with the simians than with the prosimians. Others consider the tarsiers to be merely a highly specialized group within the prosimians. Heinz Stephan, of the Max Planck Institute for Brain Research in Frankfurt, after close investigation of tarsian brains, is of the opinion that there are important considerations for including the tarsiers in the suborder Prosimiae.

The scientific name as well as the term "tarsier" refers to the special situation in the bone structure of the hind legs. The bones at the base of the foot (tarsus), consisting of the heel bone and the scaphoid, are much prolonged, the tibia and fibula being fused together. These features account for the extraordinary leaping abilities of the animals.

There are only three species living today, all belonging to a single genus *Tarsius:* the Sunda or western tarsier *(T. bancanus)*, the Philippine tarsier *(T. syrichta)*, and the Celebes tarsier *(T. spectrum)*. Essentially, all three species are inhabitants of the jungle, found chiefly in coastal forests or inland forests near bodies of water. In North Borneo, western tarsiers have been found at elevations of 3800 ft (1200 m). Dense stands of bamboo, or bamboos growing in freshly cleared forest land, also serve as a refuge. Sometimes the animals are even sighted in settlements or gardens.

For nocturnal hunters like the tarsiers, well developed senses of sight and hearing are indispensable to successful predation. In order to pick up noises and locate the source in any direction, the ears are constantly in motion, independently of each other. Even when a tarsier is disturbed in the daytime, first the external ears zero in on the direction of the source, and only then do the eyes open. Their extraordinary efficiency can be guessed from their astonishing size alone. If our eyes were as large in proportion, they would be the size of apples. The immobility of the eyeball is compensated by the ability of the head to rotate 180 degrees, left or right.

Thus, the tarsiers, like owls, are eminently equipped as nocturnal predators.

During the day, they sleep, but rarely in hollow trees; they prefer to cling to slender trunks standing in darkness, with the tail pressed firmly against the bark. Thus it acts as a support. Bristles on the under side of the tail of the Celebes tarsier, keep it from slipping; in the other two species, the same result is obtained by a special skin structure.

After dark, these little nocturnal specters attract attention by their high-pitched twittering. If there is a female ready to mate in the vicinity, the twittering and chattering is especially loud. At sundown and very early in the morning, tarsiers living in the

The large island of Borneo is the home of this western tarsier. Like its two congeners, it has excellent adaptations to the hunting of small prey by night: large ears, movable independently of each other, enormous efficient eyes, a head that can rotate to all sides, and long powerful legs for leaping. The special trait of the tarsiers, however, is the absence of a nose leather; the share this trait with this simians, and it distinguishes them from all other prosimians.

A tarsier hunting. From one of his always elevated blinds, the little hunter has located prey moving on the ground by hearing, as witness the independently moving external ears. With head rotated nearly 180 degress, he first fixedly and motionlessly eyes the victim, then carefully orients the body somewhat towards it. With lightning speed, the hind legs are extended successively and the tarsier takes off, gives himself some spin, and, rolled up into a ball, springs unerringly upon the prey. To cushion his high impact velocity, ho just before landing he extends his long jumping legs far forward. With eyes closed, as protection against defensive action, it grabs and bites the quarry into helplessness in a matter of seconds.

wild are at their most active. This is probably because many insects are about at these times of day.

Anyone who has observed jerboas will notice an outward resemblance between them and the tarsiers. Both have a rounded trunk with a long tail, and the tarsiers can hop around the area on their hind legs like jerboas. But they can leap on the ground on all fours as well, and they have a quadruped walk also. In the trees, the leap is their most rapid form of locomotion. After a sudden extension of the shanks, the tarsier flies through the air like a bullet. Arms and legs are held close to the body, while the long tail acts as a rudder. Just before landing, the tarsier extends the limbs again, the tail is elevated like a landing flap, and the animal seems as it were to be sucked tight against the target branch by the broad pads of its fingers and toes. Tarsiers living in captivity are known to be able to land on a vertical glass plate, if they can find a hold for a single finger. They can leap 6.4 ft (2 m) in distance and 1.5 ft (0.5 m) in height. In mid-leap, the tarsier can rotate up to 180 degrees around the axis of its body.

According to observations of the Celebes tarsier, there is reason to suppose that at least this species enters into a permanent pairing bond. Every group consists of a pair and its offspring. The area inhabited is about 2.5 acres (1 hectare) in extent. Adult males excrete some urine on the branches,

Second act of the hunt: With his prey, the hunter immediately starts out on the return journey to his place of concealment. Taking aim briefly, he launches himself aloft and lands in the precalculated place, the hands and feet being extended at the last moment and taking hold in that order. Only when up among the branches will he begin his meal. With his eyes closed, he bites off little pieces and masticates them carefully for fairly long intervals before swallowing. (Below) Ground-to-ground pounce. – The series of photos here shown for the first time are the achievement of Heinrich Sprankel, of Giessen, at the Max Planck Institute for Brain Research, Primatology, Frankfurt-am-Main.

▷ Western tarsier in typical vertical posture.

thereby marking the boundaries of the territory. Urine is also carried about with the broad toes. The animals also have a gland over the abdomen for scent marking. Both sexes of the western tarsier use their posterior region for marking; females ready to mate do this more frequently.

The Philippine tarsiers live in communities, and soon become tame in captivity. Whereas ordinarily a distance from conspecifics is always maintained, in the mating season there are mutual approach and grooming. When a tarsier is grooming, it licks its fur an scratches with the claws of the second and third toes. The face is not washed, but only rubbed against branches or other objects. Three Philippine tarsiers, a male and two females, kept by Osman Hill, lived peacefully together. The male obviously enjoyed being groomed by the females. They would hold him tight with their hands and lick him over thoroughly. In this species, the young are picked up and carried with the mouth.

The western tarsiers living on Borneo seem less amiable. No mutual grooming was observed, and fighting is a frequent occurrence. Forty-eight percent of these tarsiers, kept in a large outdoor enclosure for behavior studies by the Berlin zoologist Carsten Niemitz, were found to have ear, hand, foot, and lip injuries. He discovered healed fractures

Tarsiers (Tarsiiformes)

Nomenclature English common name Scientific name French German	Approximate Size Body length Tail length Weight	Distinguishing Features	Reproduction Gestation period Young per birth Weight at birth
Western tarsier *Tarsius bancanus*, 2 subspecies Tarsier de Horsfield Sundakoboldmaki	5.2 in.; 13 cm 8.8 in.; 22 cm 4.2 oz; 120 g	Large eyes; ears movable independently of each other; head can rotate 180° in either direction; very long hind legs	180 days 1, fully furred, eyes open 0.7–1.1 oz; 20–30 g
Philippine tarsier *Tarsius syrichta* Tarsier des Philippines Philippinenkoboldmaki	4.8 in.; 12 cm 9.2 in.; 23 cm to 5.1 oz; 160 g	Much like western tarsier, but tassel of tail more sparse	180 days 1; any season 0.7–0.9 oz; 20–25 g
Sulawesi tarsier; spectral tarsier *Tarsius spectrum*, 2 subspecies Tariser spectre Celebeskoboldmaki	5 in.; 12.5 cm 9.6 in.; 24 cm 4.2 oz; 120 g	Much like western tarsier, but tassel to tail longer and somewhat bushier; bristles on under side of tail; tail has scaly skin	180 days 1; usually November, December, May 0.7–0.9 oz; 20–25 g

of finger and tail bones. In this species, the babies are not carried with the mouth.

A single offspring is born with well-developed fur and eyes open. It is half the size of its mother and a quarter of her weight. Whereas at first it clings lengthwise to the mother's belly, later it hangs obliquely and is then carried crosswise. If danger necessitates rapid flight, the mother will grab the baby with her mouth and carry it. Quite young tarsiers are able to cling to branches, and at three weeks they are no longer dependent on maternal transport. In the search for food, a baby clinging to the belly is doubtless something of a hindrance. The mother will leave it sitting on a branch and depart without that burden. Babies thus parked are very good and never fidget. But if a tarsier has lost its mother, it begins to squeak, polysyllabically and at a very high pitch. This attracts the attention of the adults, but they may not be able to distinguish their own offspring from others. Probably anything the squeaks like an abandoned baby tarsier will be adopted. At six weeks, at least the young tarsiers of Borneo are ready to catch their first prey.

Births occur at any season, but in North Celebes apparently especially in the months of November, December, and May. Zoo births have been reported in Philadelphia, Pennsylvania.

In a forest region on Borneo, Niemitz collected a significant body of data on the securing of food in his large observation enclosure. Of invertebrates, the tarsiers caught spiders, roaches (possibly intermediate hosts of intestinal parasites), praying mantises, walking sticks, grasshoppers, and beetles. Among the latter were sap chafers (Cetoniidae); long-horned beetles (Cerambycidae); and even large rhinocerus beetles, of which the shell was not cracked until after a quarter of an hour's hard labor. The tarsiers also liked members of the Melolonthinae, a group of beetles including June bugs. Butterflies, ants, and cicadas were also among the smaller prey. Fast-moving snails were attractive at first sight and were promptly pounced upon and seized, but then not eaten. The tarsiers would quickly try to wipe the adhering slime from their hands on the nearest tree trunk.

Small fishes in a pond attracted the tarsiers' attention: they would watch them, but not catch them. Yet fresh-killed fish of the same species were accepted at once. Such fish were also flung out of a pot and consumed.

The tarsiers would not consider a frog meal; probably they were put off by the secretion. Small gekkos and their eggs were eaten, and once a lizard, *Draco volans*, was caught and its head eaten. Other lizards were disregarded.

Once a tarsier caught an adder, *Maticora intestinalis*, and ate it up. These adders are certainly somewhat sluggish and slow to strike, but they have a potent nerve venom that has been known to cause death in humans. The tarsiers must have an innate

COMPARISON OF SPECIES

Life Cycle Weaning Sexual maturity Life span	Food	Enemies	Habit and Habitat	Occurrence
Not known Age 12 months Over 8 years	Spiders, insects, small lizards, young birds	Feral house cats	Active at twilight and nocturnally; forest dweller near coasts and by rivers, also in plantations; quarrelsome; territory size 1.7–7.5 acres (0.7–3 hectares)	Under favorable conditions, 250–350 individuals per km^2 (1 km^2 = 0.36 mi^2)
Not known Not known In captivity, over 13 years	Predominantly insects	Not known	Active at twilight and nocturnally in rain forests; usually found in pairs; sociable; engages in mutual grooming	Endangered, despite protective measures
Not known	Insects	Not known	Active at twilight and nocturnally; rain forest dweller, also in bush and plantations; small family groups comprising one pair and latest offspring; territory size about 2.5 acres (1 hectare)	Not accurately known; endangered by destruction of habitat

technique for killing such snakes. It could not be learned, for the first mistake would be fatal to the predator.

When a three-toed kingfisher *(Ceyx erithacus rufidorsus)* flew close by a tarsier, the latter caught it with his hands. Both fell to the ground, where the bird was killed by bites in the neck. On another occasion, a nine-hued pitta *(Pitta brachyura)* was caught and killed. It took the tarsier an hour to bite open the skull; head, brain and beak were consumed. All that remained of a bulbul *(Pycnonotus brunneus)* was one wing and the legs. A lesser spidercatcher *(Arachnothera longirostris)* was consumed entirely. The animals observed by Niemitz did not even spurn bats. As we see, at least the tarsiers on Borneo have some taste for variety. The Celebes tarsier, on the other hand, is believed to live only on insects.

Of course there are also a number of animals that

can make life difficult for the tarsiers. Carsten Niemitz found tapeworms and filariae (parasitic worms in the blood) in the intestines, and ticks in the fur. Gnats will bother tarsiers, stinging them in the ears and tail, with possible transmission of filariae. Sick animals may be killed by maggots. Once a slow loris *(Nycticebus coucang)* almost caught a tarsier. The observer intervened to prevent this.

A special hazard is the feral house cat. As soon as Niemitz' subjects caught sight of a cat, they would strike up loud warnings. Once a tarsier was caught and killed by a hungry cat, but barely touched. It would seem that at least one cat did not care for tarsier.

In the London Zoo, there were lizards *(Lacerta dugesii* and *L.vivipara)*, migratory locusts, and meal beetle larvae ("meal worms") available for feeding the tarsiers. Overnight, one individual might consume up to a hundred meal worms. Of the large locusts, first the head and thorax were bitten into small pieces and eaten. The abdomen was sucked out, the hard chitinous parts and the intestine being spit out.

At the Frankfurt Institute for Brain Research, the tarsiers would repair to a place of ambush to watch for live prey. A lizard was detected first with the ear, then fixed with the eyes, and the tarsier would begin a slow approach. At this exciting juncture, the tail was curved sideways. Several quick bites along the spinal column – with eyes

Firmly clamped to a tree trunk, this tarsier is consuming a cricket.

(Opposite page) According to current popular conception, all monkeys belong "up in the trees." While not true for all species of the widely ramified and multiform suborder Simiae, it does apply to most - quite definitely to the black-handed spider monkeys shown, which spend virtually their entire life in the treetops of the forest.

closed, to avoid injury in the struggle – immobilized the lizard. The meal began at the head.

In an animal shipment from Manila, a tarsier arrived in Vienna some years ago. Before he continued his journey to the Frankfurt Zoo, where there had been some experience with the keeping of these animals, I was able to watch and observe him for a few days at the Wilhelminenberger Institute for Behavioral Research. Surprisingly enough, he was not inconvenienced in the least by photographic spotlights. He cleverly evaded the eye of the camera by always clinging to the side of a vertical limb away from the camera. But after ten minutes, curiosity overcame fear; the head with its incessantly scanning ears and enormous eyes came into view, and that was my chance to release the shutter. But when a couple of more photographs were to be taken, he refused all cooperation. He hopped into a corner, raised his arms defensively, and bared his teeth.

Nearly 300 years ago, the Jesuit Father J.G. Camel first reported a tarsier, a "small long-tailed simian on Luzon" (in the Philippines). Zoology was not yet in a position to classify the newly discovered marvel; it was thought to be a kind of jerboa. Linneus had tentatively classified the tarsier almost correctly by placing it among the simians. In a later edition of his *Systema Naturae*, he thought that was not quite right; he assigned it to the opossums, and named it *Didelphis macrotarsus*. It was not until 1777 that Erxleben discovered it to be a relative – though distant – of the lemurs.

Animals of strange appearance tend to be the subject of all kinds of stories in their homeland. The extraordinary mobility of the head – like an owl, tarsiers can turn the head 180 degrees in either direction – seemed rather uncanny to people in North Borneo. When a headhunting warrior on the march encountered a tarsier, he considered this an evil omen; it portended the loss of his own head. For the tarsier, such an encounter with a superstitious native boded no good either, and was likely to prove fatal.

On Borneo, it is also said that the tarsiers will gather in large numbers at evening to collect glowing charcoal from brush fires. Thus equipped, they are supposed to steal into the huts and set them afire. An explanation of this story is that fire attracts many insects, and insectivorous animals soon discover their whereabouts. So the fire attracts the tarsiers also. The eyes of captured butterflies reflect the light of the fire, and there you have the "glowing coals," in the reports of casual observes.

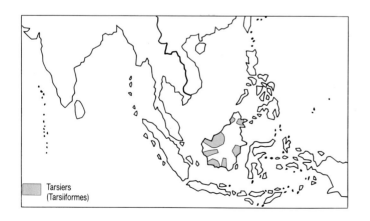

Tarsiers (Tarsiiformes)

SIMIANS

Category
SUBORDER

Classification: The traditional category of Simiae subsumes two infraorders of the order Primates: the Platyrrhini (New World primates) and the Catarrhini (Old World monkeys and apes). Together they comprise seven families. Thus, the simians include all the primates excepting the prosimians; the human family Hominidae is one of the seven families – it will be discussed separately.

Infraorder Platyrrhini (New World primates)
Family Cebidae (New World monkeys)
Family Callimiconidae (Goeldi's monkey)
Family Callitrichidae (marmosets and tamarins)

Infraorder Catarrhini (Old World monkeys and apes)
Family Cercopithecidae (Old World monkeys)
Family Hylobatidae (gibbons)
Family Pongidae (great apes)
Family Hominidae (humans)

Body length: 4.8 in. (12 cm; dwarf marmoset) to 6 ft (185 cm; gorilla)
Tail length: 0 to about 3.6 ft (110 cm)
Weight: 3.5 oz 100 g; dwarf marmoset) to about 605 lb (275 kg; male gorilla)
Distinguishing features. General appearance highly diverse; head is often rounded because of the large cranium, with more or less protruding jaw; in many species, however, there is a pronounced muzzle; eyes are consistently directed forward, and capable of depth perception; no moist nose leather; facial musculature highly developed – facial gestures are lively in higher forms especially; ears are commonly human-like; 32–36 teeth; canines are often large and pointed; two thoracic mammary glands; single uterus; in many female Old World simians, genitals swell during estrus; hands and feet generally designed for grasping, usually with opposable thumbs and great toes; prehensile tail in certain New World monkeys only; ischial callosities in Old World simians only (baboons and gibbons); the most highly developed forms (gibbons and anthropoids) are consistently tailless.
Reproduction: At any time of the year; gestation about 125–280 days; generally only one young per birth, two for tamarins; birth weight about 0.3 oz to 4.5 lb (10–2000 g).
Life cycle: Weaning between about 2 months and 4 years; sexual maturity between about 14 months and 9 years; life span over 10 to about 40 years, generally more in captivity.
Food: Diverse, according to species and habitat, but predominantly vegetable; leaves, fruits, flowers, seeds, shoots, tubers, tree saps; many species take animals either primarily or as a supplement: insects, spiders, worms, birds' eggs, nestlings, and smaller vertebrates; occasionally also small to medium-size mammals.

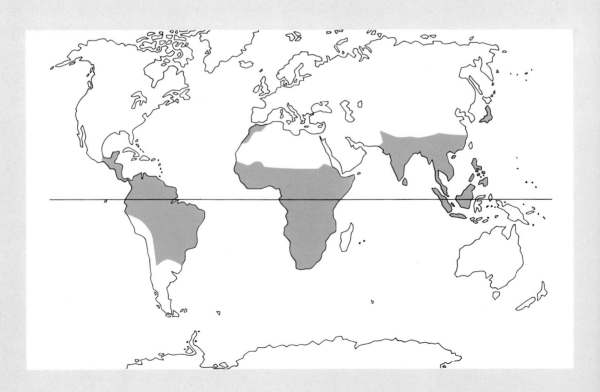

Simiae
Singes; Simiens   FRENCH
Affen   GERMAN

Habit and habitat. All species diurnal, with the exception of South American night monkeys; predominantly tree dwelling, but many species are either arboral and terrestrial or exclusively terrestrial; live chiefly in various types of forests, sometimes also in open terrain (savannas, rocky country, mountains), a few species are also found in cultivated areas and human communities, even cities; ranges of distribution almost exclusively in warmer latitudes of all continents except Australia and Antarctica; represented by only one species in Europe, the Barbary ape, namely on Gibraltar; multiform social behavior; community life in various species-typical social groups: in pairs, small family groups, or more or less stable larger groups of diverse composition; usually firm ranking and "personal" recognition of group members; mutual grooming ("nit picking") to consolidate relationships; as a rule, marked territorial behavior (occupation and defense of own turf); richly expressive behavior, especially in higher forms; comparatively long juvenile phase with strong bond between mother and offspring.

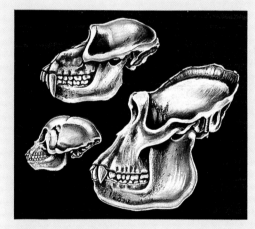

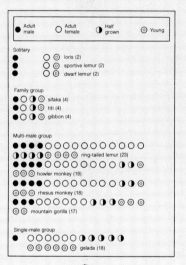

Social grouping. In a few primates only, the female and her young (except in the mating season) form the smallest social unit. Most live in larger groups. Three forms of grouping may be distinguished: (1) a family group consisting of one male, one female, and their offspring; (2) a multi-male group, composed of some adult males, about twice as many females, and their offspring; and (3) a single-male group – an adaptation to dry habitats.

Primate skull types. Characteristic features of the primate skull include a bony rim for the eye sockets and a capacious brain case. However, the absolute size of the cranium is not always apparent from the general appearance of the skull, since in non-human primates the relative size of the facial skull, especially the jaw, increases with increases body size. This is related to the fact that large primates must chew hard or fibrous vegetable food. In the male gorilla, in fact, the area of attachment of the chewing muscles is enlarged by a bony crest on the skull. Top: baboon; left; spider monkey; and right: gorilla.

Skeleton: Form and length of vertebral column, size and shape of skull, length-width ratio of pelvis and proportions of trunk to limbs and among the limbs are clues to the habits of a species.
1. tree shrew;
2. squirrel monkey;
3. baboon;
4. gibbon;
5. human.

Simians

Introduction

by Christian Welker

The simians differ from the prosimians in some conspicuous features, such as in the absence of a rhinarium, the moist nose leather adjacent to the upper lip, and in the single uterus, that of the prosimians being bicornuate. From an originally paired organ, the womb has been perfected by reduction, permitting a longer maturation of the young in the mother's body. The simians have essentially moved away from the primate ancestors, oriented principally by the sense of smell, to become visual creatures. Associated with this has been a further enlargement of the cerebrum and a further gain in grasping ability. Simians can grasp objects more precisely and gently than prosimians can, the pincer grip between thumb and forefinger being the perfected end result of the evolution of the grasping hand. In all simians, the eyes are consistently directed forward, the head is generally rounded by the enlarged cerebral volume, and the "muzzle" is comparatively short.

Every zoo visitor can recognize at a glance the common traits of a capuchin, a cebid, and a rhesus monkey, and will rightly identify them all immediately as monkeys. That impression certainly justifies contrasting simians with prosimians here, but this does not establish their phylogenetic unity. There is much to suggest that the monkeys of the New World and those of the Old World evolved independently of each other from prosimian ancestors, in which case we should actually oppose both groups to the prosimians in the narrower sense, though we did learn in the section on prosimians that their unity is also questionable. The tarsier, at least, is an entirely different sort of prosimian from the lemur and loris. The terms "prosimian" and "simian," in other words, indicate rather a difference in evolutionary level.

The monkeys of South America (Platyrrhini) differ from those of the Old World (Catarrhini) in the position of the nares, which is to say in the breadth of the cartilaginous nasal septum. In the ideal case, the nostrils point downward in the narrow-nosed monkeys (Catarrhini) and towards the side in the broad-nosed monkeys (Platyrrhini). The groups likewise differ strikingly in the number of teeth, as H.-J. Kuhn showed.

The differences of the simians in external sex characteristics and in the structure and function of the hand, foot, and tail will be fully described in the sections on the several groups. It may be as well to mention here, however, that the coiled tail attributed to monkeys in caricature is actually found in three subfamilies of New World monkeys only; no Old World primates can snatch at objects while hanging by the tail.

Unlike prosimians, among which we find many forms with an unsocial habit, virtually all simians live in communities. The simplest form of a social group is the pair, found as a social unit among the more archaic South American primates (marmosets, tamarins, night monkeys) and the more archaic anthropoid apes, the gibbons. As a rule, the young born into the group remain with their parents until attaining sexual maturity, and so do not reproduce within the group.

On the way towards larger societies, such as we find among the capuchins of South America and most Old World primates, the females have become more tolerant of other females and their own daughters. The latter remain in the group of their birth all their lives, and reproduce concurrently with the mother. Social groups of higher primates are thus the first groups of females that are stable over generations, that is, mothers with their daughters, grandchildren, and great grandchildren. Males born into the group however, leave it at sexual maturity, either voluntarily or

INTRODUCTION

under pressure from the adult male or males. It is only among the anthropoid apes and humans that the situation is otherwise. Perhaps because of their size, the anthropoid apes and humans are less dependent upon the ongoing support of other group members. Thus, as has been described for the chimpanzees and the gorilla, the females themselves may wander, and find a mate outside of the native community. Consequently, the groups are not stable across generations, but are newly formed for one generation at a time. The reason for emigration may be the excessive familiarity of individual co-members of the group. It is typical of the simians that they know and recognize each other for life and mutually support one another, but the common experience of childhood is at odds with the requisite tension between the sexes. Childhood playmates are uninteresting as sex partners.

The number of adult males in a social group is likewise dependent on the tolerance of the males for each other. Since most observers tend to be especially impressed by the size or behavior of the males in a group, they have taken the number of adult males as a criterion in the description of social groups. But we know from long-term studies that often – having been crowded out of their birth group – males may live in all-male groups. Such male groups are mutually tolerant, and will also jointly attempt to take over groups of mixed sexes. As soon as they succeed in doing this, they start fighting, and one of the males is victorious. This accounts for the alternation of single-male groups and multi-male groups, and vice versa. Hence the mere presence of a different number of adult males is no criterion. Multi-male groups in some simian species – as among the majority of the langurs – are merely transitory phenomena, limited in time. But in other species, such as the representatives of the macaques, certain adult males will definitely subordinate themselves to the group leader, so that a plurality of adult males may live together in a genuine multi-male group. So we must always inquire more closely as to the longer-term relationships among the group members before we can describe the social texture correctly.

The relationships in the large groups formed by certain species are still more difficult to interpret. They may be mere assemblages of different animals of the same species, or they may be organized troupes. If several groups, by virtue of a common previous history, have closer relationships among them than with other social groups, they collectively form a higher social unit. But in such cases, we cannot be sure that all the individuals can really positively identify each other. Probably it may suffice to have a few trusted friends with whom one lives together chiefly, and in cooperation with whom other animals are dominated.

Lastly, it should be reemphasized that all infant simians are quite helpless and dependent on their parents in the initial period of life. Insofar as other group members take a share in carrying the young during the first days of life – as in the original capuchin-like New World monkeys, the marmosets, the tamarins and the langurs – the newborn plays only a passive role in the exchange, or the exchange is prompted by the mother's rejection. Here again, the observation of care by other group members is not to be compared with the nurture of older juveniles by the Old World monkeys and the more highly developed capuchin-like monkeys of the New World. Among them, the juvenile will seek out another group member and thus get carried about. Statements about "aunt behavior" or care by "surrogate parents" among primates without specifying kind and age of the juveniles or precise descriptions of accurate observations are unsatisfactory.

▷ Only a few representatives of the New World monkeys, like this howler, have a muscular tail that serves as a prehensile and holding organ. On the under side of such a tail, there is often a tactile skin area, comparable to the palm of a hand, for sensing the nature of an object.

Gelada mother and young of different ages, illustrative of the long juvenile phase of simians.

Phylogeny
by Erich Thenius

The apes and monkeys are descended from fossil relatives of the tarsiers (Trasiiformes) that lived back in the Early Tertiary, or roughly 60 million years ago. These Tarsiiformes of the Early Tertiary, the Omomyidae, were completely or at least for the most part without the distinguishing features of present-day tarsiers, namely their enormous eyes and long hind legs for leaping. It is not, therefore, to be assumed that they were nocturnal, visual animals, hopping on two legs, with a unique capability among mammals of rotating the head, like the tarsiers of today. If only for that reason, the tarsiers cannot serve as ancestor models in the phylogeny of the simians.

In the Eocene of Europe, prosimian remains have long been known, described as Necrolemuridae *(Necrolemur, Microchoerus, Nannopithex, Pseudoloris)*, and puzzling to classify, as the generic names themselves suggest ("dead lemur," "micropig", "dwarf monkey," "false loris"). This tarsiiform tribe, commonly known today as the Necrolemurinae or Microchoerinae, are distinguished from the extant tarsiers by a longer facial skull, an archaic brain, and somewhat different dental formula – $\left(\frac{2 \cdot 1 \cdot 3 \cdot 3}{2 \cdot 1 \cdot 2 \cdot 3}\right)$, rather than $\left(\frac{2 \cdot 1 \cdot 3 \cdot 3}{1 \cdot 1 \cdot 3 \cdot 3}\right)$. The necrolemurs were tree dwellers, which, together with related forms (Omomyidae = Anaptomorphidae), flourished at that time in the northern hemisphere. Among them, the anaptomorphines (for example, *Teilhardina, Anapromorphus, Tetonius*) had dentition that was too specialized to be considered as simian ancestors. The omomyines (for example *Omomys, Hemiacodon, Navajovius*), with less-reduced dentition, are better candidates. Their dental formula is $\left(\frac{2 \cdot 1 \cdot 3 \cdot 3}{2 \cdot 1 \cdot 3 \cdot 3}\right)$, from which it would seem that the simian dentition might well have been derived. Since the omomyids of the Eocene exhibited some skeletal specializations (for example, in the base of the foot), it may be that the original simian stock is to be found among Paleocene forms, but as yet hardly any remains except those of the cranium are known. The omomyids that vanished in Europe at the close of the Eocene are represented also in the Oligocene *(Rooneyia, Macrotarsius)* and the earlier Miocene *(Ekgmowechashala)* of North America, so there is reason to believe that they left no descendants.

A question related to the phylogenetic derivation of the simians from Early Tertiary omomyids is whether the Old World (or catarrhine) and the New World (or platyrrhine) simians are to be referred to a common – Old World – parent group, or to separate Old World and New World ancestors among the omomyids. Much as in the case of the prosimians, the issue is between natural unity and mere correspondence in level of development. If the latter hypothesis proved correct, the concept of "simian" would no longer be tenable as specifying a homogeneous suborder, the Simiae.

The geologically earliest simians known – apart from some of controversial origin, and that have left very scanty remains, such as *Pondaungia* and *Amphipithecus* of the Late Eocene in Burma – were those of Early Oligocene strata in Egypt (for example *Parapithecus*), until a few years ago. Simian remains have since been described also for the Early Oligocene *(Deseadense)* of South America *(Branisella)*. On the fossil evidence alone, therefore, the above question can hardly be answered. In any event, in terms of distributional history, the hypothesis of separate origin of Old World and New World simians involves fewer difficulties than that of a common (Old World) stock. Since no South American primates are known before Oligocene times, probably the simians did not arrive there until the Late Eocene, or very early in the Oligocene. Immigration of New World monkeys and other land mammals from North America was not possible unless by "drift" (flotation, as

A predecessor of the living great apes, the giant form Gigantopithecus blacki, *which died out in the Pleistocene of Southeast Asia.*

by driftwood or tree islands), since the South American mainland was separated from Central America by open sea. The same applies to an origin in the Old World, say Africa. To be sure, according to recent information concerning the then location of the continents in relation to each other, such a drift in Eocene times cannot be entirely ruled out. The South Atlantic did not arise, as a continuous arm of the sea, until the Middle Cretaceous. Then South America and Africa drew apart as a result of sea-floor spreading until they reached their present separation of about 900 mi (1500 km) between coast. In the Eocene, the average was as much as 300 mi (500 km), but due to the coastal outline and the counterclockwise rotation of the African mainland, northeastern Brazil and West Africa were separated only by a comparatively narrow strait. Moreover, since the ocean currents were all westward, tree islands with small animals might quite well have drifted in one direction, so that in the opinion of R. Lavocat and R. Hoffstetter the possibility of a drift of small mammals (for example, rodents and primates) from Africa to South America at that time must be recognized. For the exclusively New World, originally South American relatives of the guinea pig, which were likewise first found in South America in the Early Oligocene, a derivation from Old World rodents of the porcupine type has been demonstrated by their common peculiarities. As to the platyrrhine monkeys, no final decision can yet be made. Still, certain "platyrrhine" traits in the Oligocene primates of Africa (for example, *Parapithecus*) do point to a common stock for Old and New World primates. Should this assumption be correct, the simians would be a natural unit after all.

Among the simians of today, then, we distinguish

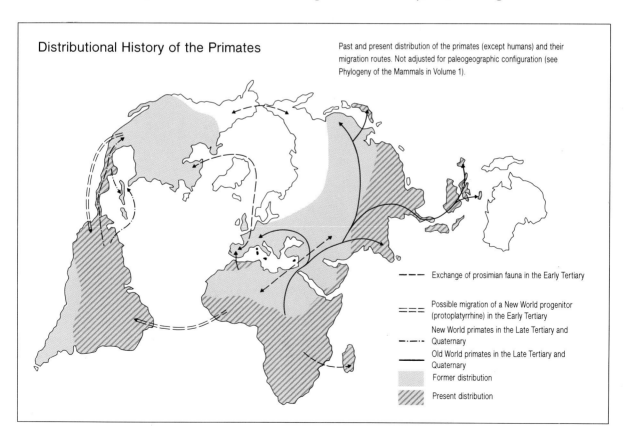

Distributional History of the Primates

Past and present distribution of the primates (except humans) and their migration routes. Not adjusted for paleogeographic configuration (see Phylogeny of the Mammals in Volume 1).

- - - - Exchange of prosimian fauna in the Early Tertiary
=== Possible migration of a New World progenitor (protoplatyrrhine) in the Early Tertiary
-·-·- New World primates in the Late Tertiary and Quaternary
——— Old World primates in the Late Tertiary and Quaternary
Former distribution
Present distribution

Evolution of the Old World monkeys, apes, and humans. The siamang *Symphalangus* has recently been assigned by some authors to the genus *Hylobates* (*H. syndactylus*).

two groups, the Old World monkeys and apes (Catarrhini) and the New World primates (Platyrrhini).

First, the New World primates, now distributed in the warmer regions of the Neotropics (Central and South America).

Here again, we must ask whether they represent a natural unit, which of them are more primitive and which derivative, and how we may classify them accordingly. Their phylogenetic unity is assured by numerous common traits (dental formula, wide nasal septum, placement of the nostrils, lack of a properly grasping hand). The usual systematic division is into the families Cebidae (New World monkeys) and Callitrichidae (marmosets and tamarins), but some assign Goeldi's monkey *(Callimico goeldii)* to a family of its own, the Callimiconidae. Certain zoologists regard the spider monkeys (for example, *Ateles, Lagothrix*) as members of a separate family. A more recent proposal merges the Callitrichidae with the Cebidae.

The Callitrichidae (marmosets and tamarins) were long regarded, on the basis of certain archaic traits, as the more original New World monkeys. But as their embryonic development shows, they are derived forms, evolved from primitive capuchin-like ancestors. Their claws are modified nails, not, that is, "true" claws. The structure of the placenta, similar to that of the capuchins, and the chromosome count are also rendered meaningful from this standpoint. Finally, the geologically oldest New World monkeys, dating from the Oligocene, are all cebids; callitrichids are first noted in the Middle Miocene of Colombia. They developed comparatively late, adapting to an ecological "niche" that was occupied on the other continents by the squirrels.

Squirrels were absent in Tertiary South America; they immigrated after a land bridge joined the two American continents.

The dental formula of the Callitrichidae, $\left(\frac{2\cdot1\cdot3\cdot3}{2\cdot1\cdot3\cdot2}\right)$, is also "derivative" compared to that of the capuchins, $\left(\frac{2\cdot1\cdot3\cdot3}{2\cdot1\cdot3\cdot3}\right)$ – farther removed, in other words, from an original state.

Goeldi's monkey is intermediate between Cebidae and Callitrichidae, and may be regarded as a primitive callitrichid.

The Cebidae, far more diverse in number of forms, are already represented by the remains – however scanty – of *Branisella boliviana* in the Early Oligocene *(Deseadense)* of Bolivia. More complete fossil finds in the Late Oligocene *(Colhuehuapiense: Dolichocebus, Tremacebus)* and the Late and Middle Miocene *(Homunculus, Stirtonia, Neosaimiri, Cebupithecia)*, belonging to the night monkeys *(Homunculus)*, the howlers *(Stirtonia)*, the capuchins *(Neosaimiri)*, and the sakis *(Cebupithecia)* prove that the principal cebid lines separated no later than the Middle Miocene. New World monkeys manifesting in some respects the one-sided development – diet of foliage, hand-over-hand climbing, tactile prehensile tail – comprise the spider monkeys. In terms of brain development, the capuchins (for example, *Cebus*) are among the most advanced of New World monkeys.

It is remarkable that not one strain of New World monkeys produced any terrestrial forms, possibly for ecological reasons, such as periodic inundation of the primeval forestland of the Amazon Basin. In the Quaternary, New World monkeys reached Central America *(Ateles* and *Alouattra)* by way of the Isthmus of Panama as far as Mexico and even the Antilles *(Ateles* in Cuba, and the extinct genus *Xenothrix* of Jamaica). As inhabitants of the tropics, however, they never advanced into North America.

The Old World monkeys and apes (Catarrhini) raise similar questions as to their origin, their classification, and the question as to which forms are primitive and which are derivative. While the history of the Old World monkeys is better documented by fossil finds than those of the New

Dinopithecus, a gorilla-size ancestor of modern baboons who lived in Pleistocene South Africa.

World, the beginnings, especially, are veiled in obscurity. The earliest known Catarrhini lived in Africa, but it is not out of the question that the original stock came from Asia. In dentition and structure of the nasal septum, modern Old World primates constitute a natural unit, to be divided into two or three groups: superfamily Cercopithecoidea (monkeys, including langurs and colobi) and a superfamily Hominoidea (gibbons, great apes, and humans). Sometimes the gibbons are placed in a superfamily Hylobatoidea of their own.

Among the modern Catarrhini, the monkeys are in many respects the more archaic forms, and the hominoids the more highly evolved. Nevertheless, the geologically older catarrhines cannot be classified as ceropithecids. Max Schlosser first described them under the names *Parapithecus* and *Apidium* for the Early or Middle Oligocene (Jung-Fayum) of Egypt, and in recent years they have been repeatedly reviewed. The dental formula of these primates, documented chiefly by remains of jaws and teeth, is $\left(\frac{2\cdot1\cdot3\cdot3}{2\cdot1\cdot3\cdot3}\right)$ according to E.L.Simons, matching that of the New World capuchins. Since certain features of the skull and the skeleton generally are reminiscent of New World monkeys, especially the squirrel monkey Saimiri, the existence of a common (proto-)catarrhine ancestral group for the Platyrrhini and the Catarrhini in Eocene Africa is not ruled out. *Parapithecus* and *Apidium* are occasionally opposed in that case as protocatarrhines to the older Old World primates, then called eucattarhines.

Propliopithecus (= Aegyptopithecus; Aeolopithecus?) and *Oligopithecus* are other primates found in the Oligocene of Egypt that, by dental formula $\left(\frac{2\cdot1\cdot2\cdot3}{2\cdot1\cdot2\cdot3}\right)$, definitely resemble modern catarrhines, of which *Propliopithecus* is classified as belonging to the Hominoidea. *Oligopithecus savagei*, on the other hand, exhibits incipient molar bilophodontia, as a trend in the direction of the Cercopithecoidea. No questionable cercopithecoids are known before *Prohylobates* and *Victoriapithecus* in the earlier Miocene of North and East Africa, respectively. The bilophodontia of the molars is less pronounced than in surviving species. The skull shows a combination of typical traits of langurs and of gibbons. Even though the phylogenetic origin of the Old World monkeys is not entirely settled, their separation from the Hominoidea in the Early Miocene is thus documented. *Mesopithecus pentelici* of the Late Miocene of Southern Europe and Western Asia was discovered in Greece as early as 1838, and is perhaps the best documented fossil colobine (langur). To judge by the skeleton of the trunk and limbs, it was predominantly ground-dwelling. It resembled modern species in the marked bilobal pattern and other structural peculiarities of the molar teeth. In the African Pliopleistocene, some langurs are known (for example, *Libypithecus*, *Paracolobus*, *Cercopithecoides*) that are closer to the Colobus group than to the Asian langurs, including *Dolichopithecus ruscinensis* of the Pliocene in Europe.

Among the Cercopithecinae, there is a considerably greater multiplicity of forms, as witness the diana *(Cercopithecus)* and patas *(Erythrocebus)* monkeys, macaques *(Macaca)*, mangabeys *(Cercocebus)*, baboons *(Papio)*, and gelada *(Theropithecus)*. The oldest cercopithecines

Two other ancestors of the great apes in the Old World: *Oreopithecus bambolii* of the Early Pliocene Monte Bamboli lignite in Tuscany (above), and *Proconsul africanus* of Miocene East Africa (below).

occur as macaque-like forms in the Late Miocene of Africa. The papionines split off later, producing not only the various baboons *(Papio* and *Madrillus; Parapapio, Dinopithecus,* and *Gorgopithecus)* but also *Theropithecus,* of which a subgenus *Simopithecus* was still prevalent throughout Africa in the Pleistocene. The geladas are highly specialized seed and grass eaters. The monkeys described as *Paradolichopithecus* and *Procynocephalus* in the Eurasian Pliopleistocene belong to the macaques. *"Cynocephalus" falconeri* of the South Asian Pleistocene is also a macaque, akin to the *Macaca thibetana* group, and not a baboon.

The various placements of *Propliopithecus* among the Hominoidea (Pliopithecidae, Hylobatidae, or Pongidae) suggest that in the Oligocene the differentiation between Hylobatides (gibbons) and Pongidae (anthropoid apes) had not yet taken place, or even that the Old World monkeys might be derived from a form close to *Propliopithecus. Propliopithecus* did have a rather anthropoid sort of jaw, but numerous archaic skeletal features, such as a tail. *Pliopithecus vindobonensis* in the Middle Miocene of Europe is not actually a direct precursor of the gibbons, yet the postcranial skeleton may be taken as a model for such an ancestral form. As H. Zapfe has shown, the forelimbs are not prolonged as is typical of the gibbons, but there are signs of incipient hand-over-hand climbing. There is a short tail as well. Various platyr-

Parapithecus frassi, a representative of the earliest African catarrhine monkeys, extinct since the Early Tertiary (left), and *Mesopithecus pentelici,* a langur forebear from the Early Pliocene of Greece (right). The background is a tropical forest scene, then and now the chief habitat of most apes and monkeys.

(Opposite page) The New World monkeys - pictured here, a young night monkey - are distinguished by a small physical feature from the Old World monkeys; they have a wider nasal septum, and are accordingly referred to as wide-nosed monkeys (Platyrrhini).

(Above) *Pliopithecus vindobonensis*, a gibbon ally at home in Europe in the Middle Miocene. (Below) *Propliopithecus haeckeli*, a pre-anthropoid of the Early Oligocene of Egypt.

rhine traits testify to the mosaic pattern of phylogenetic development. *Dendropithecus* or *Dionysopithecus* and *Laccopithecus* of the African and Asian Miocene, respectively, are known hylobatids, but nothing is known of their postcranial skeleton. *Krishnapithecus* of the Indian Miocene (Nagri) is based on a single molar, which may possibly be an anthropoid milk tooth.

For the anthropoid apes, there is better fossil documentation, but a plethora of names hint at knowledge that does not exist. In the Miocene, the *Dryopithecus* group, remains of which have been descried in Africa *("Proconsul")*, Europe *(Dryopithecus)* and Asia *(Silvapithecus)*, is the bestknown genus. These are typical anthropoids in dentition, but the skull, brain, and skeleton (for example, incipient arm-lengthening, non-hooked hand) shows some archaic traits that are absent in modern forms. They were tree-dwelling and terrestrial quadruped primates.

There has been much discussion of the position of the jaws and teeth of *Ramapithecus* of the Asian Miocene, first described in 1934 by the American G.E.Lewis as the oldest hominid remains. It has usually been accepted as the earliest representative of the human family Hominidae because of certain supposedly hominid traits, but more recently it has often been suggested that *Ramapithecus* is really a synonym of *Sivapithecus*, differing from *Dryopithecus* particularly in the nasal region and the somewhat thicker cast of the molars. In the absence of postcranial skeletal remains, definitive taxonomic evaluation of *Ramapithecus* is not possible. The relations of kinship with the orangutan *(Pongo pigmaeus)* assumed for *Sivapithecus* are highly questionable.

A separate line within the An-thropoidea consists of the genus *Gigantopithecus*, which in the giant form *Gigantopithecus blacki*, first described by the well-known paleoanthropologist G.H.R. von Koenigswald, disappeared in the Pleistocene of southern East Asia.

Oreopithecus bambolii, of the Late Miocene in southern Europe, is not a hominid but represents a line of its own among the Old World primates.

Australopithecus afarensis, in the Pliocene of East Africa (Laetoli and Hadar), age about 4.0–3.7 million years, is the earliest undoubted homonid. As seen in fossil footprints in Laetoli (Tanzania), this form walked erect, and had a cranial capacity similar to that of modern anthropoid apes. Its hominid dentition had a "simian" gap (between incisors and canines). It may be regarded as a direct form of the later hominids.

As valuable as these and other finds may be, the precise phylogenetic origin of the Hominidae remains to be clarified by further additions to the known fossil record.

One problem involving the phylogenetic origin of humans is presented by the fact that *Homo sapiens* is serologically a pongid, closer to the African anthropoid apes than to the orangutan of Asia. How is this to be reconciled with fossil history and with morphology? Far too little attention has been paid to the discovery that during phylogenetic development, protein differentiation (and this includes serodiagnostic findings) takes place for more slowly than does adaptive evolution. Many adapted forms that differ from each other widely in morphology are serologically more or less one and the same (for example, *baboon* and *gelada*).

Humans began with the erect, biped gait, a development adaptively quite different from that of (modern) anthropoid apes. Serodiagnosis is invaluable in terms of phylogenesis, because it provides evidence of common roots, but it is of limited aid to the taxonomist, or systematist.

NEW WORLD PRIMATES

Category
INFRAORDER

Classification: An infraorder in the order Primates, comprising three families. Common characteristic of New World monkeys: generally broader nasal septum, than in the Old World primates.

Family Cebidae (New World monkeys)
Family Callimiconidae (Goeldi's monkey)
Family Callitrichidae (marmosets and tamarins)

CEBIDAE
(11 genera with 31 species)
Body length: 9–29 in.; 22.5–72 cm
Tail length: 6–37 in.; 15–93 cm
Weight: 13 oz to 33 lb; 365 g to 15 kg
Distinguishing features: A highly diverse family, as the very names of its seven subfamilies suggest: Aotinae (night monkeys), Saimiriinae (squirrel monkeys), Callicebinae (titis), Pitheciinae (sakis), Alouattinae (howlers), Cebinae (capuchins), and Atelinae (spider monkeys) – all slender animals, with a generally long tail, prehensile in a number of species; three premolars; thumb only partly opposable.
Reproduction: Gestation 126–225 days; one young per birth; birth weight 2.5–8 oz (72–225 g).
Life cycle: Weaning at 5–22 months; sexual maturity, where known, at 2–8 years; life span 12–40 years.
Food: Predominantly vegetable: fruits, leaves, flowers, seeds; some species also insects, birds' eggs, and small vertebrates.
Habit and habitat: Diurnal, except the night monkeys; exclusively arboreal, in various types of forests; depending on species, either in underbrush and middle level or in treetops; live together in pairs, small family groups, or larger social groups; territory size 1–250 hectares.

CALLIMICONIDAE
(1 genus with 1 species)
Body length: 8.8 in.; 22 cm
Tail length: 11.2 in.; 28 cm
Weight: 21 oz; 580 g
Distinguishing features: Black fur; erect facial mane, tamarin-like appearance, with claws on all toes except the great toe (hence formerly and still occasionally designated a tamarin); leaps performed in erect posture.
Reproduction: Gestation 155 days; 1 young per birth; birth weight 1.8 oz (50 g).
Life cycle: Weaning at 12 weeks; sexual maturity at 14 months; life span in nature as yet not known, presumably about 12 years.
Food: Predominantly fruits and insects; incidentally, also small vertebrates (frogs, lizards, even small snakes).
Habit and habitat: Diurnal; tree dweller in tropical forests with dense undergrowth, preferably near the ground; live together in small family groups, often associated with tamarins; territory size 75–150 acres (30–60 hectares).

CALLITRICHIDAE
(4 genera comprising 20 species)
Body length: 4.8–13.4 in.; 12–33.5 cm
Tail length: 6.8–7 in.; 17–44 cm
Weight: 3.5–25 oz; 100–700 g

Platyrrhini
Singes du Nouveau Monde FRENCH
Neuweltaffen GERMAN

Distinguishing features: Squirrel-like; silky soft, often conspicuously colored coat, adorned in some species with tufts, waves, and cloaks of long shoulder or back hair; nails on all toes except claw-like on great toes; short thumb, not truly opposable.
Reproduction: Gestation 140–145 days; 2, less often 1 or 3 young per birth; birth weight 0.5–2 oz (15–55 g).
Life cycle: Weaning at 2–3 months; sexual maturity, where known, at 9–24 months; life span over 10 years, in captivity to 18 years.
Food: Tree sap, fruits, flowers, nectar, insects, spiders, occasionally small lizards, birds' eggs, nestlings, and tree frogs.
Habit and habitat: All species diurnal and markedly arboreal; predominantly horizontal running or leaping locomotion; live together in groups of 2–19 individuals; group associations, limited in time, are possible; territory size 2.5–125 acres (1–50 hectares); lower and middle tree levels preferred, more rarely at heights above (80–96 ft (25–30 m); many species prefer forest margins, with thick growth down to the ground; tamarins are comparatively well equipped to survive also in forest areas permanently altered by human intervention.

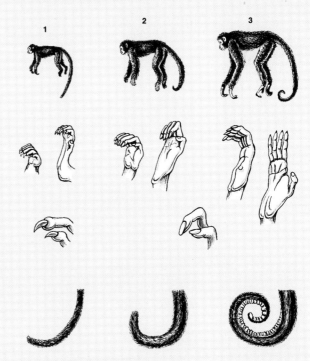

(Below) **Physical traits of New World monkeys.** The medium size forms (capuchin, 2) have trunk and limbs in balanced proportions, nails on fingers and toes, and tail adapted for assistance climbing. Probably to reach new food sources in outermost branches, the small forms (marmosets, 1) have developed dwarfism, with short limbs. Only the great toe has a nail; otherwise, all fingers and toes are clawed. The tail is not adapted as an aid to climbing. Some large forms, such as the spider monkey (3), actually exhibit some features of hand-over-hand climbers, such as a narrow hooked hand with reduced thumb. The tail is modified into a powerful prehensile organ.

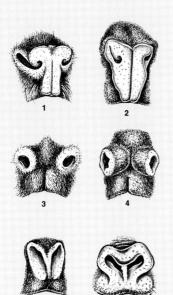

(Above) **Nasal form.** One of the most important differences among the prosimians, the New World simians, and the Old World simians lies in the structure of the nasal region. The prosimian nose (1, 2) with its nose leather resembles that of most mammals. The nostril openings of the New World monkeys (3, 4), usually rounded, are set very far apart. The nose-lip area is haired. In the Old World simians (5, 6), the nostril openings are directed forward and are close together, separated only by a narrow septum. 1. true lemur; 2. dwarf lemur; 3. Goeldi's monkey; 4. capuchin; 5. macaque; 6. gorilla.

New World Monkeys

by Christian Welker and Cornelia Schäfer-Witt

Within the New World family Cebidae, we can distinguish seven subfamilies. The most archaic are certainly the Aotinae (night monkeys), the Saimiriinae (squirrel monkeys), and the Callicebinae (titis), and they will be discussed first. In older texts, the titis are often referred to as the diurnal counterpart of the night monkeys, and the two – titis and night monkeys – are put together in one subfamily. But they are not true counterparts unless we compare active titis with sleeping night monkeys. For when the night monkeys become active, their eager movements and the spatial separation between two individuals are far more reiminiscent of teh squirrel monkeys than of the titis, which are generally slow-moving and maintain closer contact.

Traits in common between titis and night monkeys are certainly to be found; both live in small family groups, and in both cases the father takes an active part in raising the young. Such correspondences, however, are not evidence of kinship, because these forms of social and family life are original, and found in much the same form among the marmosets and tamarins. Permanent pairing, as a matter of fact, is based primarily on the intolerance of males and females for other individuals of their own sex. Males will drive off males, and females drive off females. Given that males and females are mutually attracted, the cohabiting pair is the necessary result. Again, the participation of the father (or other group member) in caring for the young is a result of the mother's endeavor to shift the burden. Among the original South American monkeys, the female will not park her baby in a forked branch or nest, as some prosimians do, but will place it in the keeping of other group members. If we compare these forms of bonding and upbringing with corresponding institutions of our own society, we easily see that our monogamy and paternal care are qualitatively quite different. In our world, males will not drive off males, or females females; rather, close relations between human beings of the same sex are typical of our society. Nor is it indispensable in our society that the father should help care for the offspring, and human mothers do not attempt to get rid of their dependent infants.

Now, whether night monkeys, titis, or squirrel

Evergreen rainforests on the Amazon are the home of the night monkeys and many other cebids.

monkeys represent the most original forms of the family Cebidae we cannot tell, but we do know that night monkeys and squirrel monkeys are earliest represented in fossil finds of South America, and we shall therefore discuss them first.

Night Monkeys (Subfamily Aotinae)

The NIGHT MONKEY *(Aotes trivirgatus)*, the sole representative of the sub family Aotinae, like all other species of the family Cebidae, has four more teeth than the Old World monkeys, namely four times three premolars instead of four times two. It is the only nocturnal simian in the world. In the order Primates, we find numerous nocturnal prosimians, but no true simian has chosen the night as its period of activity. The night monkey is adapted to nocturnal life by his large eyes. The eyes and features are suggestive of an owl, and hence the alternative English name "owl monkey."

Night monkeys awre widespread in South America and occur in nearly all forest terrains; they can even survive for short periods at temperatures below the freezing point. Owing to their nocturnal habit, they are even found in the vicinity of human settlements. Night monkeys live in small family groups consisting of father, mother, and offspring. They sleep through the day huddled together in hollow trees or dense foliage, at a height of 13 to 90 ft (4 to 28 m) above the ground. At the beginning of twilight, however, they become active, and go hunting and foraging.

As yet, Patricia Wright has been the only investigator to study night monkeys closely during their period of activity in the wild. The family she observed sleeps during the day in the dense foliage of trees 61 to 96 ft (19 to 30 m) high. Towards evening, they awaken; all family members stretch elaborately, urinate, and defecate, and then one parent leaves the sleeping tree and goes, by way of lianas at 32 to 48 ft (10 to 15 m), to the first feeding tree. The others follow, always in correct order, first parents, then children. The group studied by Ms. Wright in Peru would travel 190 to 1440 ft (60 to 450 m) within their home territory every night. Often they would visit the same fruit trees on successive nights. Throughout the period of observations, the group stayed together peaceably; Patricia Wright observed no conflict except between different groups.

Galen Rathbun and Marcelo Gache were the first to show that night monkeys also occur in Argentina. In northern Argentina, they found many pairs with one or two young in moist forestlands. Once they observed a group of no less than five at their meal in one tree. When the meal was over, however, it turned out that there were really two groups, one of two individuals and one of three. They, therefore, believe that reports by natives of as many as 15 animals living together in one group are based on sightings in favorite fruit trees. By the way, the native guides told them that the night monkey is really a diurnal animal. When they themselves, on a particularly sunny day, saw

A night monkey family stays close together. The parents live in strict monogamy and have only one offspring at a time. This is true of other simian species as well, but in one way the night monkeys differ from all their relatives: they are the only simians that sleep through the day, becoming active only at night.

night monkeys foraging in a fruit tree, they began to doubt whether the night monkey nocturnal is after all.

Such doubts are strengthened by behavioral studies done by Gerald Jacobs, according to which night monkeys can distinguish colors quite as well as the diurnal squirrel monkeys. Also, Hans Erkert observed in Colombia that the "locomotor activity" of night monkeys, that is to say, the frequency of leaping, climbing, and running, diminishes in extreme darkness. The night monkeys seemed most comfortable in bright moonlight, when they were consistently active. Erkert proceeded to check these observations under controlled laboratory conditions. He was able to verify scientifically that the four individuals he tested under various lighting conditions were most active on nights of full moon. In nearly total darkness, they would shift some of their activity into the later and earlier hours of daylight.

Should we say, then, that the night monkey is rather crepuscular than nocturnal? That was the question raised by Erkert's observations. To answer it, he again kept his specimens under constant conditions with regular alternation of light and darkness. As we have seen, the night monkey is the only nocturnally active simian, sleeping during the bright phase and seeking food during the dark phase. In such experiments – as in the wild, or otherwise in captivity – it will sleep in the bright phase and be active in the dark phase. Now in the bright phase, Erkert lighted the cages as if at full moon, the light that the animals prefer, and in the dark phase, as at new moon, in which their activity is reduced. Would they prefer their favorite light for their activities, and sleep in pitch darkness? By no means; they were inactive in the full-moon phase and active in the new-moon phase, choosing their period of activity in the time of greater relative darkness. Thus, it was demonstrated that the night monkey is truly nocturnal.

Night monkeys often fail to breed satisfactorily in captivity, especially in the second generation. The reason for this was discovered by the Dutch zoologist Leobert de Boer, who many years ago investigated the chromosome counts and forms of many primate species and subspecies. Among the night monkeys, he found seven distinguishable types by number and structure of chromosomes. The differences are so great that progeny of unlike types would in all probability the sterile. Poor breeding results were thus explained; in the zoos, individuals of different types had been paired. Leobert de Boer made a standing offer to examine any night monkeys for their keepers. The examination is perfectly safe. A blood sample is carefully taken and closely examined under the microscope. (Since zoo keepers will certainly be interested in making up their pairs correctly, they should apply to him by letter. He works at the Rotterdam Zoo and will cooperate fully.) We should heed his warning against pairing animals of different geographical provenance, not only night monkeys but any South American primates, lest we inadvertently imperil the survival of the species.

Whether all these types are to be regarded as a single species we cannot as yet decide. We have given only one specific name, *Aotus trivergatus*, but would emphasize that it very likely covers more than one species.

How can we explain the occurrence of so many forms, not only of night monkeys but also of the other New World simians? In the first place, the forest areas of South America are not so coherent as they might seem at first glance. The Amazon forest is laced with wide rivers, some of which represent natural barriers between species or subspecies. On opposite banks, representatives of one species may enter upon different directions of development; the appearance of different subspecies is easily accounted for in this way. In the second place, the present forest region is comparatively recent. According to the zoogeographer Paul Müller of Saarbrücken, displacements of vegetation occurred during the last glaciation; the continuous rainforest came to be interrupted by wide stripes of savanna, constituting a natural distributional barrier. This beating back of the connected forest last took place about 10,000 years ago. Probably such alternations between dry and moist phases occurred repeatedly in the past, thus accounting for the present wealth of species and the large number of differently colored subspecies. As in the case of the night monkeys, of course, it is not always easy to decide whether animals of similar appearance represent different species or only different subspecies. In principle, we speak of species when several similar representatives of the same genus occur side-by-side in the same habitat without mingling, and of subspecies when similar representatives are separated in space.

Precise identification of species is not important, however, for the purpose of preservation in captivity. Night monkeys from Colombia and Bolivia, for example, will behave differently; the Bolivian individuals are far livelier in the first hours of their period of activity than the Colombian, maintaining more social contacts and grooming themselves more frequently. They are also differentially susceptible to various parasites; only Colombian night monkeys contract malaria, and Bolivian ones do not. Hence a physician seeking the right animal "model" for developing a vaccine must pay attention to accurate identification.

At our primate facility, we have been keeping a small breeding group of the Bolivian night monkey since 1978; it has now grown from 6 to 18 in number. As with all our births, we record twice each day where the baby happens to be. In the first week of life, we almost invariably find the newborn night monkey nestling in his mother's groin. At first we thought it was there because it was still too weak to climb on the mother's back by its own efforts – that it had, as it were, gotten stuck in midpassage on the way to its mother's back. In the context of the displacement of an older offspring by the mother on the day of the birth, we learning that this assumption was false. The newborn little one was perfectly able to hold on anywhere in the mother's fur, and would crawl around on her back and belly while she was threatening her elder daughter. As the mother leapt about wildly, the baby was again and again set in motion. We had the impression that it crawled about more or less aimlessly until it found the only part of the body where the mother's movements did not disturb it – the groin. Whether the restless female climbed up or climbed down, or ran along the boughs and back down, the baby now rested motionless in the one secure and warm place. We learned that it was not at all helpless to climb up on the mother's back, but had found the best place on her body the very first day of his life.

Beginning with the second week, the father takes a share in carrying the baby (also at the groin). Beginning with the third week, the baby is carried

The charming squirrel monkeys, whose ominous German name of "death's head monkey" refers to their facial markings, are highly athletic forest dwellers. They can run along even slender branches with great speed and agility.

by the father almost exclusively, and beginning with the second month mostly on his back, in a jockey-like posture. In our rounds of observation, that is where we often found the young night monkey until the age of 4 or 5 months. The other group members, namely siblings, take virtually no part at all in the carrying. The night monkeys born in our facility are regularly seen alone beginning with the fourth month. Thus a small night monkey will change its "chief person of reference" at an early age. A phase of complete dependence on the mother is followed by a phase of increased contact with the father, and then by a phase of increasing integration with the group. During the second half-year, the young night monkey increasingly plays with siblings, and once again relates more to the mother than to the father.

As previously mentioned, night monkeys are strictly monogamous. Hence it is not possible to keep more than one male and one female together, if quarrels are to be avoided. On the other hand, the pair are very tolerant of their offspring, who can remain in the group into adulthood. Our largest groups consist of the two parents and four of their offspring. The latter do not breed within the group, not even if they are already completely grown up and sexually mature.

We were especially impressed by loud "choral singing", which at first we had never been able to hear. We first became aware of this on the occasion of the aforementioned displacement of the elder daughter. At the beginning of our observations it seemed that the air-conditioning system was starting to whirr. We soon realized that there was nothing wrong with the system; for the first time, we were hearing the long-distance calls of our night monkeys. Our groups of night monkeys ordinarily have no eye contact with each other; but now that we are keeping two groups in one space, and they see each other all the time, we hear their songs readily at the beginning of their period of activity, when the "master clock" has switched to nighttime duty. Often the entire group participates in the calling.

Squirrel Monkeys (Subfamily Saimiriinae)

The SQUIRREL MONKEYS *(Saimiri sciureus* and *S. oerstedii)*, assigned to their own subfamily, Saimiriinae, are about the same size as the night monkeys. These handsome beasts, often referred to in German as "death's head monkey", owe their macabre name to their peculiar facial markings, which are distantly reminiscent of a human skull. Unlike the night monkey, they are regularly encountered in social groups of considerable size. Such groups may comprise over 300 individuals, but in that case, we believe, they are not truly social groups but "aggregations" – temporary associations of many smaller groups. This belief, however, is unproven, as there have been very few opportunities for observation of squirrel monkey groups in nature. It is in fact quite difficult to distinguish the individual members of a group from each other. To date, this has been done only by R.W. Thorington for a group of Colombian squirrel monkeys and by John and Janice Baldwin for two groups of the red-backed squirrel monkey; however, these were small groups. Thorington reports that all the animals sleep together at night. At dawn, they wake up and move off to the feeding trees in small subgroups. He was able to distinguish three typical subgroups from each other: adult males, grown females, and females carrying young room separately; older juveniles

Using their long tails as balancing rods, squirrel monkeys leap about surefootedly in the branches of the trees.

NEW WORLD MONKEYS

generally attach themselves to the female subgroups. Individual subgroups never stay long in one tree, but change from tree to tree frequently; on the way, they hunt for insects and spiders. They never "plunder" one tree completely. Thorington observed that one subgroup after another would visit the same tree, or return again and again to the same tree. The Baldwins found no such definite division into subgroups among the squirrel monkeys, though here also the adult

males kept to the periphery of the group. As a rule, squirrel monkeys live in the middle forest level, but they are occasionally found on the ground or in the undergrowth. In all three of the groups closely studied, the number of infants and juveniles was grestest, while the number of adult females exceeded the number of adult males. Squirrel monkeys will cover 1.5 to 2.5 mi (2.5 to 4.2 km) a day. Disputes within the group are rarely observed; if in wandering about the animals discover a new source of food, they will often all rush in at once and try to get the fruit, climbing all over each other. The home territories of different groups overlap, so that there are no exclusive private territories.

More precise details have been reported on a group of squirrel monkeys kept under seminatural conditions in a wooded area of 4 acres (1.6 hectares) in southern Florida. Here the Baldwins made a close investigation of social life. They verified that the groups are in fact divided into subgroups, and they collected extensive information concerning reproductive behavior and juvenile development. They, thus, laid the foundation of our present knowledge of squirrel monkeys in the wild.

In Germany too, a group led by Detlev Ploog at the Max Planck Institute for Psychiatry in Munich has studied squirrel monkey behavior. Most of the pioneering work on squirrel monkey behav-

ior was done here. Detlev Ploog was the first to point out that the behavior profile of each group member is unique. Today, 25 years later, this statement seems obvious, but it was a striking discovery at that time, when many researchers still believed that one monkey was like another, and one squirrel monkey even more so, and was a starting point for nearly all later studies of squirrel monkeys. Here Sigrid Hopf accomplished the labor of compiling the first complete behavior catalogue.

Detlev Ploog was likewise the first to describe and interpret the habit of "genital display." Animals approach each other with erect penis, or clitoris, and raised leg. Ploog realized that this is a distinct dominance or "ranking" signal among squirrel monkeys. He never observed reciprocal display between two males. "This mode of behavior proved a very effective social signal, contribution substantially to establishing a hierarchical order

(Left) A squirrel monkey will not lose its balance swaying on twigs and at the same time grasping food with the hands. (Right) Squirrel monkey mothers carry their babies on their backs from the first day of life. This is the beginning of a close, long-lasting mother-to-offspring bond, consolidated especially by regular nursing. The little ones cling so tight to the mother's fur that they are not dropped even when the mother executes long leaps.

within the group." However, in his very first studies, Ploog found that "the traditional concept of a linear ranking ... [proves] inadequate to represent group structure."

Doubtless, genital display by one individual to another is a clear sign that the former is in fact dominant. For a description of the social texture of the group, however, this one behavior pattern is insufficient. In the first place, as the Baldwins repeatedly observe, relations of dominance between adult males change rapidly and frequently, and in the second place, females do not as a rule manifest any clear relations of dominance.

As already noted, observations in the wild and in captivity confirm that males have close physical contact primarily with males, and females primarily with females. Groups of squirrel monkeys thus resolve into single-sex subgroups. Adult females occupy the center; adult males live on the periphery, and often maintain but little contact with each other as well. Now this must not be understood to the effect that the adult females form a close subgroup; on the contrary, in larger social groups we can also distinguish several female subgroups. Results in our breeding group suggest that a female usually has one close female friend with whom she chiefly maintains contact. This friend then becomes the preferred adult social partner, next to the mother, of the growing infant. If a female fails to attach herself to a subgroup, she must fall back upon contact with a male, and this may account for the mixed-sex subgroups that are also reported.

This hypothesis is supported by experiments of the American psychologist Lynn Fairbanks. She studied a group consisting of two males and six females and found that the females formed two closed subgroups of three individuals each, in each of which subgroups two females maintained especially close relations. She was able to show by an interesting experiment that the subgroups are in fact closed. She divided the cage into two areas, thus partitioning off the subgroups, each with one of two males. In one experiment, Fairbanks placed a female of subgroup B in the cage of subgroup A. The newcomer did not make any connection with the female group, but formed relationships with the males of subgroup A only, thus forming a mixed-sex subgroup with them. After reunion of the eight squirrel monkeys, in the second section of the experiment, this female was readmitted to her old female subgroup. The control experiment (female from subgroup A to subgroup B) gave the same result.

Squirrel monkeys, then, apparently do not form mixed-sex pair groups such as we described for night monkeys. The American psychologist William Mason has documented this by extensive experiments. He has kept his squirrel monkeys consistently in pairs, so that the two animals are perfectly familiar with each other. Then if a strange female is added to such a pair, the two fe-

In the first months of life, squirrel monkeys are totally dependent on their mothers.

males immediately make close contact, and the stranger is even preferred to the familiar male. The intimate relationship between females is the most important factor in the social structure.

However, the social organization of the group is also affected by the annual reproductive cycle. In the natural environment, all the females have their young at about the same time. Therefore, we can divide the year into a mating season, a birth season, and a rearing season. During the mating season, the adult males, ordinarily living at the margin of the group, become very active and excitable, and fight among themselves. At the same time, at least in some habitats and in some captive colonies, they gain a good deal of weight, and seek the company of the females, but are quire often rejected by several adult females in common. Courtship behavior – such as we shall describe for the hooded capuchins – has not yet been observed.

The same manifest temporal uniformity over the course of the year was observed in our facility also. However, it was not reported for all colonies alike; in many instances, births are distributed

around the year. We found no explanation for this variation in behavior until after moving our animals into the new monkey house. In the new quarters, we formed new groups. To our astonishment, the animals paired successfully enough, but at the "wrong" time. The change of scene, or more likely the change of partner, had led to a spontaneous commencement of the cycle. We therefore surmise that this in the "secret" of the large colonies in which new groups are formed constantly and there are no agreements of seasonal rhythm in reproduction. When squirrel monkeys are consistently kept together after importation, as is the rule in smaller colonies, they keep to the annual cycle.

Squirrel monkeys are generally born at night. In the first few weeks, the mother and other group members show little interest in the infant, which is carried on the mother's back from the very first day. Yet this part of life seems to be important to the formation of the close mother-to-offspring bond. Matt Kessler and Hector Martinez, of the Caribbean Primate Center in Puerto Rico, observed in a total of three instances that a female would successfully raise, in addition to her own young, a second baby born on the same day as her own. In each case, the change occurred at the end of the first week of life. Attempts to return the babies to their mothers were unsuccessful. The three "good" mothers proved that squirrel monkeys are quite able to raise twins, though no twin births of squirrel monkeys have yet been reported. The weight curve was normal in each case, though it rose somewhat more slowly than for single offspring. Observations of a juvenile born in our colony indicate that it is not the carrying but the nursing of the young that controls development of the mother-to-offspring bond. A female giving birth for the first time was much weakened afterwards, and generally lay motionless on one of the platforms. Since the baby showed no signs of weakness, and the adoptive mother stayed in proximity to the natural mother, we did not take the infant away from her. This decision proved the right one, for evidently the baby was nursed by the natural mother and taken back by her at the end of the first week of life.

In the third and fourth weeks, the youngster becomes more mobile and attracts the attention of other females, who may then also participate as "aunts" in carrying the infant about. This "aunt behavior," however, like the carrying of young night monkeys by their father, is not always ob-

Squirrel monkeys manifest a special attitude of rest. Normally they lie head down with the tail over the shoulder from the front. Only at high temperatures or after vigorous movement will they lie on their belly with arms and legs stretched out sideways, presumably to cool off.

served; in our colony it is in fact rarely seen; many young have grown up without having been even once taken over by an aunt. In the fifth to tenth weeks, the young finally begins to leave the mother's back occasionally, explore the surroundings, and take some solid food. From the eight week to the fourth month, mother-offspring contacts become rarer and rarer. From the fifth to the tenth month, the juvenile becomes increasingly independent of the mother.

Social play, according to studies at the laboratory in Munich and under seminatural conditions in Florida, begins at the end of the second month. The baby squirrel monkeys often tussle with each other. During the first year, both male and female youngsters take part in fighting games. In the second year, however, the young females more often seek out the adult females of the group, and start playing when they reach sexual maturity, while young adult males still skirmish frequently.

Childhood for squirrel monkeys lasts until about the end of the eleventh month, and adolescence from the twelfth to the thirtieth month. The females are now adult, but the males are still passing through a phase of maturation until the end of the fourth or the middle of the sixth year of life.

The squirrel monkeys are remarkable for a special and typical resting position. Ordinarily, like the dormouse, they will lie head down with the tail over the shoulder from the front, but at high temperatures or after periods of intense physical activity they will lie on the belly with arms and legs outstretched to the sides, probably to cool off. Possibly another bit of remarkable behavior is also related to temperature control: squirrel monkeys will "wash" in urine. They will urinate on the hand and rub the hand against the sole of the foot, moistening the surface. Sometimes they then proceed to rub the tail as well. More often, the tail is rubbed with food; possibly squirrel monkeys feeding in the same territory acquire a "group odor" in this way. Urine washing probably does not serve for marking. Incidentally, similar behavior may be observed on the part of many South American species of monkeys, including the night monkeys, though among them it has never been properly studied.

Independent of the general problem of hybridization that we have already touched upon, anyone engaged in squirrel monkey research soon finds that it is not a matter of indifference which subspecies is selected. One squirrel monkey is not like another. Since the several subspecies have only been confusingly described, researchers have classed their specimens according to the shape of the eyebrows, as "romanesque" (round) or "goth-

ic" (pointed). They have also been accustomed to specifying the points from which their specimens were shipped. Unfortunately, however, animals in laboratories and zoos are often incorrectly identified. By chromosome analysis, a research group at Michigan State University has shown that especially among animals from Peru, numerous crosses have taken place. The difficulties that may arise through neglect of careful subspecies identification may be illustrated by an occurrence at our facility. Doris Merz, observing squirrel monkeys in connection with her thesis, found that our Bolivian squirrel monkeys regularly engage in mutual grooming, as specimens from British Guiana have never been seen to do. Now if someone interested in the study of social grooming were simply to place an order for "squirrel monkeys," he might never observe any social grooming, and would search in vain for the reason, in the absence of precise identification of the specimens.

A swamp titi, also called a red or dusky titi, in the Peruvian rainforest.

Titis (Subfamily Callicebinae)

By contrast with the squirrel monkeys and night monkeys, which climb, run, and leap at high speed through branches an lianas without apparent effort, the movements of the Callicebinae, namely the dusky or RED TITI *(Callicebus moloch)*, the WIDOW MONKEY *(Callicebus torquatus)*, and the MASKED TITI *(Callicebus personatus)*, are slow and cautious. They can actually leap well and accurately, but they prefer a quadruped gait and prefer climbing to leaping. These three species make up the subfamily Callicebinae. Like the night monkeys, titis live in small family groups. In the early morning hours, their song – extraordinarily tuneful compared to the night monkeys – can be heard from far off. The natural community life of the dusky titi was observed by the American psychologist William Mason, and we are informed concerning the life of the widow monkey through several field studies by the zoologist Warren Kinzey. We are likewise indebted to him for the first observations of masked titis in their natural habitat.

All titis have a fixed territory and will defend it against other family groups; however, the preferred habitats and forms of defense vary from species to species. Dusky titis live in dense underbrush of frequently or constantly flooded forests, and will travel to the boundaries of their territories each morning to "sing." Then they return to the interior of the territory. Their calls reaffirm the boundaries between groups daily. Masked titis and widow monkeys prefer more open forest terrain, have larger territories, and make their presence known to other families by loud singing from the center of their territories. Thus, the groups augment the distance separating them, and their calls serve rather to maintain that distance.

The three species of titis likewise differ in the composition of their diet. Whereas dusky and masked titis live principally on fruits and leaves, the widow monkeys require insects as well. The latter constitute 20 percent of their diet and thus equals the leaf component of the other two species.

Pairs of titis seem literally inseparable. When resting – their main "activity" – they generally sit close together and entwine their tails, maintaining the most intensive contact possible. In family groups, the tails of all the members are often intertwined. The young born into the group – like night monkeys – do not reproduce within the group, even when they are themselves full-grown.

Titis are only very rarely found in zoological gardens; they are considered extremely difficult, if not impossible, to keep. Our dusky titis have now been living under our care for more than nine years in some instances. What maximum age they

The rainforests in northern South America, which are frequently or even constantly inundated, are the preferred habitat of the swamp, or dusky, titi. These animals prefer to frequent dense undergrowth. There, the closely integrated family groups occupy their territories. Each morning they travel to the boundaries of their home area, to determine the boundary lines with neighboring groups by their songs. After the morning singing session, they return so the interior.

may actually attain we do not yet know. When we acquired these animals nine years ago they were at least 4 years old, but still show no signs of senility.

The successful care and keeping of titis goes back to William Mason, who – as has been mentioned – observed them in Colombia for an aggregate eleven months in 1964 and 1965. He had the imagination and the courage to found a colony of this almost unknown species at the Delta Regional Primate Center in Covington, Louisiana. Once the colony had at last been established, with very great initial difficulty, Mason moved to Davis, California, with some of the animals; those left behind were later brought to West Germany by the German zoologist Rainer Lorenz, Mason's collaborator, and found a temporary home in the anthropology department of the University of Göttingen. It was from Rainer Lorenz that we obtained our original colony. Since he had to return his other animals to Mason in the United States, there are two successful colonies in existence today, at Davis and here at Kassel. Our colony now consits of 19 individuals, some of whom represent the third generation in captivity.

Since titis are possibly among the most delicate of all simians to keep, we would like to report our experience more fully. To be sure, cage size and equipment and the feeding program are merely important prerequisites. Perhaps of greater impor-

tance for sensitive primates in the daily visit to each individual by us, the scientists, so that we can note even minor changes in behavior and respond to them promptly. We have come to regard these daily visits as mandatory. At first, quite often we had the experience of seeing one of our titis appear sickly, then lose weight, and finally die, after lingering for months, despite our efforts. It did indeed seem that titis could not be kept. We tried in vain to solve our problems by careful variation of the feeding program. Finally, Reina, who was born in Kassel on November 5, 1977, gave us the answer. At the age of 3 years, she became sickly, then weaker and weaker, and eventually, practically lay dying. We were especially grieved about Reina, the first titi to have been born into our colony. We took her out of the cage, put her in a very small carrier cage, and brought her into our study. We did not want to find her dead on the

In their wide range of distribution, the dusky titis have developed several subspecies and color varieties. As a rule, their long, soft fur is gray to red-brown. The face is generally dark, but may also be light, as in this subspecies.

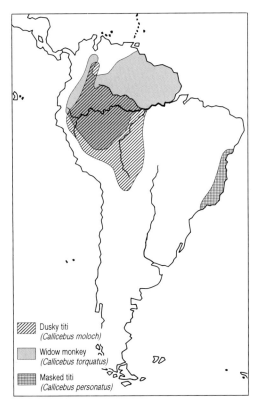

floor the next morning, having dropped from the climbing tree. Some baby macaques were living in the study at that time. It was a place of work, with smoking and loud talking – highly unsuitable, it would seem, for the keeping of such sensitive animals. But next morning, Reina was not dead; she eagerly accepted the special porridge offered her, and the chronic diarrhea from which she had been suffering seemed to have ceased. From day to day she got better. Apparently the unusual environment did not trouble her; she recovered completely. When one sees Reina leaping about vigorously in her cage today, there is nothing at all to suggest a narrow escape from death.

The question we had to answer was why Reina had suddenly recovered. We soon found the solution; we had separated Reina from other titis that had been living in the same room with her. We reviewed all past fatalities, and found to our surprise that deaths had always occurred in only one of the two groups kept in one room. Evidently the titis were oppressed by the presence of their conspecifics, kept in another cage, without eye contact. With their loud cries, they would threaten individual members of the group so seriously that the latter became mentally ill and gradually lost all will to live. Now we had the solution to our problem. As soon as any of our titis suffers from diarrhea for any length of time, it is taken out of the cage and placed in another room. Not one member of our colony has died since!

In the morning, we prepare a porridge for our titis; at noon, toast, potatoes, and egg; and in the afternoon, fruits and vegetables in season. Bananas, apples, beets, peppers and onions are offered daily. During the evening rounds, we add pressed pellets specially developed for marmosets, but all our monkeys like it. Preparation of the feed porridge, which by the way is fed to our night monkeys also, is quite expensive; we do not even know whether this porridge is ideal, nor can we give convincing scientific reasons for all ingredients. However, since it has served well for years, we will disclose its composition. For 37 monkeys (18 night monkeys and 19 titis), we take a cup of powdered mashed banana baby food, a teaspoonful each of grape sugar, wheat bran, and calcium, and add a pinch of salt, vitamin D-3 (15 drops), vitamin B-complex (10 drops), a capsule of medicinal brewer's yeast, a tablespoonful of cod liver oil, and a quarter-pound of whey. The porridge is mixed with boiled water in which a bit of gelling agent for making marmalade has been cooked, stirred, and thickened with baby food rice flakes until the monkeys can readily pick it up with the hand. We have learned from our animals that it must not be too soft and not sticky. When it is sticky, the titis – manifestly disgusted – will shake it off with outstretched fingers. Each titi and each night monkey gets half a heaping tablespoonful of this porridge daily.

Our colony also indulges in the characteristic

Titi families are inseparable. The members sit side by side close together on a branch during their long rest periods, and in token of community, they often wind their long tails about each other. Here, from left to right, is an adult son, the father with a baby at the back of his neck, and the mother, all in a row.

NEW WORLD PRIMATES

The masked titi may be recognized by a black face mask, contrasting plainly with the brown or gray body fur.

song every morning, reminding one of the loud calls of howler monkeys. About noon, the titis take an extended nap. In the forenoon and afternoon they are active; their activity resembles the inactivity of squirrel monkeys or night monkeys. Whereas in groups of these last-mentioned species, differences in activity may be observed among individual group members, apparently mood transmission between titis is highly developed. All family members are either active or inactive, adjusting their behavior to each other completely.

However, the members of pairs that were put together by us are not always compatible, so that new pairs have to be formed. Here we have frequently observed behavior at first meeting. The male chirps, arches his back, shakes himself, and follows the female around the cage, examining her genital region. As a second stage, sniffing by the male is followed by sniffing of the male sex organs by the female. Then (stage 3) the partners sit in close physical contact, groom each other (stage 4) and twine their tails together (stage 5). Stage 3 is usually reached on the first day, stage 5 not until several days later. If the pair do not wind their tails together, they are not goint to get along.

When pairing has been concluded, the two will often sit in close physical contact for long periods, side by side with tails intertwined. Tail winding is more often initiated by the female than by the male. Sitting next to the partner, one titi will move its tail towards the partner, grasp the partner's tail with the proximal third of its own tail, pull the partner's tail nearer and wind around it several times. "Sitting in contact" is the chief form of social contact between the members of a pair. Mutual grooming is less often observed, the male grooming more frequently than the female.

In family groups, this fixed pair bond is loosened, though the partner remains more attractive to both parents than any of the young. Independently of the persistence of this close pair bond, the father is the most attractive group member in all the family groups we have studied. All the juveniles prefer to sit with him rather than with the mother.

Newborn titis – always only one born at a time – are carried on the back, crosswise of the body, in the first weeks of life. The carrying is shared in by all group members, first and foremost by the father. Often the mother carries her baby only in the first few days of life; then she picks it up only to nurse, then returns it to the father. The newborn titi must actively climb onto its father's back by itself. We often observed how the young are held and supported with the hand when changing carriers. As soon as the young titi is hungry, it begins to cry and crawl around restlessly on the father, at which time the mother will pick it up for a while. After nursing, the baby cries again, whereupon the father hurries over and willingly takes charge of his youngest. Once on its father's back, the baby stops crying immediately.

At the age of one month, the young begin to explore their surroundings, and clearly take an interest in the environment. At five weeks, they abandon their carrier from the waste up, but continue to cling to the fur on their feet. At that age, they also take their first short leaps from one adult to another. In their first independent climbing trials (end of the seventh week), the juveniles are still very unsure. We saw only short leaps in the cage at an age of 65 days. Increasing independence is matched by an increasing rejection of the young by the parents. As a rule, the young are more or

The third species of titi is the widow monkey. Its special trait is the wide black band framing the light-colored face.

less independent after the twentieth week of life, and become attractive as playmates for parents and siblings. Fighting games, in which the adults try to hold the juveniles tight, squeeze them against bars or bite them playfully, become common. Older juveniles play mostly with the father: they leap at him, pull his tail, or bite him around the trunk. Even past the age of one year, they promptly seek proximity to the adults at any suspicion of danger.

Siblings tend to be unskillful carriers, and try to shake or pull off a baby that clings to them. In such situations, it would often seem as though the mother meant to punish her elder offspring. She would jump in and bite or strike the older sibling. But invariably, neither parent will pick up a crying baby. Only in a real crisis, such as a fall, will both hurry to the youngster's side to get it on their backs. It was also often observed that the mother would pick up the baby and bring it to the father. The father enjoys the highest rank among the titis.

As emphasized at the outset, the distribution of roles between males and females is no model for our own society; at each change, the young are in actual peril, and are far less sheltered than baby monkeys that are raised by the mother only, at least in the first phase of life. Carrying by the father is not more highly evolved behavior than carrying by the mother exclusively. But the transfer to the father is surely a further development of the "baby parking" that we may observe among many prosimians, even on the first day of life.

Sakis and Short-tails (Subfamily Pitheciinae)

Our fourth subfamily Pitheciinae comprises the sakis and short-tails, today divided into three genera of eight different species. At first glance, one might not recognize the connection between the sakis and the short-tails, but if we abstract from the obvious differences and look closely at their faces, we see that in all members of this subfamily the nostrils are exceptionally far apart, the incisors in the upper and lower jaws are directed forward, and their canine teeth are remarkably large. The two genera of sakis, *Pithecia* and *Chiropotes* (bearded sakis), resemble each other in tail structure; in the skull and teeth, we note common feature between the bearded sakis and the short-tails, or uakaris, of the genus *Cacajao*. All representatives of this subfamily – as of those previously discussed – lack the prehensile tail, and between the index finger and middle finger we find a cleft that functionally corresponds to the space between our thumb and index finger. The three genera, then, to constitute a natural unit.

Probably the more primitive representatives of this subfamily are the sakis, among which we distinguish four different species, namely the white-faced saki *(Pithecia pithecia)*, the red-bearded saki *(Pithecia monachus)*, the black-bearded saki *(Pithecia hirsuta)*, and the buffy saki *(Pithecia albicans)*. Since the subdivision of the shaggy monkeys into three species in quite recent – for reasons already mentioned in the section on night monkeys – we must here report on the shaggy monkeys in general and on the whitefaced saki.

White-faced sakis live in small family groups, eating chiefly fruits and seeds, but systematic observations of their life in the wild are as yet lacking. Once, Van Roosmalen was able to follow a white-faced saki for three hours. In that time, the animal traversed 290 ft (90 m). That saki thereby became the longest-observed white-faced saki ever, in the

The sakis, of which we distinguish three species today, are sometimes called shaggy monkeys because of their unkempt fur. The picture shows a black-bearded saki.

wild! We learn from John Fleagle and Russell Mittermeier that the white-faced sakis, like the titis, are remarkably passive animals, resting often and hiding from observers. But they also discovered that these animals are leapers; the white-faced saki will indeed run and climb on all fours like the species already described, but it leaps much oftener. Three-quarters of all movements recorded were leaps. Mittermeier also reports a 'characteristic hopping (with no support from the hands). The leaps may sometimes be longer than 32 ft (10 m), for which reason the natives commonly refer to the animals as flying monkeys. The sounds uttered by the white-faced saki remind one of the chirping of birds, and when excited, they emit shrill cries.

The handsome white-faced sakis, with the strikingly colored males, are much kept in zoos, and sometimes bred as well. Concerning the second birth to her white-faced sakis, Uta Hick of Cologne reports: "Our female always used to sleep in her house at night with her firstborn. A few days before giving birth, she alone changed to the other house. The birth took place in the evening; she was heard, but not seen. At the age of 16 days, the baby was climbing around on the mother's back. Early in June, we saw it for the first time on the father's back." Since the baby was born on the 19th of May, that change may have been observed at about the age of 14 days. Additional observations are due to Renate Claussen of the Krefeld Zoo. There, the young – she describes two births – climbed onto the mother's back independently by the tenth day of life. The photographs accompanying Claussen's report are consistent with the supposition that previously – as among the night monkeys – the groin had been the preferred location on the mother's body. In one of the family groups, there remained an older sister, who was very much interested in the baby but who was always kept off by the mother. "When the male baby was scarcely three weeks old, it was first observed that the sister was allowed to pick up her little brother. From that time on, she took charge of the little one more and more frequently." In the white-faced saki's case also, then, it would seem that other group members take part in child-rais-

(Above) Red-bearded saki. (Left) Male white-faced saki, belonging to the same genus as the bearded sakis. (Below) The black sakis stay for the most part in the treetops of the primeval forest. Unlike their close relatives, they prefer to walk or run rather than leap or climb.

ing. But here the mother-to-child bond seems far closer than in the species previously discussed, which live in small families. Renate Claussen tells us also of the time of color change: "The young males are at first almost black, with strikingly light-colored, hairless faces. About the fourth week of life, they change color, and they then resemble the adult females. Much later, at the age of two months, according to our earliest observation, the typical white face mask of the male appears." The females become sexually mature at perhaps two years, the males at three.

We know still less about the bearded sakis than about the white-faced sakis. They too live in small family groups, and have been kept for years in captivity – at least in the old "small mammal" house of the Frankfurt Zoo. Bearded sakis are known to include ants in their diet in their natural habitat. This would be a difference from the white-faced sakis, which hav hitherto been observed to consume only vegetable food in the wild, though they will accept insects in captivity.

Unlike the sakis, which prefer the underbrush and the middle levels of the forest for habitat, the bearded sakis prefer the treetops of the tropical rainforests. Here, they live chiefly on fruits and seeds; the seed component of the diet is the largest reported for any South American simian species. Bearded sakis differ entirely in their behavior from the sakis discussed so far. They generally run or scurry through the treetops; climbing and leaping are seldom observed. Also, they live in large social groups, with several adult males. During the day, a group will traverse up to 1.8 mi (3 km). When disturbed, they move their tails back and forth excitedly. The tail is carried forward over the body in running and in resting.

Bearded sakis have been satisfactorily kept only by Uta Hick of the Cologne Zoo. She reports her recipe for success: "I would say from my experience that all sakis are extremely sensitive animals, and that it would be a good idea in captivity to put them in the hands of a keeper who will maintain close contact with them. I have observed again and again that the condition of sakis improves dramatically when they are kept as close as possible, so that they often see the same familiar face. Big cages with more room for exercise are less important factors than a close relationship with their keeper."

When Uta Hick had kept her female white-nosed saki for one Year, she had already scored a record. She would not then have thought that she would be able to keep Bella successfully for 18 years. Only Uta Hick has so far had real success in keeping bearded sakis of various species and subspecies (black saki, red-backed saki, white-nosed saki), and breeding them as well (crosses between white-nosed saki female and red-backed or black saki mali), thanks to her personal dedication. By this time there have been a number of successful further breedings at Cologne. We may again quote Uta Hick on the development of young bearded sakis: "As I enter the lemur facility on the morning of the 20th of April, Bella is holding a baby in her arm. The youngster is already licked clean and dry ... The little one – a male – is firm-

The adult male black saki wears a remarkably long beard. Representatives of the genus are, accordingly, referred to as bearded sakis.

ly seated in the depression formed by the mother's thigh and abdomen. He is surprisingly large – the head especially – and just like his mother ... As soon as the baby utters a soft squeaking sound, Bella reacts with a shrill cry and helps im find the nipple, lifting him to the breast with her hands ... Once, when the baby gives a loud scream and Bella becomes much excited, Ringo hastens to the scene. Both parents feel the baby over, lift the tail, touch the eyes, the head, the whole body ... At the age of five days, the baby will fix objects with his eyes and starts to reach for them ... At 24 days, the young saki for the first time is climbing towards the mother's back. When he is just four weeks old, I see him on his mother's back for the first time. A striking feature of the juvenile is

the vestigial prehensile tail. The tail will curl around the human hand or the mother's body. This ability has been lost in the adult." At the end of the second month – the clinging reflex is no longer to be observed at 61 days – the saki begins to climb about: "By August, his independence is so far advanced that now, as soon as the dish is placed in the cage, the young saki leaves the mother's back to go and eat the food." There is more information about the sounds uttered by the bearded sakis: "The voice of the white-nosed saki has a faint chirping sound. When excited, the animal utters strange cries, somethin like the whistle of a bird." And as to expressiveness, we read: "Whenn the red-backed saki is angry and excited, he stands up, shows his teeth and vibrates his beard."

The uakaris also live in large groups. In view of their preferred habitat – swamp forests alongside the rivers – observations in the wild are not available for either the UAKARI *(Cacajao calvus)* or the BLACK UAKARI *(C. melanocephalus)*. Common to both species is the short tail, not to be found in any other South American primate. An especially striking featur of the uakaris is that the face, the sides of the head, and the frontal portion of the scalp are naked, and, moreover, pink to bright red in color. Because of these traits, uakaris are often described as particularly ugly, often meeting with prejudice and rejection. But upon closer acquaintance, we find that they are unique and no less fascinating then the sakis we find so attractive. There have been no observations in the wild, but their

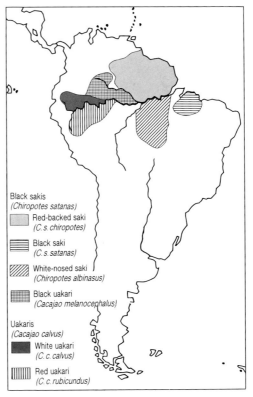

Black sakis *(Chiropotes satanas)*
Red-backed saki *(C. s. chiropotes)*
Black saki *(C. s. satanas)*
White-nosed saki *(Chiropotes albinasus)*
Black uakari *(Cacajao melanocephalus)*
Uakaris *(Cacajao calvus)*
White uakari *(C. c. calvus)*
Red uakari *(C. c. rubicundus)*

The uakaris, with their naked, pink to brilliant red "old man's face" and "high forehead," are not beautiful. However, they are peaceable, social animals, apparently free from both food competition and dissension over rank.

▷ The swamp forests of the Amazon Basin are the home of the uakaris. They are usually associated in fairly large social groups with two or more adult males.

"Scarlet face", an apt name for a red uakari like this. Note the cleft between the index finger and middle finger.

community life under seminatural conditions has been studied in detail at the Monkey Jungle in Florida, in the same area where the Baldwins studied their squirrel monkeys.

The American zoologist Roy Fontaine quickly realized that the RED UAKARI *(Cacajao calvus rubicundus)* that he studied manifests no relationships of dominance ("ranking") to speak of. If an attempt is made to determine the "rank" of an individual, it is soon observed that fighting over food is not a good criterion of the position of an individual in the group, because the uakaris do not quarrel over food; also, judging by the outcome od other observable confrontations, a number of adults must have been of "equal rank." Like all other South American primates that we have dis-

cussed, they live together very peaceably in their group. Fontaine frequently observed mutual grooming. Adult females participate more than do other group members.

A number of uakaris were born in the Monkey Jungle; the births took place at night. At first the young is carried at the hip, alongside the abdomen. The mother seems to pay scant attention to the baby, but allows it to climb about and nurse at will; she herself reduces contact with other group members. During the third and fourth months, the youngster becomes more independent and is now carried on the back. In the sixth of seventh month, it begins to play with others, and to approach or climb around upon the adult male. The male is very indulgent and will sometimes accept it, or carry it, over short distances. At the age of six months, the yount take their first solid food, but they remain dependent on their mothers. Even at the age of one year, they will sleep on the mother, but they are no longer carried. In the second year of life, the young are completely weaned, but the mother will permit nursing until the birth of the next child, or until the age of 20–22 months. Even in the third year of life, the close mother-child relationship is preserved. Children and juveniles play frequently and will also challenge adult group members, not always successfully, to join in the game.

Fontaine also reports that upon any suspicion of danger, the male uakaris he observed will shake branches and perform a dance of intimidation, beating with their hands and at the same time urinating so as to wet the entire ventral surface with urine. Adult uakaris are physically powerful and may be aggressive and dangerous.

In Uta Hick's care, representatives of both the white and the red uakari have been living at the Cologne Zoo for nearly two decades, breaking records with every day that passes. Uta Hick is also the keeper of the only living black uakari outside of South America. He was presented on March 15, 1974 to the Mayor of Greater Cologne, Theo Burauen, by Pedro Trebbau, director of the Caracas Zoological Garden; in 1986, it still enjoyed the best of health.

Howler monkeys (Subfamily Alouattinae)

The remaining three subfamilies of capuchin-like New World monkeys, the Alouattinae (howlers), the Cebinae (capuchins), and the Atelinae (spider monkeys) have as their common feature an exceptionally muscular tail, which serves them, to some extent, as a fifth hand. They use it as a sort of safety anchor. In the case of the howlers and the spider monkeys, we also find, on the under side of the end of the tail, a naked area with well-developed tactile skin, which they can use, as we use the palms of our hands, to find out what something feels like. We believe the howler monkeys to be the more original representatives. Therefore, despite the well-developed prehensile tail that associates them with the spider monkeys, we shall discuss them before we go on to the capuchins.

The first field studies of living monkeys were studies of the howler monkey. As early as 1931, the American primate researcher Clarence Ray

The black uakari differs from its red congener not only in color but also by the abundant growth of hair on its scalp. Both species, however, have the short bushy tail, a unique trait among the New World primates.

Carpenter acted upon his insight that we must get acquainted with our nearest relatives in their natural habitat in oder to gain a better understanding of them and of ourselves, and made his observations on Barro Colorado. Barro Colorado is a small island – originally a mountain – artificially created in 1914 as a "by-product" of the Panama Canal. Carpenter at that time had to develop for himself all those methods of field research that are a matter of course today, and his classical work made him a pathfinder of modern primatology. His howler monkeys are perhaps the most thoroughly investigated South American primates in the wild. Through numerous surveys made on Barro Colorado in the meantime, research on the movement of population density and long-term changes in group composition have been made possible as well.

As Carpenter realized, howler monkeys are characteristically friendly animals; overt conflict is hardly ever to be observed. Non-comprehension manifests itself simply in cautious evasion. Thus an observer may often pass several hundred hours with them and not see any open confrontations. Most howler monkeys live in large social groups; then there are individual males, either of advanced age or just full-grown, who seem to be seeking an attachment to another group. Usually the number of adult females in the group is larger than that of males. Howler groups have fixed territories of residence, but they do not defend them by fighting; they defend only the place where they happen to be. Territories of different groups often overlap. Each morning, groups of howlers will notify neighboring groups of their whereabouts by means of their loud, penetrating "songs." Thus, the howling serves primarily to maintain distance between groups. Immigration of strangers is prevented by peaceful means. Their loud calls are often to be heard in th evening also, and they will howl if danger threatens.

Howler monkeys rarely leap. In moving along, they anchor themselves by the tail, and so can negotiate most of their routes by climbing. As in the case of the sakis, their hands have a cleft between index finger and middle finger that affords a secure functional grip. So ist is not surprising that the howlers are very rarely met on the ground; they prefer the safety of the high crowns of the trees, where their locomotion is usually slow. But when threatened, groups of howler monkeys can retreat very quickly. Another characteristic of the howlers is that they will rest for as much as 80 percent of their period of activity, and hence they are possibly the least active of all monkeys. They exhibit still another peculiarity: they are dietary specialists, subsisting primarily on leaves. However, they will also eat fruits, especially ripe fruits.

Howler monkeys, though, are not exclusively peaceful. Carpenter himself notes that there must have been fights among adult males, as witness the scars often seen on the faces of the males especially. He surmises that these contests simply do not take place in the observer's presence, but, are as it were, inhibited by him. It is certain that the males who grow up in the group are expelled on reaching sexual maturity, or else they leave of their own accord. Accordingly – as careful population studies have shown – most fatalities happen to males at the age of five to seven years. Occasionally a dead male howler is found whose head bears traces of previous combat. In connection with confrontations in displacement of males, sometimes the most vulnerable members of the community, the babies, are injured or even killed. In this respect, the howler monkeys also have taken on a more "human" image. As in all natural populations, death rates are high before sexual maturity is attained; thus we know today – thanks to Barro Colorado – that nearly 90 percent of the males and 65 percent of the females die before their fifth year. But once they have passed that point, the males will live to an average of 17 and the females to 16 years of age, and may even survive past 20.

We do not yet know what actually determines the

Mantled howler monkey (left) howling, and red howler monkey (right) at rest. The howler monkeys are well named; they have a very loud voice, and every morning, in chorus, each group advertises its location to the neighbors. They will often perform their howling songs again in the evening, or in case of danger.

All howler monkeys have a long, powerful prehensile tail, which they use as a safety anchor in climbing.

▷ The howler monkey is not much inclined to rush madly about in the forest. Between extended rest periods, it prefers climbing, using the strong prehensile tail - its "fifth hand."

density of a population. It was long believed that the habitat would not "support" more animals, and that food supply determined population density. Comprehensive and detailed studies by an American team under the leadership of Anthony Coelho have not produced any confirmation.

Encouraged by the interesting mantled howler findings on Barro Colorado, researchers are turning their attention to other howler species. We know by this time that the past observations apply in principle to all howler monkeys. But we have also learned that the large groups of mantled howlers – with an average of about 18 individuals in a group – are not typical. Red howlers and brown howlers generally form groups only half that size, and Guatemala howler groups only a third. In these species, accordingly, the proportion of males to females varies, the number of females per group being smaller than among the mantled howlers. Females are less tolerant among themselves, so it is not surprising that females also change groups quite often, thereby modifying group composition.

Red howlers groom each other far more frequently than mantled howlers. Mutual social grooming is engaged in especially by females; they groom young females in preference to adolescent females, but the latter are preferred in turn to fully adult females as grooming partners. Mutual grooming of males and females is observed as well. As a rule, in fact, females groom adult males more frequently than they groom adult females.

Among the species that live in smaller groups, all-male groups are found more often. Individual males – this again applies to all species – seek to join existing groups. Without combat, they will attach themselves to a selected group, and, even as a new male, such an individual may become the dominant male of the group. A striking feature of howler groups is that one of the adult males, usually the biggest, is always clearly dominant. His superiority manifests itself primarily in the high degree of attraction he exerts on other group members. Observation suggests that probably one of the females likewise dominates the other females. We shall study a social fiber like this in more detail when we come to the capuchins.

We are indebted to Carpenter for still other insights into the courtship behavior of howler monkeys. A female will select a particular male, look at him, and move her tongue rhythmically. Often the male will respond with similar behavior. As in the case of our capuchins, however, sometimes the female will get no response, and in that case must content herself with a less attractive partner. As a rule, female howlers pair with a male of their own group, but on one occasion Robert Horwich saw a female wooing the male of another group of the same species, in fact actually moving into the territory of the other group, and pairing with him there. Her regular male was running about excitedly all this time, but was unable to prevent the "infidelity".

The mothers and any children born into the

Guatamala howler (*Alouatta villosa*)
Red howler (*Alouatta seniculus*)
Black-and-red howler (*Alouatta belzebul*)
Brown howler (*Alouatta fusca*)
Black howler (*Alouatta caraya*)
Mantled howler (*Alouatta palliata*)

group are highly attractive to all group members. They are extremely gentle with the little ones, and even adult males have no objection to a youngster climbing all over them. Newborn howler monkeys are at first quite helpless, and are carried at the mother's belly. In the case of red howlers, we know that the young use their prehensile tails beginning in the fourth week. They lay it across the mother's back or aroung the base of her tail and thus anchor themselves securely. In this phase of life, the mother seems to pay no attention to her offspring at all. She is passive and merely permits the clinging and nursing, without giving the baby any assistance.

Most present information relates to the juvenile development of mantled howlers. In their case, the several age groups are easily distinguished, because the small mantle howlers change color several times as the grow up. The same Baldwins who gave us much information about the squirrel monkeys have made a close study of the development of young howlers. In the case of the mantled howlers specifically, it is difficult to distinguish young males from young females, because the testes do not descend and become visible until about the 24th month (visibility of the testes, by the way, is a good criterion for distinguishing between mantled and Guatemale howlers, who occupy the same habitat). In the first six months – the mantled howler baby is gray-brown, and carried on the belly and the back – no grooming of the young is observed; the mother goes about her business seemingly without concern for the baby, which simply hangs on by the mother's tail. Even in nursing, there is no apparent coordination between mother and offspring. Contacts between the infant and other group members are infrequent or indirect, limited, that is, to those between the mother and other group members. In the second half-year of life – the young are now brown-back – they become more active and more mobile. They mostly ride on the mother's back, and will now even sometimes run a yard or two away from her.

At this point, the mother is observed to pay attention to the youngster after all; if she intends to move on, she waits until the baby has taken hold. If the mother is resting, the little one is occupied chiefly in exploring the immediate surroundings; it will observe butterflies, birds, squirrels, and also begin to use its tail for prehension – it is now able to hang entirely by its tail for as long as three minutes. When the mother becomes more active, it immediately hurries back to her. In the early morning and late evening hours, it will still snuggle against the mother's belly. It is still being nursed regularly (every 1–3 hours). Contacts with other group members increase; the young one is allowed to crawl around over them unrebuked. At the age of one year to a year and three-quarters – the coloration is now black – the young will stay away from the mother for 4 or 8 hours at a time, running about independently. At any time, however, she will permit it to come aboard, and never rejects it. The juveniles will now follow the group alone, but it may happen that the distance between two trees is too great for them, and one may be left behind alone. In such situations, the Baldwins often observed that the mother would come back, fetch her child, and cross the gap with it. At this

Brown howler monkey in the Brazilian rainforest.

age, the young are still allowed to nurse. In the course of the next year, the young become more and more independent, and differences between the sexes become apparent; females are grown up at the end of the third year, and males in the middle of the fourth.

During these observations in Barqueta, the Baldwins observed exploration and play in all age groups; exploration decreased with increasing age. In the second half-year of life, the little howlers begin to play rather cautiously, quite often with infants, juveniles and their own mothers. In the second year, play becomes more vigorous. In their fighting games, the players will often hang by their tails only, leaving their hands and feet free for sparring. They interrupt their play only if one of the adult males approaches. It is quite noticeable that juveniles at this age do not require as much rest. They will play through rest periods, while older adolescents and adults will take part in play only after sleeping. With increasing age, play becomes rougher; chasing games as well as fighting games are now observed. Over all, juvenile howlers play most (15 to 30 minutes a day) between the ages of 18 and 30 months, while older ones, 30 to 36 months, play less. Adult females are usually only passive playmates of their young, but females may be observed playing with each other; adult males rarely take part in play. It remains to be added that howler monkeys, especially adults of both sexes, also sometimes rub themselves with urine. The reason for this is not yet clear.

Capuchins (Subfamily Cebinae)

The true capuchin monkeys, the sixth subfamily (Cebinae) of the capuchin-like New World monkeys, have a curling tail, but its use as a fifth hand is quite limited. They are indeed able to hang by it and to pick up sticks or food with it, but they use it mostly as an anchor in traveling through their habitat. Quadruped running, climbing, and leaping as well as biped running and hopping can be observed. The animals are impressive as acrobats.

Capuchins live in all forest regions of South America; they are the most robust of the monkeys to be found there, and therefore they have been seen for many years in the menageries and zoological gardens of Europa. Because of their liveliness, their intelligence, and their agility, they best typify the popular conception of a monkey.

Capuchins live in large social groups. Usually the number of females is greater than the number of

Although the brown fur color predominates, all of these are black howler monkeys, a mother and baby at the left and an adult pair at the right. In this species, only the males are solid black, while the coat of the females is consistently of a lighter, olive color.

males, and half the group will comprise juveniles of various ages. Capuchin groups are cohesive; males will change groups occasionally, but basically the entire group is defended, the males fighting with the males of other groups. Capuchins are active throughout the day. In the morning twilight, the young are the first to leave the sleeping trees; until evening, all are busy foraging – with the interruption of a good midday nap. Capuchins find food in every corner of their habitat, and so are to be found on the forest floor as well as in the highest crowns of the trees. In the search for food, the group spreads out – especially in pursuit of insects – within a radius of 320 ft 100 m, often separating some individuals from the group entirely.

Representatives of different species meet in this way, and may associate with each other temporarily. Capuchins are encountered in the company of squirrel monkeys, howler monkeys, spider monkeys, and woolly monkeys. Social contacts between the various species are also observed regularly. Sometimes, however, individuals of different species will quarrel.

Like the squirrell monkeys, capuchins also wash with urine and erect the penis or clitoris when excited, though the case would not seem to be one of intimidating display.

A large colony of the HOODED, or BROWN CAPUCHIN *(Cebus apella)* has been kept at our primate center for about a decade. Since our largest social group of more than 30 individuals is not only the largest group of capuchins ever kept in captivity, but also the only large social group of cebine New World monkeys that has been under regular (daily) and uninterrupted observation for years, we should like to give some further account of their social life.

Our first four capuchin monkeys were received in 1974. In 1975, we acquired Bubi, then 17 years old, and imported eleven more animals, of whom one female, Teufel, was clearly dominant. At first we kept all the capuchins in one group, but we soon found that this was not feasible. Repeatedly, certain individuals would get bitten and be driven off, so that it seemed impossible to maintain a single large group. Six capuchins acquired in 1977 did not get along either, and could not be kept together. We observed, however, that the quarreling animals would help each other, and were loud in support of each other even through screen partitions, when the adversary was a newly introduced animal – or in the case of the newly introduced, if the adversary was an already acclimated individual. Apparently the capuchins in their own shipment were more familiar to them than the others. We utilized this situation to construct a large capuchin group. We put two displaced newly imported animals with a female of the original group. Then, as the balance of social relations seemed to dictate, we would add either settled or newly imported individuals, and were thus able to maintain a more or less stable group.

Before we discuss our own research finding in detail, it should be emphasized that even today, a

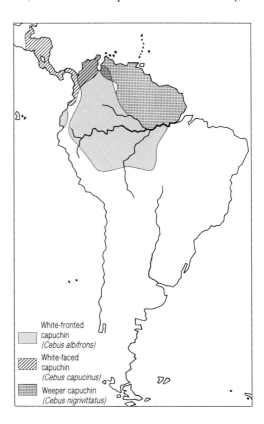

The white fur of the white-faced capuchin extends over the shoulders to the upper arms. Like all capuchin monkeys, this is a forest dweller, traveling on all levels as well as on the ground - climbing, running, leaping, and hopping. Because of their acrobatic feats, their intelligence, and their hardiness, capuchins were imported to Europe very early.

NEW WORLD MONKEYS

preference for individuals of the same transport class is still perceptible, though less pronounced than formerly. Likewise, our procedure of constructing balanced groups is still the same. Any rejected individuals are placed in association with others that are less familiar, in common with still other familiar ones. On the basis of experiments of recent years, we can now predict with fair confidence what will happen in the new combination.

It is a conspicuous trait of our capuchin monkeys that the females court the males. They approach the selected male with eyebrows raised high and beckon him to follow them with gestures and typical sounds. This is often a dubious enterprise, as the males may not be interested in the courting females. But when it is successful, the males will respond with the same mime and gesture, following the females and mating with them. Mating is commonly preceded by prolonged mutual handling.

Newborn capuchins are quite helpless, and for the first three weeks they remain more or less motionless, crosswise on their mother's back. Their only activities are nursing, looking around, scratching, climbing on the mother, and standing up on her. Nevertheless, capuchins only three weeks old will climb on the backs of conspecifics resting together with the mother, but these others will not yet carry them. From teh second month of life – the young are now mostly borne lengthwise to the body, and are themselves more active – we regularly see them carried by other individuals, "friends" of the mother (elder siblings, female or male). In the second month of life, the young begin to sit beside the mother. By the end of this month they are alone regularly.

Independent climbing and eating of solid food are now see for the first time. The juveniles actively begin to make contact with other conspecifics in the third month. The close bond with the mother continues, independently of increasing maturity, until the end of the six month of life; for sleeping and for safety, the young keep returning to the mother's back. They are also occasionally nursed as late as the eleventh month.

Now if we consider the relations of the young with other group members, we find that owing to their helplessness and constant dependence on the mother, as we might except, capuchin babies are at first merely "recipients" of social contacts. In the first months, siblings, especially sisters, are particularly interested in the little capuchin. In the third month, the youngsters themselves seek contact with their siblings. At four months, they become attractive to other young ones, whose company they themselves will seek out deliberately in the fifth month. From that point on, even adult males will take an interest in the adolescents. As a rule, one particular male will take up relations with a particular youngster. This male need not be the father; as a rule, according to our observations, it will be the mother's "friend." Then, in the sixth month of life, the youngster will seek contact with just this male. In this same month, for the first time, the young capuchins themselves will find

▷ Were it not for the long prehensile tail, one might take the little fellow with his dark "hood" and "cap" for a capuchin monk, hastening to the kitchen with two bananas. Actually, the monkey is only trying to get the precious fruit to a safe place, and is thereby impelled to rapid, almost human-like locomotion on two legs.

(Left) With the sureness and skill of a tightrope walker, this capuchin is running along a smooth, slender rod. (Right) Two capuchins engaged in social grooming, an activity to which these monkeys devote much care and patience.

additional social partners, who will be group members a year older, and not related.

An especially remarkable feature of capuchin groups is that there are no rankings such as we find among many Old World primates. As we have noted before, the group depends entirely on a balance of social relationships.

In each capuchin group, there is one dominant male and one dominant female. The dominant male isolates himself from the rest of the group. Nonetheless, all other group members are especially interested in contact with him. The dominant female also seeks to maintain distance from the others. Her object is to build up special relationships with the dominant male and disperse any others who approach him. The dominant capuchin male, however, does not police or "supervise" the group; unlike the alpha (highest-ranking) male of the macaques, who exerts an influence on all relationships, and composes differences, he has little influence over group events. However, duties definitely devolve upon him in defending the group against conspecifics. Accordingly, the respective dominant males will sit on the outskirts of the group and watch what is happening in the other groups, often making loud "comments." The criterion of superiority is obviously not physical strength; rather, what is important is "feeling" superior and "being regarded as" superior. Dominant males go about with no fear of others, do not give ground in danger, and do not respond to threatening by others.

The dominant female has no such paramount position. She is the female who decides all disputes in her favor. In contradistinction to the great popularity of the dominant male, other animals tend to avoid her. Our observations indicate that females too must feel superior in order to dominate.

Each dominant female seeks to secure the dominant male to herself alone. She will threaten females who court "her" male. Only when the courting female has succeeded in getting the dominant male to pursue her will this threatening begin. It is our impression that dominant females seek to build up a mating relationship, but, because of the indifference of the dominant male, they do not succeed.

Besides the presence of the two dominant individuals, we soon noticed that a capuchin group is divided into several subgroups. In our colony, at first we observed only subgroups based on one-to-one relationships. These were and are still today very stable. If an individual enters into several one-to-one relationships at the same time, larger subgroups grow up. Eventually, one individual comes to be so esteemed as a social partner that several partners are recruited through one-to-one relationships. We have been able to observe several forms of such relationships in the past. One-to-one relationships may be based on common social experience (pursuing or being pursued) or on real friendships. These alone are highly stable. Augmentation of the groups by additional births has

The alert and astonishingly intelligent capuchins have for decades been "show" specimens for animal psychologists, or behavior researchers.

Brown capuchin
(Cebus apella)

shown us that the young grow into their mother's subgroup and become members of that subgroup. Kinship thus has special importance.

Examining individual social partners more closely, we find that females are prone to maintain contact with females and males with males. These preferences become more pronounced with advancing age. However, they can be detected even at the age of six months. Social contacts among males and among juveniles, or between males and juveniles, are primarily play contacts. Males and juveniles will also play with males and juveniles of other subgroups. Fighting games between males are based on one-to-one relationships, where one individual can play with two others at the same time. When two individuals exhibit the same behavior and play with different partners at a given time, larger play groups may be formed.

In addition to this form of play, we also find what we call the "free-for-all," sometimes organized in the evening hours by our strongest male, Don. He will hop about on the floor with outstretched arms and be attacked by more or less all of the juveniles, or he will attack them himself. When he, a heavy, full-grown capuchin male, jumps up and down on a one-year-old several times, the observer may get the impression that the youngster is being abused. The young capuchin himself appears to think otherwise. Once he has managed with difficulty to escape from Don, he will spring at Don once more as if to recommence the game. The evening "free-for-all," with often more than ten participants, is one of the most impressive shows put on by the capuchins.

Whereas males, then, mostly engage in play, females engage in other social contacts. They prefer mutual grooming and close seated contact with other females. Grooming between females of different subgroups is very rarely observed. We have learned repeatedly that even small capuchin females are more inclined to sit together and groom each other than are young males. They do indeed often play, but the sex differences are apparent by the end of the first six months of life.

According to our observations so far, kinship relationships are maintained throughout life. As a result, even the newborn capuchin ha a special built-in relationship not only with its siblings but also with its grandmothers, its uncles, and its aunts. The newborn is in their company more often than in that of other group members of the same age. Kinship, age, and sex determine the network of relationships of the individual capuchin.

We are indebted to our male capuchins, especially Bubi, for the interpretation of a cry seldom heard in everyday capuchin experience. We have referred to it as the cry of greeting. Both Angelo Nolte's famous Pablo and our Bubi use it especial-

When capuchin monkeys want to quench their thirst, they sometimes use their hand for a cup.

▷ A brown capuchin in the rainforest of Peru. The characteristic the black skull cap of this species, consisting of erectile hairs, gives the monkey its name.

ly in greeting familiar or attractive persons. We now know what this utterance means. It is commonly heard when, after some separation, males have their first opportunity for physical contact. Simultaneously with the loud cry of greeting, we observed that they touch each other's bodies and erect sex organs. Females too may participate in this greeting. We can induce such an outcry artificially by giving two groups with no screen contact access to neighboring cages. Evidently mere visual contact has a very special quality.

Capuchins exhibit certain innate behavior patterns that were observed thirty years ago by the above-mentioned Münster zoologist Angela Nolte, again in the case of Pablo. In eating, they will perform the maneuver of "making a table"; that is, they place their lower arms close together to intercept any morsels that may be dropped. They crack nuts by impact, lay the head back when squeezing fruits, and will rub fragrances all over the body. Their handling of tools has also been described by her in detail. The capuchins in our colony daily confirm Angela Nolte's interpretations of the behavior patterns she first observed and described. She is rightly considered the pioneer in laboratory researches on South American monkeys.

A weeper capuchin has wrapped its strong, fully furred prehensile tail decoratively around a tree trunk. This species also wears a skull cap, though it is narrow and wedge-shaped.

Spider Monkeys (Subfamily Atelinae)

The last of the subfamilies of capuchin-like New World monkeys that we are to discuss is the Atelinae, the spider monkeys. Their common feature, as the name suggests, is a prehensile tail providing a versatile grasping organ and differing from the tail of the capuchins in having a sensory area on the under side of the tip. It bears skin ridges of different configuration in each individual. The tip of the tail is hairless and resembles the ball of a finger; such a tail is also found in the howler monkeys. In outward appearance, however, the spider monkeys are distinguished by a rounder head, the structure of the lower jaw, and other cranial features. We shall distinguish three genera here, the spider monkey *(Ateles)*, the woolly monkey *(Lagothrix)*, and the muriqui *(Brachyteles)*.

Spider monkeys are surely the most charming and attractive of South American primates. It is surprising, then, that we know so little about them. The first field study, also by Carpenter, was done more than fifty years ago. Carpenter recognized that the black-handed spider monkeys he studied were associated in subgroups. Soon, however, this insight came to be forgotten. In the wild, therefore, groups of spider monkeys were always described as remarkably small.

Now that the zoologists Lewis and Dorothy Klein of Canada have observed long-haired spider monkeys in the wild for a total of 1100 hours and identified all individuals concerned, we know something of their social structure. The study area, 3 mi^2 (8 km^2) in extent, was inhabited by three groups of more than 20 members each. These groups, however, fell into small subgroups, one subgroup consisting of only two or any number up to twenty individuals. The composition of the subgroups varied from observation to observation. Through their accurate acquaintance with all individuals, however, the Kleins were able to show that certain individuals regularly meet each other again and again within the area, so that they are really members of one group. Yet the distance between group members may be 515 to 4100 ft (180 to 1500 m); sometimes certain animals may be separated for days at a time.

Spider monkeys often wander about in single-sex subgroups, consisting, that is, of males only or females only. Female subgroups are commonly larger than male or mixed subgroups. No disputes are observed between adult males, but females are sometimes displaced by males and by females, and – when kept in captivity – driven off or threatened. Males especially will defend their group in case of danger.

Carpenter reports that the males and sometimes also adult females uttered peculiar sounds like the barking of a terrier at his approach, and would come down within 38 ft (12 m) of him. Their barking became intensified in his vicinity. They would climb down to the ends of the branches and begin to shake them with the hands or feet, or both, while holding on by their tails. Carpenter frequently observed scratching in association with this shaking. They would also break off branches and drop them, often quite close to him. (Such branches, weighing as much as 8 or 10 lb (4 or 5 kg) when dropped from a height of 38 ft (12 m), are a real hazard.) At the same time, they would urinate and defecate, in manifest reference to the observer.

Spider monkeys are tree climbers; they travel chiefly by swinging, anchoring themselves with their prehensile tail or using it as a climbing tool in its own right. Their hands are specially adapted to this mode of locomotion. The thumb is completely or almost completely lost – a non-func-

tional stump – and their other four long fingers serve as a short of hook with which they can anchor themselves securely in the branches.

Spider monkeys are able to run upon two legs on the ground, but for the most part they stay in the crowns of the trees. The long-haired spider monkeys observed by the Kleins, for example, were never encountered on the ground; they would come down to a minimum height of 38 ft (12 m) while foraging. Their preferred diet was ripe fruits (83 percent of 2156 recorded meals) rather than leaves and buds (5 percent); quite often, they would eat bark or even wood (10 percent). Finding food and eating it accounted for almost a third of their period of activity. They rest more than half the day, and otherwise spend their time traveling, covering more than one-half mile (1 km) per day.

Young spider monkeys are carried on the abdomen in the first four weeks of life. According to observations by Franz Podolczak at the Frankfurt

Three black spider monkeys looking at their observe – distrustfully, amiably, and defensively.

Zoo, the mother will often support her offspring with the hand during transport at this time of life. Carrying at the belly is replaced by carrying on the back as time goes on. The baby will lie on the lower back with his prehensile tail coiled around the base of the mother's tail. Older children play, as we know from Carpenter, with their own bodies, with sticks, or with other youngsters. He also suggests that the spider monkeys probably have no fixed season for births, as he was able to observe babies and young of all sizes at one time.

Although large numbers of spider monkeys are kept in zoological gardens, their preservation is by no means assured, given the large number of species and subspecies exhibited and the extensive hybridization that has taken place. Populations in the wild are steadily declining. It is not only that they are coveted game. Perhaps because of their special dietary need (ripe fruits), they are highly dependent on undisturbed forest terrain, and consequently they are the first monkeys to disappear in disturbed areas. Hence their protection is especially important.

We should mention that spider monkeys in captivity perform remarkable feats with their tails. If they are unable to reach food outside the cage with their hands, they will deftly retrieve it with their "fifth hand," which they will also use for the purpose of begging.

The WOOLLY MONKEY *(Lagothrix lagotricha)* lives in more compact social groups in the tropical rainforest. According to ovservations by the Japanese zoologist Akisato Nishimura and Kosei Izawa, the living territories of different groups overlap. There are usually several adult males living in a group. Like the spider monkeys, they are hardly ever found on the forest floor. They generally remain in the upper level at a height or 38 ft (12 m), but they will come down to 22 ft (7 m).

In undisturbed groups, the grown males will approach the observer and shake the branches, with a loud barking like that of the male spider monkey. Woolly monkeys are peaceful among themselves; no quarreling over food is observed. They often groom each other. Youngsters play with each other, mostly around midday. Adolescent and adult females also participate in the games.

Young born within the group are carried at first on the abdomen, later on the back. Michael Kavanagh and Lawrence Dresdale, who have been responsible for some of these field researches, observed a baby being transferred to another female. More details of youthful development are given by Leonard Williams, Herbert Schifter, David Mack, and Helen Kafka, who observed the development of young woolly monkeys in captivity. The babies are able to hold thight to the mother's

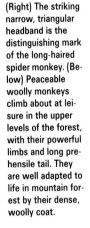

(Right) The striking narrow, triangular headband is the distinguishing mark of the long-haired spider monkey. (Below) Peaceable woolly monkeys climb about at leisure in the upper levels of the forest, with their powerful limbs and long prehensile tail. They are well adapted to life in mountain forest by their dense, woolly coat.

back from an early age, but they are regularly carried at the abdomen in the first month of life. In the next six weeks, they climb to the mother's back more and more frequently, and are only very rarely found on the ventral side from the tenth week on. Young woolly monkeys first leave their mothers after eight weeks, and they become increasingly independent from the fifth month on.

Both in their natural habitat and in captivity, a remarkable rubbing of the chest is sometimes observed; in the Zurich Zoo, it is conspicuously when moving into a new cage, on a climbing tree. Schifter says of the dominant male, "After sniffing, often but not always followed by repeated licking of the spot in question, the chest is pressed against the substrate at the level of the nipples, and cyclic movements are executed. The operations described are usually performed several times successively in the same order. Again and again, the nose is applied to the marked spot, and it too may execute circular rubbing movements." In this way, the chest and the rubbed surface are wetted with saliva. Observers at the Zoo often interpret this as marking behavior, but the field researchers' observations do not confirm that interpretation.

The second species of woolly monkey, the YELLOW-TAILED WOOLLY MONKEY *(Lagothrix flavicauda)*, has been just recently "rediscovered." It was Alexander von Humboldt who first discovered it, in 1802. He described it, from a hide being used as a saddle blanket, as a new howler species, *Simia flavicauda;* he never saw a live yellow-tail. About 120 years later, Watkins (1925) and Hendee (1926) together "collected" five yellow-tailed woolly monkeys. The three woolly monkeys sent to the British Museum of Natural History in London by R.W. Hendee were recognized as representing an independent species, described as the yellow-tailed woolly monkey. The identity of all six finds was recognized at last by J. Fooden in 1963.

Together with Hernando de Macedo-Ruiz, Anthony Luscombe, and John Cassidy, an expedition was undertaken in 1974 by Russel Mittermeier to learn whether this species still existed. While sill on their way to the locality of Chachapoyas, they met a hunter with a yellow-tailed woolly monkey shot six days before. From him they obtained three more hides and two skulls of this rare species; the man had been shooting the specimens for food. Finally, a day before their departure, children in Pedro Ruiz took the naturalists to the home of a soldier who was keeping a young male yellow-tail as a pet, and was willing to part with him. Forty-eight years after the last report, the species had finally been rediscovered. That young yellow-tailed woolly monkey was and still is the only one of his kind in captivity.

Only two years later, in August 1976, the American zoologists Gary R. Graves and John P. O'Neill had the great good fortune and opportunity to observe this species in the wild. On the 23rd of

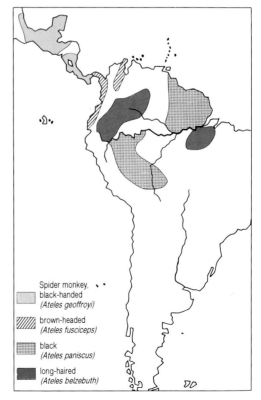

▷ The "gentle temperament" of the woolly monkey conforms to its diet, consisting exclusively of vegetable fare.

▷▷ A black-handed spider monkey takes a walk with her baby. The tail is used for support. The young are carried at first on the hip, as shown here, and later on the back.

▷▷▷ (Left) Black-handed spider monkey. (Right) Woolly monkey.

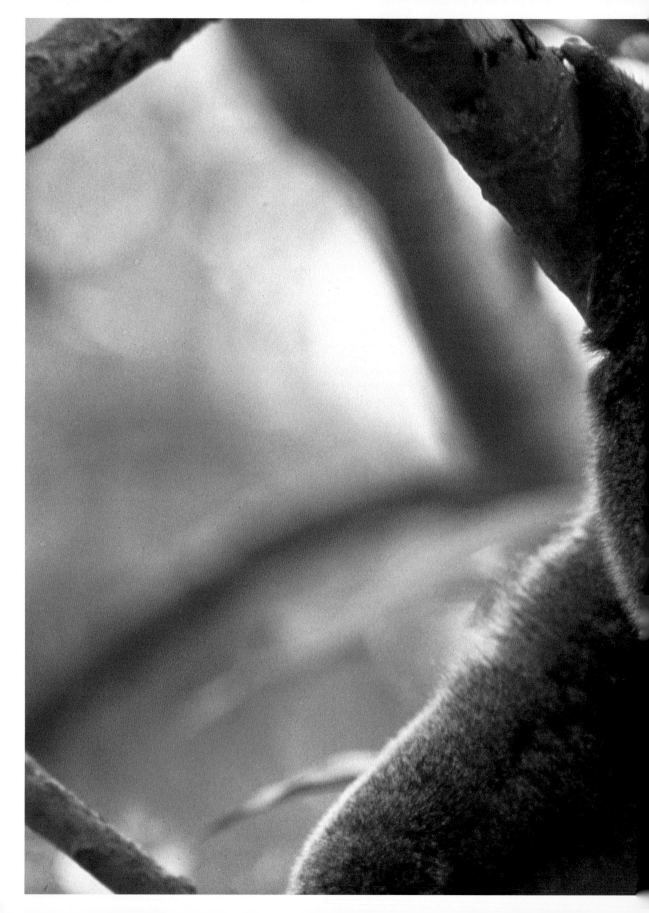

August, O'Neill heard their loud calls. Some days later, Graves, with his companion M. Sanchez Semba, 3 mi (5 km) from their main camp, at an elevation of 5340 ft (1670 m) in the evening sunlight, saw an adult male, five adult females, and several young ones 80 ft (25 m) up in the epiphyte-rich treetops. The reported: "To entice them to come closer, Graves 'squeeked' loudly on the back of his hand. The grown male came down calmly, to a horizontal branch about 26 ft (8 m) above the ground and watched Graves addressing him. The deep mahogony brown and black fur, the white-gray muzzle, the straw-yellow tail stripe, and the straw-yellow scrotal tuft made a striking pattern in the filtered sunlight. The dark for color, relieved by bright markings, provided perfect camouflage in the contrasting light of the leaf canopy. After a few minutes, the male lost interest in the sound and returned to the group, which then moved away from us."

Since that time, yellow-tails have been observed regularly. As with other woolly monkeys, the dominant male advances to meet danger, displays his genitals, moves his hips, and shakes the branches, sometimes urinating and defecating at the same time. Excited males bark outright, like spider monkeys. Today, the Peruvian government in association with the WWF is educating the population through motion pictures and placards as to the uniqueness of this monkey. The Peruvian researcher Mariella Leo Luna distributes posters, post cards, and T-shirts picturing the yellow-tailed woolly monkey to school children. It is to be hoped that the species still may survive.

Prospects are by no means as good for the WOOLLY SPIDER MONKEY or MURIQUI *(Brachyteles arachnoides)*, the largest of the South American primates. Only between 300 and 1000 individuals of this species live scattered over a range, more than twice the size of West Germany, in the southeast of Brazil, the region of highest population density on the continent. Hence the muriqui is possibly the most endangered primate species alive today: 85 muriquis live on the Fazenda Barreiro Rico, a large cattle ranch in the State of São Paulo that includes 7500 acres (3000 hectares) of undisturbed forest. Fortunately, the owners, the Magalhães family, are confirmed conservationists, and take care that nothing happens to the muriquis. Nevertheless, the prospects are not very good. The forest land is not continuous, but subdivided into five rather small areas with farmland in between and no communication. There are muriquis still living in three of these patches of forest.

Muriquis are peculiar animals. Their nostrils are large and, like those of Old World monkeys, pointed downward. At first glance, they may remind one of spider monkeys; like them, they lack a thumb, their tail is prehensile, and they travel hand over hand, hanging and swinging. However, a distinguishing trait of the spider monkeys is that the females have a much elongated clitoris, while the male testicles are almost unnoticeable, so that males and females are often mistaken for one an-

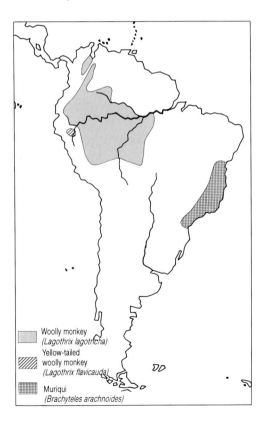

Woolly monkey
(Lagothrix lagotricha)
Yellow-tailed woolly monkey
(Lagothrix flavicauda)
Muriqui
(Brachyteles arachnoides)

other. For the muriquis, this situation is "corrected"; the female sex organs are inconspicuous, and the males have quite large testicles.

A field study done by Katharine Milton is our source of the information that muriquis live in large social groups, which, however, as in the case of the spider monkeys, are divided into several subgroups and may range as much as one-half mile (1 km) apart. Considering that muriquis travel only 2600 to 3800 ft (800 to 1200 m) a day, they are therefore separated by "a day's march." Subgroups often consist of individuals who are all of the same sex. The composition of a subgroup varies continually, but mother and infant always stay together. Small subgroups composed of females only are often observed, and may be joined at times by several males for a short time.

When a female is ready to mate, she will usually utter a certain regular call, whereupon as many as nine males will approach. They will follow her for a day or two and mate with her repeatedly.

Muriquis also may be observed "washing" with urine. Usually only one hand is moistened; sometimes muriquis will then scratch themselves at the point of the foot, or touch the sole of the foot, or the breast with the wet hand.

In complete contrast to the eating habits of the spider monkeys and the woolly monkeys, the muriquis' diet consists chiefly of leaves. Like the howler monkey, they are definitely specialists. The greater part of the day, they rest. They spend most of their time in the treetops, but muriquis have been seen to come down to the ground for water or to rob papaya fields.

In this chapter, we have not mentioned experience with capuchin-like New World monkeys kept individually. In the past, there was but little success. As we have mentioned before, most are very delicate to keep and care for, and their stocks are much threatened by destruction of primeval forest and by hunting. Most of them practice wetting themselves with their urine, and are not very acceptable as pets in a private home for that reason. Besides, adult males of the species living in larger groups grow very big and strong.

Since none of these species have systems of ranking, they will not subordinate themselves to a human "master" either. Their behavior is "specific" to them, and so they are mistakenly stigmatized as "vicious" and unpredictable. The species of this family are indeed charming companions, but they remain so only if they receive constant attention and respect, and are not asked to conform to human expectations.

(Top) The yellow-tail woolly monkey, now rare, is believed to have been extinct, and rediscovered only a few years ago. (Bottom) The better-known woolly monkey, is also threatened by increasing destruction of habitat.

CEBIDS

Cebids (Cebidae)

Nomenclature English common name Scientific name French German	Approximate Size Body length Tail length Weight	Distinguishing Features	Reproduction Gestation period Young per birth Weight at birth
Owl monkey; night monkey *Aotus trivirgatus*, 10 subspecies Singe de nuit Nachtaffe	males 11.8–19 in.; 29.5–47.5 cm females 10.8–15.6; 27–39 m: 10–17.2 in; 25–43 cm f. 11.6–17.6; 29–44 m. 1.8–2.3lb; 825–1025 g f. 1.6–2.5; 750–1120	Tail not prehensile, equal to body length; rounded head; conspicuously large eyes; small ears; nostrils closely spaced; white face markings bounded by three black stripes; back of head, back, and legs gray-brown to gray-olive; under parts lighter; posterior third of tail black	126–133 days 1 or very rarely 2 3.2–3.7 oz; 90–105 g
Common squirrel monkey *Saimiri sciureus*, 14 subspecies Sapajou jaune Totenkopfaffe	m. 10–14.8 in.; 25–37 cm f. 9–14.8; 22.5–37 m. 14.4–18.6 in.; 36–46.5 cm f. 14.8–18; 37–45 m. 1.2–2.5 lb; 550–1135 g f. 0.8–1.6; 365–750	Tail long, not prehensile; fur short and dense; white face, throat, and ears: head, back, and legs yellow to gray-green; under parts lighter yellow to white; mouth area and tip of tail black; scalp sometimes black	152–172 days 1 2.5–5.1 oz; 72–144 g
Red-backed squirrel monkey *Saimiri oerstedii*, 2 subspecies Sapajou à dos rouge Rotrücken-Totenkopfaffe	m. 11.8–13.2 in.; 29.5–33 cm f. 11–12.2; 27.5–30.5 m. 15.4–17.4 in.; 38.5–43.5 cm f. 15.4–16.4; 38.5–41 Not known	Tail long, not prehensile; appearance similar to common squirrel monkey, but with black hood and orange to red back; often the Bolivian subspecies of the common squirrel monkey is mistakenly exhibited in zoos as *S. oerstedii*	Not known 1 About 8.5 oz; 100 g
Dusky titi; red titi *Callicebus moloch*, 7 subspecies Callicèbe roux, Callicèbe arabassu Sumpfspringaffe; Roter Springaffe; Grauer Springaffe	10.4 in.; 26–41 cm 11.8–20.8 in.; 29.5–52 cm 1.5–2.9 lbs; 700–1320 g	Tail long, not prehensile; long, soft fur; short canine teeth; head and back gray to red-brown: belly and tail differently colored in some subspecies; head sometimes with face markings of strikingly different color	136 days 1 About 2.5 oz; 70 g
Widow monkey *Callicebus torquatus*, 3 subspecies Callicèbe à fraise Witwenaffe	m. 15.2–18.4 in.; 38–46 cm f. 14.8–18.4; 37–46 m. 19.2–20.4 in.; 48–51 cm f. 17.6–20.4; 44–51 2.2 lb; 1 kg	Tail long, not prehensile; long, soft fur; long canine teeth; wide black band around white face; red-brown to black-brown body; black forearms, legs, and feet; white or yellow hands; dark brown or black tail	Not known 1 Not known
Masked titi *Callicebus personatus*, 3 subspecies Callicèbe à masque Maskenspringaffe	m. 15.2–16.8 in.; 38–42 cm f. 13.2–16; 33–40 m. 16.8–22 in; 42–55 cm f. 15.8–21.2; 39.5–53 2.2 lb; 1 kg	Tail long, not prehensile; long, soft fur; black face mask; gray or brown body, sometimes with long tan hairs; black hands and feet; tan arms and legs; red-brown tail	Not known 1 Not known
White-faced saki *Pithecia pithecia*, 2 subspecies Saki à tête blanche Weißkopfsaki	m. 13.2–18.4 in.; 33–46 cm f. 12–18; 30–45 m. 14–17.8 in.; 35–44.5 cm f. 13.2–17.4; 33–43.5 m. 4.2–4.6 lb; 1.9–2.1 kg f. 3.3–4.2; 1.5–1.9	Long, bushy tail; males consistently black, with short white hair on face, except nose, mouth area, and frontal midline; female brown to gray-brown, under parts lighter, with white to light red stripe from eye to corner of mouth	163–176 days 1 Not known
Red-bearded saki *Pithecia monachus* Saki à perruque Rotbärtiger Mönchsaffe	m. 15.8–17.6 in.; 39.5–44 cm f. 13.4–16.2; 33.5–40.5 m. 18–19 in.; 45–47.5 cm f. 15.2–18; 38–45 m. 5.5 lb; 2.5 kg f. 5.5; 2.2	Long, bushy tail; both sexes uniformly gray-brown to gray; hands and feet lighter, chin beard and under parts tan or red; head hair of females long and loose, males short and stiff	Not known 1 4.3 oz; 121 g
Black-bearded saki *Pithecia hirsuta* Saki à perruque Schwarzbärtiger Mönchsaffe	m. 15.2–17.6 in.; 38–44 cm f. 15.2–18.4; 38–46 m. 17.6–20.4 in.; 44–51 cm f. 16–19.2; 40–48 Not known	Long, bushy tail; both sexes uniformly gray-brown to gray; hands and feet lighter; black chin beard, and under parts	Not known 1 Not known
Buffy saki *Pithecia albicans* Saki à perruque Schwarzrückenmönchsaffe	m. 16–16.4 in.; 40–41 cm f. 14.6–16.2; 36.5–40.5 m. 16.8–17.6 in.; 42–44 cm f. 16.2–18.2; 40.5–45.5 Not known	Long, bushy tail; both sexes tan to red; back and tail mostly black	Not known 1 Not known
Bearded saki; black saki *Chiropotes satanas*, 3 subspecies Saki noir Satansaffe	m. 16–19.2 in.; 40–48 cm f. 15.2–18.4; 38–46 m. 14.8–16.8 in.; 37–42 cm f. 14–16.8; 35–42 m. 6.4 lb; 2.9 kg f. 5.7; 2.6	Long, bushy tail; comparatively short, soft black fur; back and shoulders light yellow-brown to dark brown; pronounced beard growth in adult males	About 5 months 1 Not known
White-nosed saki *Chiropotes albinasus* Saki à nez blanc Weißnasensaki	15.2 in.; 38 cm 16.8 in.; 42 cm m. 6.8 lb; 3.1 kg f. 5.5; 2.5	Long, bushy tail; black fur, except for red nose area which is covered with white hair; beard less pronounced than in bearded saki	About 5 months 1 Not known

COMPARISON OF SPECIES

Life Cycle Weaning Sexual maturity Life span	Food	Enemies	Habit and Habitat	Occurrence
Age 5–12 months Age about 2 years 12 years	Fresh, young leaves, flowers, seeds, also insects (including caterpillars), birds' eggs	Not known	Nocturnal; found in pairs or small family groups (parents and 2–4 young) in variety of primary and secondary forest terrains, from evergreen rain forest to dry thorn woods of Paraguay, at elevations up to 10,240 ft (3200 m); territory size 7.7 acres (3.1 hectare)	Not yet endangered due to wide distribution, despite destruction of habitat; status of some subspecies uncertain
Age 5–10 months Age 3–5 years 21 years	Fruits, insects	Birds of prey	Active by day; found in fairly large social groups, with several adult males (10–50, sometimes up to 300 individuals), in variety of forest terrains (tropical rain forests mangrove swamps, river forests, as well as dry forests), chiefly in the middle tree level; territory size 36.7–325 acres (14.7–130 hectares)	Status of some subspecies uncertain due to habitat destruction and capture for research purposes
Age 5–10 months Age 3–5 years Not known	Fruits, insects	Not known	Acitve by day; found in fairly large social groups, with several adult males (10–35 individuals); dense, wet tropical forests; territory size 43.7–100 acres (17.5–40 hectares)	Threatened with extinction, especially because of deforestation and use of insecticides
Age 5 months Not known Over 13 years	Fruits, leaves	Birds of prey, humans	Active by day; found in pairs or small family groups (parents and 2–3 young) in wet or inundated forests, especially in dense underbrush, and in gallery forests; territory size 1.7–12.5 acres (0.5–5 hectares)	Habitat being steadily diminished by destruction of biotope
Age 5 months Not known Not known	Fruits, insects	Birds of prey	Active by day; found in pairs or small family groups (parents and 2–4 young at high elevations in forests with sandy soil; remote from rivers; only rarely in underbrush; territory size 1.5–50 acres (3–20 hectares)	Habitat being steadily diminished by destruction of biotope
Not known	Fruits, leaves, flowers	Birds of prey	Active by day; found in pairs or small family groups (parents and 2–4 young) in variety of forests, including secondary forests, at elevations up to 3200 ft (1000 m)	All subspecies much threatened by destruction of habitat
Not known Age 4 years Over 14 years	Fruits, seeds, flowers	Humans, birds of prey	Active by day; found in pairs or small family groups in tropical rain forests savanna forests and mountain forests at high elevations, including secondary forests, predominantly in underbrush and middle levels of trees; territory size 10–25 acres (4–10 hectares)	Habitat being steadily diminished by destruction of biotope
Not known Not known In captivity, over 14 years	Fruits, seeds, flowers, leaves, ants	Humans, birds of prey	Acitve by day; found in pairs or small family groups (parents and 2–4 young) in tropical rain forests; also in flood forests and farther from rivers; only in primary forest, predominantly in underbrush and middle levels of trees	Habitat being steadily diminnished by destruction of biotope
Not known	Fruits, seeds, flowers, leaves, ants	Humans, birds of prey	Acitve by day; found in pairs or small family groups (parents and 2–4 young) in tropical rain forests; also in flood forests and farther from rivers; primary forests only, predominantly in underbrush and middle levels of trees	Habitat being steadily diminished by destruction of biotope
Not known	Fruits, seeds, flowers, leaves, ants	Humans, birds of prey	Acitve by day; found in pairs or small family groups (parents and 2–4 young) in tropical rain forests; also in flood forests and farther from rivers; primary forest only, predominantly in underbrush and middle levels to trees	Habitat being steadily diminished by destruction of biotope
Not known Age 4 years Over 18 years	Fruits, seeds, flowers, leaves	Humans	Active by day; found in large social groups with more than one adult male (8–18 individuals); only in high forest growth (rain forests, high savannas, gallery), predominantly in crown level of trees; territory size 500–625 acres (200–250 hectares)	Endangered by destruction of habitat; *Ch. s. satanas* is threatened
Not known Age 4 years Over 17 years	Probably much like bearded saki	Humans	Acitve by day; found in fairly large social groups with more than one adult male (8–30 individuals); in flood forests and farther from rivers; mainly at crown level of trees	Seriously threatened by increasing destruction of habitat

CEBIDS

Nomenclature English common name Scientific name French German	Approximate Size Body length Tail length Weight	Distinguishing Features	Reproduction Gestation period Young per birth Weight at birth
Uakari *Cacajao calvus*, 2 subspecies Ouakari Uakari	males 1.5–1.8 ft; 43.5–56 cm females 1.2–1.9; 36.5–57 m. 6–7.4 in; 15–18.5 cm f. 5.6–6.6; 14–16.5 m. 9 lb; 4.1 kg f. 7.7; 3.5	Distinctly short, bushy tail; little or no fur on face and forehead; pink to red face skin; long, thick white or red body fur	Not known 1 Not known
Black uakari *Cacajao melanocephalus* Ouakari à tête noire Schwarzkopfuakari	1.6 ft; 50 cm 6.8 in; 17 cm Not known	Distinctly short, bushy tail; black face skin; scalp fully furred; long black coat; hind legs and tail sometimes chestnut brown	Not known
Red howler *Alouatta seniculus*, 9 (?) subspecies Hurleur roux Roter Brüllaffe	m. 1.6–2.4 ft; 49–72 cm f. 1.5–1.9; 46–57 m. 1.6–2.5 ft; 49–75 cm f. 1.6–2.4; 50–71 m. 14.3–17.8 lb; 6.5–8.1 kg f. 9.9–14; 4.5–6.4	Prehensile tail with hairless area on under side; howling voice; prominent muzzle; hyoid bone and larynx greatly enlarged, housing cavernous vocal apparatus; orange-brown to red fur	186–194 days 1 Not known
Brown howler *Alouatta fusca*, 3 (?) subspecies Hurleur brun Brauner Brüllaffe	m. 1.6–2.2 ft; 50–65 cm f. 1.4–1.8; 44–54 m. 1.6–2.2 ft; 48.5–67 cm f. 1.6–1.9; 48–57 Not known	Prehensile tail with hairless area on under side; howling voice; prominent muzzle; hyoid bone and larynx greatly enlarged, housing cavernous vocal apparatus; dark brown fur	Not known 1 Not known
Black howler *Alouatta caraya* Hurleur noir Schwarzer Brüllaffe	m. 1.7–2.2 ft; 52–65 cm f. 1.6; 50 m. 1.8–2.2 ft; 55–65 cm f. 1.8–2; 54.5–60 m. 14.3 lb; 6.5 kg	Prehensile tail with hairless area on under side; howling voice; prominent muzzle; hyoid bone and larynx greatly enlarged, housing cavernous vocal apparatus; nostrils close together; male black, female olive	187 days 1 4.5 oz; 125 g
Mantled howler *Alouatta palliata*, 5 (?) subspecies Hurleur à manteau Mantelbrüllaffe	m. 1.6–2.3 ft; 47–67.5 cm f. 1.2–2.1; 36–63 m. 1.6–2.2 ft; 50–66 cm f. 1.8–2.2; 56–67 m. 17.2–22 lb; 7.8–10 kg f. 14.5–19.8; 6.6–9	Prehesile tail with hairless area on under side; howling voice prominent muzzle; hyoid bone and larynx greaty enlarged, housing cavernous vocal apparatus; black or brown fur; long golden or brown fringes at the flanks; sexual parts of males inconspicuous until sexual maturity	186 days 1 Not known
Guatemalan howler *Alouatta villosa* Hurleur de Guatemala Guatemala-Brüllaffe	m. 2–2.1 ft; 60–64 cm f. 1.6–1.8; 50–54 m. 2.2–2.4 ft; 66.5–71 cm f. 2.1–2.2; 64–67 m. 23.9 lb; 10.9 kg f. 19.8; 9	Prehensile tail with hairless area on under side; howling voice; prominent muzzle; hyoid bone and larynx greatly, enlarged, housing cavernous vocal apparatus; soft, dense black fur; hair red-brown at root	Not known 1 Not known
Black and red howler *Alouatta belzebul*, 5 (?) subspecies Hurleur à mains rousses Rothandbrüllaffe	m. 1.9–2.1 ft; 56.5–63 cm f. 1.3–2.2; 40–65 m. 1.8–2.3 ft; 55.5–69 cm f. 1.6–2.3; 47–69 m. 15.8 lb; 7.2 kg f. 12.3; 5.6	Prehensile tail with hairless area on under side; howling voice prominent muzzle; hyoid bone and larynx greatly enlarged, housing cavernous vocal apparatus; black or black-brown fur; hands, feet, and tip of tail yellow to red or red-brown	Not known 1 Not known
Brown capuchin *Cebus apella*, 6 subspecies Sapajou apelle Gehaubter Kapuziner	m. 12.8–22.6 in; 32–56.5 cm f. 12.8–20.4; 32–51 m. 15.2–22.4 in; 38–56 cm f. 14.4–18.8; 36–47 m. 7.7 lb; 3.5–3.9 kg f. 5.5–6.6; 2.5–3	Prehensile, fully-furred tail; fur light to dark brown or yellowish-brown, sometimes lighter on belly; arms, legs, and tip of tail often darker than body; face lighter; head hooded black; erect hairs on head form horns or skull cap	153–161 days 1 8.9–10.3 oz; 250–290 g
White-fronted capuchin *Cebus albifrons*, 13 subspecies Sapajou à front blanc Weißstirnkapuziner	m. 14–21.2 in; 35–53 cm f. 13.2–16.8; 33–42 m. 16.2–19.8 in; 40.5–49.5 cm f. 16.4–20; 41–50 f. 4.6 lb; 2.1 kg	Prehensile, fully-furred tail; fur light brown or cinnamon to dark red-brown; hands, feet, and tip of tail usually lighter than body; broad cap of smooth brown hair at back of head; face, forehead, neck, shoulders, forearms, and breast creamy white	Not known 1 6.5–8.9 oz; 184–251 g
White-faced capuchin *Cebus capucinus*, 5 (?) subspecies Sapajou capucin Weißschulterkapuziner	m. 13.2–18.4 in; 33–46 cm f. 12.8–16.2; 32–40.5 m. 16–20 in; 40–50 cm f. 16.8–18.2; 42–45.5 m. 5.3–8.4 lb; 2.4–3.8 kg f. 4.4–5.9; 2–2.7	Prehensile, fully-furred tail; trunk, arms, legs and tail black; broad cap of smooth black hair at back of head; face, forehead, neck, shoulders, upper arms, and breast white	Not known 1 Not known
Weeper capuchin; wedge-capped capuchin *Cebus nigrivittatus*, 4 (?) subspecies Sapajou brun Brauner Kapuziner	m. 15.2–18.4 in; 38–46 cm f. 14.4–15.6; 36–39 m. 17.6–19.6 in; 44–49 cm f. 15.2–18; 38–45 m. 6.4 lb; 2.9 kg f. 4.6; 2.1	Prehensile, fully-furred tail; fur medium to dark brown; hair on sides, arms, legs, and tail ringed black and brown; hands, feet, and tip of tail darker than body; narrow wedge-shaped brown cap on back of head; face and forehead light gray-brown	Not known 1 Not known

COMPARISON OF SPECIES

Life Cycle Weaning Sexual maturity Life span	Food	Enemies	Habit and Habitat	Occurrence
Age 13–22 months Not known Over 18 years	Fruits, leaves	Humans	Active by day; found in fairly large social groups with more than one adult male (5 to over 30 individuals); probably only in swampy forests	Endangered by destruction of habitat; white uakari (*C.c.calvus*) believed threatened
Not known Not known Over 12 years	Probably much like uakari	Humans	Active by day; found in fairly large social groups with more than one adult male (15–25 individuals); only in swampy forests	Seriously threatened by increasing destruction of habitat
Not known	Leaves, fruits	Humans	Active by day; found in social groups of several adult males and females (4–17 individuals, average 6–9); in a variety of forest terrains (tropical rain forests, wet, high savannas, gallery, also deciduous and mixed forests); territory size 2.5–55 acres (1–22 hectares)	Habitat being steadily diminished by destruction of biotope
Not known	Leaves, fruits	Humans	Active by day; found in social groups of several adult males and females (up to 11 individuals, some groups males only); in wet mountain forests at elevations of 2400–3800 ft (750–1200 m)	Existence threatened according to recent surveys
Not known	Leaves, fruits	Humors	Active by day; found in social groups of several adult males and females	Habitat being steadily diminished by destruction of biotope
Age 1.5–2 years Females 4–5 years, males 6–8 Over 20 years	Leaves, fruits	Humans	Active by day; found in social groups of several adult males and females (5–31 individuals average about 18); in evergreen rain forests, often on river banks; territory size 12.5–150 acres (5–60 hectares)	Habitat being steadily diminished by destruction of biotope
Not known Not known Over 20 years	Leaves, fruits	Humans	Active by day; found in social groups of several adult males and females (6–7 individuals); in tropical rain forests and mixed woods, also in flood forests	Probably the most endangered species of howler monkeys
Not known	Leaves, fruits	Humans	Active by day; found in social groups of several adult males and females	Existence of some subspecies definitely threatened
Age 6–12 months Not known Over 40 years	Fruits, insects, small vertebrates	Humans, sometimes birds of prey	Active by day; found in large social groups with more than one adult male (3–40 individuals); in nearly all forest terrains (from tropical rain forests to deciduous forests) at elevations up to 8600 ft (2700 m); territory size 0.11–0.15 mi² (0.3–0.4 km²)	Some subspecies at least *C.a. xanthosternos*) are threatened
Not known Not known Over 40 years	Fruits, insects	Not known	Active by day; fairly large social groups with more than one adult male (7–30 individuals); in a variety of forest terrains at elevations up to 6700 ft (2100 m); territory size 100–175 acres (40–70 hectares)	Habitat being steadily diminished by destruction of biotope
Not known Females 4 years, males 8 In captivity, nearly 47 years	Fruits, insects, flowers, young birds, small mammals, lizards	Humans, possibly of prey birds	Active by day; fairly large social groups with more than one adult male (2–24 individuals); in nearly all forest terrains, from wet lowland forest on the Atlantic to dry forests on the Pacific, and in mangrove forests, at elevations up to 4800 ft (1500 m); territory size 80–200 acres (32–80 hectares)	Habitat being steadily diminished by destruction of biotope
Not known Not known Over 30 years	Fruits, insects	Humans, possibly of prey birds	Active by day; fairly large social groups with more than one adult male (10–33 individuals); in a variety of forest terrains, including gallery forest or small stands; only in fairly low-lying regions	Habitat being steadily diminished by destruction of biotope

CEBIDS

Nomenclature English common name Scientific name French German	Approximate Size Body length Tail length Weight	Distinguishing Features	Reproduction Gestation period Young per birth Weight at birth
Black spider monkey *Ateles paniscus*, 2 subspecies Singe-araignée noir Schwarzer Klammeraffe	males 15.2–23.2 in.; 38–58 cm females 16–24.8; 40–62 m. 2.1–2.8 ft; 63–85 cm f. 2.1–3.1; 64–93 m. 17.1–20.9 lb; 7.8–9.5 kg f. 19.3; 8.8	Distinctly long, slender legs; long, prehensile tail with hairless area on under side; black fur; face light red and partly hairless; female has exceptionally long clitoris, male genitalia small	210–225 days 1 Not known
Long-haired spider monkey *Ateles belzebuth*, 3 subspecies Singe-araignée à ventre blanc Goldstirn-Klammeraffe	m. 16.8–20 in.; 42–50 cm f. 13.6–23.6; 34–59 m. 2.3–2.7 ft 69.5–82 cm f. 2.1–2.9; 61–88 m. 7.3–12.7 lb; 3.3–5.8 kg	Distinctly long, slender legs; long, prehensile tail with hairless area on under side; black or dark brown fur; underbelly, hind legs, and base of tail white to light brown; yellow-brown triangular blaze on forehead; female has exceptionally long clitoris, male genitalia small	210–225 days 1 Not known
Brown-headed spider monkey *Ateles fusciceps*, 2 subspecies Singe-araignée à tête brune Braunkopf-Klammeraffe	m. 14.8–23.6 in.; 37–59 cm f. 18–22; 45–55 m. 2.1–2.4 ft; 63–72 cm f. 2–2.7; 60–81 Not known	Distinctly long, slender legs; long, prehensile tail with hairless area on under side; black trunk; brown or black head; black face; female has exceptionally long clitoris, male genitalia small	210–225 days 1 Not known
Black-handed spider monkey *Ateles geoffroyi*, 9 subspecies Singe-araignée aux mains noires Geoffroy-Klammeraffe	m. 15.2–19.8 in.; 38–49.5 cm f. 13.6–20.8; 34–52 m. 1.9–2.7 ft; 59–82 cm f. 2.3–2.8; 70–84 m. 16.5–17.8 lb; 7.5–8.1 kg f. 16.7–18.5; 7.6–8.4	Distinctly long, slender legs; long, prehensile tail with hairless area on under side; fur gold-brown or red to dark brown; black hands and feet; female has exceptionally long clitoris, male genitalia small	210–225 days 1 12.1 oz; 340 g
Woolly monkey *Lagothrix lagotricha*, 4 subspecies Singe laineux; lagotriche Wollaffe	m. 16.4–26 in.; 41–65 cm f. 15.6–23.2; 39–58 m. 1.7–2.6 ft; 53–77 cm f. 2–2.4; 60–73 m. 14.3–23.7 oz; 6.5–10.8 kg f. 12.1–17.8; 5.5–8.1	Massive build with heavy trunk; powerful limbs; large, round head; coat soft and woolly; trunk gray to olive-brown, dark brown, or black; head often darker, nearly black; prehensile tail	207–211 days 1 Not known
Yellow-tailed woolly monkey *Lagothrix flavicauda* Singe laineux à queue jaune Gelbschwanzwollaffe	m. 16–20.8 in.; 40–52 cm f. 20.8; 52 m. 22.4–23 in.; 56–57.5 cm f. 25.2; 63 Not known	Same size as woolly monkey, with shorter limbs; brown face with light nose spot; fur deep mahogany brown; yellow hair on scrotum and under side of tail up to the hairless prehensile portion; white fur at nose and mouth	Not known 1 Not known
Muriqui *Brachyteles arachnoides*, 2(?) subspecies Eroïde Muriki	m. 18.4–25.2 in.; 46–63 cm f. 18.8–24; 47–60 m. 2.2–2.5 ft; 65–74 cm f. 2.5–2.8; 74–84 m. 26.4–33 oz; 12–15 kg f. 20.9; 9.5	Long, prehensile tail; dense, light gray to brown fur; nostrils large and pointed downward; female has short clitoris; male has large testes; canine teeth barely larger than incisors	Not known 1 Not known

The muriqui is the largest and also the most threatened New World primate. Its large nostrils, directed downward like those of Old World monkeys, distinguish them from other "spider" monkeys.

COMPARISON OF SPECIES

Life Cycle Weaning Sexual maturity Life span	Food	Enemies	Habit and Habitat	Occurrence
Not known Age 4–5 years Over 20 years	Fruits, leaves, flowers, buds	Humans	Active by day; social groups break up into subgroups (exact group size difficult to determine); only in rain forests free from inundation, predominantly at crown level of trees	The most effectively protected of all spider monkey, this species; both subspecies locally eliminated by hunting
Not known	Fruits, seeds, leaves	Humans	Active by day; social groups break up into subgroups; exact group size difficult to determine (more than 20 individuals); in rain forests free from inundation, predominantly at crown level of trees	Threatened by hunting and destruction of habitat, especially subspecies *A. h. marginatus*
Not known	Fruits, seeds, leaves	Humans	Active by day; social groups break up into subgroups; group size difficult to determine; only in rain forests free from inundation, predominantly at crown level	Much endangered by habitat destruction and hunting, especially subspecies *A. f. fusciceps*
Not known	Fruits, seeds, leaves	Humans	Active by day; social groups break up into subgroups (possibly up to 80 individuals to a group); only in rain forests free from inundation or rain forest stands within mixed forest regions, predominantly at crown level of trees	More threatened than the other three species; only two of nine subspecies protected
From 5th months Age 4–5 years In captivity, over 25 years	Fruits, leaves, buds, flowers, bark, nuts	Humans	Active by day; found in large social groups with more than one adult male (up to 70 individuals); in rain forest and wet hill horests, at elevations up to 8000 ft (2500 m); territory size 400–1100 hectares	Considerably threatened by increasing destruction of biotope
Not known	Fruits, flowers	Humans	Active by day; found in large social groups with more than one adult male (4–14 individuals); live high in wet forest rich in epiphytes, at elevations between 5300 and 8600 ft (1670 and 2700 m)	Extremely endangered by very limited occurrence and by hunting and deforestation
Not known	Leaves, fruits, seeds, berries, and flowers	Humans	Active by day; large social groups, with more than one adult male (10–50 individuals), break up into smaller subgroups; in remote hill forests	The most endangered now-world monkey; complete extinction avoidable only by strictest protective measures

Remote mountain forests in southeastern Brazil are the home of the rare muriqui, no less strange than it is rare. Few students have yet succeeded in observing and photographing these animals, which live widely scattered in small groups, in their natural environment.

Goeldi's Monkey

by A. George Pook

The small, black, unassuming GOELDI'S MONKEY, or CALLIMICO *(Callimico goeldii)* was the last primate species to be discovered. It was not until 1904 that the Swiss naturalist Emilio Goeldi described it for the first time and sent a specimen to Europe. Until then, the inaccessibility of its habitat, its shyness, and its inconspicuous behavior, as well as the fact that it was not of any economic value, had enabled Goeldi's monkey to live in seclusion, for the most part overlooked even by the native population.

Once discovered, however, it not only created a considerable stir in the classification of New World monkeys, but also changed the general conception of their phylogenetic origin. Until then, all New World monkeys had been assigned to two fixedly defined families of separate phylogeny, the family Callitrichidae (marmosets and tamarins) and the family Cebidae (capuchins and allies). The marmosets were regarded as primitive primates, representing a phylogenetic stage prior to the evolution of true simians.

The callimico was immediately recognizable as an intermediate form. It was small, and like the callitrichids, bore claws on almost all digits, but it had the dental formula and skull shape of the cebids. Ever since, many systematists have been unable to agree whether the two main families are really so different, and whether the callimico is closer to one than to the other, or whether it should be placed in a separate family. The provenance of the family Callitrichidae also cam into question; perhaps their small size was a later development, and they had evolved from a larger, more simian-like ancestor. These problems still await solution. Strictly speaking, the creature is neither marmoset-like nor capuchin-like, and the name Goeldi's "tamarin" is misleading, since it is not really a tamarin either. It may best be called simply a callimico.

The callimico. This small, unimpressive monkey was first discovered in 1904, and to this day its systematic position is not completely settled. It is the sole representative of its family.

Notwithstanding the scholarly debate, little more was learned about Callimico in the ensuing 70 years. It is only since the mid-1970s that it has been variously kept in captivity and observed in the wild, where it is quite shy. Observers interested primarily in other species rarely sight it. Although it is fairly widespread in the upper reaches of the Amazon, it would appear that the Callimico does not occur anywhere in great numbers. When we first attacked the problem of selecting a suit-

able location for a study in the wild, no one was able to recommend an area in which we could be certain of success.

When many specimens suddenly appeared on the market in Europe, it became important to discover their source, for two reasons. First, it must be a place where the species was reasonably abundant; and second, there was the need to suppress this traffic. In the meantime, concern over the survival of Callimico had become so important that it had been included, as a rare species, in the International Red List of Endangered Species. The source turned out to be northeastern Bolivia, a country where, as in many comparatively poor countries of the world, there are indeed laws against the exportation of endangered species, but very limited means of enforcement. That is where we finally traced this illusive creature.

The forests in that area are not typical Amazonian primeval forest. The crown canopy is patchy and irregular, and the height of trees ranges from 32 to 96 ft (10 to 30 m). Thus, the sun's rays can reach the lower branches and consequently, there is an abundance of undergrowth, with a maze of bamboo thicket beneath it. Except in the few rainiest months of the year, the soil is well drained by many natural streams and the rolling conformation of the terrain.

Throughout its range of distribution, the callimico is associated with various callitrichid species. In some respects, their habits are similar: a family group will spend the night in dense underbrush or in a hollow tree, and spend most of the day going from one fruit-bearing tree to another.

In the dry season, the choice of fruit trees is very limited, which means that the group will visit the same tree perhaps every day, and on some days more than once. A number of trees thrive here that produce soft fruits in the rainy season, but the time during which a particular tree will bear fruit may sometimes last only a few days. Hence, when various species of small monkeys travel in groups, their collective knowledge of which trees are bountiful at given times may be advantageous to all. Often, the entire group will suddenly take off for a particular tree as far away as 0.3 mi (0.5 km). Utilization of trees in such a manner gives the appearance of being planned, and, at any rate, is not accidental.

When a suitable fruit tree is reached, there is plenty to eat for all. In fact, much fruit falls to the forest floor and is lost. Hence, there is not much competition among species for the fruit portion of their diet. Surprisingly, there are hardly any open interrelationships between individuals of different species at close quarters.

The living territory of such a mixed-species group may be about 75–100 acres (30–40 hectares) in area. During the rainy season, the group will roam more or less in a circle, traversing an average of 1.2 mi (2 km) per day, and sometimes covering a large part of the territory in a short time. Many of the main feeding trees stand on the edges of the territory, where they must be shared with neighboring callitrichid groups. A callimico family will spend about half the day traveling about, and eating with other species or resting nearby.

The other main component of the callimico's diet consists of insects and smaller vertebrates. In seeking food, individual monkeys will often leap to the ground, where they have been observed to catch and consume lizards, frogs, and even small snakes. In this hunting activity, the animals are more successful as individuals; hunting seldom involves several species acting together. One of the greatest differences in behavior between the callimico and the other clawed monkeys should be

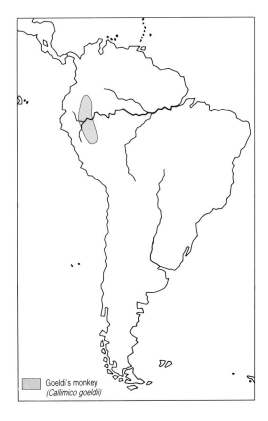

Goeldi's monkey (*Callimico goeldii*)

New World Primates

Callimiconids (Callimiconidae)

Nomenclature English common name Scientific name French German	Approximate Size Body length Tail length Weight	Distinguishing Features	Reproduction Gestation period Young per birth Weight at birth
Goeldi's monkey *Callimico goeldii* Tamarin de Goeldi Springtamarin; Callimico	8.8 in; 22 cm 11.2 in.; 28 cm 1.1 lb; 500 g	Completely black coat; face surrounded by erect main; clawlike nails on all toes except great toe; snub nose; some prosimian features in musculature	155 days 1 1.8 oz; 50 g

Callimicos prefer a vertical posture, and they do not abandon their erect posture even in making long leaps. With their powerful legs, they launch themselves upright from a trunk, execute a turn about their longitudinal axis in mid-air, and land feet forward, again vertical, on the next tree.

noted. The latter spend their time in the middle and upper levels of the trees, where branches are numerous. In locomotion, they generally run along the branches and leap from the twigs at the ends, from one tree to the next. This mode of locomotion betrays the presence of the animals (to their enemies as well) even when they are not seen, because it makes a great deal of noise and sets the twigs and branches into wild motion.

By contrast, the callimico might be described as a vertical climber and leaper. Although it will climb and leap about without restraint in the crown of a tree while eating fruit, it will regularly descend nearly to the forest floor when about to move on. Most of its time is spent at heights of less than about 10 ft (3 m), but it climbs around in trees of any height. At the lower levels, vegetation consists almost exclusively of vertical tree trunks and saplings, as well as a dense growth of bush and bamboo. The callimicos move about by leaping in an erect posture from one vertical support to the next, with the aid of their powerful hind legs, turning while still in mid-air, and landing vertically, feet forward.

The leaping ability of the callimico is impressive. I have seen one of these animals leap a distance of about 13 ft (4 m) horizontally without losing height; and this was done only about 12 in. (30 cm) above the ground, so that it should actually have been easier to drop to the ground and run.

This mode of locomotion is preferred to such a degree that the animals will make a detour in a circle on the ground whenever they encounter an obstacle, such as a creek or an unexpected human being, whereas most arboreal animals will hasten up into the trees. It is a comparatively silent mode of locomotion, and causes little disturbance in the vegetation. When I was following a callitrichid group, I often met callimico monkeys whose presence I had not noticed.

Whatever advantages the callimico may gain by this form of locomotion – foraging for food at ground level being in my opinion the most important – it is far slower than the quadruped run-and-jump locomotion of the callitrichid monkeys. When traveling between fruit trees in mixed groups, a callimico family tends to drop back, and may be up to ten minutes later in arriving than the callitrichid family.

The typical day for a callimico is divided between traveling, eating and resting. Three breaks, each of 30 to 90 minutes duration, are usual in the course of the day. At these times, they separate from other species. A callimico family of 4 to 10 individuals will usually rest not more than about 6.5 ft (2 m) from the ground in dense bush or on the inclined trunks of fallen trees, which by falling may have created a small clearing for Sun bathing.

GOELDI'S MONKEY

Life Cycle Weaning Sexual maturity Life span	Food	Enemies	Habit and Habitat	Occurrence
Age 12 weeks Age 14 months In captivity, over 9 years	Chiefly fruits and insects, also small vertebrates	Possibly ocelot, tayra, and birds of prey	Diurnal in seasonally dry forests with dense undergrowth; station near the ground, preferred; found in small family groups; territory size 75–150 acres (30–60 hectares)	Rare and endangered; population very limited

Usually, the animals rest crowded together in a cat-like crouch. Pairs, or small groups, will also engage in grooming, which may last as long as 15 minutes. One individual will give another a grooming invitation by stretching out flat right in front of the other one. Younger animals sometimes take part, but they make an unstable impression and are quite restless during a break. However, the young do not seem inclined to play very much.

Another important aspect of the callimico's social behavior involves their manifold vocal utterances. These comprise a number of expressive sounds, ranging from soft piping tones between individuals sitting close together to loud, piercing series of cries in a descending tones, whereby they maintain contact with each other as well as with callitrichids over distances of 320 ft (100 m) or more. In connection with feeding, they make a chirping sound and a loud "chuck" call in response to frightening or threatening situations. Young submissive individuals make squeaking and choking sounds. There are mixed vocalizations between many of these utterances; some extend into the ultrasonic range.

Callimicos often communicate by scent signals. They have more scent glands than any callitrichid. Some of these glands are located on the belly. In perhaps the most striking behavior pattern in this context, they stretch their limbs, arch their backs, and thrust the coiled tail under their body, moving it back and forth over the ventral surface, which is thereby moistened with urine and scents. Sometimes the nose, face, and other parts are rubbed on branches.

Visual expression and "body language" include grimaces, hard stares, and bristling of the fur, attended by a stiff-legged gait. The meaning of some of these expressions is still unclear; some are employed in altercations within the group, while others may relate to sexual behavior. Unlike callitrichids, the callimico hardly uses such devices for making territorial claims. Callimico groups seem to live widely separated, which is very unusual for a simian species. We estimate that in the area we studied, one of comparatively frequent occurrence, there is about one group in 1.4 mi² (4 km²). Since each group seems to spend all its time inside its own territory of about 0.1 mi² (0.3 km²), it would seem that about 90 percent of the range is not utilized.

On the other hand, the callitrichid species with which the callimico associates divide their entire

Even in climbing, the callimico prefers the vertical posture. It prefers to climb on straight trunks and saplings.

range of distribution into an unbroken network of overlapping territories. We do not yet know with certainty why the callimico practices this peculiar distribution. A possible explanation might be that a territory must meet very definite requirements in order to provide the callimico with special sources of nourishment. It seems to prefer low-lying terrain that has ample underbrush, but is not prone to inundation. Kosei Izawa, a primate researcher at the Japan Monkey Center and a contributor to this volume, believes that geological processes slowly eliminated the available habitat for the callimico. In addition, there is the increasing competition of successful callitrichid species of the genus *Saguinus*. This might imply that the callimico is destined to disappear in the long term, and is just hanging on in a region where his occurrence was formerly abundant.

The comparative remoteness of groups from each other has a significant impact on the reproduction of the species. If the groups exchanged members only very infrequently, the genetic pool would deteriorate, and problems of inbreeding might result if a group depended upon only one breeding pair. Therefore, it is a very important observation that a callimico female, by contrast with typical callitrichid reproductive habits, will have only one young at a birth, but the group may have more than one breeding female. Whether there will also be more than one breeding male is not known. In fact, however, this system ensures greater genetic variety within a group.

In the first two weeks after birth, and when not nursing, an infant spends most of its time clinging tightly to its mother's shoulder. In the third week of life, the mother begins to push the young one away, and the father gradually assumes the main burden of carrying. In the fourth week, other group members sometimes take turns carrying the baby. The experience gained by juveniles in this way benefits them when they come to have young of their own. About the same time, the youngster will begin to climb around independently for short periods, and at the age of seven weeks it can run and leap about very well. After about four weeks, the young begin to take solid food from their parent's hands, sometimes while still being carried. When the young are about ten weeks old, the adults become less and less willing to share food with them in this way, and the young gradually begin to fend for themselves.

Both sexes attain maturity at about 14 months. Although matings have seldom been observed, breeding pairs generally show clear signs of sexual activity about 7 to 10 days after the female has given birth. Pregnancy lasts about 155 days, so that a female may reproduce twice in a year. Field observations indicate that this happens in the wild also.

Under suitable conditions, in which they are undisturbed, secure, and well fed in ample variety, the callimico seems to respond well to life in captivity. Since the mid-1970s, several zoological gardens, notably the Jersey Wildlife Preservation Trust on the Isle of Jersey and the Brookfield Zoo near Chicago, have kept breeding groups that are growing steadily and span several generations.

We can but hope that these will not remain the only places where these shy, modest, and charming little monkeys can be observed, as their remnant in the wild is increasingly threatened by destruction and exploitation of their rainforest home.

Callimico monkeys live in family groups. Unlike callitrichids, where twin births are not uncommon, they have only one baby at a time. The young is carried about by the mother for the first three weeks and chiefly by the father thereafter. In the fourth week, other group members will help with carrying the baby.

Marmosets and Tamarins

by Jürgen Wolters and Klaus Immelmann

The marmosets and tamarins are a remarkable group of primates. Natural selection has reduced them to a small size, and they exhibit some close parallels to the smaller species of birds, namely the sparrows. They are the smallest true monkeys; their lightest representative, the pygmy marmoset, weights about 3.5 to 4.2 oz (100 to 120 g) when full grown, which is hardly more than a dormouse. The lion tamarin, the largest member of the family, tips the scales at no more than about 25 oz (700 g). Even among the prosimians, there are only two species, the mouse lemur and the dwarf galago, that are smaller than the pygmy marmosets.

The Callitrichidae comprise a great multiformity of species in the South to Central American rainforests. With a total of 20 species and not less than 39 subspecies, they are among the most speciated groups of primates. We divide the family Callitrichidae into four genera. Cebuella and Callithrix together make up the MARMOSETS. A common feature of these forms is the length of the incisors in the lower jaw, about equaling the canines. The TAMARINS, collectively of the genera *Saguinus* and *Leontopithecus*, have canines in the lower jaw that are distinctly longer than the incisors, as do the other true monkeys.

The origin of the name tamarin, like that of many animal names, may lie in a linguistic misunderstanding. The French explorer A. Binet had supposed that Indians in Cayenne would call the red-handed tamarin *Saguinus midas* a "tamarin." He was mistaken, but the name afterwards came to be accepted nevertheless.

The etymology of "marmoset" is not as abscure. It derives from the French *marmouset*, originally, more or less meaning "shrimp" or "dwarf", and hence an apt designation for these creatures.

Since we have no fossil finds of direct callitricid ancestors, the evolutionary history of these animals is still in dispute. It was long supposed that the ancestors of the callithricids were possible even smaller than the pygmy marmosets of today. This view was related to the assessment that the callithricids, on the basis of some structural traits, are very primitive true monkeys. The American evolutionary biologist John Eisenberg was one of the first to suspect that they evolved to their present size later, being descended from larger ancestors. Their light weight enables the marmosets and tamarins to feed in the thin outermost twigs and branches, and thus gives them access to food sources that are less readily available to the larger

The physical feature that was the basis for naming the squirrel-like "clawed" monkeys is the modified, claw-like nail borne on all digits except the great toe. The photograph shows an emperor tamarin.

cebid monkeys. That suspicion has been confirmed in recent years by some field studies. However, there has been only conditional confirmation of another premise of Eisenberg: that the callithricids took over the habitat of arboreal squirrels in the tropics of the New World; the squirrels did not immigrate into South America until a few million years ago.

The occasional application, even today, of the term "squirrell monkey" to the callithricids alludes to some similarities, say, in locomotion, at least between certain marmosets and the squirrels. More recent field observations have made it clear that the callithricids are in only limited competition with arboreal squirrels for the same habitat, and have evolved another highly specialized habit of life that puts the seemingly primitive physical traits of the marmosets and tamarins in a different light.

First, mention must be made of the claws formed on all the fingers and toes of these monkey except the great toe. Only the great toe bears the characteristic primate nail in all species. The American anatomist E.E. Thorndike has been able to show that the claws are only laterally compressed nails. In other words, the claws are not a vestigial trait of phylogenetic ancestors, but a newly acquired, specialized feature.

The clawed fingers characterize the marmoset and tamarin hands, which are otherwise of a quite primitive structure. For "handling" objects, or for climbing and hanging hand-over-hand, they are doubtless less appropriate than the extremities of other primates. Thus, the absence of a saddle joint at the base of the thumb precludes "opposability", that is, the ability to oppose the thumb to the other fingers in grasping an object.

Locomotion and Foraging

The callithricid food is comparatively elongated and shows a distinct adaptation to a predominantly horizontal, running or leaping locomotion. Accordingly, we classify the callithricids as leapers in their motor behavior. The term perhaps best applies to the tamarins however. They will repeatedly interrupt their quadruped progress over the branches with leaps, which are usually rather short. In our callithricid facility, we have found that the cotton-top tamarin, for example, can negotiate horizontal leaps of about 10 ft (3 m) or more without much difficulty, landing safely in the twigs and branches. There is no apparent deficiency in attaining a secure grasp when climbing through the twigs.

The lion tamarin resembles the other tamarins in mode of locomotion, but at the same time its movements suggest more of a climbing type of locomotion, like that of the squirrel monkeys among New World primates.

The other extreme is represented by the pygmy marmosets. Their gait is rather mouse-like; they

The callithricids are the smallest true simians. But even among the smallest, there are considerable differences in size. The largest species, the lion tamarin (left), may attain a height of more than 12 inches (30 centimeters); and the smallest, the pygmy marmoset (right), measures about 6 inches (15 centimeters), making it the world's smallest monkey. Due to their minute size and light weight, the marmosets and tamarins can utilize the thin outermost branches in the primeval forest as additional habitat and food source.

avoid long leaps, and are well adapted to vertical progress on tree trunks, like arboreal squirrels. Representatives of the genus *Callithrix* are comparable in locomotion to the tamarins but show a marked tendency towards the pygmy marmoset type of behavior. Still, no callithricid matches the ability of a squirrel to run up and down a vertical tree trunk.

The preferred horizontal locomotion of the tamarins in particular does not explain why the nails have been transformed into claw-like structures. This became more comprehensible when, in recent years, we became more familiar with the feeding habits of the marmosets and tamarins in the tropical rainforest. It turned out that the marmosets especially consume considerable quantities of tree sap. The evolution of claw-like nails may well be closely related to this special dietary component.

The monkeys use their claws to establish a firm grip on the vertical trunk and gnaw little holes in the bark. This causes the trees to produce more of the sap that is used by plants to seal injuries to the bark, and perhaps also to discourage insect pests. Clearly, the claw-like nails afford a firm position during the gnawing activity. The tamarins employ a different strategy. They do not actually make any tap holes; rather, they utilize tree sap found in the trees as a result of the action of other animals. Hence, it is not surprising that the marmosets that deliberately tap the trees are the most able trunk climbers of the family, and that their lower incisors are of much the same lengtht as the canines. The Peruvian primatologist Pekka Soini, who studied the life of the pygmy marmosets in nearly twenty different groups for 16 months in the rainforest of Peru, discovered that these animals spend the better part of the day gnawing holes and harvesting tree sap. The wildlife biologist Marleni Flores Ramirez has observed that these animals will make fresh holes in the bark just before nightfall; since sap will collect abundantly overnight in the fresh incisions, this may be goal-directed behavior.

Tree sap supplies the clawed monkeys with valuable carbohydrates and minerals. Possibly the consumption of tree sap is directly related to another important source of nourishment, namely insects. More than most other primates, the callithricids utilize animal protein, especially in the form of locusts, beetles, butterflies, and their larvae. Spiders, snails and ants also contribute to the food supply, as well as occasional small lizards, birds' eggs, nestlings, and tree frogs.

In the hunt for insects, their small physical size enables the little monkeys to stalk large insects by stealth. Their specialization in catching and consuming insects may well be related to a peculiarity of dentition that was formerly thought of as a relic of the primordial level of development: this is the conformation of the molars in the upper jaw. In all true simians, these have four cusps, whereas in the callithricids there are only three, as in many prosimian species.

The diet of callithricids is supplemented with fruits, flowers, and nectar. In fact, for the tamarins,

A pygmy marmoset with twins. Whereas single children are the rule for all other simian species, the marmoset and tamarin females generally give birth to two offspring, both of which are lovingly reared and cared for.

apart from insects this is the mainstay of the food supply. The American wildlife biologist John Terborgh spent a year in the Manu National Park of Peru (one of the still largely untouched rainforest areas of South America) comparing the habits of various New World monkeys. He learned that two of the clawed species, the brown-headed tamarin an the emperor tamarin, use different species of three than cebid monkeys for gathering fruits and blossoms. Whereas the larger New World monkeys prefer trees with spreading crowns, he found tamarins almost exclusively in trees with small crowns, less than 48 ft (15 m) in diameter. Another characteristic of the callitrichids' fruit trees was that only a few fruits would ripen at the same time, possibly not enough for the needs of cebid monkeys, which generally roam about in larger groups. Other field researchers have observed that tamarins will specialize in one or just a few tree species, as the fruit ripens. It is only with this information about special feeding habits of the callitrichids that we can properly understand their role in the network of the complicated biocommunity of the tropical rainforests. We may now assume that the physical features of these animals are not primordial, but are adaptations to the occupation of a particular ecological niche, to a special habitat. This adaptation has enabled the callitrichids to spread so successfully into nearly all tropical regions of the New World.

Genus Cebuella (Pygmy Marmosets)

The most extreme dietary specialist among the clawed monkeys is undoubtedly the pygmy marmoset *(Cebuella pygmaea)*, sole representative of its genus. The life of these animals revolves around a few trees and climbing plants, which they regularly tap for their coveted sap.

Territories are not much larger than 2.5 acres (1 hectare). The previously mentioned primatologist Soini found territories with fixed boundaries, at most 16,000 ft^2 (5000 m^2) in extent (a few measured only 10,200 ft^2 or 1000 m^2) in his area of investigation in Peru. The animals made thorough use of only about one-third of the area. Here stood the preferred sap trees, often peppered with round and along gnawed holes; as many as 1300 such incisions were found in 10 ft^2 (1 m^2).

As a rule, the trees seem to survive exploitation by the monkeys for a long time without injury. It the flow of sap gives out in the pygmy marmosets' core territory, they will move on or even temporarily leave their territory entirely. In the rainforest region studied by Soini, this was altogether feasible, since the territories of neighboring groups were not contiguous. Perhaps for this reason, Soini found no signs of confrontations with neighboring groups, as among other callithricids.

In the comparatively large total range of distribution in the Upper Amazon region, the pygmy marmoset occupies quite diverse forest types. However, it seems to prefer the marginal zones – densely grown all the way to the ground – of the tropical rainforest, as for example along the banks of rivers. If suitable feeding trees are available, the pygmy marmoset will not avoid the secondary forests created through human intervention. Therefore, despite destruction of on natural habitat after another, the adaptability of the pygmy marmoset may help save it from extinction.

Undoubtedly, however, trees and climbing plants with useful sap are indispensable to survival. Fruits do not appear to be especially important in the diet of the pygmy marmosets. They lie in wait

The pygmy marmoset, like all marmosets, has an unusual mode of feeding: it holds on to tree trunks by its claws, gnaws little boles in the bark, and taps the emerging sap.

for insects, their second important food source, near their tree incisions, where emerging sap attracts butterflies in particular.

The pygmy marmoset lives in rather small social bands: Soini found an average group size of six individuals. Groups comprise two or more adults and several young of various ages. Soini also found solitary individuals, chiefly young adults, but evidently their "vagabond" phase was of brief duration. Occasionally he encountered animals living in pairs. He was not able to determine definitely whether such pairs are a typical foundation for the formation of new groups. Nor was it clear whether the larger groups are purely family associations, that is, a pair of parents with descendants of various ages.

As in all other well-known callitrichid species, only one female at a time will reproduce in a group; other adult females are excluded from procreation. With an interval of only five to seven months between births, and a gestation period of about 140 days, this breeding female will generally produce twins. Births are distributed throughout the year in the wild, with a slight peak at about the middle and the end of the year.

The young, weighing only about 0.5 oz (15 g) at birth, are nursed up to the age of three months. At only one year, they have attained the weight of an adult, and within the second year of life they can reproduce.

As in other callithricids, males and females do not develop any so-called secondary sex characteristics, so that they are outwardly indistinguishable, except for their genitals.

By contrast with all other true simians, twin births are the rule among callithricids. From breeding in captivity, we know that single births are less than half as frequent as twin births, on the average. Also, especially marmosets breeding in captivity are increasingly bearing triplets, and very occasionally quadruplets. A mother of quadruplets cannot possibly supply all the babies adequately with milk. In rare cases, triplets have been raised successfully by the mother. In our facility, we have had two opportunities to observe this situation in the case of white-fronted marmosets. In the phase during which the fast-growing youngsters require a good deal of milk and are not yet taking much solid food, we witnessed serious tussling and even biting between nurslings for the mother's nipples, which incidentally, in adaptation to the simultaneous rearing of several young, are located near the shoulders. The injuries sustained by the young marmosets in these exchanges, especially on the hands and forearms, left scars that were clearly visible in adulthood. Even though we do not yet know just why marmosets breeding in captivity tend to have triplets, it is to be assumed, according to our observations, that a high frequency of such multiple births is not the natural condition.

In captivity, the raising of twins is usually as free of problems as the raising of a single young. In the wild, it seems that this may not be the rule. Thus,

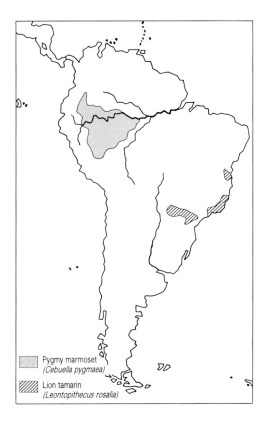

in observing pygmy marmosets, Soini noted many losses of young during the nursing period. Meanwhile, however, from field studies of other species we know that both twins have good chances of growing up healthy.

Whereas mothers of other primate species receive comparatively little help from other group members, including close relatives, in raising their young, the marmosets and tamarins have evolved a system of division of labor for the simultaneous raising of twins.

Genus *Callithrix* (Marmosets)

The second genus of marmosets, *Callithrix*, embraces seven species, according to the classification of the primatologists Adelmar Coimbra Filho and Russel Mittermeier. This includes the silvery marmoset *(Callithrix argentata)* and the white-shouldered marmoset *(C. humeralifer)*, which live south of the Amazon in central Brazil; the silvery marmoset is also found in eastern Bolivia. The other representatives, the common marmoset *(C. jacchus)*, the white-eared marmoset *(C. aurita)*, the buff-headed marmoset *(C. flaviceps)*, the white-fronted marmoset *(C. geoffroyi)*, and the black-penciled marmoset *(C. penicillata)* all live in eastern Brazil, especially in the coastal rainforest of the southeast.

The American biologist Philip Hershkovitz regards all East Brazilian marmosets as merely subspecies of a single species, the common marmoset. The scientific debate over the correct classification revolves most particularly on the question of whether the

Things happen in captivity that are not possible in the wild: Two very rare specimens of different callithricid species, a buff-headed marmoset (left) and a white-eared marmoset (right), seem to get along very well living in intimate contact with each other.

Marmosets *(Callithrix)*

natural ranges of distribution of the marmosets of East Brazil overlap, and whether there is, accordingly, a natural mingling of the various forms. It is of some concern that such questions can no longer provide a clear answer because the white-eared marmoset, the buff-headed marmoset, and the white-fronted marmoset, which occur only south of the coastal rainforest, can no longer occupy wide areas of their original habitat.

Of the original coastal rainforest of southeastern Brazil, less than five percent has been preserved in its original state. The remaining stands of semi-intact primeval forest are broken up like islands, separated from each other by agricultural land and human settlements. The marmosets and tamarins living here share the fate of hundreds of animal and plant species that evolved in isolation from the great rainforest of the Amazon. By the year 2000, it will have been decided whether protection of nature will afford adequate habitat to this unique flora and fauna.

It is astonishing how little we knew until recently about the life of marmosets in the wild. Even in the case of the common marmoset, which is able to adapt still better than the pygmy marmosets to an environment modified by human hands, and might almost be called a camp follower of civilization, our knowledge has relied largely on anecdotal accounts, until very recently. The species, originally occurring only in northeastern Brazil, was introduced by humans in the south of eastern Brazil, where it has become so naturalized that one may observe it regularly in the park-like marginal districts of Rio de Janeiro.

Today we know that the common marmoset resembles the pygmy marmosets in many aspects of behavior, such as preference for tree sap. The availability of useful trees appears to exert a considerable influence on the size of individual group territories, which may occasionally extend over 12.5 acres (5 hectares).

Groups frequently numbering up to 13 members live here. When two neighbor groups encounter each other in areas of mutual overlap, there are sometimes exchanges between them. Serious conflict arises especially if members of one group encroach upon the tapped trees of the neighbors.

In this context, it seemed an interesting observation that the common marmosets, unlike the pygmy marmosets, regularly mark new tap sites with urine and a special secretion produced by glands in the animals' genital region. To do this, the monkeys will crouch with the posterior above the gnawed hole and rub the genital region repeatedly across the bark sidewise. They thus strip the secretion off and also add a few drops of urine.

Such behavior suggests the gnawed holes are marked to deter neighbor groups from using the same trees; observations in the wild soon showed that neighbor groups are not impressed. Tapped trees in use by three groups at the same time were even found. Also, there was no indication that the scent tended to be used for marking at the boundaries of the territory; instead, marmosets dropped their secretion in the middle of the territory at least as often as at the edge. These observations do not rule out territorial marking, but indicate that the secretions play at least an equally important role in olfactory communication within the group. We are much indebted to the behavior researcher Gisela Epple for a great deal of our knowledge of the possible functions of scents among the callithricids.

Epple studied not only the common marmoset but especially various species of tamarins. To ascertain the information content of scents, she made skillful use of the virtually automatic callithricid habid of immediately marking with secretion any objects that are newly introduced into their cages. The secretions thus collected from donor animals and placed on test plates were then presented to other individuals. When it was found

(Left) Black-penciled monkey, with black ear tufts. (Right) The similar, common marmoset, perhaps the best-known callithricid species, and formerly popular as a household pet. Its ear tufts are white and project outward from the head.

that the secretions of males provoked stronger reactions than those of females, it was to be concluded that evidently an animal's sex was identifiable by the secretion alone. Epple performed a large number of different experiments, and also had the secretions chemically analyzed. She learned that the secretions are not only species-specific; subspecies or races may produce scents of different composition, thus distinguishing the animals from each other. It was also learned that, in addition to the ability to discriminate sex, individuals apparently can be recognized by their scents. There were even clear indications that information concerning the social position of the transmitter was received by scent.

In observing callithricid groups, it is soon noticed that individuals of high rank mark more frequently than the those of lower rank. Also, since the common marmosets tend to apply the secretion to tap holes, it is virtually assured that low-ranking group members will catch the scent of the higher-ranking when feeding.

Although we do not yet know just what behavior patterns inhibit the reproduction of low-ranking females, the British biologist John Hearn and his associates have quite impressively shown the operation of the mechanism. Hearn placed several adult female common marmosets in a cage and observed their hormonal cycles by means of regular blood tests. Whereas the highest-ranking female maintained a regular cycle, in the other two individuals the cycle was so disturbed within a few days that ovulation was precluded. As soon as the highest-ranking individual was taken out of the group, the individuals of second rank promptly achieved a normal cycle.

There is probably some comparable mechanism that regulates the reproduction of all callithricids. An involvement of scents in suppressing reproduction of low-ranking females may have developed in all forms, for marking of the habitat with secretions is a habit of all marmosets and tamarins. The callithricids have glandular fields in the genital region, and also on the lower abdomen and near the sternum. Marking with the sternal glands, whereby the upper trunk is rubbed across a branch in the direction of locomotion, may also be observed in some species.

Our knowledge of the fabric of relationships within social groups of common marmosets, as well as other callithricids, has so far been derived essentially from behavior studies of animals in captivity. As a rule, stable social associations are formed only if the line is begun with a single pair and the descendants are left in the parental family. The behavior researcher Hartmut Rothe of Göttingen was thus able to build groups of up to 20 common marmosets on this basis. With the exception of the young, all group members were placed in two rankings separated by sex, and headed by the father and mother respectively.

(Right) Common marmoset. (Below) white-shouldered marmoset (left) and silvery marmoset (right).

The mechanism of suppression described above prevents the lower-ranking daughters – individuals other than the mother – from reproducing. Sexual relations between grown sons and the mother or between grown daughters and the father are prevented also by intervention of the parent of the opposite sex; sexual relations between siblings is probably prevented by an incest barrier as well.

Aggressive confrontations, given the separate rankings, take place almost exclusively between individuals of the same sex. Rothe discovered that the relationships of individual animals are intricately interwoven, and that the father especially, together with several sons, will form subgroups characterized by extensive contact, the stability of which is important to the coherence of the entire community. Rothe was able to demonstrate experimentally the special importance of the father. If he removed the mother, the group remained stable; but if he took out the father, the entire social fabric became quite unstable, with the most severe altercations taking place among the sons, resulting in the exclusion of several individuals from the group.

Rothe was able to catalogue a total of 240 different forms of behavior in his animals' social encounters. Because of their rapidity of motion and small physical size, the animals must be observed very closely. Contrary to a widespread impression these callithricids have a repertory of very subtle mimetic expressions, supplemented in the common marmoset by erection and movement of the ear tufts.

An especially impressive example is the "arched back" run of the marmosets, a behavior pattern equally pronounced among the lion tamarins. The animals run with back arched high, tripping along a branch, all body hair erect. This typical threatening and intimidating attitude makes the animals appear almost twice their actual size. No less remarkable is another form of marmoset display behavior, that is, "genital presentation." An individual will aim its posterior at another particular individual, extending the tail almost vertically upward. This displays the genitals, which are conspicuously colored in many species.

Our white-fronted marmosets practice genital presentation towards their keepers, and especially towards human strangers, but never towards members of their own family group. It would seem that human beings are equated to conspecifics who are not members of the group. Anthony Rylands, a Brazilian primatologist, observed this behavior regularly when two groups of the white-shouldered marmosets that he studied came into contact at the common boundary of their territories. This imposing behavior, in which both males and females participate, evidently reinforces claims to the possession of a particular territory, and obviates unnecessary, serious confrontations. After an interval of intimidation, each group goes its separate way.

Many callithricids thrive in captivity and become quite tame. Formerly, private individuals tended to keep single animals, but in modern scientific institutions efforts are made to create larger breeding groups, like that of these white-fronted marmosets.

Much as in the common marmosets, Rylands found overlapping territories of neighboring groups of white-shouldered marmosets, comprising between 4 and 13 individuals. Some of these territories, exceeding 25 acres (10 hectares) in area, were more than double the size of the common marmoset's territory. White-eared marmosets, in groups of 5 to 8 individuals, had similarly large territories. Both species were rarely – the white-eared marmoset, in fact, not at all – observed to consume tree sap. It seems possible that these two species, like the tamarins, stress the hunting of insects, and have deemphasized the consumption of fruits. This would account for their comparatively large territories.

The black-penciled marmosets and the buff-headed marmosets, according to our present knowledge, resemble the common marmoset to a large extent. For want of comprehensive field studies, there are few available observations concerning the ecology of the silvery marmoset and the white-fronted marmoset. If we judge by the observations at our callithricid center, we may assume food habits similar to those of the common marmoset for both species. At any rate, special tree trunks or branches set up on end in their cages will be gnawed all over, even though no sap can be obtained from the incisions.

In their reproductive behavior, the white-fronted marmosets and all other *Callithrix* marmosets resemble the pygmy marmosets. About 150 to 190 days after mating, our white-fronted females have their first babies, followed by births at intervals of 150 to 160 days. In contradistinction to their natural state, births do not peak seasonally. The young, weighing 0.7–1.2 oz (20–35 g), will voluntarily dismount from their carriers at an age of only 3 weeks, and at about 50 days they are weaned, for the most part.

After that they are only occasionally nursed or carried by other group members. In nearly half of all cases, the mothers would turn their burdens over to other group members from the first day of life. Several females would not allow their young to be picked up by others until three days after birth, others not until after a week, and until that time all offers of assistance, even the father's, were energetically rebuffed.

Once the first surrender has occurred, especially in families of five or more members, the elder siblings assume most of the burden of carrying the little ones. The mother will then limit herself to nursing. The father too, in larger family groups, is

Pied tamarin, carrying a baby. All tamarins in a group participate in child care.

usually restrained in carrying the young, but performs an important function as a relay in distributing the task among other group members, and he supervises the youngsters closely in critical phases of development, for example at the time of their first independent excursions into the branches in the cage.

Tamarins
(Genera *Saguinus* and *Leontopithecus*)

The genus of New World monkeys comprising the greatest wealth of forms is *Saguinus*, which together with the lion tamarin *Leontopithecus*, a genus of only one species, forms the group of the tamarins. Dozens of different scientific names for the tamarins pervaded the scholarly literature until, in the 1970s, the confusion was dispelled, notably by the zoologist Hershkovitz. Today we divide the genus *Saguinus* into 11 species comprising 28 subspecies.

Most tamarin species inhabit an enormous area, extending from the Amazonian region of Ecuador, Peru, and Bolivia in the West to the region of the mouth of the Amazon on the Atlantic. The Amazon and the Rio Madeira form a fairly rigid limit of distribution on the South. The red-handed tamarin has definitely transgressed this limit, only in the region of the mouth of the Amazon.

The Central Amazonian range of distribution is inhabited primarily by representatives of the moustached tamarins. Their most varied representative, the BROWN-HEADED, or SADDLE-BACK TAMARIN *(Saguinus fuscicollis)*, has given rise to no less than 14 different subspecies, which today live separated from each other by the tributaries of the Amazon. Although clawed monkeys are tolerably good swimmers, they shun the water under all circumstances, and thus, among the brown-headed tamarins, subspecies were formed that we would doubtless hardly recognize as representing one and the same species by their markings. The sole common trait is the dirty-white whiskers adorning the upper and lower lips. Some subspecies have a dark brown background coloration with variously marbled fur on the back, and another subspecies, *Saguinus fuscicollis melanoleucus,* is white over its entire body.

The moustached tamarins further include the BLACK-AND-RED TAMARIN *(Saguinus nigricollis)*, the MOUSTACHED TAMARIN *(S. mystax)*, the rather gaudy RED-BELLIED TAMARIN *(S. labiatus)*, the magnificiently whiskered EMPEROR TAMARIN *(S. imperator)*, and the previously mentioned RED-HANDED TAMARIN *(S. midas)*.

At home here as well is the little-known MOTTLE-FACED TAMARIN *(Saguinus inustus,* the sole representative of its subgroup. This form is characterized by a regular sprinkling of light, nonpigment spots on its otherwise dark face. Fairly recently, by the way, the mottle-faced tamarin was regarded as a natural hybrid (cross) of the black-and-red tamarin with the red-bellied tamarin. However, the comparative lack of facial hair characterizes it rather as an independent connecting link between the moustached tamarin and the bare-faced tamarin. The sole representative of the latter in the Central Amazonian region is the nearly bald, PIED TAMARIN *(S. bicolor);* much like the mottle-faced tamarin, it may be characterized by rather more irregular nonpigmented skin areas in the otherwise black face.

The other three representatives of the bare-faced tamarins – GEOFFROY'S TAMARIN *(Saguinus geoffroyi)*, the COTTON-TOP TAMARIN *(S. oedipus)*, and the WHITE-FOOTED TAMARIN *(S. leucopus)* today live separated from the other tamarins at the northernmost tip of South America, Geoffroy's tamarin is the only callithricid with a range that extends into Central America, about as far as Costa Rica.

We know very little of the habits of the WHITE-FOOTED TAMARIN *(S. leucopus)*, and it is doubtful whether we shall learn anything more. Its range on the Rio Magdalena in the north of Colombia is very small, and has already been largely destroyed by timbering.

NEW WORLD PRIMATES

Much like the white-footed tamarin, the other species are quite adaptable forest dwellers, inhabiting very diverse types of forests. They are rather rarely found either in the topmost level above 80–96 ft (25–30 m), or as frequently as the marmosets in the very densely grown forest margins.

Eating chiefly fruits and insects, the tamarins are distinctly more dependent on a varied diet than most marmoset species that live on tree sap. The rather large territories of up to 125 acres (50 hectares) may be a direct consequence.

Territories 75 to 125 acres (30 to 50 hectares) in size, but with considerable overlap, were discovered by the Japanese primatologist Kosei Izawa when he observed the black-and-red tamarin in Colombia. In his field of study, he found a total of 10 distinct groups, averaging six to seven members. The course of their day was very substantially determined by several hunts, lasting up to two hours, for exceptionally large grasshoppers. The monkeys traversed several hundred yards in each instance, and quite often abandoned the trees. On the ground, they would track the insects, but seldom stray farther than about 10 ft (3 m) from the tree trunks.

An especially interesting observation of Izawa's is that the groups cultivated regular relationships, which must be called amicable, with neighboring groups. By the same token, there were no signs of territorial behavior. The members of different groups would wander for a time through the area of territorial overlap, eat together, and sometimes associate intensively with each other. It was very noticeable that the high-ranking animals of different groups kept their distance from one another.

Occasionally, more than two groups were involved in these encounters, in which case up to 40 individuals were thus assembled. Still, it was clear that some groups seemed to avoid meeting under any circumstances.

Amalgamations of groups, as among the black-and-red tamarins, seem to occur in other tamarin species as well. There have been observations of

Representative of a rare subspecies, the red-handed tamarin *(Saguinus midas niger)*.

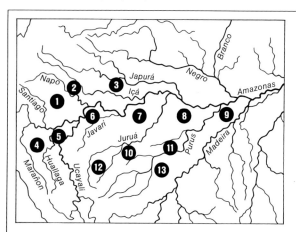

An example of the terrain-related formation of subspecies: the brown-headed, or saddle-back tamarin *(Saguinus fuscicollis)*.

① *S. f. lagonotus:* gray-brown coat on back, red-brown limbs, with black hair on feet and hands. ② *S. f. tripartitus:* gray-brown coat on back, extending to the legs; light brown shoulders and arms, white blaze on the forehead. ③ *S. f. fuscus:* marking like *S. f. lagonotus*, although the rust-brown areas of *S. f. lagonotus* are coffee-colored in *S. f. fuscus;* black fur on head. ④ *S. f. leucogenus:* gray-brown coat on back; red-brown legs with black feet, anterior back, and arms. ⑤ *S. f. illigeri:* similar to *S. f. fuscus,* but brown fur areas have a reddish tinge; black feet and hands. ⑥ *S. f. nigrifrons:* similar to *S. f. fuscus,* but brown fur on head, and black hands and feet. ⑦ *S. f. fuscicollis:* dark brown, with yellow-brown coat on back. ⑧ *S. f. avilapiresi:* similar to *S. f. fuscus,* but all brown parts are deep dark brown; black hands. ⑨ *S. f. crandalli:* red to light brown legs, otherwise, with some mingling of posterior coat of back, silvery gray throughout; silvery gray tail. ⑩ *S. f. melanoleucus:* fur almost all white. ⑪ *S. f. cruzlimai:* except dark brown tail and mixed coat on back, red to light brown throughout; white headband. ⑫ *S. f. acrensis:* coloration similar to *S. f. crandalli;* ochre to light brown legs; base of tail beige. ⑬ *S. f. weddelli:* coloration like *S. f. leucogenys,* but medium brown legs, and white headband.

this kind, for example, in the case of brown-headed tamarins, Geoffroy's tamarin, and red-handed tamarins, likewise leading to brief assemblies of more than 25 individuals. At the same time, in some of the same species, clear signs of strict territorial behavior were repeatedly found. We have as yet no conclusive explanation for the differentially peaceable behavior of neighbor groups towards each other, or corresponding differences in various ranges of distribution.

Details of relationships among the animals in the groups became accessible to field investigators only when they began individual tagging of the callithricids, which are quite difficult to observe in the wild. The American biologist John Terborgh tagged a total of 50 brown-headed tamarins with small neck chains and observed their behavior, with his associates, for several years. First it was learned that the individual groups generally contain several sexually mature individuals, in particular, usually two adult males. This finding became truly remarkable when it emerged that the manifestly highest-ranking female in such a group would mate with several males. We happen to know from brown-headed tamarins in captivity that two adult males may be kept together with a sexually mature female for some time. Nevertheless, there is much to indicate that the callithricids are typically monogamous primates, that is, a species that practices monogamy rather rigidly. This is just what the brown-headed tamarins in Peru did not do; indeed, neither of the sexually mature males showed any tendency to interfere with the other's sexual relations with the female. Terborgh's further observations yielded an apparently quite plausible explanation of this social system, extraordinary not only among the primates. These observations related to the rearing of the young, in which the sexually mature males especially participate, along with the younger group members who act as porters.

Animal behavior, at least, is only rarely quite altruistic, and a callithricid lacking some assurance of its paternity would be little inclined to transport the offspring of some other male. Perhaps evolution has given the females a chance to utilize just this uncertainty of fatherhood. Since she mates with both males, each can presume to be the father of the young. Consequently, both males as experienced baby carriers will invest in the rearing of the offspring, thereby doubtless enhancing the probability of their being successfully brought up.

This context will also fit Terborgh's observation that territories left vacant are usually occupied jointly by two males, who may derive from different groups. If the animals mate with a female recruit, it is assured that several experienced child-raisers will be on hand from the beginning. Terborgh's findings have recently been corroborated by similar observations of moustached tamarins. Consequently, at least for most tamarin species, we should accept the possibility that in the wild, straight family groups consisting of mother,

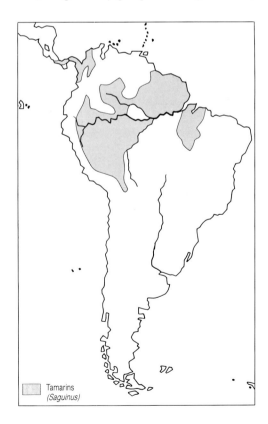

Tamarins (*Saguinus*)

father, and offspring may be the exception. This is argued also by the numerous observations according to which individual animals will change groups rather frequently, and male or female, sexually mature or immature individuals may participate.

Some quite extraordinary behavior was found in the case of the emperor tamarin. Groups of these animals inhabited the same territory with one group each of brown-headed tamarins for several years; not only were their various activities highly coordinated, but their territory was jointly defended against neighboring groups of similar composition. Individuals of the two species do not continually traverse the territory together, but they maintain voice contact by prolonged sounds. Such alliances between groups of different species have been observed for other tamarin species as well, as between the red-bellied and brown-headed tamarins, or the moustached and the brown-headed tamarins. What advantages the different species thereby gain has not been definitely ascertained. At least we know from field studies that the two species are not competitors for food to the extent that one might at first suppose. Whereas the lion tamarin finds its diet of insects chiefly in the foliage of the trees and captures it by lightning attacks, the brown-headed tamarin deliberately searches the trunks and branches, especially the knot holes, for insects, as well as lizards. Similar observations have been made for various groups of birds, such as titmice.

It has already been mentioned that the callithricids often behave quite intolerantly towards unrelated conspecifics in captivity. This applies especially to the cotton-top tamarins. In our colony, we therefore kept these animals almost exclusively in family groups, which we allowed to increase with a little interference as possible over the years. In so doing, we gained valuable insights into the social behavior of these animals, supplementing our knowledge of their life in the wild.

The cotton-top tamarins, like the marmosets, live in rankings separated by sex. To settle disputes about rank, they have a pattern of behavior that we call "heckling." The challenger first stares at a partner of the same sex from a distance, uttering long drawn-out, very melodious sounds. He will keep up this "heckling" until the other responds in kind. The result is a sort of duet, which may last for several minutes. The animals' fur is on end. In the further course of events, they approach each other, with a somewhat stilted gait, closer and closer, until finally direct bodily contact takes place. At that point, the two touch each other only with their open mouths, which they advance towards each other. They hold that position almost motionless for a short time until seemingly one of the two can no longer stand the stress-laden situation and takes flight. "Heckling," in other words, is a sort of psychological war of comment, an altercation in which the stronger is identified without danger of injury. To be sure, when this behavior fails to establish relations between two individuals, the cotton-top tamarins will enter into serious, injurious combat, ultimately expelling one individual from the group.

Closer study of such combat has yielded highly interesting results. In more than 50 cotton-top families, not even once did we observe the father of a group to be involved in such a confrontation. This means that to a large extent the sons decide among themselves who may remain in the group and who must leave. It is otherwise with the mothers, who were directly involved in more than half of the serious contests, and accordingly initiated the expulsion of certain descendants from the group. The outcome of the occasional serious

Emperor tamarin. The species owes its name to the facial hair of the Emperor William II, although the points of his moustaches were directed upward. The little monkey had been discovered only in 1907, and when the taxidermists saw dead specimens, they mistakenly twirled the long, drooping moustaches upward.

tamarins is quite harmonious. The mother is usually indulgent even towards her grown daughters, and has good reason to be so, for in larger groups the older children do the main job of carrying the young.

The length of time the young cotton-tops are carried, 70 to 80 days, is a good deal longer than among the marmosets. When single offspring are raised, especially in groups with an elderly and very tolerant father, it may sometimes be 100 days.

In studying the young, our associate Gerlinde Mika found that important steps towards weaning are initiated by the youngsters themselves. When young cotton-tops about 18 to 24 days old make their first excursions into the branches, as a rule they do so of their own accord, and not, as the seemingly anxious cries continually uttered would suggest, upon compulsion by their carriers. The cries of the young perhaps serve rather to attract the attention of the family members to themselves and prepare them to lend aid if necessary. Other group members consistently refuse to carry the young only when their motor skills are so far developed that they can move surely among the branches without assistance.

(Left) Cotton-top tamarins "heckling." The challenger stares at his opponent until he gets a response (top). The two stalk towards each other (center) and physical contact occurs (bottom), until the less resolute can stand the tension no longer. (Right) Cotton-top monkey. These monkeys are named (in German) for the celebrated composer Franz Liszt, because of their shock of white hair.

conflicts was an average group size of 7 to 8 individuals, with a balanced sex ratio and a staggered age composition. This agrees surprisingly well with the structure of groups in the wild, where on the other hand the make up is determined in part by immigration and emigration of individuals.

All the same, in general the group life of these

MARMOSETS AND TAMARINS

Marmosets and Tamarins (Callitrichidae)

Nomenclatur English common name Scientific name French German	Approximate Size Body length Tail length Weight	Distinguishing Features	Reproduction Gestation period Young per birth Weight at birth
Pygmy marmoset *Cebuella pygmaea* Ouistiti mignon Zwergseidenaffe	4.8–6 in.; 12–15 cm 6.8–9.2 in.; 17–23 cm 3.5–4.2 oz; 100–120 g	Smallest of the callitrichid family; fur uniformly yellow; tail indistinctly dark-ringed	About 140 days 1–2, rarely 3 About 0.5 oz; 15 g
Common marmoset *Callithrix jacchus* Ouistiti à toupet blanc Weißbüschelaffe	7.6–8.8 in.; 19–22 cm 11.8–14 in.; 29.5–35 cm 10.1–12.8 oz; 300–360 g	Strikingly erect, white ear tufts; white blaze on forehead; head fur generally dark brown; gray-brown back fur with light transverse striping; pronounced transverse striping of tail	140–145 days 1–2, rarely 3; triplet births frequent in captivity About 1.1 oz; 30 g
Silvery marmoset *Callithrix argentata*, 3 subspecies Ouistiti mélanure Silberaffe	7.2–11.2 in.; 18–28 cm 10.6–15.2 in.; 26.5–38 cm 10.7–12.8 oz; 300–360 g	Body white to silvery gray, except for black tail; *C. a. argentata* naked ears and flesh-colored face; *C. a. leucippe* fur very light, tail beige; *C. a. melanura* fur nearly light brown, face dark	140–145 days 1–2, rarely 3 About 1.1 oz; 30 g
White-shouldered marmoset *Callithrix humeralifer*, 3 subspecies Ouistiti à camail Weißschulterseidenaffe	8–12 in.; 20–30 cm 12–16 in.; 30–40 cm 10.7–12.8 oz; 300–360 g	Large, silvery ear tufts; general body color brown, anterior back silver and black mixed, increasingly dark towards tail with streamer-like white stripes; predominantly dark tail, with light striping	140–145 days 1–2, rarely 3 About 1.1 oz; 30 g
White-eared marmoset *Callithrix aurita* Ouistitit oreillard Weißohrseidenaffe	8.8–10 in.; 22–25 cm 12–14 in.; 30–35 cm To over 14.2 oz; 400 g	White ear tufts; whitish-yellow stripe traverses black-brown head fur from a white blaze on the forehead; back predominantly dark gray-brown, especially around arms, with yellow-beige interspersed	140–145 days 1–2, rarely 3 About 1.1 oz; 30 g
Buff-headed marmoset *Callithrix flaviceps* Ouistiti à tête jaune Gelbkopfbüschelaffe	8.8–10 in.; 22–25 cm 12–14 in.; 30–35 cm 10.7–12.8 oz; 300–360 g	White-ocher fur in wreath-like arrangement on ear and cheek region border a light face of the same hue; ocher back fur finely striped transversely in dark gray to black; generally darker tail transversely striped as well	140–145 days 1–2, rarely 3 About 1.1 oz; 30 g
White-fronted marmoset *Callithrix geoffroyi* Ouistiti à tête blanche Weißgesichtseidenaffe	8.8–10 in.; 22–25 cm 12–14 in.; 30–35 cm 10.7–14.2 oz; 300–400 g	Black ear tufts border white face; back fur more or less conspicuously marbled black and golden-yellow; gray-white and black striped tail	140–145 days 1–2, rarely 3; triplet births more frequent in captivity About 1.1 oz; 32 g
Black-pencilled marmoset *Callithrix penicillata*, 2 subspecies Ouistiti à pinceau noir Schwarzpinselaffe	7.6–9 in.; 19–22.5 cm 10.8–13.4 in.; 27–33.5 cm 10.7–12.8 oz; 300–360 g	General coloring similar to that of common marmoset but black-brown ear tufts *(C. p. penicillate)*; whitish gray-brown crown of head and distinctly white cheek fur *(C. p. kuhlii)*	140–145 days 1–2, rarely 3; triplet births more frequent in captivity About 1.1 oz; 30 g
Black-and-red-tamarin *Saguinus nigricollis*, 2 subspecies Tamarin rouge et noir Schwarzrückentamarin	8.4–10 in.; 21–25 cm 12.4–14.4 in.; 31–36 cm 12.5–16 oz; 350–450 g	Dirty white, short-haired beard around mouth and nose; head, nape, arms, and anterior back dark to black; posterior back striped red and black lengthwise; tail black, red at base *(S. n. nigricollis)*; *S.n. graellsii* interspersed with gray on anterior back, posterior back more vivid brown	140–145 days 1–2, rarely 3 About 1.6 oz; 45 g
Brown-headed tamarin; saddleback tamarin *Saguinus fuscicollis*, 14 subspecies Tamarin à tête brune Braunrückentamarin	7–10.8 in.; 17.5–27 cm 10–15.2 in.; 25–38 cm 10.7–13.5 oz; 300–380 g	White fringe beard; fur predominantly dark brown with fine light brown-beige marbling on back, dark brown-black tail *(S. f. fuscicollis)*; some subspecies vary greatly in coloration	140–145 days 1–2, rarely 3 About 1.4 oz; 40 g
Moustached tamarin *Saguinus mystax*, 3 subspecies Tamarin à moustaches Schnurrbarttamarin	9.4–11.2 in.; 23.5–28 cm 14.6–17.4 in.; 36.5–43.5 cm 12.5–16 oz; 350–450 g	Broad white area of fur around nose and mouth, extending laterally far across the cheeks; otherwise black to dark brown *(S.m. mystax)*; *S. m. pileatus* forehead and crown rust-red; *S.m. pluto* back hair wavier	140–145 days 1–2, rarely 3 About 1.6 oz; 45 g
Red-bellied tamarin *Saguinus labiatus*, 2 subspecies Tamarin labié Rotbauchtamarin	9.4–12 in.; 23.5–30 cm 13.8–16.4 in.; 34.5–41 cm 12.5–16 oz; 350–450 g	Only a small strip of white moustache; black head hair; brown-black back; trunk reddish-orange on under side *(S. l.labiatus)*; *S. l. tomasi* upper part of breast black	140–145 days 1–2, rarely 3 About 1.6 oz; 45 g
Emperor tamarin *Saguinus imperator* Tamarin empereur Kaiserschnurrbarttamarin	9.2–10.4 in.; 23–26 cm 14–16.6 in.; 35–41.5 cm 10.7–14.2 oz; 300–400 g	Long drawn out moustaches; black hands, feet, and cap; back gray, mixed with fine yellow hair; rust-red scattering of color on breast; tail black, except for rust-brown base *(S. i. imperator)*; *S. i. subgrisiscens* has ragged whiskers on upper and lower lip	140–145 days 1–2, rarely 3 About 1.2 oz; 35 g

COMPARISON OF SPECIES

Life Cycle Weaning Sexual maturity Life span	Food	Enemies	Habit and Habitat	Occurrence
Age 2 months Age 18–24 months Over 10 years	Chiefly tree sap, also insects and some fruits	Small cats, birds of prey, snakes	Active by day; found in groups of 2–15 individuals in various kinds of forests; at lower and middle tree height; preference for thickly grown tree margins; territory size rarely more than 2.5 acres (1 hectare)	Still abundant, but habitat dwindling
Age 2 months Age 14–18 months Over 10 years	Chiefly tree sap, insects, spiders, fruits, flowers, nectar, also small lizards, birds' eggs, nestlings, frogs	Same as pygmy marmoset	Active by day; found in groups (2–13 individuals); in wide variety of forest types, even plantations; originally only in the extreme northeast of Brazil, but introduced by humans into the south of the East Brazilian coastal rain forest; territory size 2.5–12.5 acres (1–5 hectares)	Original habitat in northeast Brazil destroyed; abundant in newly opened habitat
Much like common marmoset	Much like common marmoset	Same as pygmy marmoset	Active by day; found in groups (2–8 individuals) in both evergreen primary and secondary hill forests and, in Bolivia, in forests with marked dry seasons	Subspecies *C. a. leucippe* threatened in its small range; other 2 subspecies still fairly abundant
Much like common marmoset	Much like common marmoset	Same as pygmy marmoset	Active by day; found in groups (2–15 individuals) in primary and secondary forests of Central Brazil, south of the Amazon; territory size over 25 acres (10 hectares)	Endangered by severe threat to habitat
Much like common marmoset	Like common marmoset, but possibly tree sap less important	Same as pygmy marmoset	Active by day; found in groups (2–8 individuals) in primary and secondary forests of the southeast Brazilian coast (states of São Paolo, Minas Gerais, and Rio de Janeiro); territory size about 25 acres (10 hectares)	Directly threatened with extinction; natural habitat largely destroyed
Much like common marmoset	Much like common marmoset	Same as pygmy marmoset	Active by day; found in groups (2–8 individuals); like most callitrichid species, preference for marginal forest along rivers, now also along roads; in hill forests of the Brazilian coastal state of Espírito Santo, also in neighboring state of Minas Gerais	Very small range of distribution, already widely cleared; threatened with immediate extinction
Much like common marmoset	Much like common marmoset	Same as pygmy marmoset	Active by day; no accurate knowledge of habits in nature; preference for bushy margins of rain forest on water courses; range similar to that of buff-headed marmoset, but only below elevations of 1300 ft (400 m)	Still considered quite abundant in the early 1980s, but now facing extinction
Much like common marmoset	Tree sap, like pygmy marmoset, otherwise like common marmoset	Same as pygmy marmoset	Active by day; found in groups (2–9 individuals); comparatively wide range of distribution in western continuation of Brazilian coastal rain forst to about the Mato Grosso state line; *C.p. kuhlii* south of Bahia state; territory size over 5 acres (2 hectares)	*C. p. penicillata* still abundant; *C. p. kuhlii* endangered
Age 2–3 months Age 16–20 months Over 10 years	Insects, spiders, fruits, tree sap; also flowers, leaves, nectar, lizards, birds' eggs, nestlings, tree frogs	Same as pygmy marmoset	Active by day; found in groups (2–12 individuals); in primary and secondary forests, also in more altered habitats near human settlements; distribution from eastern Ecuador to junction of Peru, Colombia, and Brazil; territory size up to 125 acres (50 hectares)	Both subspecies still fairly abundant
Much like black-and-red tamarin	Much like black-and-red tamarin	Same as pygmy marmoset	Active by day; found in groups (2–17 individuals); habitats similar to those of black-and-red tamarin; largest range of all callitrichids; from the foothills of the cordilleras to central Amazonia, from south Colombia to central Bolivia; territory size to 87.5 acres (35 hectares)	All subspecies probably still fairly abundant
Much like black-and-red tamarin	Much like black-and-red tamarin	Same as pygmy marmoset	Active by day; found in groups (2–16 individuals); apparent preference for primary forests or undisturbed old secondary forests; range of distribution in northeastern Peru and a large adjoining territory in northwestern Brazil; territory size up to 87.5 acres (35 hectares)	None of the subspecies are known to be endangered
Much like black-and-red tamarin	Much like black-and-red tamarin	Same as pygmy marmoset	Active by day; found in groups (2–13 individuals); habitat similar to moustached tamarin; range of distribution a strip from the middle west of Brazil to the foothills of the cordilleras near the Peru-Bolivia border; territory size up to 87.5 acres (35 hectares)	Both subspecies still fairly abundant
Much like black-and-red tamarin	Much like black-and-red tamarin	Same as pygmy marmoset	Active by day; found in groups (2–8 individuals, possibly up to 15); in various forms of light, dry-bottom forests to densely-grown forest margins in seasonally inundated regions; territory size up to 100 acres (40 hectars)	Regarded as endangered or threatend in Brazil and Peru

Nomenclature English common name Scientific name French German	Approximate Size Body length Tail length Weight	Distinguishing Features	Reproduction Gestation period Young per birth Weight at birth
Red-handed tamarin *Saguinus midas*, 2 subspecies Tamarin aux mains rousses Rothandtamarin	8.2–11.2 in.; 20.5–28 cm 12.6–17.6 in.; 31.5–44 cm 14.2–19.6 oz; 400–550 g	Black coloration all over except for sharply contrasting red-gold to orange (*S. m. midas*) or dark brown (*S. m. niger*) hands and feet	140–145 days 1–2, rarely 3 About 1.6 oz; 45 g
Mottle-faced tamarin *Saguinus inustus* Tamarin à face marmoréenne Marmorgesichtstamarin	8.4–10.4 in.; 21–26 cm 9.2–16.4 in.; 23–41 cm 12.5–16 oz; 350–450 g	Black face sprinkled with unpigmented light spots; face hair largely absent; body hair nearly black all over	140–145 days 1–2, rarely 3 About 1.6 oz; 45 g
Pied tamarin *Saguinus bicolor*, 3 subspecies Tamarin bicolore Mantelaffe	8.4–11.2 in.; 21–28 cm 13.4–16.8 in.; 33.5–42 cm 12.5–16 oz; 350–450 g	Front of head, including ears, hairless, with black, partly flesh-colored, freckled skin; back white to behind armpits; rest of back ocher; upper side of tail sometimes black (*S. b. bicolor*); *S. b. martinsi* dark brown back, lighter shoulders and under parts; *S. b. ochraceus* pale all over	140–145 days 1–2, rarely 3 About 1.6 oz; 45 g
Cotton-top tamarin *Saguinus oedipus* Pinché Lisztaffe	8–11.6 in.; 20–29 cm 12.4–16.8 in.; 31–42 cm 12.5–16 oz; 350–450 g	Long, mane pure white, tapering more or less to a point of the forehead; under parts of trunk, arms, and shanks white; back dark brown; upper half of tail rust-red to brown, lower half black	140–145 days 1–2, rarely 3 About 1.6 oz; 40 g
Geoffroy's tamarin *Saguinus geoffroyi* Pinché de Geoffroy Geoffroy-Perückenaffe	8–11.6 in.; 20–29 cm 12.4–16.8 in.; 31–42 cm 12.5–16 oz; 350–450 g	Coloration similar to cotton-top tamarin, but stronger dirty-white marbling on black; head adorned only by short, strip of white hair	140–145 days 1–2, rarely 3 About 1.4 oz; 40 g
White-footed tamarin *Saguinus leucopus* Pinché aux pieds blancs Weißfußaffe	8.8–10.4 in.; 22–26 cm 14–16.8 in.; 35–42 cm 12.5–16 oz; 350–450 g	Bearing short silvery hair; on front of head and crown; back of head brown; arms and legs white, back fur brown-caramel, tail dark to black	140–145 days 1–2, rarely 3 About 1.4 oz; 40 g
Lion tamarin *Leontopithecus rosalia*, 3 subspecies Singe-lion Löwenaffe	8–13.4 in.; 20–33.5 cm 12.6–16 in.; 31.5–40 cm 13.5–25 oz; 380–700 g	Golden-yellow fur with occasional lighter or darker areas, especially in tail region (*L. r. rosalia*); *L. r. chrysomelas* black with reddish-yellow face and arms; *L. r. chrysopygus* dark-black, shanks light reddish-brown up to buttocks and to base of tail	132–145 days 1–2, rarely 3 About 1.9 oz; 55 g

The comparatively long period of time for which the young cotton-tops are carried is not typical of all tamarins. The pied tamarins, for example, are carried only for a full 60 days, and their rate of development is more like that of the marmosets.

In the wild, the pied tamarins are seriously threatened with extinction, and are still largely unresearched. One adverse factor is their comparatively small range of distribution, which unfortunately coincides in large part with the steadily growing city of Manau.

The LION TAMARIN *(Leontopithecus rosalia)*, the only representative of the tamarins in the coastal rainforest of southeastern Brazil, has almost reached the critical point on the brink of extinction. All three subspecies of this largest and perhaps handsomest callithricid are limited in occurrence to a few remnant areas in the Brazilian East Coast states of Rio de Janeiro, Bahia, and São Paulo. Populations of each of the three subspecies were under 500 individuals at the beginning of the 1980s. The lion tamarin strongly resembles the other tamarins in its feeding and other habits, but

COMPARISON OF SPECIES

Life Cycle Weaning Sexual maturity Life span	Food	Enemies	Habit and Habitat	Occurrence
Much like black-and-red tamarin	Much like black-and-red tamarin	Same as pygmy marmoset	Active by day; found in groups (2–6 individuals); very large range of distribution in the northeast of South America, from Guiana in the north far into the Brazilian state of Pará; westward, as far as the junction of Colombia, Brazil, and Venezuela; territory size about 25 acres (10 hectares)	Generally still very abundant; regionally threatened by destruction of habitat
Much like black-and-red tamarin	Probably like black-and-red tamarin	Same as pygmy marmoset	Active by day; no accurate knowledge of habits in nature; range of distribution from northwestern Brazil as far as the Río Guaviare in southeast Colombia	Survival in the field probably still quite secure
Much like black-and-red tamarin	Much like black-and-red tamarin	Same as pygmy marmoset	Active by day; no details concerning natural habits; appears to have some limited ability to survive even in permanently altered habitats; quite small range of distribution on the north of the Amazon, also in the west of Pará state in Brazil	Ranked as threatened because of small range of distribution in a habitat greatly stressed by human occupation
Much like black-and-red tamarin	Much like black-and-red tamarin	Same as pygmy marmoset	Active by day; found in groups (2–13 individuals); rarely found deep in original primary forests; preference for secondary forests and forest margins; range of distribution in northwestern Colombia; territory size over 25 acres (10 hectares)	Threatened with extinction by extensive destruction of habitat
Much like black-and-red tamarin	Much like black-and-red tamarin	Same as pygmy marmoset	Active by day; found in groups (2–19 individuals); habitat similar to cotton-top tamarin; northernmost range of all callotrichids, from frontier country of Colombia and Panama to Costa Rica; territory size up to 80 acres (32 hectares)	As yet comparatively abundant, at least in Panama
Much like black-and-red tamarin	Much like black-and-red tamarin	Same as pygmy marmoset	Active by day; found in groups (2–15 individuals); preference for terrains on edge of primary forests, but will not avoid secondary forests; comparatively small range of distribution in northern Colombia between the Río Magdalena and the Río Cauca	Threatened with extinction by deforestation
Much like black-and-red tamarin	Much like black-and-red tamarin	Same as pygmy marmoset	Active by day; found in groups (2–8 individuals); originally in primary forests, but apparently in secondary forests also; minute remaining ranges in Brazilian state of Rio de Janeiro (*L. r. rosalia*), in southwest of Bahia state (*L. r. chrysomelas*) and in two wooded areas of São Paulo state (*L. r. chrysopygus*)	All subspecies threatened with extinction by destruction of habitats

seems to be even more dependent than they are on the intact primeval forest.

In view of the ominous decline of the lion tamarins, a model breeding project was founded under the direction of zoologist Devra Kleiman of the National Zoo in Washington State. Early in the 1970s, when fewer than 100 individuals of the golden lion tamarin subspecies *Leontopithecus rosalia rosalia* were living in captivity, she created an international studbook in which all these animals were registered. Their breeding was coordinated worldwide, but for several years failed to progress.

Today, there are already more lion tamarins in captivity than in the wild, where their population has declined to fewer than 500 individuals. The precious nurslings must be fed in the same variety to which they are accustomed in nature. Their main diet consists of insects and larvae, but a small mouse will not be refused.

▷ The lion tamarins, with their gorgeous manes which bristle when they are excited, are considered the handsomest callithricids. Unfortunately this species is threatened with extinction in its South American home.

OLD WORLD PRIMATES

(Opposite page) Old World monkeys are also referred to as "narrow-nose" monkeys, because the narrow nasal septum - plainly seen in the photograph - is common to them all. Here are two male barbary apes sitting together and studying the baby's genitalia. In that context, they are able to approach each other peaceably.

It was only when details of the special habits of the lion tamarin had been learned through scientific research and applied to its care and keeping that the number of individuals in captivity began to increase noticeably. By the early 1980s the number of lion tamarins being kept in captivity outnumbered those still at large. The survival of the golden lion tamarin in captivity is assured today. Similar projects have meanwhile been started not only for the other two subspecies of the lion tamarin but also for some other very severely threatened callithricids.

The long-range objective of such projects is of course to release animals raised in captivity for the support of wild populations that have become too small, or even to resettle habitats in which the animals have already died out. The breeding program to preserve the golden lion tamarin had reached that stage by the mid-1980s. In cooperation with the exemplary primate center in Rio de Janeiro, Devra Kleiman and her associates have returned a about a dozen lion tamarins to Brazil. In preparation for release, the monkeys had to go through an extensive training program, in which they had to relearn the art of getting their food independently. After a quarantine, the animals were then gradually released in the newly created Poco das Antas preserve, located 60 mi (100 km) from Rio de Janeiro. The release was successful; a respectable proportion of the zoo-born animals became acclimated to their new, natural habitat, and within a year the new generation was beginning to be born.

The lion tamarin is quite a skillful trunk climber, although it prefers to run and leap in a horizontal direction.

OLD WORLD PRIMATES

Category
INFRAORDER

Classification: An infraorder in the Order Primates, comprising two superfamilies, Cercopithecoidea and Hominoidea, with a total of four families. Only the first mentioned superfamily will be dealt with here; synopses of the three families of Hominoidea – the Hylobatidae (gibbons), the Pongidae (large apes) and the Hominidae (humans) – are given later.

Superfamily Cercopithecoidea (Old World monkeys)
Family Cercopithecidae (Old World monkeys)
Subfamily Cercopithecinae (macaques and allies)
Subfamily Colobinae (langurs and colobi)

CERCOPITHECINAE (9 genera comprising 45 species)
Body length: 10–44 in.; 25–110 cm
Tail length: 0–3.5 ft; 0–105 cm
Weight: 1.5–121 lb; 0.7 to 50 kg

Distinguishing features: Narrow nasal septum ("Catarrhini"); wide variation in body build; larger, sturdy terrestrial forms have a moderately or completely reduced tail, and smaller, more slender, chiefly arboreal forms have a longer tail; cheek pouches; single stomach; ischial callosities; posterior of some species is brightly colored, generally red; broad hands and feet; plantigrade locomotion, or wrists and ankles slightly elevated; thumbs are well developed and opposable to fingers, permitting a precision grip; menstrual swellings in females of many species living in multi-male or harem groups.
Reproduction: Menstrual cycle 26–35 days; fertile throughout the year or in a limited season only; gestation 5–6 months; one young per birth, twins are rare.
Life cycle: Weaning at about 3–12 months; sexual maturity at 3–5 years, males full-grown some years later; life span 25–45 years.
Food: Diverse; vegetable (fruits, leaves, seeds, nuts, flowers, buds, sprouts, roots, mushrooms, husks, grass, rosin) or animal (insects, spiders, crustaceans and other invertebrates, and small vertebrates, such as lizards, bird and birds' eggs, rodents, young antelopes).
Habit and habitat: Highly diversified; terrestrial and arboreal; live in dense, moist tropical forests, dry and mountain forests, savannas, rocky terrain, etc.; social; group size generally 20–60 individuals; multi-male, single-male and harem groups; territory size varies widely (0.004–36 mi^2 or 0.01–100 km^2).

COLOBINAE (5 genera comprising 24 species)
Body length: 1.5–43 ft; 45–108 cm
Tail length: 3.3–3.6 ft; 100–109 cm
Weight: 9–52 lb; 4–23.5 kg
Distinguishing features: Narrow nasal septum; reduced thumb; compartmented stomach with abundant bacterial flora for digestion of cellulose in diet; molar teeth with cusps connected by transverse crests (bilophodontia); no genital swellings (except *Procolobus* species); conspicu-

Catarrhini
Singes de l'Ancien Monde — FRENCH
Altweltaffen — GERMAN

ous infant coat of newborn, with fur coloration in sharp contrast with that of adults, in the context of an "infant transfer" system, that is, babies passed around among group members (except *Procolobus* species); far-reaching cries and associated intimidating leaps of males, early-morning calling "concerts" in some species.

Reproduction: Seasonal or throughout the year; gestation about 200 days; reproductive behavior not as yet much studied, but details are available for some species; usually one young per birth, twins are rare.

Life cycle: Not much reliable information; weaning, at 7–15 months where known; sexual maturity usually at 3–6 years; life span probably about 20 years.

Food: Vegetable, with highly diverse proportions of leaves, flowers and fruits; only occasionally insects, worms, or birds' eggs.

Habit and habitat: Predominantly arboreal, some species more or less frequently met with on the ground; in nearly all zones of climate and vegetation: rainforests, mangrove, gallery, coastal, and mountain forests, bush, savannas, etc.; monogamous, single-male and multi-male groups, as well as all-male groups; in many species, male replacement (expulsion of grown male by male from an outside group), often attended by infanticide.

Normal and defense formations of baboon troupe. Especially in the case of terrestrial monkeys, for example the baboons, who have many enemies to contend with, social ranking has a vital function. It is dictated in part by the great difference in size between males and females. But among the males also, there are dominat and submissive individuals. As a group of baboons travels through open grassland, the subordinate males form the vanguard and the rear, to protect the company, and the dominant males are in the center, keeping order. If the group encounters an enemy, the dominant males assume a posture of defense on the periphery of the group, and the other members gain time to retreat.

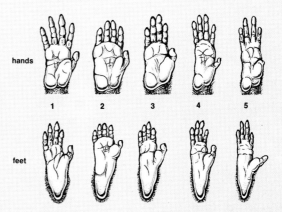

Hands and feet of Old World monkeys are similar. Both are grasping organs, eminently adapted to life in the treetops. Note especially the broad flat palm, the long fingers and the outspread thumb. There are of course adaptations to differing conditions of life, for example the foot-like hand of the baboon (2), adapted to travel on the ground, or the hand of the colobus (5) with reduced thumb, adapted to hanging and swinging in the branches. 1. Macaque; 3. Guenon; 4. Langur.

SUBFAMILY Cercopithecinae

Macaques and Allies

by Jan A.R.A.M. van Hooff

Only one simian species, the Japanese red-faced macaque (above), has adapted to cool northern latitudes with deep Winter snow, and only comparatively few species live a terrestrial life in open country, such as the anubis baboon (below), foraging through the wide African savannas. But they too, after the day's march, seek out trees or rock walls, in which to pass the night undisturbed.

Since ancient times, the human-like appearance of this group of animals has excited scientific curiosity. Galen (129–199), the great physician of antiquity, dissected barbary apes to examine their internal structure. The same was done by medieval physicians, when it was forbidden to open human cadavers. Long before there was any knowledge of phylogeny or evolution, it was assumed that the *pithekoi* (apes and monkeys) were constructed on the same pattern as humans.

Compared to the round faces of the South American monkeys and the langurs, the cercopithecines have a dog-like muzzle, or "snout." This feature is most pronounced in the terrestrial representatives of the macaques and in the baboons. The longer jaws accommodate more powerful molars, and with their stronger chewing muscles, these species can consume harder and more fibrous nourishment than their fruit- and insect-eating cousins. Basically, they are true omnivores, partial to animal as well as vegetable fare.

Their variability and adaptability has enabled the representatives of this group to occupy a multiplicity of ecological "niches." While some African monkeys and mangabeys and a few Asian macaques (Java monkey, lion-tailed macaque) remain in their original habitat, namely the branches of the tropical forests, others such as the Asian pig-tailed macaque and the African baboons came down to the forest floor. The green monkey and the patas monkey, or red guenon, venture out into the low shrub and grass expanse of the treed savanna. This is still more widely traversed by the baboons. The hamadryas baboon and the gelada feel at home even on the open dry steppe and the rocky hills. A similar enlargement of habitat has been experienced by some macaques (for example, rhesus monkey, Barbary ape). Some species (Barbary ape, rhesus monkey, Tibetan and red-faced macaques) are widespread in areas where they must endure snow and cold during parts of the Winter.

The guenons and other members of the genus *Cercopithecus* itself, who still follow the original way of life in the treetops of the tropical forest, are admirably adapted to that environment, having strong hind legs which to launch themselves into a leap. The long tail is an important balancing organ. The arms are slightly shorter, so that the animals walk in a stoop. The posture of the ground dwellers is blocky; their arms are longer, and their actual length is augmented because in walking the flat of the hand is not put to the ground, but only the fingers, with the palm held nearly vertical.

With their fully opposable thumbs, the hands are very well suited to the deft "handling" of objects, to plucking and to picking, even more so than those of the anthropoid apes.

To keep the hands free and utilize their dexterity, cercopithecines sit a great deal. Thus, all species have more or less pronounced ischial callosities.

Another special adaptation involves the cheek pouches. An extensible, sac-like excrescence in the lower part of each cheek will instantly accommodate a whole meal. This enables them to spend little time in dangerous feeding places. When they have returned to a place of safety, they force the food out of the pouches back into the mouth by pressing with the back of the hand, and then they consume it at leisure.

The cercopithecines are highly visual animals. Studies in sensory physiology have shown that they have excellent powers of color discrimination. A mangabey, for example, may actually surpass humans in this respect, for the structures of both the retina of the eye and the centers of vision in the brain are more elaborate and more complex than ours.

Like nearly all other primates, the cercopithecines live together in communities with a strict ranking. Individuals regulate their relations with the other members of the group by means of a large repertory of behavior patterns and expressive movement, the meaning of which arises from the particular context.

An example is genital "presentation." In the narrow sense, this is an expression of sexual readiness and functions as an invitation, triggering readiness to mate in the males. This effect may also be employed to placate hostile members of the group and thus procure indulgence or at least sufferance.

Another example is "nitpicking," or mutual "grooming." The immediate purpose is cleaning of the fur and skin, which the subject finds agreeable. Therefore, and because the procedure presupposes a friendly attitude, it may serve indirectly to influence the social setting, either to appease the higher-ranking or to reassure the lower-ranking.

Friendly "smacking" of the lips is an example of the lively facial pantomime engaged in by these animals. This facial gesture may be reinforced by retracting the scalp, raising the eyebrows, and laying back the ears. All these movements can be performed to an extent far surpassing human abilities.

There are also lasting features that are to be interpreted as messages or "signals." Thus, the females of some species experience a conspicuous swelling or reddening of the genital region at times of fertility. This evidently affects their sexual attractiveness. The red coloration may spread to other naked skin areas, for example to the face as in the red-faced macaques or to the breast as in the gelada.

Likewise lasting are the deep blue skin colorings exhibited by many monkeys, especially males, on the buttocks or the scrotum, and in the face of the mandrill.

Likewise remarkable is the contrasting fur and

Trees are the proper habitat of the cercopithecines. Java monkeys or long-tailed macaques (top) are inhabitants of the dense, moist tropical forest. The red-faced macaques (below) that have come to be known as "snow monkeys" are also dependent on trees, especially in Winter, when the bark serves them as an emergency ration.

skin coloring of cercopithecines in the first months of life. The skin of the face, hands, feet and buttocks is pale in young baboons and juveniles of many other species, whereas it is generally black or brown in older individuals. The fur of the newborn, on the other hand, is black, contrasting sharply with the gray, greenish or reddish-brown coat of adults. The newborn of the dark-haired stump-tailed or bear macaque and of the Celebes crested macaque are more or less white in whole or in part.

Macaques (Genus *Macaca*)

The numerous species of macaques all belong to the one genus, *Macaca*, but they exhibit a wide variety of forms. This variation is most conspicuous in the length of the tail. In some species the tail is actually longer than the body (for example, Java monkey), in others it is completely reduced (for example, Barbary ape). There are all intermediate degrees in between.

Of all primates, the macaques, next to humans, have adapted best to widely diverse environmental conditions. With the exception of the Barbary ape, living in the Atlas Mountains of northwestern Africa, they are all indigenous to Southeast Asia. There they are found in tropical rainforests, and also in villages, temples, even in large cities, in dry treed savannas and even in mountainous country. The short-tailed macaques especially have largely adopted a terrestrial way of life. In the northernmost parts of their range, in China and Japan, they have a snow-covered landscape to contend with in winter.

The JAVA MONKEY or LONG-TAILED MACAQUE *(Macaca fascicularis)*. Within the genus *Macaca*, the Java monkey represents quite an original type. Not only does it more closely resemble the African guenons and mangabeys in its long tail and elongated figure; it is also markedly arboreal in much of its range. This comprises almost all of the Southeast Asian archipelago, and is therefore split up into many separate parts. This accounts for the variety within the species. Today about 21 subspecies are recognized, differing in fur color (red-brown to green-yellow and gray), facial coloring (pale to dark brown), and scalp hair (from bald to pig-tailed).

The name "crab-eating monkey" for this species suggests an occurrence primrily in the vicinity of coasts or rivers. The species is found even in mangrove forests. The monkeys swim and dive very well, and will feed on crustaceans and other invertebrates. But they are also widespread inland.

The Java monkeys, at home not only on the island of Java but in almost all of Southeast Asia, live in family groups, centered in the currently reproducing mother and her offspring (left). A striking feature, unusual for monkeys, is their fondness for water (right).

For example, in the hill country of the Gunung-Leuser reservation in Atjeh, Sumatra, covered with tropical rainforest, where conditions almost undisturbed by humans still prevail, their groups are found at intervals of about 1600 ft (500 m) along the rivers and streams, where their sleeping trees stand. Each evening, they return to their sleeping tree, doubtless because they are safest in the branches overhanging the water. During the day, they travel through the branches of the forest. Either they go straight to trees which, evidently on the basis of earlier expeditions, they know to be bearing edible fruit at the time, or they move slowly through dense brush on a broad front, looking for insects, caterpillars, pupae, etc. Their behavior exhibits a fixed pattern. After a period of traveling and foraging in the morning, they rest for about an hour towards noon; the older individuals doze or engage in grooming, while the younger ones play. In the early afternoon, the group will undertake another tour in search of food. When food is abundant, they may return to the river as early as about 4 p.m. Usually there will then be some more play. If a small beach has formed at a bend in the river, the animals will come down and romp around. When the brief tropical twilight sets in at about 6:00 p.m., they will have returned to their sleeping trees. In the gathering darkness, there will be a few more little altercations before each has found his sleeping place, and quiet reigns.

The Java monkey may serve as an example of the typical cercopithecine social habit. Group size varies between about 20 and more than 60 members. The groups are matriarchal in structure, that is, within a group family units may be distinguished, each with a reproducing female at the center. Whereas the males almost without exception leave the group of their birth when they are grown up, the females remain for life in the group into which they were born. They there attain a position of rank corresponding to that of their mothers. At first the young individual behaves very submissively towards larger and stronger group members, even if their family rank is below that of the young one's mother. But since its relatives, especially its mother, can support it effectively, the growing youngster also gradually acquires a self-assured attitude towards these fellow members. If the juvenile gets into a dispute with members of higher-ranking families, the members of its own family will rarely lend aid, perhaps not unless the emergency is acute. Then the mother primarily will assume the risk of being chastised herself if she stands by her offspring. But if the young one belongs to a powerful family, it will often even be supported by non-family members, who will take advantage of the conflict to improve their own position towards this opponent through the expected approval of the young one's relatives. Such "opportunistic" assistance is of course not forthcoming for the young of low-ranking families. However, the strategy can be successful only if the individuals have exact knowledge of the relations between other individuals, and this tells us something about their mental faculties. Some scientists in fact advocate the view that the social forms of behavior practiced by the

Java monkeys inhabit especially the forests of the coasts and river banks. They are excellent swimmers and divers, and have even included crustaceans and mollusks in their diet. Because of this habit, they are sometimes called "crab-eating monkeys".

monkeys in their complex communities have supplied the selective pressure for the evolution of their intelligence.

The males generally move about from one group to another. The Netherlands researcher Van Noordwijk studied Java monkeys for a long time in the primeval forest of Sumatra. She discovered that there are two types of migration. The first "immigration" into a neighboring group takes place about the 5th year of life, when the male is not yet full grown. It is accomplished "discreetly": the immigrant carefully insinuates himself into the new group from the periphery and takes over a lowly position. This operation is repeated perhaps once more before the young monkey is about 9 or 10 years old. At that age, he follows a more solitary way of life for a time. During this period he grows big and strong. When he now reenters his previous group or a new one, he does so as an "aggressive" immigrant. Either he succeeds, and remains in a position of high rank for several years, or he is compelled to wander off again. Then he must try again in another group. Once he makes it, he goes through two role phases. During his first year as a comparatively high-ranking male he associates with the fertile females but takes no interest in the infants. After a year or two his behavior changes. Now he attaches himself to the mothers and becomes the protector of their little ones, and in all probability, he is their father. After a few years, he may then wander off again and find a new group.

Our detailed knowledge of the environmental conditions and their influence on the social life of Java monkeys is due to long-term researches on Sumatra by Van Schaik. Animals of a given species seek to create a social environment in which the most effective and least dangerous performance of the task necessary for self preservation and reproduction is possible. Certain primates such as the orangutan, for example, or nocturnal species, evidently do this best if their life is for the most part solitary. Most diurnal primates evidently manage best in larger communities, whose organization may nevertheless be widely diverse.

Now the question is in what respect community life is advantageous to the individuals. Some zoologists surmise that animals can better utilize a limited food supply in a larger group because they are able to drive out other smaller groups or individuals. This view contrasts with another according to which life in groups involves a great disadvantage because these animals more often take food away from each other, and they must "work" longer and travel farther in order finally to get the necessary amount of nourishment together; this disadvantage, however, it is argued, is offset by an even greater advantage, in that animals in larger groups are far better protected against enemies. In the case of the Java monkey, these are chiefly felines, such as the tree-climbing cloudy leopard, and snakes, such as the python. Detection of such enemies will be timelier in a larger group. As a matter of fact, Van Schaik and Van Noordwijk found that smaller groups can be approached more closely than larger ones in the forest before the animals will take flight. On the other hand, careful measurements of the time spent on various forms of activity in groups of different sizes have

In the "yawn of menace", grown male Java monkeys bare their terrifying dentition with its dagger-like canines. This picture was taken on the island of Bali, where this species is abundant.

shown that members of larger groups are longer engaged in seeking food, are more frequently involved in disputes with each other, and make a longer day's march.

Hence the energy balance – output over input – is less favorable to the individual in larger groups. This is confirmed by the fact that in many species living in smaller groups, more young are born per female. On the other hand, in areas where there are predators, more young survive per female in larger groups. Living in smaller groups, then, is less taxing but more hazardous. This would mean that enemy-avoidance would be the chief selective pressure favoring social life among primates. This pressure prevails when it is greater than the opposing pressure exerted by competition.

Java monkeys, which still occur in large parts of their extensive range of distribution, soon become accustomed to humans. The tourist in Bali, for example, meets with them there primarily as temple monkeys, who leap on the visitor's shoulder and search his pockets for delicacies. On the Hindu island of Bali, the animals were deemed sacred, and gifts of food are placed for them in temple courtyards and in the woods. It is because of the uncommon adaptability of the Java monkeys that they are not highly esteemed everywhere. In some areas, they are regarded as pests because they raid the crops. However, humans also profit by these animals in ways not always beneficial to them. In past years, tens of thousands of them have been shipped to medical institutions, where they are among the most-used laboratory animals, second to the rhesus monkey.

The RHESUS MONKEY or RHESUS MACAQUE *(Macaca mulatta)*. The rhesus monkey is one of the best-known simian species. It is not only found in many zoos but is also one of the most-used experimental animals. It serves for biological and medical as well as psychological research, with special emphasis on the study of perception, learning, and memory.

This species has given its name to the so-called rhesus factor, discovered in its blood in 1940. It is a hereditary antigen factor, found also in human blood. It may cause dangerous defense reactions if the blood of a person with positive Rh-factor enters the bloodstream of someone with negative Rh-factor, for example in blood transfusions, or in the last phase of pregnancy, if an Rh-negative mother gives birth to an Rh-positive baby. As a

(Top) Philippine macaque, member of a Java monkey subspecies, eating a locust. (Bottom) Java monkey mother with baby, grooming an adult companion.

result of the knowledge gained through studies of rhesus monkeys, such complications can now be avoided and controlled.

Within its wide range of distribution, the rhesus monkey exhibits considerable adaptability. It occurs in low-lying flatland but also in the north of India and Pakistan, at the foot of the Himalayas, at elevations up to 9600 ft (3000 m). Both hot, dry summers and cold winters, with snow and temperatures below the freezing point, are tolerated. The rhesus monkey is encountered in forests as well as in dry and desert areas. However, the religiously inspired indulgence of the Buddhist and Hindu populations has had the effect that most of the animals have long been living near human settlements, especially in temple precincts, marketplaces, and railway stations. Obviously, this can become a nuisance. The monkeys will not only pick up anything edible that human beings discard; occasionally they will also raid fields and gardens and steal provisions from dwellings and from shops. Attempts have therefore been made to transfer groups out of human settlements back into nature. But this is not without its difficulties. The "citified" monkeys are not at home in their new environment. They migrate until they find an urban milieu that meets their requirements and expectations.

The diversity of environmental conditions under which the species may be found suggests that its dietary habits will also vary. In some areas, the animals eat chiefly herbs, roots, and buds; elsewhere, much of their diet consists of fruits. Thus,

(Right) Rhesus monkeys in a forest region of the northwest Indian state of Rajasthan. (Below) A group of rhesus monkeys engaged in social "nitpicking." Better referred to as mutual grooming of the fur or person, this is an important component of social behavior in most primate species. This activity serves not only to clean the fur and skin but also to consolidate group cohesion. It presupposes a friendly inclination between participants, mollifies higher-ranking and soothes lower-ranking group members.

rhesus monkeys observed in the mountain forests of northern Pakistan live primarily on clover, available in superfluity on open ground, in Summer. In Winter, when snow covers the earth for several months, they resort to food of lower nutritive value and higher fiber content, such as oak leaves and pine needles. Mortality is no higher in the Winter months, but the animals lose a good deal of weight. In this region, with its marked seasonal fluctuations, the young are born a few weeks after the beginning of the Spring thaw, in April and May. Matings occur in September and October. In more southern regions, where seasonal differences are less pronounced, the scatter is much wider; the female cycle of ovulation during which fertile matings are possible extends over several months. Under uniform laboratory conditions, the females have a steady menstrual cycle of about 26 to 28 days. This cycle, as in some other

primate species, is characterized by visible changes in the genital region. In the rhesus monkey, however, the swelling of the genital region is minor; it occurs chiefly in young adult females, and is attended by a conspicuous red coloration of the genital region, sometimes extending over the tail and the thighs.

A remarkable feature is that in areas where there is a mating season, not only the females but also the males exhibit alterations in reproductive physiology. In this period, the testes become enlarged, about twofold. At the same time, the naked skin of the posterior takes on a redder color. In this phase aggressiveness

increases, owing to competition among the males for access to the fertile females. It is intensified because the males then have a greater inclination to switch to other groups; this in turn leads to uncertainty as to the social relations in the group, and hence to altercations.

As in othe macaques, the size of the testes is notable. At about 2.5–2.8 oz (70–80 g), the weight is greater than in many a healthy human male – about 0.7% of total body weight. This heavy weight is characteristic of primate species living in so-called multi-male communities, where the males cannot monopolize access to the females, or not completely. When females pair with several males, those males are more successful in selective competition for reproduction which are able to mate several times, and quantitatively surpass the genetic contribution of other males. Remarkably enough, in just these species the females exhibit a

more conspicuous fertility signal and impress everyone with their sexual attraction.

Thus, the rhesus monkey lives in multi-male communities, the adult males having as a rule immigrated from elsewhere, and the group being composed of matriarchal kinship units such as have been described for other macaques. The social life of the rhesus monkey, apart from field research in the wild, has become known to us through observations of a "population" (animals of one species in a bounded land area) on the island of Cayo Santiago, to the southeast of Puerto Rico.

An important finding related to the choice of sex partners. While the animals often engage in alternative sexual intercourse and generally have a preference for higher-ranking representatives of the opposite sex, there are also certain relationships of preference within the group between particular males and females.

However, the contrary is also true; certain individuals avoid sexual intercourse. Both with the red-faced macaques and with the rhesus monkeys of Cayo Santiago, it is found that pairings between mother and son are rare, although in other respects the individuals have a good relationship and may for example support each other in alliances. It turns out that this absence of sexual attraction extends also to other members of the mother's kinship group. The attractive effect of females in neighbor groups may be one of the reasons why males abandon the group of their birth, those of low rank tending to change more than the high-ranking (or young males with a high-ranking mother). Several factors are thus involved in this process.

How important the matriarchal kinship structure in a macaque group can be is made clear when

An unusal phenomenon is this albino rhesus monkey.

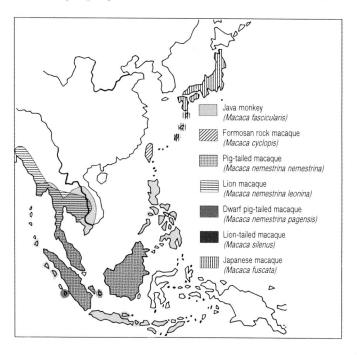

groups split up. In the case of the rhesus, a split may occur if the number of group members rises to about 80–100 individuals. The females of a matriarchal subgroup will then detach themselves progressively from the other members until there can be said to be two groups.

Like other macaque species, rhesus monkeys are not territorial. This means that a group will not defend a particular area and claim it exclusively. The groups do have their own living areas, but the territories of neighboring groups may overlap in large part. When groups meet, generally the weaker avoids the stronger. In certain cases, for example if there is a shortage of water in a dry season, different groups may form assemblages of 200 individuals or more. When confrontations occur, they are initiated by uncertainty regarding mutual relationships.

Group size and consequently the density of occupancy depends greatly on the particular environment. In dense forests, territories of 6 mi^2 (16 km^2) have been measured, with a density of 5 to 15 individuals per square kilometer (1 km^2 = 0.36 mi^2). In urban surroundings, territories are much smaller, and densities up to 750 individuals per square kilometer have been counted. Under these circumstances, conflict between neighbor groups is far more frequent. The character of the animals growing up under these conditions is of course affected. Thus, experiments in which town and country individuals were compared indicated that the town animals are bolder and more enterprising, and easily win out in altercations with forest-bred individuals.

The rhesus monkeys were once very abundant in India, and also in southern China and Tibet. Whereas it was recently supposed that in the Indian state of Uttar Pradesh, where the predominantly Hindu population was no threat to the rhesus monkeys, there were many millions of them, later estimates arrive at figures of only half a million to a million. Exportation of animals for experimental purposes has contributed substantially to this decline. Little care was taken for their welfare. Protective measures and restriction have resulted in some improvement, leading to greater care in the treatment of captured animals.

The RED-FACED MACAQUE *(Macaca fuscata)*. This species is described by Kosei Izawa in a separate chapter, following the present article, on the subject of so-called simian "cultures."

The FORMOSAN ROCK MACAQUE *(Macaca cyclopis)*. This simian, with its rounded head and flat muzzle, is intermediate in size and appearance between the rhesus monkey and the Java monkey. It differs from both species in that the females, during estrus, exhibit a conspicuous genital swelling. This may be an indication that the animals live in groups including several sexually active males. However, details of their social organization in the wild are not yet known.

Reports dating from the previous century mention that this animal traveled with great agility over virtually inaccessible rocks on the seacoast, and that it lived in part on crustaceans and mollusks. Today the species has been driven back into the central hill country of Taiwan. The Formosa macaque is a ground dweller, comfortable in terrain with few trees or none. However, it is not shy, and sometimes visits the fields of the hill people, where it digs sweet potatoes and peanuts. It is

Rhesus monkeys are at home both on the ground (above) and up in the boughs (right). Their diet is correspondingly varied; they find herbs, roots, and field fruit on the ground, and sprouts, buds, and fruits in the trees.

The remarkably adaptable rhesus monkey is one of the few primate species that have long since lived within the perimeter of human settlements, even in the center of large cities. They are tolerated by the Buddhist or Hindu population, although their importunity and greediness may become a nuisance.

▷ (Left) Ceylon toque macaque eating a large fruit. A foot is helping to hold the food. (Right) In the south of India, where the rhesus monkey is absent, we find the Indian bonnet macaque, identifiable by his radiating crown of hair.

hunted for this reason, and also for its flesh; at the same time, the species is threatened. The Formosa macaque has been settled on some Japanese islands and in certain spots on the Japanese mainland, where the animals can maintain themselves and reproduce.

The INDIAN BONNET MACAQUE *(Macaca radiata)*. The most striking feature of this macaque is its head hair. From a point at the apex of the midline, long hairs radiate to all sides, so that it looks like a crown; the animal also has a sparse, wrinkled forehead.

In habits and social structure, the bonnet macaques are like the others. Groups vary in size but comprise at least some dozens of individuals. They include several adult males, who tolerate each other to a large extent. Although they occur in the tropical forest, they easily become accustomed to the presence of humans. The bonnet monkeys likewise live in or on the outskirts of human settlements, roam through markets and waste heaps, or beg from temple visitors. An omnivores, however, they also incur displeasure by raiding fields and gardens.

Large predators, such as tigers and leopards, have disappeared from the range of distribution of the bonnet monkeys. They have been long familiar with smaller carnivores such as jackals, and are not much disquieted by their approach. Dogs in the vicinity of settlements, however, are feared and dangerous enemies. When the monkeys feel threatened, they warn each other with barking cries of alarm and take refuge in trees and on buildings.

Bonnet monkeys engage in vivid mimicry. A friendly attitude is expressed by smacking of the lips, fear and submission by an exaggerated grin with the corners of the mouth drawn back and the teeth bared by retracting the lips. A blend of these two expressive gestures – tooth chattering – expresses an intermediate mood, a guarded friendliness.

The females do not exhibit the swelling or discoloration of the genital region in estrus that is common in other species. Nevertheless, the males can detect mating readiness of the females, which secrete a thick, colorless, scented mucus in the fertile period.

The CEYLON TOQUE MACAQUE *(Macaca sinica)*. This specis is somewhat smaller than the Indian bonnet macaque. It is also distinguishable by a somewhat different headdress, and often by redder fur, from its Indian relative. The Ceylon toque macaque inhabits the large island for which it is named, now known as Sri Lanka. It occurs both in the moist forests and in the central hill country, and in dry, less wooded areas in the east of the island. It does not frequent the vicinity of human settlements as much as its mainland cousin.

Extensive observations of a simian population in

Humans have long used rhesus monkeys for their own purposes. Despite all protective measures, these higher animals are still unduly exploited, both in their homeland and in research facilities of the western world. (Left) Animal experiment in a scientific laboratory. (Right) Monkeys as "entertainers" on a street in the Indian capital of New Delhi.

the dry northern part (near Polonnaruwa) by W. Dittus have given us an understanding of the manner in which sharply fluctuating conditions of life affect the composition of the population.

In dry periods, it is chiefly the juveniles that perish of undernourishment, most especially the smaller females. The older and stronger individuals are able to drive the weaker away from food sources. And although the animals' food is usually so dispersed that it cannot be completely monopolized by the high-ranking individuals, these regular displacements nevertheless have the result that the low-ranking individuals need far more time to get a given quantity of food.

Nevertheless, the frequency of acts of aggression did not increase in time of food shortage, but rather diminished. The animals were fully occupied in the search for food, and traveled widely separated, accepting the consequent greater hazards.

In a four-year period of observation, it was found that 85% of the females and 90% of the males failed to reach adult age. Mortality was highest among the females in the first three years of life; thereafter, that of the males was higher, especially between the 3rd and the 5th year. This concentration of fatalities is consistent with the higher risks incurred by the males when they move from one group to another; they perish under attacks by enemies (dogs) and in combat when penetrating a strange group. As a result, there are twice as many adult females as adult males.

The ASSAMESE or HIMALAYAN MACAQUE, or HILL MONKEY *(Macaca assamensis)*. The Assamese macaque has traits in common with the rhesus monkey and to a greater extent with the toque macaque, but is considerably larger. As in the bonnet macaque, the head hair radiates more or less. The Assamese macaque lacks the feature of bare red skin on the posterior.

Scientifically speaking, this species is still something of a blank page, since it is rare in captivity. The name "hill monkey" suggests that it is not be met with in the most accessible of terrain. It lives in the forest of the Himalayas at elevations up to at least 7200 ft (2250 m). Although the Assamese macaque is not a camp follower of civilization like the rhesus monkey and the bonnet monkey, it will occasionally invade maize fields. It has been observed in groups of as many as 58 individuals. The few known observations in captivity report an expressive behavior similar to that of the bonnet monkey.

The TIBETAN STUMP-TAILED MACAQUE *(Macaca thibetana)*. The first European explorer to see the Tibetan macaque was the celebrated French missionary Père Armand David. Before he saw the first live Tibetan specimen, more than a century ago in Western China, he had already seen its footprints in the snow.

Although available measurements are inadequate, this short-tailed macaque is undoubtedly one of the largest, if not the largest, representatives of the genus. Its dense and long gray-brown fur enables it to withstand the severity of Winter in the hill country of western China and eastern Tibet, where it is found at elevations up to 6400 ft (2000 m). Another adaptation to the cold is the

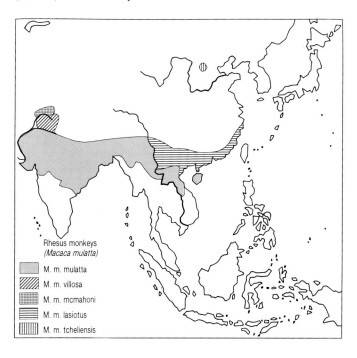

thick fur about the ears. Characteristically, these macaques have a well-developed gray beard, and the head hair forms a crown by radiating from a center. In this way, these monkeys resemble the bonnet macaques. The Tibetan macaque is today regarded as related to the toque and bonnet macaques and to the Assamese macaque, rather than to the stump-tailed or bear macaque as formerly believed. In their habits, the Tibetan macaques probably resemble the Assamese macaques, whose range of distribution they adjoin in the northeast.

The BARBARY APE *(Macaca sylvanus)*. The Barbary ape is the sole representative of the genus *Macaca* in Africa. It is also the only simian species to occur in northwestern Africa and even in Europe. From time immemorial, this species has lived on the Rock of Gibraltar. It has been surmised that these monkeys may be the last representatives of a population that once inhabited large parts of Europe. But it is more likely that the present Barbary apes of Gibraltar are the descendants of animals brought to the Peninsula from North Africa by humans, perhaps as early as the beginning of our Era. And not for the last time; since the Rock has been held by the British, the stock has been repeatedly replenished with animals from Morocco, whenever it declined. Since 1915, British soldiers have been charged with the care and feeding of the apes. The close association with humans of course has its disadvantages. The animals often inconvenience the inhabitants by entering gardens and even houses. In addition, their health is endangered by the unsuitable food offerings of tourists. The growing numbers of vacationers since the opening of the frontier between Spain and Gibraltar will aggravate this situation, and necessitate special accommodation in a monkey "park."

In our time, Barbary apes have again been brought to Europe to live there in large groups under nearly natural conditions, namely in "monkeylands" created for mass tourism, as at Kitzheim in Alsace, Salem in South Germany, and Rocamadour in southeastern France.

(Top and far right) The Indian bonnet monkeys are indeed original inhabitants of the tropical forest, but like the rhesus monkeys, they frequent human settlements and temple grounds, where they become quite tame. They beg for charitable donations, but can inflict painful bites if they feel threatened. (Below) Very similar to the Indian bonnet monkey, but somewhat smaller, the Ceylon toque macaque lives on the island now named Sri Lanka.

Originally, the Barbary apes occurred throughout North Africa. Today they are limited to certain regions in Algeria and Morocco, where they live in cedar, pine, and oak forests of the central Atlas Mountains. In Summer, the food supply there is plentiful and varied. Herbs, fruits, seedlings, seed pods, roots are all part of the diet, and occasionally invertebrates. But in Winter, when snow lies in the hills, the monkeys must content themselves with poorer fare, leaves, needles, bark and buds, more difficult to digest. However, they are fairly well adapted to the cold, growing a very thick coat in Winter.

The Barbary ape is considered a threatened species. Although perhaps more than 20,000 individuals still exist, the prospects are dim because the hill forest habitat is seriously endangered. More and more grazing of cattle, sheep, and goats impedes renewal of the forest and causes erosion, the loss of soil. Destruction of the forests by timbering is no less a matter for concern.

Like other macaques, he Barbary ape lives in hierarchically organized communites, incorporating several adult males. However, the Barbary ape differs from the other macaques and from the baboons in that the adult and adolescent males expend much effort in tending the young. This begins only a few days after the baby is born. The males like to carry the little ones about at the breast or on the back, hold them on their laps, groom them and play with them. Often the youngster is invited to interact by a peculiar facial expression of the older male: the lips are drawn back, as well as the corners of the mouth, and then the teeth chatter in a very fast tempo. Whereas in many other species the complete baring of the teeth expresses submission or readiness to take flight, in the case of the Barbary apes it must be interpreted as having primarily a friendly significance.

Contacts between adult male and infant may be one-to-one as described, or else two-to-one, one male introducing the youngster to another male. There is then a thoroughgoing exchange of "chattering" during which the little one is held between the two males, and they regard both the child and each other, making gestures of embracing.

The interpretation of these continually repeated behavior patterns has been the subject of much discussion in the technical literature. At first, they were supposed to represent an "agonistic" buffering: one male feeling threatened by another seeks to utilize the disarming influence of the sight of a baby to mollify his possible adversary. In this way, for example, subordinate males might contrive to associate with higher-ranking males. The infant is

(Left) A group of Ceylon toque macaques engaged in mutual social grooming. (Right) The Tibetan macaque's "menacing" yawn exposes his massive canine teeth to view.

OLD WORLD PRIMATES

as it were utilized as an "implement" in regulating relations with other males.

Careful observations of this behavior by the American primatologist David Taub in Morocco have led to another interpretation (not, by the way, necessarily ruling out the first in all cases). Taub found that each male addresses his behavior to one or more "preferred" young ones. Thus, a little one may have one or several males paying it special attention; most interactions were of the one-to-one type. Now the question is why a male should accord such care to particular children. Such behavior would of course enhance his reproductive success if he addresses it preferentially to his own descendants, and of course also if these child-rearing efforts, so far from interfering with the development of the young, actually improve their chances of survival. Now paternity is generally difficult to prove among primates in groups where several males are present. This certainly applies to the Barbary apes. When the females are in their fertile phase, they actively seek mating opportunities with numerous males. Nevertheless, certain males will have a substantially greater share in the conceptions of a particular female than do others. And it is in fact these males only who pay special attention to the offspring of the female in question.

In the two-to-one configuration also, males with one and the same preferred youngster may seek each other out. In this behavior, then, there is clearly an emphasis on predilection or protection,

and one might reasonably speak of paternal behavior.

However, still another advantage may result, namely when the males in question are close to adulthood. Such males also take a large part in these activities, often even more so than the adult males who are presumably the fathers. Since the Barbary females presumably play an important part in the choice of mates, a young male might enlist a female's preference by taking care of her offspring, thus serving his own reproductive interests in the longer term.

The PIG-TAILED MACAQUE *(Macaca nemestrina)*. This macaque of sturdy build takes its name from its little tail, carried high and more or less curled. The pig-tailed macaque is a typical dweller in the rainforest, responding to danger in an unusual way. Whereas other primates living in the forest will take flight high up in the trees, pig-tailed macaques throw themselves to the ground and sprint off at high speed. The LION MACAQUE *M. n. leonina*, a subspecies, is more like the other macaques in this respect, intermediate between its "brother the pig-tailed macaque and its "cousin" the lion-tailed macaque.

This Barbary ape, on the street in a Moroccan town, must entertain the public in this unnatural way.

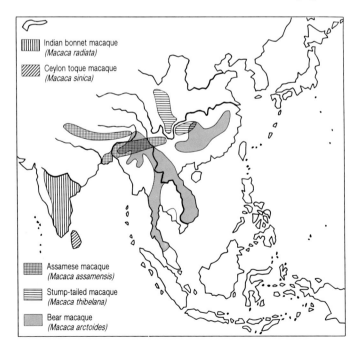

The pig-tailed macaques travel in large groups on the forest floor from food source to food source, and may traverse long distances. Ripe fig trees attract them especially. However, they choose not only vegetable fare such as fruits, seed pods, roots etc., but also caterpillars, locusts, spiders and spider webs, termites, and even vertebrates, such as, for example, young birds. Thus, the pig-tailed macaque has created a niche of its own in the forest. Through the boughs overhead ranges another macaque species, the Java monkey, likewise chiefly fructivorous and insectivorous, but able to forage and hunt at greater heights because of its lighter weight. The other fructivores are the gibbons and siamangs, and on Borneo and Sumatra the orangutan also. Then there are the langurs, but as specialized leaf-eaters they do not compete with the macaques to the same degree.

In the large groups or troupes of pig-tailed macaques, the adult males are easily distinguished: they are considerably larger than the females. This is one of the *Macaca* species with the most conspicuous sex differences (sexual dimorphism). Old males are quite often found to be solitary.

In its language of gesture, the pig-tailed macaque employs a remarkable facial expression shared only with the lion-tailed macaque in this genus. The facial expression of "greeting" may be observed in zoos also, when familiar human beings approach the cage. The monkey lifts its head, nose in air, raises the eyebrows and retracts the scalp, and at the same time the lips are extended forward over their entire width. It then looks down its nose at the visitor.

Primarily, this is, of course, an intraspecific form of communication. It expresses a peaceful inclination, and will thus reassure a diffident companion. The same facial expression is often observed when an adult male seeks to approach a female to mate with her. The female is then very likely to assume the attitude of "presentation," thus offering herself.

The origin of this peculiar facial gesture has not been entirely clarified. It resembles the movements and gestures we ourselves use when we carefully sniff something. The male pig-tail will in particular perform the gesture when he is sniffing the genitals of a female in heat. Thus, the behavior pattern may originally have been a means of expressing sexual interest, but it has evolved into a sign of a general friendly attitude.

Many other macaques, the baboons, and many simian species show a friendly social inclination by smacking the lips, or derivative gestures. The origin of this behavior is fairly clear. Monkeys groom each other as a form of social aid. Of course this is done only by individuals who are inclined to be friendly. In the course of this mutual grooming, they will eat recovered particles, smacking their lips. Lip smacking in thus associated with a benevolent attitude towards the groomed party. Hence it can be used as a message

▷ Barbary apes - a mother and baby - are the only simians that occur in Europe. Many centuries ago, they were already inhabiting the Rock of Gibraltar. There they are under the protection of the British military, and are a very popular tourist attraction. Probably this population derives from North Africa, where the species occurs in some parts of the Atlas Mountains.

This confrontation is not so unfriendly as it looks. For the Barbary apes, baring the teeth completely does not convey the usual threat, but counts as an amicable gesture.

Course of a "triadic" interaction among Barbary apes. As a rule, two males and their "favorite" youngster are involved. The baby is cared for, caressed, and carried about by the two jointly. This behavior evidently serves both to ally the "foster fathers" with each other and to benefit the infant, affording the younger male reproductive opportunities if he gains favor with the mother in this way.

addressed to another at a greater distance, and giving expression to this benevolent attitude. Among the pig-tailed and lion-tailed macaques, this friendly smacking of the lips has been replaced to a large extent by the facial gesture of protruding the lips.

A zoo visitor who has seen the other evidently lively and "hot blooded" species of monkeys may perceive the pig-tailed macaque as a somewhat stoical creature. But are there in fact objectively ascertainable differences in character between closely related species? The American psychologists Kaufman and Rosenblum have compared pig-tailed macaques in certain peculiarities of their behavioral development and social conduct with bonnet macaques kept under the same conditions. Whereas bonnet macaques, when resting, will creep close together and seek physical contact, pig-tailed macaques rest separately. Also, they "converse" chiefly with near relatives. Mother pig-tails always attend their young themselves, whereas the more sociable bonnet macaque mother will allow other females to pick up the baby. One is therefore inclined to regard the difference in sociability as an inherited trait. Probably this is correct in the present instance, but one must always allow for the possibility of a "tradition," a behavior pattern that has spread and been passed on within a given population, and may be very persistent because individuals are exposed to it from earliest youth.

The LION-TAILED MACAQUE *(Macaca silenus)*. With its glossy black fur and conspicuous gray-white crown and beard, this macaque is assuredly one of the handsomest representatives of the genus. It is a close relative of the pig-tailed macaque, though it has a tail of medium length. It shares the pig-tailed macaque's peculiar behavior trait of a facial expression with protruding lips, not found in other *Macaca* species. The lion-tailed macaque also uses this gesture as a friendly act of greeting. The expression is a good example of peculiar forms of behavior that may indicate kinship between species. Another trait common to both species is the female genital swelling. Though not quite so large as in the pig-tail, it is spread broad and high, so that the posterior is bounded by the margin of the swelling, nearly exceeding the height of the back line and surrounding the base of the tail in a kidney shape.

Lion-tailed macaques live in moist, evergreen tropical hill forests of the Western Ghats. They inhabit great trees, 95 ft (30 m) tall. If they make an exception and come down to the ground in search of food, at the slightest disturbance they will immediately take flight into the treetops. In this respect their behavior is quite different from that of the pig-tails, which are typical inhabitants of the forest floor. The lion-tails eat blossoms, fruits, nuts, and buds, and also insects and caterpillars, which they find in the branches and among the leaves.

(Top) Thorough grooming is important to the well-being of all simians. (Left) A pair of Barbary apes in the typical simian mating posture.

This species forms comparatively small groups, numbering 10 to 20 individuals. Information about social structure is as yet conflicting. Some professional observers report multi-male societies, while the German researchers M. Herzog and G. Hohmann assert that the animals live in single-male groups. This difference may not be so fundamental as it seems. In a small group, a male may try to keep other males away and thereby gain exclusive access to the females. In a larger group, the outlay of time and energy that this would require may be out of proportion to the results obtainable. It will then be obligatory to tolerate other males. The alternative strategy for successful reproduction will be to mate as often as possible with as many fertile females as possible, as the rhesus monkeys reportedly do.

The lion-tail exhibits a behavior trait that distinguishes it from the other macaques. The leading male occasionally utters a very loud call. It is heard in particular when the group is setting itself in motion, and thus promotes cohesion within the group, but its great volume suggests that the call also serves to make the group's presence known to others.

The bonnet macaque of India occurs in the same region. Groups of the two species occasionally mingle. As a rule, however, the lion-tails avoid other macaques. The lion-tail is very shy, and reluctant to encounter human beings. In this respect it differs from other *Macaca* species, which tend to adapt to an environment altered by humans. The shyness is surely in part attributable to the fact that the lion-tail has been intensively hunted, not only for its flesh but also for its fur. Since the mountain forests where it is at home have been much diminished by timbering, it is the most endangered macaque today. Only a few hundred individuals remain in the wild. Fortunately, there are about another 300 individuals in various zoological gardens. A controlled breeding program is designed to secure optimum matings and groupings and avoid weakening by inbreeding.

The STUMP-TAILED OR BEAR MACAQUE *(Macaca arctoides)*. The bear macaque is of a massive and powerful build, with a short tail betokening that it is more at home on the ground than up in the trees. Zoo visitors are less attracted by these animals than for example by the more teddy-bear-like Barbary ape. This is certainly related to the fact that older animals become bald on the scalp and forehead. Also, the pale skin color of the juveniles is altered with advancing age. Brilliant red or purple patches of color then become visible on the

face and on the genital and posterior regions. These alternate with dark-pigmented patches. These monkeys lack a dense, woolly coat, and are thinly clad in long, coarse brown hair.

Stump-tailed macaques inhabit the dense tropical forest, and are very similar in habits to the pig-tailed macaques, which appear to occupy the same niche in more southeasterly parts of Asia. They eat chiefly fruits, young leaves, buds and occasionally small prey. In captivity, at least, it has been observed that they are sclever at catching birds. Where opportunity offers, they can leave

In their home country, pig-tailed macaques are often recruited to help with the coconut harvest. The half-tame animal is fitted with a collar (above), allowed to climb the coconut palm on a long leash (left), and then it remains to wait for the heavy fruit to be picked and dropped to the ground (right).

the forest to raid fields and gardens for maize, sugar cane, rice, papayas, and mangoes.

They live in multi-male groups, comprising from 20 to presumably as many as 50 individuals. In zoos, even larger groups are kept. The sex life of the bear macaques seems to have a special social significance, because mating is less restricted to the period of female fertility than in other primates. After ejaculation, the male is unable to detach himself from the female immediately, much as in canines. This is due to a swelling at the entrance to the vagina. The female will sit "tight" on the male's lap for a time. Both are then rather defenseless, and the male partner particularly must accept excited mock attacks by younger individuals; they will hit him in the face, or perhaps pull his hair. The biological function of this behavior is not known.

The bear macaques have a lively repertory of gestures. Chattering the teeth is a striking part of their facial behavior. As in the case of the Barbary apes, it is a friendly, conciliatory or inviting token.

The bear macaque is considered to be seriously threatened. Unfortunately it is at home in a region where political unrest and active hostilities have prevailed for decades. Thus, humans have become a more serious threat than the original natural enemies, the leopard and the python.

The CELEBES MACAQUES. The Island of Celebes (Sulawesi), which on the map looks more like an assemblage of peninsulas, is a fine example of multiplicity of species due to geographical isolation. A number of *Macaca* species live here, each confined to a certain part of the island. After much scientific debate, it is now agreed that these species derive from a common stock, a subgenus that must have been closely related to the pig-tails living farther to the west in the archipelago (Borneo, Sumatra, Malacca).

The first simian species on Celebes to become known did not look so very much like a macaque at all. This was the black ape, or crested macaque, of the extreme northeastern peninsula. It has a much prolonged snout and pronounced bony cheek ridges. Thus it resembles a baboon or – with hardly any tail – a drill or mandrill.

I was impressed with the resemblance to the last-mentioned genus when I became acquainted with the crested macaque in the London Zoo, where I made a comparative study of the facial expressions of Old World monkeys, early in the 1960s. An adult male greeted me each morning by looking at me with a broad grin. The lips were drawn up and the corners of the mouth pulled back. The bared teeth made a sharp contrast with the coal-black skin of the face. The resemblance to the manner in which I was greeted by a male drill was astonishing. This showing of the teeth as a gesture of greeting distinguishes these species from most other macaques, baboons, and mangabeys, which are more likely to respond with lip smacking or chattering in such a case. The crested macaque reinforces the gesture of greeting by raising the eyebrows very high. This flattens the crest towards the rear, with a very striking signal effect.

The second species to become known in Europe was the moor macaque, from the extreme southwest of the island. It has an unmistakably macaque-like figure. Its black face and black-brown fur do not conceal its resemblance to the pig-tails and lion-tails. Thus it was reasonable to suppose that Celebes harbored two genera. The northeastern, baboon-like crested species was named *Cynopithecus niger* ("black dog ape"), and the southwestern macaque *Macaca maura*. The possibility that *Cynopithecus* belonged to a branch paralleling the macaques was seemingly corroborated when fossil remains of large simians, given the generic name of *Procynocephalus*, were found in India. Today, however, these too are considered to be macaques. An independent descent of *Cynopithecus* seemed unlikely also when more

This yawning, stretching and scratching bear macaque, also known as the stump-tailed macaque because of the short, nearly hairless tail, lives in dense tropical forest, but chiefly on the ground. The survival of this species is seriously endangered.

species from other parts of the island were described. It turned out that the species in intermediate geographic locations occupy an intermediate position in their characteristic also. Thus for example, as one travels from southwest to northeast the ratio of nose length to breadth keeps increasing from one species to the next. The same applies to biochemical traits, as recent studies of blood proteins have shown. Today, therefore, it is hardly doubted that the Celebes simians all proceeded from the same *Macaca* stock (for which reason many now call the crested macaque *Macaca nigra*). Animals similar to the pig-tailed macaque today may once have reached Celees from Borneo when there were land bridges between the islands. They then probably developed differently on Celebes, in more sharply separated habitats.

The macaques of Celebes have not yet been thoroughly studied. From the reports of various observers, however, it is clear that all species live in multi-male societies, like other macaques. The groups are often considerably smaller. Although groups of more than 25 individuals have been sighted for all species, usually groups of about 6 to 15 members are mentioned. In these smaller groups, there may often be only one adult male. This presumably explains why the native population believed that the crested macaque was monogamous.

The natural home of the Celebes macaques is the forest which still covers large parts of the mountain ranges on the island. They feel especially at home in rocky terrain. It has been observed that they like to retreat into caves and crevices in the rock face, not only when they feel threatened but apparently also to rest.

In the southwestern and northeastern parts of the island, where the human population density is considerably greater than elsewhere, the primeval forest has for the most part disappeared. There, the moor macaques have become raiders of cultivated crops, especially maiz. The farmers of course make every attempt to discourage the monkeys.

Celebes monkeys in captivity are quite affectionate. They lack the irritability that renders other macaques, such as rhesus and Java monkeys, so difficult to keep. They are consequently popular

as household pets. A traveler on Celebes will often see them in villages and towns. With a strap or ring about the loins, on a cord or chain, they will sit somewhere outside near the house, often with no shelter from the Sun or rain.

The crested macaque and the moor macaque, both of which live in the most severely transformed environment, are also the most endangered. The other macaques of Celebes are as yet comparatively unthreatened.

The lion-tailed macaque, with its fine full beard, is surely the handsomest of the genus. Unfortunately it is also the rarest, being intensively hunted for its flesh and fur. The population in the wild is estimated at only about 400 individuals today.

It is no simple matter to catch one of the shy lion-tailed macaques in front of the camera in the moist tropical forests of southwestern India. Their favorite abode is in the mighty trees, from which they seldom descend to the ground.

Macaques (*Macaca*)

Nomenclature English common name Scientific name French German	Approximate Size Body length Tail length Weight	Distinguishing Features	Reproduction Gestation period Young per birth Weight at birth
Crab-eating monkey; long-tailed macaque; Java monkey *Macaca fascicularis*, 21 Subspecies Macaque de Buffon Javaneraffe; Langschwanzmakak	males 16–22 in.; 40–55 cm females 16.2–20; 38–50 m. 17.2–26 in.; 43–65 cm f. 16–22; 40–55 m. 11–19.8 lb; 5–9 kg f. 6.6–13.2; 3–6	Long-tailed macaque is the most monkey-like species of macaque; short arms and legs; fur yellow-green, gray-green, brown-green to reddish-brown; muzzle dark; skin above eyes white; head hair smooth or tossled; blue abdominal skin; menstrual swelling highly variable	167–193 days 1 About 12.5 oz; 350 g
Rhesus monkey *Macaca mulatta*, 5 subspecies Rhésus; macaque rhésus Rhesusaffe; Rhesusmakak	m. 19.2–25.6 in.; 48–64 cm f. 18–22; 45–55 7.6–12.8 in.; 19–32 cm m. 14.3–26.4 lb; 6.5–12 kg f. 12.1; 5.5	Tail of medium length; face pale not heavily pigmented; fur brown, olive-brown, and yellow-brown; large area of naked skin on buttocks; no marked menstrual swelling, but skin of buttocks is red during fertile period	135–194 days 1; twins very rare About 16 oz; 450 g
Japanese macaque *Macaca fuscata* Macaque de Japon Rotgesichtsmakak	m. 2.9–3.2 ft; 88–95 cm f. 2.5–2.8; 79–84 4 in.; 10 cm m. 22–30.8 lb; 10–14 kg f. 17.6–22; 8–10	General coloring brown or gray; face and hind parts naked, red in adults; no menstrual swelling; limited period of pairing	about 173 days 1 About 17.8 oz; 500 g
Formosan rock macaque *Macaca cyclopis* Macaque de Formose Formosamakak; Rundgesichtsmakak	m. 18–22 in.; 45–55 cm f. 16–20; 40–50 10.4–18 in.; 26–45 cm 8.8–17.6 lb; 4–8 kg	Brown face with hairless forehead; beard; abdominal skin slightly blue; fur predominantly dark olive-brown; pronounced menstrual swelling around the buttocks, vulva, and base of tail	about 165 days 1 About 14.2 oz; 400 g
Bonnet macaque *Macaca radiata*, 2 subspecies Macaque toque Indischer Hutaffe	m 20–24 in.; 50–60 cm f. 14–18.8; 35–47 m. 20–27.2 in.; 50–68 cm f. 14–22.8; 35–57 m. 12.1–19.8 lb; 5.5–9 kg f. 7.7–9.9; 3.5–4.5	Crown of hair on head; naked, wrinkled forehead; face light brown; predominant color gray-brown; orbital ridges; no menstrual swelling	150–170 days 1 About 14.3 oz; 400 g
Toque macaque *Macaca sinica*, 3 subspecies Macaque bonnet chinois Ceylon-Hutaffe	m. 16–21.2 in.; 40–53 cm f. 14–17.2; 35–43 m. 19.2–24 in.; 48–60 cm f. 16–22.4; 40–56 m. 8.8–12.1 lb; 4–5.5 kg f. 5.5–9.9; 2.5–4.5	Smallest macaque species; crown of hair on head; unpigmented face pale brown; fur gray; olive-gray or reddish-brown; no menstrual swelling	Same as bonnet macaque
Assamese macaque; Himalayan macaque; hill monkey *Macaca assamensis*, 2 subspecies Macaque d'Assam Assammakak; Assamrhesus; Bergrhesus	m. 20–29.2 in.; 50–73 cm f. 20–24.8; 50–62 7.6–15.2 in.; 19–38 cm m. 22–31.9 lb; 10–14.5 kg f. 17.6–26.4; 8–12	Pronounced snout and eyebrows; face pale; suggestion of crown of hair on head; fur yellow to dark brown; no menstrual swelling in young adult females	Presumably, much like bonnet macaque
Tibetan stump-tailed macaque *Macaca thibetana* Macaque de Thibet Tibetmakak	2 ft; 60 cm 2–3 in.; 5–7.5 cm m. 26.4 lb; 12 kg	Long, dense gray-brown fur; ears haired; crown of hair on head	Presumably, much like bonnet macaque
Barbary ape *Macaca sylvanus* Magot; singe de Berberie Berberaffe; Magot	15.2–30.4 in.; 38–76 cm 0 11–28.6 lb; 5–13 kg	Thick, dense fur; yellow-ocher all over; tailless; face pale, and covered with many short hairs; moderate menstrual swelling, with light purple-blue skin set with small, fine hairs	About 7 months 1, rarely 2 About 16 oz; 450 g
Pig-tailed macaque *Macaca nemestrina*, 3 subspecies Macaque à queue de cochon; macaque maimon Schweinsaffe	m. 17.2–30.8 in.; 43–77 cm f. 18.8–22; 47–55 4.8–10 in.; 12–25 cm m. 9.9–29.7 lb; 4.5–13.5 kg f. 7.7–15.8 3.5–7.2	Predominantly olive-brown, yellow-brown, or reddish-brown; short curly tail; white fringe beard; extensive menstrual swelling with red skin in adults	162–186 days 1 About 17.8 oz; 500 g
Lion-tailed macaque *Macaca silenus* Ouandérou Bartaffe; Wanderu	m. 18–24.4 in.; 45–61 cm f. 16–18; 40–45 9.6–15.2 in.; 24–38 cm m. 11–22 lb; 5–10 kg f. 6.6–13.2; 3–6	Black fur; tasseled tail; black face with erect, silver-white fringe main; eyelids lighter; pronounced menstrual swelling, with pink skin	162–186 days 1 About 16 oz; 450 g
Celebes black ape; Celebes crested macaque *Macaca nigra*, 2 subspecies Pagion nègre; papion huppé Schopfmakak; Schopfaffe	m. 20–24 in.; 50–60 cm f. 17.6–22; 44–55 0.8 in.; 2 cm Not known	Long muzzle with high, bony cheek ridges; large retractile crest; predominantly gray-black to deep black; black face; pronounced menstrual swelling, with light pink skin	Presumably, like pig-tailed macaque
Moor macaque *Macaca maura* Macaque maure Mohrenmakak	m. 2–2.3 ft; 60–70 cm f. 1.8; 55 0.8–1.6 in.; 2–4 cm Not known	Predominantly gray to gray-black; broad, dark grey face; projecting round muzzle; cheek ridges and crest absent; pronounced menstrual swelling, with light pink skin	Presumably, like pig-tailed macaque

COMPARISON OF SPECIES

Life Cycle Weaning Sexual maturity Life span	Food	Enemies	Habit and Habitat	Occurrence
Age 6–12 month Females about 3.5 years, males about 4.5 years In captivity, to 38 years	Fruits, buds, sprouts, leaves, insects, spiders, crustaceans, shellfish, also cereals	Clouded leopard, panther, possibly also golden cat; python	Tree dweller in dense primeval forest, often along rivers; also in secondary or mangrove forests; readily descends from trees to ground; swims well; social group includes several males (20–60 individuals); territory size 100–250 acres (40–100 hectares)	Abundant and widely distributed
Age 6–12 months Females about 3.5 years, males about 4.5 years Over 30 years	Fruits, seeds, roots, buds, sprouts, greens, leaves, small invertebrates, inner bark, resin, cereals	—	Highly adaptable and widely distributed in forest from tropics to temperate zone; in mountain forests to 9600 ft (3000 m); also in human settlements, even large cities; chiefly grounddwelling; social groups include several males (10–100 individuals); territory size 12.5 acres (5 hectares; in town) to 4000 acres (1600 hectares; hill forest)	Still fairly abundant, but diminishing
Age 6–12 months Females about 3.5 years, males about 5 years Over 30 years	Fruits, berries, nuts, young leaves, grasses, greens, bark, buds, insects, spiders, snails, crayfish, birds' eggs, mushrooms	Lynx, coyote	Terrestrial and arboreal in highlands and along coast; northernmost simian species, sometimes in snow-covered terrain ("snow monkey"); social groups average 20–100 individuals; territory size 7.5–11.5 mi² (20–30 km²), often smaller	Total population estimated at 35 000–50 000; diminishing and threatened
Presumably like rhesus monkey	Fruits, berries, buds, sprouts, roots, insects, crustaceans, shellfish	Clouded leopard	Terrestrial on sparsely treed or treeless coastal rocks and in the hill country of Taiwan; social groups presumably include several males	Displaced into remote regions and threatened
Age 6–12 months Females 2.5–4 years, males 4–7 years Over 20 years	Fruits, nuts, seeds, buds, sprouts, flowers, leaves, insects, cereals	Dogs, leopard	Arboreal and terrestrial in woods and open terrain up to elevations of 5100 ft (2200 m) around and within settlements and temples; social groupc include several males (7–76 individuals); territory size 100–625 acres (40–520 hectares)	Fairly abundant
Much like bonnet macaque	Much like bonnet macaque	Leopard	Terrestrial and arboreal, in close or open, wet or dry woods as well as on shrubby hills; social groups include several males (7–76 individuals); territory size 42.5–287 acres (17–115 hectares)	Endangered; 0.3–100 individuals per km² (1 km² = 0.38 mi²); subspecies *M. s. opisthomelas* seriously threatened
Presumably, like bonnet macaque	Fruits, young leaves, insects and other small prey, also cereals	Not known	Predominantly terrestrial in woods of Himalayas at elevations up to 7200 ft (2250 m); social groups include several males (10–100 individuals); habits largely unknown	Not known
Presumably, like bonnet macaque	Omnivorous	Presumably, leopard	Chiefly terrestrial dweller in hill forests of eastern Tibet at elevations between 2500 and 6400 ft (800 and 2000 m)	Not known
Age 6–12 months Age 3–4 years Over 20 years	Leaves, inner bark and seeds of cedars, berries, greens, buds, sprouts, roots, invertebrates	Not known	Predominantly terrestrial dweller in cedar and oak forests of central Atlas Mountains; usually at elevations between 4800 and 6400 ft (1500 and 2000 m); social groups include several males (12–40 individuals); territory size 0.4–0.6 mi² (1.2–1.7 km²)	Population about 25 000; seriously threatened by hunting and habitat destruction
Age 6–12 months Age 2.5–4 years Over 30 years	Fruits, seeds, nuts, young leaves, sprouts, buds, greens, mushrooms, invertebrates	Tiger, leopard, python	Terrestrial and arboreal dweller in tropical rain forests and humid secondary forests; social groups include one or several males (5–50 individuals); adult males often solitary; territory size about 1250 acres (500 hectares)	Stock decreasing rapidly; dwarf pigtailed macaque (*M. n. pagensis*) most severely threatened
Age 6–12 months Age 2.5–4 years Over 20 years	Fruits, flowers, nuts, young leaves, buds, sprouts, cardamon pulp, invertebrates	Not known	Arboreal dweller in dense, wet tropical forests; social groups include one or several males (4–35 individuals); territory size 250–1250 acres (100–500 hectares)	Most severely threatened macaque species; population estimated at only 405 individuals in 1975
Presumably, like pig-tailed macaque	Fruits, flowers, young leaves, buds, sprouts, insects, caterpillars	Not known	Arboreal and terrestrial forest dweller; social groups include one or several males (about 20 individuals)	Black crested macaque (*M. n. nigra*) seriously threatened by habitat destruction and hunting
Presumably, like pig-tailed macaque	Fruits (especially figs and bamboo seeds), buds, sprouts, invertebrates; also cereals (maize)	Not known	Arboreal and terrestrial in dense stands in mountainous regions, near or on limestone rock; refuge in caves; social groups include one or several males (about 20 individuals)	Habitat much fragmented; destroyed for plundering crops, hence threatened

MACAQUES

Nomenclature English common name Scientific name French German	Approximate Size Body length Tail length Weight	Distinguishing Features	Reproduction Gestation period Young per birth Weight at birth
Booted macaque *Macaca ochreata*, 2 subspecies Macaque à bras gris Grauarmmakak	males 20–24 in.; 50–60 cm females 16–20; 40–50 1.2–2.4 in.; 3–6 cm Not known	Slightly prolonged muzzle; poorly developed cheek ridges; face black; fur predominantly black-gray or brown; buttocks, inside of thighs and upper arms, shanks and forearms slightly ocher-gray; pronounced menstrual swelling	Presumably, like pig-tailed macaque
Tonkean black ape; Tonkean macaque *Macaca tonkeana*, 2 subspecies Tonkeanamakak	m. 21.2–27.2 in.; 53–68 cm f. 16.8–24; 42–60 1.2–2.4 in.; 3–6 cm Not known	Shorter muzzle than Celebes crested macaque, and less-developed cheek ridges; small crest; face black; fur predominantly gray-black; buttocks and legs light gray; menstrual swelling	Presumably like pig-tailed macaque
Stump-tailed macaque; bear macaque *Macaca arctoides*, 3 subjects Macaque à queue en trognon Bärenmakak; Stumpfschwanzmakak	15.2–28 in.; 38–70 cm 0.6–3.2 in.; 1.5–8 cm m. 19.8–28.6 lb; 9–13 kg f. 13.2–26.4; 6–12	Dark brown, yellow-brown, or reddish-brown fur; short, nearly hairless tail; forehead wrinkled, becoming bald in older males and in females; face skin pale in juveniles, later spotted black and red; no menstrual swelling	166–185 days 1 10.1–21.4 oz; 350–600 g

The long snout with prominent cheek ridges, the blackish face, and especially the large crest, which is lowered at will, are characteristic of the crested macaque or ape, found only on the Island of Celebes (Sulawesi).

The baboons (Genera *Papio, Mandrillus, Theropithecus*)

"Baboon" is a collective term for a fairly integral group (the genera *Papio*, *Mandrillus* and *Theropithecus*) to which the largest and heaviest cercopithecids belong. With their sturdy arms and legs, which are of equal length, they stand firmly on the ground. They are thus certified to be confirmed terrestrials. The animals of this group have been called "dog apes" with reason, for they have a distinctly canine shape of the head. This is emphasized by a long, blunt snout with an angular outline and a dog-like nose at the tip. The small eyes are deep set under prominent superorbital ridges. There is a very marked difference in size between the sexes; the adult males are nearly double the weight of the females. When gaping, the males bare their enormous canine teeth. The sharp posterior edge renders these dagger-like teeth dangerous. To make room in the closed mouth for the great canines, they slide past each other, the lower canines fitting into a gap in front of the upper canines.

The baboons are closely related to the African mangabeys and to the predominantly Asian macaques. We distinguish two principal genera: *Papio*, comprising the yellow baboon and hamadryas baboon, which live in open country, and *Mandrillus* with two species of ridge-cheeked baboons living on the floor of the forest. We also include the gelada, in its own genus *Theropithecus*. How closely these genera are related is illustrated by the fact that many scholars, for example the mammalogist Theodor Haltenorth, place them all in a single genus *Papio*.

The YELLOW BABOONS (*Papio cynocephalus* and allies). The general term "steppe baboon" embraces at least four species inhabiting different parts of Africa but so much alike that many systematists have recently regarded them as mere subspecies of the best-known representative, the YELLOW BABOON (*Papio cynocephalus*).

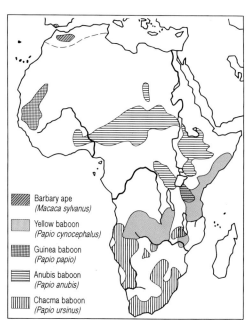

Barbary ape (*Macaca sylvanus*)
Yellow baboon (*Papio cynocephalus*)
Guinea baboon (*Papio papio*)
Anubis baboon (*Papio anubis*)
Chacma baboon (*Papio ursinus*)

COMPARISON OF SPECIES

Life Cycle Weaning Sexual maturity Life span	Food	Enemies	Habit and Habitat	Occurrence
Presumably, like pig-tailed macaque	Much like moor macaque	Not known	Occurs in sparsely populated regions of southeastern Celebes; habits not much studied, probably similar to moor macaque	Relatively unthreatened
Presumably, like pig-tailed macaque	Much like moor macaque	Not known	Dweller in dense, hardly accessible hill forests in sparsely settled areas of central Celebes	Relatively unthreatened
Age 6–12 months Males 6–8 years, females 4 years Over 30 years	Preference for fruits and (young) leaves; occasionally small prey	Leopard, python	Predominantly terrestrial in dense forests of plains and mountains at elevations up to 6400 ft (2000 m)	Seriously threatened by deforestation, and especially by warfare within range of distribution

Retaining the traditional division into species, there are the following three to be mentioned besides the yellow baboon proper: the GUINEA or SPHYNX BABOON *(Papio papio)*, the ANUBIS or GREEN BABOON *(P. anubis)*, and the CHACMA or BEAR BABOON *(P. ursinus)*.

Due to their accessibility, these baboons have been studied for a long time by numerous zoologists. They are ground dwellers, mostly in open country – open woodlands, savannas, and steppes, as well as semi-deserts and rocky terrains. Visitors to the Amboseli National Park in Kenya sometimes happen upon a troupe of baboons with a couple of behavioral scientists in their midst, accompanying the animals on their day's march. That means showing up by sunrise at the sleeping trees, to be on the spot in time for the first assembly. The daily progress ends in the late afternoon when the animals arrive at the same or other sleeping trees. They may have traversed from 3 to 12 mi (5 to 20 km).

The baboons in the Amboseli Park sleep in tall acacia trees. They use the same trees only once, or at most for a few days in succession. Then they move on to new sleeping places, not returning to the starting point until after quite a long time. In this way they are best protected against reinfestation by parasites excreted in the feces.

In regions where sleeping trees are scarce, baboons retire at night into difficulty accessible rock terrain. The anubis baboons in Kenya, for example, are very well protected in the vertical rock

Single-male group of the Ethiopian hamadryas or sacred baboon on the troupe's sleeping rocks. The females and juveniles stay close to the large male, with the silver-gray mane.

walls to be found everywhere in the Rift Valley of East Africa. For these animals, it is still more important to avoid contaminated camp grounds until they have been "biologically cleaned."

The first stage of the day's march is covered quickly, but gradually the animals will fan out over the terrain to seek food. Some pick flowers, pods, berries and buds; others pull up grasses. They will eat the tender juicy stalk parts, or strip the seeds from the blade. In times of drought, they find nourishment on the steppe, where no green meets the eye, by digging bulbs, root stocks, tubers, and sprouts of grasses and herbs from under the ground. They will often find water by digging a hole in a dry river bed. Every stone or elephant dropping is turned; there will probably be something edible crawling under it.

A comparison of diets in different regions shows how adaptable the baboons are in their choice of food. In the Cape Province, the chacma baboons of the coastal rocks go out on the beach at ebb tide to catch crustaceans, mollusks, and snails in pools and crevices. In the hard Black Rocks of Angola, the local yellow baboons subsist chiefly on lichens growing on the rock. Where opportunity presents, young birds and young mammals such as hares and gazelles may be caught and eaten. In the Gilgil, however, a form of behavior seems to be developing that looks like deliberate hunting. Researchers Strum and Harding report how certain animals, chiefly males, would traverse an area at a fact pace, some distance from each other, thereby considerably improving the chance of encountering young animals. This may have been an example of a locally evolved tradition, made possible because the original predators of this agricultural region have been killed off; in such a niche, the baboons do not have much to fear from venturing into open country in a small group.

In their community life also, the yellow baboons are somewhat flexible. The basic pattern is in each instance a multi-male group of from about 30 to over 100 individuals; occasionally, hordes of up to about 200 individuals have been observed. The size and cohesion of the group apparently depend on various factors, notably the quantity and distribution of food and the presence of lions or leopards. Thus, cohesion is strengthened noticeably when a troupe passes through dangerous terrain. In the case of chacma baboons, it has been found that extreme food shortage will cause indi-

(Above) With space-devouring stride, this anubis baboon travels on the ground, the principal abode of all baboons. (Right) A small troupe of yellow baboons passing through the Amboseli National Park in Kenya.

Male hamadryas baboon's "yawn of menace." His dagger-sharp canines are very formidable weapons, used in intraspecies confrontations, in repulsing enemies, and in the hunt for small vertebrates.

After roaming the open steppe, a troupe of anubis baboons finds shelter in sleeping trees at evening, but will use them only once or at most for a few nights at a time.

viduals to forage singly or in small groups, apparently accepting the heightened risk involved.

The importance of the predator hazard as a factor of group cohesion is illustrated also by the situation in Suikerbosrand, a preserve to the south of Johannesburg, where there have been no predators for many baboon generations past. A horde of chacma baboons living there exhibits much loosening of the group bond. Frequently, "single-male" groups forage independently. This resembles the situation among hamadryas baboons with a very important difference. One male will not "guard" certain females, as a harem-keeping hamadryas would do. Consequently, membership in these groups is not fixed.

As in the case of the closely related macaques, the female baboons form the permanent core of the group. They remain for life in the group of their birth. The center is the matriarch, surrounded by siblings with their descendants. Close relatives also support the growing youngsters of their tribe when they seek to enlarge the limits of their influence and discover which possible confrontations with other individuals will succeed and which will not.

As a rule, the males leave their birth group upon reaching adult age, and may change from group to group later on as well.

Behavioral researchers have noted that among the yellow baboons in the Amboseli National Park, males in the prime of life will often transfer to a new group. Remarkably, such a newcomer usually attains the highest rank very quickly. As a result, males in general will hold a lower rank, the longer they have remained in their group, even though they are maintaining more firm reltionships. A baboon male seeking to join a strange group will encounter a richly graduated texture of relationships. Especially if he enters a horde of considerable size, he will seek to form a bond with the females of a particular tribe. From such an attempt, lasting connections may arise; the animals will often rest and travel together – and the male will protect the females and their young from attacks by other group members. That he is thus protecting offspring that are not his own, and do not represent his genetic interests, may be accepted if, as a result, a female will prefer him as a mating partner on the next occasion.

This means that the females influence the mating choice. When a female is in heat, the adult males will compete for a so-called husband relationship to that female. Such a relationship has the appearance of connecting the pair as if by a rubber band. The male will dog the female's footsteps. The two will search for food, rest, and engage in grooming together. They will prefer to travel somewhat apart from the horde. During this relationship, matings occur. The partnership may last for one or a few hours, or for several days. Usually the males will change places as "husband." Field researchers have repeatedly observed that the high-ranking males in the group have the best prospects. They will strive for a "husband" relationship especially in the few days of maximum female genital swelling, which indicates fertility. Immediately thereafter, the swelling recedes rapidly, and the female loses her attractiveness. The competition is keenest in the maximum phase. And it is intense, because it is unlikely in a group of medi-

Not only trees, but also steep cliffs, difficult of access to enemies, serve as sleeping quarters for the hamadryas baboon.

▷ A young anubis baboon riding on his mother's back during the long day's march. This is the typical mode of baby transport for the baboons of the steppe.

um size that several females will be fertile at the same time. Most of the females are either pregnant or nursing.

On the day when the swelling is at its peak, usually the highest-ranking male, while still in the sleeping tree, will establish a husband relationship. But the process is not without tension. The other males cannot be induced to ignore the pair, and they will try to follow at some distance. This makes the husband somewhat nervous. He keeps looking around at the other males, frequently yawns violently to display his canine teeth, and urges the female, insofar as she will cooperate, away from the other males.

But the rivals are excited also. Certain individuals embrace and mount each other more and more frequently, vigorously smacking their lips. These are conciliatory expressions, confirming association and portending the formation of an alliance. Eventually there may be a brief struggle in which the previous husband may have to yield and the leader of the alliance takes over the husband relationship. My associate Ronald Roë, who has studied these alliances and mating exchanges in the yellow baboon of Kenya, found that certain low-ranking males become comparatively successful in assuming husband roles by entering into coalitions. The low-ranking males may have succeeded in this because they have been in the group for a long time and maintain relationships of confidence with certain other males.

The mating partnerships, then, depend on relations of influence and on successful male alliances.

But it would be a great mistake to suppose that the female herself would not influence the result. Her willingness to mate with a particular male probably in large measures determines his ability to start a husband relationship and maintain it against others. Rivals can tell by the female's behavior whether she can be easily "alienated." Hence good relations with the female are manifestly very important to the males.

Relationships between groups of baboons are not particularly friendly, but not markedly hostile either. The group that has strayed farthest from the center of its own territory and entered the territory of a neighboring group will give way to the other. Territories are evidently so large that it does not pay to defend them against intruders.

Now and then, however, combat will occur, serving to stake out the groups' spheres of action. Once I had the good fortune to witness a contest between two very large groups, each of more than 100 members. In the company of the American

(Right) Heated altercations like this between two male baboons usually pass off harmlessly. (Below) Two adult male yellow baboons have formed an alliance and together are driving off a rival.

primate researcher Shirley Strum, I was following one of her specimen groups in the Gilgil of Kenya. All at once, far ahead, we heard loud barks and cries. As we approached, we saw the two groups on a long front of about 320 ft (100 m) facing and threatening each other in the bush savanna. In so doing, the animals would keep glancing about at the members of their own group and occasionally make sallies towards an adversary. Now and then, this individual activity would escalate to somewhat broader combat. If any opponents gave way, several would chase them with loud barking. This would immediately set the whole front in motion, members of one group charging down upon the fleeing members of the other. After 160 ft (50 m) or so, another halt would be made, as the pursuers gradually fell back. The cycle repeated itself as soon as the foremost pursuers in turn retreated. Real fighting took place only briefly here and there. We did not get the impression that there were any serious casualties. After about half an hour, the zeal of the combatants cooled, and the intruding group withdrew without pursuit.

Although the baboons of the steppe are affected by the restricting influence of humans, they still occur in large parts of Africa, where their mode of life renders them highly adaptable. Their most important natural enemy is the leopard. Lions are feared as well. Where the two big cats are no longer found, humans and their dogs are the major hazard.

(Left) He follows tirelessly in her tracks, as though the two partners were joined by a rubber band, in the phase of female readiness to mate. The considerable difference in size between the sexes is characteristic of all baboons. (Bottom) Group separation at close quarters: three hamadryas baboons have rounded up their harems.

The HAMADRYAS OR SACRED BABOON *(Papio hamadryas)*. The English named him the "sacred baboon" because of the special significance of the hamadryas baboon in ancient Egyptian religion. Since the fourth millennium B.C., it was worshipped as a guardian and representative of Thoth, the god of science and the art of writing.

As among other baboons, the males are much larger than the females. They are distinguished also by their fur color and the imposing mane on head and shoulders.

Like the yellow baboons, the hamadryas is a pronounced ground dweller. In the Ethiopian and southwestern Arabian parts of his range, it chiefly inhabits dry, treeless semi-desert and rocky terrain up to elevations of about 6400 ft (2000 m). Its distribution is limited by the circumstance that in the course of a day's march it must be able to reach at least one watering place, even in the dry season. The home territory of a group or "band" (see below) measures about 11 mi² (30 km²) in extent.

The hamadryas baboon receives much scientific attention primarily because of an extraordinary social organization, compared with what we know of other baboons and the macaques. We owe our knowledge especially to the Zurich zoologist Hans Kummer and his associates. Within the "troupe" of several hundred animals studied in the Danakil semi-desert of eastern Ethiopia, three further levels of organization may be distinguished. Groups of about 60 individuals, called "bands," by Kummer, travel about together and may jointly enter into confrontations. Each band in turn is composed of several "clans" whose grown males are often so much alike that one would suppose them to be close relatives. Now it appears to be the fact that the hamadryas differs from other species in that it is not the females, but the males, who remain members of their native clan.

A harem-keeper's life is not an easy one. The hamadryas male must be constantly on guard to keep his females together. If they will not mate or stray too far away, he may become violently infuriated.

The most striking feature of the social texture of the hamadryas baboons is the presence of single-male harem groups. Within a clan, there will be one or two powerful males, each of which possesses some of the adult females of the clan. That is to say, each steadily guards one or more females; he prevents relationships with other males, and compels his females to always remain in his immediate vicinity and to follow him. Hence he is also the undisputed father of the young in his harem.

A clan will often include an older male that has lost his consort to his descendants, as well as one or more "successor" males – younger, nearly adult, and soon to attempt to establish harems of their own.

A harem owner gets his females to follow him by threatening them if they stray farther than he likes. Often he will run upon such a female and bite her in the back of the neck. The females know that the only way to escape such chastisements is to keep as close to the male as possible. Apparently the females must learn this. Female yellow baboons, for example, would shun males that abused them. Kummer performed an experiment in which he released female anubis baboons in a hamadryas group. Certain males immediately claimed them for themselves, and then tried to keep them in train with bites in the nape. The yellow baboon females soon discovered that attempts at flight only made matters worse, and within about an hour they had learned to follow their master. Conversely, hamadryas females in a yellow baboon group quickly learned that no male would demand this obedience.

Whereas the female role in the adaptation to the social structure is thus a matter of experience, it is otherwise with the male behavior. Where hamadryas and anubis ranges adjoin, hybrid groups are to be found – groups in which crosses of the two species are represented. Presumably they owe their genesis to the fact that it is sometimes easier for young adult hamadryas males in this region to abduct a female from a neighboring group of anubis baboons than to win a female within their own horde.

If the possessive attitude of the males towards the females and the attendant modes of behavior such as guarding had been acquired by learning in the group tradition of the hamadryas baboons, then the features of male hamadryas behavior should be uniformly expressed in groups with hybrids as well. But this is not the case. The observers determined that some males exhibit pronounced guarding behavior, others not, or to only a minor extent. The extent to which they exhibited such behavior seemed to match the extent to which they displayed physical hamadryas traits. This is a very strong indication that the difference in male behavior that accounts for the difference in social structure between hamadryas and yellow baboons has a hereditary (genetic) base.

An adolescent hamadryas male finds to his frustration that all females in his clan are "bespoke." Violent displacement of a male whose strength has begun to wane is quite possible, but not a suitable enterprise for untried young males. Instead, a "successor" male will try to found a relationship with a young virgin of his own clan. By following her, "mothering" her, carrying her about on his back, and grooming her, he gradually accustoms not only her but the other group members to the relationship. Because he begins when the female is still far from sexually mature, he meets with little

The bite in the back of the neck is the male hamadryas baboon's most effective method of subduing a refractory female.

or no resistance from other adult males. Young males may also abduct young females more violently. In either case, with a little patience, young males may thus acquire their first consorts.

It is remarkably how patiently fellow members of a band – males, that is, who are well acquainted with each other – accept ownership of females among themselves. Kummer conducted experiments in Ethiopia with temporarily captive individuals. When a strange female was placed in a cage with a male, he would immediately try to exact submissive behavior. Now if a fellow member who had been able to observe this process at some distance was also admitted to the cage, he would exhibit conspicuous lack of interest in the female. He avoided any attention that might be taken as threatening. And he behaved thus even if he was the other male's superior and could have appropriated the female by force. This "respect" for the liaisons of fellow members averts a rivalry that might jeopardize community life.

This the more so insofar as the males remain with their male relatives, so that it is also in their own genetic interest for the clan to maintain itself successfully within the group.

That the members of the band attach value to joint action is shown by a remarkable process that cannot be interpreted in any other way than as a consensus concerning the route of the common journey in the daily search for food. To these animals, which must range far over dry, barren country to obtain at least the minimum of food necessary to sustain life, a correct course that will moreover bring them to a watering place about midday is very important. Before starting out early in the morning, the males regularly greet one another with conspicuous anal "presentation," that is, a display of the posterior. Meanwhile, occasionally one of the males will make a brief excursion in a certain direction. If the others do not follow, he returns and sits down again. Then another male may try a start in a different direction. After a few such attempts, the entire band may start off on a certain course. If this happens, and only then, all clans of the band will be found in the afternoon at a common resting place, lying in the direction jointly adopted in the morning, even though soon after the start the clans went their own ways and lost contact with each other, arriving at the rendezvous after long detours. It must be concluded

The hot-headed hamadryas baboon males can also be very tender in "loving."

from this that the first, hesitating starts are for the purpose of comparing opinions as to the proper course and submitting them for consideration. But the suggestions can be compared only if each gives the beholder a "mental image" of what lies beyond the horizon in the indicated direction. The result is therefore highly relevant to the controversial question whether animals such as these

have an active power to imagine things not present "here" and "now." This would be different from the modest faculty of recognizing objects perceived on an earlier occasion when encountered anew.

The MANDRILL *(Mandrillus sphinx)* and DRILL *(M. leucophaeus)*. These two stump-tailed baboon species, with their nearly identical build, are very different in appearance. The drill, with its coal-black face encircled by a white wreath of hair, is a wholly different apparition from the mandrill, with its facial skin colored brilliant red and blue, framed by a yellow-orange beard on the chin and cheek. In their behavior and ecology, however, the two are very much alike.

Both are ground dwellers in the high, moist forests on the Bight of Biafra. The grown males spend most of their time on the ground. Although the far smaller females and the young climb trees somewhat more frequently to pick the fruit, they also spend more time on the ground. There they collect fallen fruits, turn over stones, fallen leaves, and branches in search of insects and other small prey, pick up seeds and nuts, and dig up tubers, roots, and herbs. It has been reported that mandrills occasionally visit the fields of the native farmers to collect manioc and the fruit of the oil palm. For sleeping, mandrills and drills climb trees, but the grown males remain on the lower level.

Since visibility on the forest floor is often no more than 32 to 64 ft (10 to 20 m), rather little has yet been learned about the habits of these two species. We do know that there are two types of group formation. Single-male groups averaging about 14 members are led by an adult male of enormous stature compared to the others. Then there are

Baboons live chiefly on vegetable fare but do not dislike animal food sources. While one anubis baboon has killed a small gazelle and is bringing it to a place of safety (top), his fellow plays the part of an ostrich-egg thief (bottom).

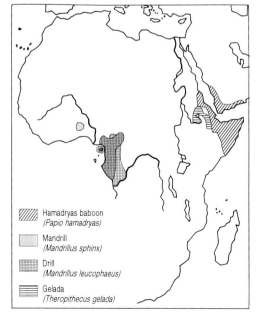

//// Hamadryas baboon *(Papio hamadryas)*
▦ Mandrill *(Mandrillus sphinx)*
▦ Drill *(Mandrillus leucophaeus)*
≡ Gelada *(Theropithecus gelada)*

▷ Leaning close to the harem owner, whose gray coat can be discerned in the background, the mothers nap side by side. Only one of the babies is awake.

multi-male groups that may number up to 200 individuals. Hoshino and his associates have observed how these larger communities occasionally arise through the coalescence of smaller, multi-male or single-male groups, and how they will split up again into these components. It is not known whether the females are attached to a particular adult male, or whether they change membership as groups unite and divide. Active "guarding" would seem hardly feasible in the obstructed environment. Evidently the females "voluntarily" follow a male, who constantly announces his whereabouts on the move by uttering an unmistakable two-phase snarl. It is frequently sounded also in answer to crowing sounds of group members who have ventured out of sight. That is, they regard this male as the permanent focus of their attention. They are aided by the color contrasts in the face and posterior of the male, conspicuous in the confused light and shade of the forest floor.

The amalgamation into larger multi-male groups is evidently seasonal. From October to March, many widely scattered trees in the forest are heavy with fruit. Effective utilization of this food supply requires independent ranging in smaller units. The rest of the time, when the forest floor must be searched for anything edible, larger groups can do this just as well; and may be preferable for reasons of safety. It is not known to what extent drills and mandrills are preyed upon by enemies or what protection their social organization affords.

At first glance, the grown males especially have a ferocious appearance. This impression is reinforced by a remarkable expressive gesture in which the corners of the mouth are drawn back and the lips are pulled up over the canines, while remaining closer together in the center. The effect is reminiscent of a human face contorted by wrath. The terrifying impression is intensified when the beast shakes its head as though about to devour one.

But an interpretation based on superficial formal resemblances can be very misleading. In the Arnhem Zoo, there lived for many years a mandrill who would greet me enthusiastically when he saw me coming. As I would pass through the zoo with visitors, they were very much impressed at my extending my arm to the supposedly raging wild beast. Instead of attacking, he would begin to "groom" my arm eagerly, with loud smackings. For in reality, this behavior is a modification of the conciliatory baring of the teeth that we observe in other monkeys also. It is a gesture of friendly greeting. The horizontal shaking of the head is the opposite of the exaggerated vertical nodding of the head and upper body that the drill and mandrill engage in when they are threatening an adversary – a ritualized incipient pounce. The shaking of the head, in other words, literally negates the presence of aggressive intent.

Both species, the mandrill and the drill, are seriously threatened in nature, by deforestation as well as by being hunted for their flesh.

The GELADA *(Theropithecus gelada)*. Although some researchers believe that the gelada is closer to the macaque, and also has much in common with *Cercopithecus*, the animal shows a striking resemblance to the baboons, especially the hamadryas baboon,

(Top) "Presentation" is a conciliatory gesture between males not intimate enough for mutual grooming. (Bottom) Female with genital swelling "presenting" to a male.

which also lives in the Ethiopian region. The resemblance relates not only to size and shape and the possession of an imposing mane by the male, but also to certain features of community life. The similarity is a convergence, that is, the matching features were evolved independently in response to comparable selective pressures. However, we shall note important differences as well.

Geladas live in the treeless world of rocks at mountain elevations from 6400 to 16,000 ft (2000 to 5000 m). At night they sleep in large troupes on the face of cliffs, where they are safe from predators. During the day, they forage on alpine meadows. Their diet consists almost exclusively of grasses, of which they consume the seeds, the blades, and the dug-up sprouts and root stocks. They strip off the seeds with their teeth or between thumb and forefinger. Fruits, leaves, tubers and roots as well as small prey supplement their frugal fare.

They wander about timidly throughout the day, never straying far from their cliff wall, to which they retreat when danger threatens.

Geladas live in great hordes sometimes comprising more than 600 individuals. As in the case of the hamadryas baboons, there are bands of 50 to 300 individuals that usually range together. Within these bands, single-male harem groups and bachelor groups can be distinguished in turn.

Whereas the hamadryas male holds together a harem of females, usually not related, by means of threats, and the males are united in kinship societies, the gelada way is entirely different. Here it is again the females whose kinship forms the fixed, matriarchally organized core. An adult male may attempt to join such a little group and repulse the attempted approaches of other males.

The expressive behavior of geladas includes a peculiar element, namely the so-called "lip flip." They draw back the corners of the mouth and flip the upper lip over the nose, exposing a shiny, bright pink surface of mucous membrane. This gesture can be seen from afar. It was at first interpreted as threatening. We now know it to be derived from conciliatory tooth baring. Like the corresponding gesture of the mandrill, it has developed into a calming and encouraging gesture of greeting.

A special feature of this species is the bright red hourglass-shaped patch of naked skin on the throat and breast of both males and females. In the females, it is encircled by a string of "pearls" – small blisters that swell during the fertile phase of the cycle. At the same time, the bare skin takes on a deep red color. Thus, much the same process takes place on the breast of the gelada female as in the genital region, where blisters also appear and the skin reddens. Only the process is not so easy to observe in the genital region, because the animals are usually sitting or crouching.

The German behavioral scientist Wolfgang Wickler first pointed out that the expressive forms of the genital region are imitated on the breast. Especially the unusual location of the nipples, close together near the ventral midline, reminds one of the structure of the vagina and clitoris.

The gelada is considered to be seriously threatened. Perhaps about half a million individuals re-

The mandrill is one of the primate species that are at home primarily on the ground.

▷ The male mandrill constantly wears a kind of "war paint." His powerful teeth, too, show that he is not to be trifled with.

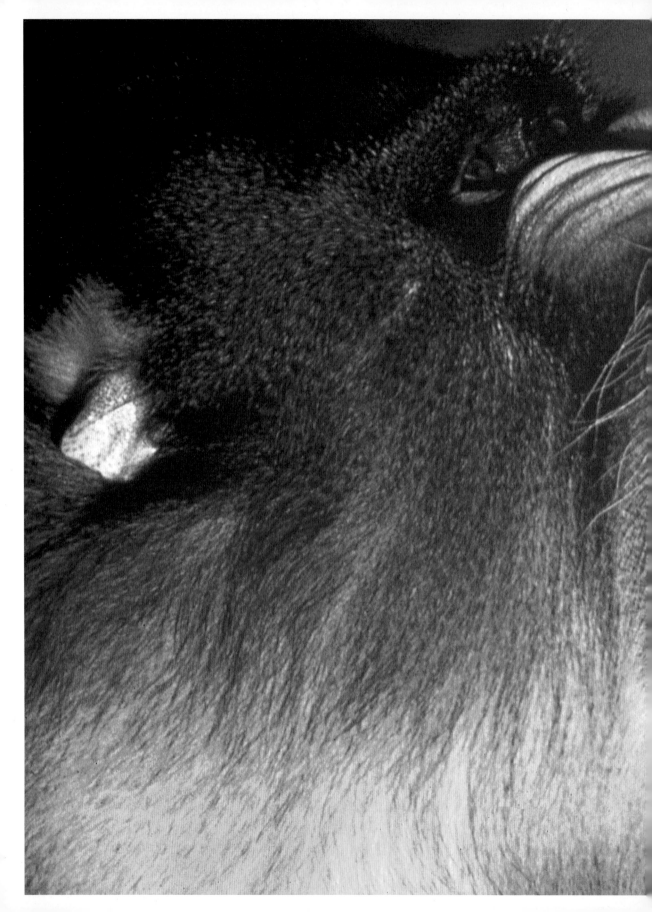

Brilliant colors with a signal effect. Only the male mandrill has such a colorful face; the female markings are far less conspicuous. This is related to the social life of the species; in the larger single-male groups, the male leader is the very center, and all female glances are directed at him. In the confused light and shade of the forest, this high-contrast coloration facilitates visual contact.

main. They are not only being displaced by human intrusion, but seriously interfered with in their social structure as the adult males are shot down. Their fur with its magnificent mane is highly prized.

Mangabeys (Genus *Cercocebus*)

The visitor to the African rainforest soon notices how varied the primate fauna is compared to that of the open country, such as the wooded savanna. There, from one to three species will be found in a given location. But if one counts the simians living, say, in the primeval forest of Makokou in Gabon, the figure is at least seven: talapoin monkey, crowned guenon, moustached monkey, spot-nosed or greater white-nosed monkey, De Brazza's monkey, gray-cheeked mangabey, and agile or crested mangabey.

In the variegated three-dimensional world of the forest, each of these species has defined its own niche. This only overlaps partly with the niches of other primates, so that they can exist side-by-side without being displaced by competition. The wealth of *Cercopithecus* forms is especially overwhelming. These animals, with their extremely long tails, are clearly typical tree dwellers.

The mangabeys of the genus *Cercocebus* (the somewhat misleading common name derives from the town of Mangabé on Madagascar, where these simians do not in fact occur) are at first glance very similar to *Cercopithecus*. They too are arboreal acrobats, moving effortlessly in the third dimensions with their slender bodies and long tails. Upon closer examination, however, the superficial re-

semblance is found to conceal important differences, raising the question of how to account for these differences phylogenetically.

Unlike *Cercopithecus*, the mangabeys do not live in single-male groups. There are often several males present, rather peacefully disposed towards each other. Groups roam over large territories, and do not exert themselves to expel others. Females exhibit menstrual swellings; males have absolutely and also relatively larger testes than *Cercopithecus*. This suggests a social order like the one we found among the macaques.

These differences between *Cercocebus* and *Cercopithecus* doubtless reflect differences in the evolutionary potential inherited by them, which controlled their specialization in forest life. The mangabeys, as we now know, are closely related to the baboons and the macaques.

Among the mangabeys *(Cercocebus)*, two subgenera

(Left) Gelada male with long shoulder mane and naked red patch on breast, fringed with white hairs, for which this species has been called the red-breasted baboon. The animal's seemingly threatening posture is actually a conciliatory and inviting gesture of greeting. (Right) In this gelada interaction, the animal at left is expressing submission by "lip lifting."

BABOONS AND MANGABEYS

Baboons (Papio, Mandrillus, Theropithecus) and Mangabeys (Cercocebus)

Nomenclature English common name Scientific name French German	Approximate Size Body length Tail length Weight	Distinguishing Features	Reproduction Gestation period Young per birth Weight at birth
Yellow baboon *Papio cynocephalus*, 4 subspecies Papion, babouin, cynocéphale Steppenpavian; Gelber Pavian	males 1.8–2.8 ft; 55–84 cm females 1.1–2; 35–60 m. 1.7–2.2 ft; 53–66 cm f. 1.1–1.8; 35–56 m. 44–55 lb; 20–25 kg f. 17.6–35.2; 8–16	Slender with comparatively long limbs; round head, domed skull, short muzzle; silver-white sideburns; black face; predominantly yellow-green fur; tail carried oblique upward, conspicuously arched, with last two-thirds turned downward; menstrual swelling with red skin	173–193 days 1, twins rare Not known
Guinea baboon *Papio papio* Cynocéphale rouge; babouin de Guinée Guineapavian; Sphinxpavian; Roter Pavian	1.6–2.8 ft; 50–83 cm 1.5–2.3 ft; 45–70 cm Not known	Smallest baboon species; reddish-brown fur; black face; yellow-brown sideburns; nape and shoulder hair long in adult males, forming light hood; tail carried in round arc; menstrual swelling	Much like yellow baboon
Olive baboon; anubis baboon *Papio anubis*, 7 subspecies Papion anubis; babouin doguera Anubispavian; Grüner Pavian	m. 2.3–3.2 ft; 70–95 cm f. 1.6–2.6; 50–80 m. 1.5–2 ft; 45–60 cm f. 1.3–1.5; 38–45 cm m. 66 lb; 30 kg f. 33; 15	Heavy build; long muzzle; black face; dark olive-green fur; first third of tail directed upward, then hangs down; naked violet-brown, skin on buttocks of males and females not in heat, pink in pregnancy	Much like yellow baboon
Chacma baboon *Papio ursinus*, 8 subspecies Chacma; cynocéphale chevelu Bärenpavian; Tschakma	m. 2.3–3.5 ft; 70–114 cm f. 1.6–2.6; 50–80 1.1–2.8 ft; 35–84 cm m. 74.8 lb; 34 kg f. 37.4; 17	Largest baboon species; very long, narrow muzzle; face black; no sideburns; very dark brownish-gray fur; tail like that of yellow baboon	Much like yellow baboon
Hamadryas baboon; sacred baboon *Papio hamadryas* Hamadryas; tartarin; papion à perruque Mantelpavian	m. 2–3.1 ft; 60–94 cm f. 1.6–2.2; 50–65 1.2–2.1 ft; 35–61 cm m. 39.6 lb; 18 kg f. 22; 10	Marked sex difference in size and coloring; males, long silver-gray shoulder mane; females, gray-brown; pink face and naked skin of buttocks; pronounced menstrual swelling	170–173 days 1 Not known
Mandrill *Mandrillus sphinx* Mandrill Mandrill	m. 2.2–3.2 ft; 65–95 cm f. 1.6–2.2; 50–65 2.8–4.8 in.; 7–12 cm m. 44–61.6 lb; 20–28 kg f. 24.2; 11	Pronounced muzzle; bony ridges on either side of bridge of nose, with light blue skin grooved lengthwise; nose and lips red; penis bright red; scrotum red; naked skin of buttocks red, purple and blue; small, but conspicuous, pear-shaped menstrual swelling	167–176 days 1 Not known
Drill *Mandrillus leucophaeus*, 3 subspecies Drill Drill	Same as mandrill	Pronounced muzzle with bony ridges; face coal black, framed with white ring of hair; dark brown fur; striking coloration of buttocks in red, violet, blue, and green tones; penis bright red; scrotum blue and violet	Same as mandrill
Gelada *Theropithecus gelada*, 2 subspecies Gélada Dschelada; Blutbrustpavian	m. 2.3–2.5 ft; 69–74 cm f. 1.6–2.2; 50–65 m. 18–20 in.; 45–50 cm f. 12–16.4; 30–41 m. 46.2 lb; 21 kg f. 30.8; 14	Short skull; round muzzle; nostrils directed laterally; brown fur; male with long shoulder mane; naked, patch shaped like an hourglass on throat and breast, encircled by white hairs in the male, and white vesicles in the female, which swell during estrus; eyelids light in color	6 months 1 Not known
Collared mangabey; sooty mangabey *Cercocebus torquatus*, 2 subspecies Mangabey à collier blanc Halsbandmangabe	m. 1.5–2.2 ft; 46–67 cm f. 1.4–1.8; 42–56 m. 1.6–2.5 ft; 50–76 cm f. 1.5–2.1; 46–64 m. 11–19.8 lb; 5–9 kg f. 8.8–15.4; 4–7	White or light gray sideburns; black face with bright white eyelids; large cheek pouches; predominantly gray fur; tail with sharp bend carried forward over back; small, pear-shaped menstrual swelling	about 170 days 1 Not known
Agile mangabey; crested mangabey *Cercocebus galeritus*, 3 subspecies Mangabey à ventre doré Haubenmangabe	m. 1.4–2.1 ft; 42–62 cm f. 1.3–1.7; 40–52 m. 1.8–2.5 ft; 55–76 cm f. 1.5–2; 45–60 m. 15.4–28.6 lb; 7–13 kg f. 9.9–15.4; 4.5–7	Fur brown with sprinkling of yellow hairs; crest; otherwise like collared mangabey	Much like collared mangabey
Gray-cheeked mangabey *Cercocebus albigena* Mangabey à joues grises; mangabey à gorge blanche Mantelmangabe	m. 1.8–2.4 ft; 54–72 cm f. 1.6–2.1; 50–61 m. 2.4–3.3 ft; 72–100 cm f. 2.3–3; 68–90 m. 13.2–24.2 lb; 6–11 kg f. 8.8–15.4; 4–7	Fur predominantly black; brown-gray shoulder mane of long hair; projecting eyebrows; throat pouch in male; otherwise like collared mangabey	Much like collared mangabey
Black mangabey *Cercocebus aterrimus* Mangabey noir Schopfmangabe	m. 1.5–2.1 ft; 45–62 cm f. 1.3–1.7; 40–52 m. 2.2–2.8 ft; 65–85 cm f. 1.8–2.5; 55–75 8.8–24.2 lb; 4–11 kg	Glossy black fur; high black crest; broadly projecting sideburns; tail carried pointing up and curved forward; otherwise like collared mangabey	Much like collared mangabey

COMPARISON OF SPECIES

Life Cycle **Weaning** **Sexual maturity** **Life span**	Food	Enemies	Habit and Habitat	Occurrence
Age 6–8 months Age about 5 years 30–45 years	Grass, greens, seeds, fruits, tubers, roots, leaves, nuts, cereals; also, invertebrates, young birds, small mammals	Leopard; lion, hyena dog, python, eagle	Chiefly terrestrial in open woods, tree and bush savannas, and steppes, near gallery woods or rock hillocks; social groups include several males (10–150 individuals); territory size varyies widely (0.7–19 mi^2; 2–50 km^2)	Abundant
Much like yellow baboon	Much like yellow baboon	Much like yellow baboon	In dry forests, gallery forests and adjoining bush savannas or steppes; otherwise much like yellow baboon	Abundant
Much like yellow baboon	Much like yellow baboon	Much like yellow baboon	In tree, bush, or grass savannas and steppes, also in forests; otherwise much like yellow baboon	Abundant
Much like yellow baboon	Same as yellow baboon; on coast, also crustaceans and shellfish	Much like yellow baboon	Chiefly terrestrial dweller on grass and bush savannas or rocky terrain, from the seacoast to elevations of 9600 ft (3000 m); otherwise much like yellow baboon	Abundant
Age 8 months Age about 5 years Over 37 years	Grass, roots, tubers, seeds, nuts, fruits; invertebrates, also small vertebrates	Same as yellow baboon	Terrestrial dweller in semi-desert regions, steppes or rocky terrain; single-male harems within groups of about 60 individuals; average territory size 11.5 mi^2 (30 km^2)	Abundant
Age 6–12 months Age about 5 years In captivity, up to 46 years	Fruits, leaves, seeds, nuts, roots, greens, invertebrates	Leopard	Dwells in low levels of trees, and especially the forest floor; single-male harems of up to 20 individuals, often united in bands of 200; territory size about 19 mi^2 (50 km^2)	Seriously threatened
Much like mandrill	Much like mandrill	Same as mandrill	Much like mandrill	Seriously threatened by hunting and by destruction of forest
Age 1.5–2 years Males 8 years, females 3.5 years Over 20 years	Almost exclusively grass	Leopard, humans, dog	Ground dweller in rocky mountain terrains of central and northern Ethiopia, at elevations of 6400–16 000 ft (2000–5000 m); graze on high meadows; sleep on steep cliffs; single-male harems combined in troops of over 600 animals; territory size 0.5–0.9 mi^2 (1.5–2.5 km^2)	Threatened by habitat destruction and hunting
Age not known Age 5–7 years Over 30 years	Fruits, seeds, palm, nuts, young leaves, grass, mushrooms, invertebrates	Leopard, eagle	Arboreal and terrestrial dweller in lower levels of rain forests; found in social groups (10–25 individuals) including several males	Seriously threatened by deforestation and hunting
Much like collared mangabey	Fruits, nuts, seeds, grass, mushrooms, buds, invertebrates	Leopard, eagle	Arboreal and terrestrial dweller in lower levels of rain, marsh and river forests; found in social groups (10–35 individuals) including several males; territory size 0.07–0.7 mi^2 (0.2–2 km^2)	Tana crested mangabey (*C. q. galeritus*) endangered; remains only in 1825 acres (730 hectares) of river forest
Much like collared mangabey	Fruits, nuts, flowers, seeds, buds, invertebrates, small vertebrates	Leopard, cat, golden eagle	Dwells in upper levels of trees in of rain forests; found in social groups (6–30 individuals) including several males; territory size 0.09–1.5 mi^2 (0.25–4 km^2)	Threatened by destruction of habitat
Much like collared mangabey	Fruits, seeds, sprouts, buds, leaves, invertebrates	Same as gray-cheeked mangabey	Much like gray-cheeked mangabey	Much like gray-cheeked mangabey

Long-tailed monkeys I (Cercopithecus)

Nomenclature English common name Scientific name French German	Approximate Size Body length Tail length Weight	Distinguishing Features	Reproduction Gestation period Young per birth Weight at birth
Mona monkey *Cercopithecus mona* Mone Monameerkatze	1.1–1.8 ft; 35–55 cm 2.3–3 ft; 70–90 cm 5.5–13.2 lb; 2.5–6 kg	Long, dense sideburns; flesh-colored mouth parts; dark eyebrow band; white forehead band; back red-brown; no menstrual swelling	5 months 1 Not known
Campbell's monkey *Cercopithecus campbelli* Mone de Campbell Campbells Meerkatze	Much like mona monkey	White-gray side whiskers; blue face; pink mouth parts; predominantly gold-green upper parts; no menstrual swelling	Much like mona monkey
Crowned guenon *Cercopithecus pogonias* Mone pogonias Kronenmeerkatze	Males 1.6–2.2 ft; 50–66 cm Females 1.2–1.5; 38–46 m. 2–2.9 ft; 60–87 cm f. 1.6–2.3; 50–68 m. 6.6–13.2 lb; 3–6 kg f. 3.9–6.6; 1.8–3 kg	Yellow side whiskers; black face; yellow-green upper parts; black midline stripe; no menstrual swelling	Much like mona monkey
Wolf's monkey *Cercopithecus wolfi* Mone de Wolf Wolfs Meerkatze	Much like mona monkey	Dark gray side whiskers; otherwise much like Campbell's monkey	Much like mona monkey
Dent's guenon *Cercopithecus denti* Mone de Dent Dents Meerkatze	Much like mona monkey	Predominantly brown back; otherwise similar to Campbell's monkey	Much like mona monkey
Moustached monkey *Cercopithecus cephus* Moustac Blaumaulmeerkatze	1.3–2 ft; 40–60 cm 1.6–2.6 ft; 50–80 cm 6.6–14.3 lb; 3–6.5 kg	Bushy side whiskers; blue face with white stripes on upper lip; green-brown upper parts; no menstrual swelling	5 months 1 Not known
Black-cheeked white-nosed monkey; redtail monkey *Cercopithecus ascanius* Ascagne Rotschwanzmeerkatze	Much like moustached monkey	White side whiskers; blue around the eyes; white or yellow spot on nose; posterior end of tail red-brown; no menstrual swelling	Much like moustached monkey
Red-eared monkey *Cercopithecus erythrotis* Moustac à oreilles rousses Rotohrmeerkatze	Much like moustached monkey	White to yellow side whiskers; blue around the eyes; red ears and nose; no menstrual swelling	Much like moustached monkey
Red-bellied monkey *Cercopithecus erythrogaster* Hocheur à ventre rouge Rotbauchmeerkatze	Much like moustached monkey	White or white-yellow side whiskers; blue around the eyes; white spot on nose; red-brown or gray breast and back; white throat stripe; no menstrual swelling	Much like moustached monkey
Lesser white-nosed monkey *Cercopithecus petaurista* Hocheur blanc-nez Kleine Weißnasenmeerkatze	Much like moustached monkey	White side whiskers; gray-blue, hairy face; white nose spot, throat stripe, and ear margin; no menstrual swelling	Much like moustached monkey
Blue monkey *Cercopithecus mitis* Singe à diadème; singe argenté Diademmeerkatze	1.3–2.2 ft; 40–70 cm 2.2–3.4 ft; 70–102 cm 8.8–26.4 lb; 4–12 kg	Fur predominantly blue-gray, yellow-gray or reddish-gray; white stripe on forehead; dark face; no menstrual swelling	5 months 1 About 14.3 oz; 400 g
White-throated guenon; Sykes monkey *Carcopithecus albogularis* Cercopithèque à collier blanc Weißkehlmeerkatze	Much like blue monkey	Fur predominantly olive-gray; dark face; white chin and throat, not sharply marked off from surrounding gray; no menstrual swelling	Much like blue monkey
Spot-nosed monkey; greater white-nosed monkey *Cercopithecus nictitans* Hocheur Große Weißnasenmeerkatze	Much like blue monkey	Fur uniformly gray to yellow-gray; dark face; nose spot of very short, dense, white hairs; no menstrual swelling	Much like blue monkey

are distinguished: Torquatus, the COLLARED MANGABEY *(Cercocebus torquatus)* and the AGILE or CRESTED MANGABEY *(C. galeritus);* and ALBIGENA, the GRAY-CHEEKED MANGABEY *(C. albigena),* and the BLACK MANGABEY *(C. aterrimus).*

Recent immunogenetic studies have led to the hypothesis that these subgenera split off independently from an original baboon-macaque complex, and have since undergone separate development. Accordingly, some systematists have proposed regarding the Albigena subgenus as a genus, *Lophocebus,* in its own right.

COMPARISON OF SPECIES

Life Cycle Weaning Sexual maturity Life span	Food	Enemies	Habit and Habitat	Occurrence
Age about 1 year Age 2–3 years Over 30 years	Fruits, sprouts, young leaves, insects, invertebrates	Crested eagle; also python, leopard, and golden cat	Dwells in upper levels of trees in rain forests; apparently lives in single-male social groups, sometimes united in larger troops (up to 50 individuals); territory size 5–50 acres (2–20 hectares)	Still quite abundant
Much like mona monkey	Much like mona monkey	Same as mona monkey	Much like mona monkey	Much like mona monkey
Much like mona monkey	Much like mona monkey	Same as mona monkey	Much like mona monkey	Much like mona monkey
Much like mona monkey	Much like mona monkey	Same as mona monkey	Much like mona monkey	Much like mona monkey
Much like mona monkey	Much like mona monkey	Same as mona monkey	Much like mona monkey	Much like mona monkey
Age about 6 months Age 2–3 years Not known	Fruits, young leaves, sprouts, flowers, invertebrates	Crested eagle, leopard	Dwells in lower and middle levels of trees in of rain forests or secondary forests; social groups include one male; territory size 0.007–0.4 mi² (0.02–1.2 km²)	Still quite abundant
Much like moustached monkey	Much like moustached monkey	Same as moustached monkey	Much like moustached monkey	Much like moustached monkey
Much like moustached monkey	Much like moustached monkey	Same as moustached monkey	Much like moustached monkey	Seriously threatened
Much like moustached monkey	Much like moustached monkey	Same as moustached monkey	Much like moustached monkey	Seriously threatened
Much like moustached monkey	Much like moustached monkey	Same as moustached monkey	Much like moustached monkey	Still fairly abundant
Not known	Fruits, leaves, flowers, sprouts, buds, invertebrates, small vertebrates	Leopard, crested eagle, golden cat, python	Tree dweller in rain forests, hill forests and secondary forests; in single-male social groups of 15–40 individuals; territory size 25–175 acres (10–70 hectares)	Still fairly abundant
Not known	Much like blue monkey	Same as blue monkey	Much like blue monkey	Much like blue monkey
Not known	Much like blue monkey	Same as blue monkey	Much like blue monkey	Much like blue monkey

A close relative of the mandrill is the very similar drill, which lacks the colorful face mask. As among most of the terrestrial primates, the female is considerably smaller and less conspicuous than the male.

The species within a subgenus are geographically separated. Representatives of the two subgenera occupy different ecological compartments, and so may occur in the same area, as, say, crested and gray-cheeked mangabeys in the Dja forest of Cameroon and the Makokou forest of Gabon.

OLD WORLD PRIMATES

The two representatives of the Albigena mangabeys, the GRAY-CHEEKED MANGABEY *(Cercocebus albigena)* and the BLACK MANGABEY *(C. aterrimus)*, are definitely arboreal, staying almost exclusively in the upper layers and treetops of the forest. They are often encountered in association with other monkeys, especially *Cercopithecus* species. But the mangabeys draw upon their own food supply, because they can crack hard nuts and seeds with their teeth, as well as fruit stones that *Cercopithecus* cannot use. This ability is betokened by their strong jaws and comparatively large incisors.

The mangabeys live in groups of 6 to 30 individuals, including about the same numbers of adult males and adult females. The males of a group, especially the highest-ranking males will utter a loud two-syllable call, sounding like "hon-hon." Resonance sacs in the throat lend great volume to this cry. It is produced only a few times a day, especially at the morning concert. Not only members of the same group react to it by assembling; other groups will respect it by selecting an alternative route. Thus, the call tends to even out the distribution of groups in the forest.

The CRESTED MANGABEY *(Cercocebus galeritus)* and the COLLARED MANGABEY *(C. torquatus)* are not so restricted to arboreal life as their cousins the Albigena mangabeys. They live in the lower strata of the forest, especially in swamp forests, and like alternative travel on the ground. They will sometimes venture into cultivated fields, where they particularly esteem peanuts.

The crested and collared mangabeys are extraordinarily mobile animals, exhibiting lively expressive behavior. Eye play commands special attention, the movements of the eyebrows and upper eyelids being very conspicuous. In many simian species, the pigmentation of the upper eyelid is somewhat less pronounced than that of the surrounding facial skin, being somewhat protected from the Sun's rays. The trait is grotesquely exag-

The tropical rainforest of Africa is the home of the mangabeys and of most *Cercopithecus* species.

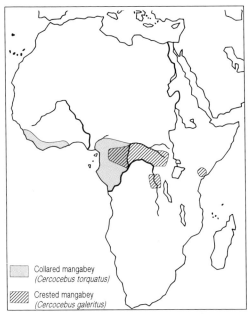

gerated in the collared mangabey, the area being wholly unpigmented. As a result, two contrasting spots are seen in an otherwise deep black face. If one animal looks at another, and raises its eyebrows, the other cannot ignore this; two eye spots will immediately "light up."

Guenons (Genus *Cercopithecus*)

Within the subfamily Cercopithecinae, two supergenera may be distinguished. On the one hand there are the genera *Cercocebus* (mangabeys), *Macaca* (macaques), *Papio* baboons), *Mandrillus* (cheek-pouched baboons), and *Theropithecus* (geladas). These are closely related simians, fairly large, which – except for the mangabeys – have evolved

towards terrestrial life; the three genera last mentioned, in fact, are markedly terrestrial. The other group comprises the genera *Cercopithecus*, *Miopithecus* (talapoin), *Allenopithecus* (Allen's swamp monkey or guenon), and *Erythrocebus* (patas monkey or red guenon), which agree in some traits. The patas monkeys are exceptional in this group, because, as long-legged, swift forms, they have adapted to life on the savanna.

The (Dutch and German) word *Meerkatze* is interpreted as referring to "cat-like" monkeys that have reached us from "overseas" *(übers Meer)*. In fact, it is a corruption of the Sanskrit word *marka-ti*, meaning simply ape or monkey, and also the source of the modern generic name Macaca. Norwegian *marekatt* and Netherlands *meerkat* (also archaic English) show the origin of the word more clearly. Dutch settlers in the Cape Province of South Africa were also misled by the "cat" sound and applied the name to the mongoose-like suricate *Suricata tetradactyla.* The Dutch word *meerkat* has since been borrowed into English for a second time in this sense.

The *Cercopithecus* monkeys are called in English "guenons" (a French word for she-monkey). The *Cercopithecus* monkeys are a fairly integral genus. Still, it contains 21 species with a total of more than 70 subspecies, nearly all of them living in or near the forest regions of central Africa. The division into species probably took place in geologically recent times, when in the dry phases of the Ice Age the rainforests shrank into isolated patches, so that the several populations developed separately from each other for a long time.

(Left) The green monkey, perhaps the most familiar representative of the genus *Cercopithecus*. (Below) Mangabeys at a glance. Gray-cheeked mangabey (top), collared or red-headed or sooty mngabey, and black mangabey (center); crested mangabey and moor mangabey, a subspecies of the collared mangabey (bottom).

OLD WORLD PRIMATES

Owing to their comparatively short snouts and moderate superorbital ridges, the *Cercopithecus* monkeys have neat rounded heads. The face is mostly covered with short, dense hair in some species, often leaving a clear space around the eyes only. Varicolored facial markings asre notable.

Cheeks, chin and neck are sometimes adorned with fringes of hair and beards, contrasting in white, gray, yellow or reddish tones with the gray, greenish or black background color of the fur. These are manifestly signal colors, associated with the identification of species. As we shall see, groups of some guenon species often roam about for considerable periods in company with groups of other species living in the same region. Whatever the advantages of such an association may be, it involves the risk of hybridization. Yet crosses are extraordinarily rare. When several species live in the same area, one is sure to displace the others in the long run, if all occupy the same environmental niche, that is to say if they have exactly the same habits. Closely related species can live together stably in the same station only if they are differently specialized, thus avoiding competition for limited supplies of goods. But when this is the case, a hybrid will generally be less successful in maintaining itself in either of the species-specific realms than the pure species.

The multiplicity of species seems at first glance chaotic, but closer inspection shows that (following Osman Hill and John Napir), they may all be reduced to about eight subgenra:

These subgenera reflect the principal ecological specializations (environmental adaptations). Species of the same subgenus never occur in the same area (except for a few overlappings of evidently recent origin). In other words, they are species that have developed on either side of a geographic barrier. On the other hand, representatives of different subgenera may well occur together in the same area.

The danger of extinction is drawing nearer and nearer for some species. The animals in the West African rainforest belt especially are endangered. They are not only shot to be sold on the market for meat or in order to protect fields and gardens; the forest itself is vanishing at a dismaying pace. At least, the diana monkey, the red-bellied monkey, the red-eared monkey and L'Hoest's monkey must be regarded as seriously threatened.

The MONA MONKEY *(Cercopithecus mona)* and allies. The representatives of this subgenus are all fairly small and are typical rainforest dwellers. They occupy the uppermost and intermediate levels of the foliage. They run along the branches and take flying leaps from the outermost twigs of the crown into the branches of the next tree, where they land in a vertical posture, grasping with all fours to regain a hold. Occasionally one will miss its aim. As a rule, however, the fall will be broken sufficiently so that no harm is done. The American zoologist I.T.Sanderson once observed a whole troupe of mona monkeys in flight, dropping one after an-

In Africa south of the Sahara, green monkeys (above and right) are met with nearly everywhere; unlike most of their relatives, they do not live hidden in dense forest but prefer open terrain, on the margins and in the adjoining savanna. They move with equal speed and skill in the branches and on the ground.

264

Subgenus	Species	Common Name
Mona	*Cercopithecus mona*	Mona monkey
	C. campbelli	Campbell's monkey
	C. pogonias	Crowned guenon
	C. wolfi	Wolf's monkey
	C. denti	Dent's guenon
Cephus	*C. cephus*	Moustached monkey
	C. ascanius	Black-cheeked white-nosed monkey, redtail monkey
	C. erythrotis	Red-eared monkey
	C. erythrogaster	Red-bellied monkey
	C. petaurista	Lesser white-nosed monkey
Mitis	*C. mitis*	Blue monkey
	C. albugularis	White-throated guenon, Sykes monkey
	C. nictitans	Spot-nosed monkey, greater white-nosed monkey
L'Hoesti	*C. l'hoesti*	L'Hoest's monkey
	C. preussi	Preuss's monkey
Hamlyni	*C. hamlyni*	Owl-faced monkey
Neglectus	*C. neglectus*	De Brazza's monkey
Diana	*C. diana*	Diana monkey
	C. dryas	Dryas monkey
Aethiops	*C. aethiops*	Grivet, savanna monkey
	C. sabaeus	Green monkey
	C. pygerythrus	Vervet, vervet monkey

more quickly. Even more than the other guenons, the little monas must fear the crested eagle, which often outwits their watchfulness. Like many birds of prey, this eagle lives in pairs, male and female often hunting together. One will fly over and draw all frightened monkeys' eyes to itself, while the other bird comes up from behind and snatches a victim.

Mona monkeys are noisy; as they travel through the foliage, they keep uttering expressive sounds. It has been thought that the name "mona" refers to these "moaning" calls, but it is almost certain that it derives from the Moorish word "mona," designating long-tailed monkeys.

In the Mona group, we also find an exception to the rule that species of the same subgenus cannot occur in the same area. The mona monkeys spread out on the one hand into the range of CAMPBELL'S MONKEY *(Cercopithecus campbelli)* and on the other hand into that of the CROWNED GUENON *(C. pogonias).* In the areas of overlap, the two species are gradually acquiring different habits. Mona monkeys like river banks, where crowned guenons are less often to be met with.

The MOUSTACHED MONKEY *(Cercopithecus cephus)* and allies. Monkeys of the subgenus *Cephus* are of medium size. Their highly diverse coloration and marking of the face are noteworthy. Blue, red, yellow, white, and gray hairs are scattered over their faces in fantastic patterns.

▷ The black face and white headband, as well as the conspicuous red penis and blue scrotum, are distinguishing features of the green monkey. Its fur is not actually green, but possibly yellow-green or even gray.

Guenons are temperamental animals, but the squabbles between group members are for the most part harmless.

other into a stream that they intended to cross. "They all uttered loud screams, but swam ashore as readily and quickly as dogs, and climbed back up into the branches." Of all members of the genus, they eat the least leaves and the most insects. Most of all, they eat fruits. They accept sprouts and young leaves only when fruits are scarce.

Mona monkeys are mobile, lively simians, observed in groups of up to 50 individuals. Presumably such larger groups, in which several adult males may be found, are composed of single-male groups, temporarily combined. Smaller groups almost invariably include one adult male only. It may be advantageous for the animals to unite in larger bands because many eyes will detect danger

All species of this subgenus are tree dwellers of the original rainforest, although they are encountered also in secondary forests. They are swift, and generally streak about in the lower levels of the leaf cover. They eat fruits and insects as well as young leaves, blossoms, sprouts, and shoots.

Moustached monkeys differ from the other members of the subgenus in that they have adapted to a particular kind of food. Their principal diet is the flesh of the oil palm *(Elaïs)*. This tree bears fruit the year round. The monkeys remain in the vicinity of where the palms grow, and their home territories there are considerably smaller than in the primeval forest, where the palm is absent.

Cephus monkeys live in groups of about 20 animals, led by an adult male. Single-male groups are characteristic of the forest-dwelling monkeys. However, a study of the REDTAIL MONKEY *(C. ascanius)* has shown that occasional penetration of the group by other males is not ruled out. Such was the case in one group of 26 individuals. It happened when the females came into heat. The invaders participated in sexual intercourse, and may have begotten young. This took place not without strenuous resistance on the part of the ruling male; nevertheless, they succeeded in remaining in the group for some time.

The remarkable point about this situation is that there are two strategies for reproductive success. In one, the object is to strive for exclusive access to the females; in the other, to consummate a mating occasionally. It would be natural to suppose that a superior male would have the greatest reproductive success. In the case cited, it was actually verifiable that a certain high-ranking male had the readiest access to the females. On the other hand, the "cost" was manifestly high. Overexertion and injuries soon compelled it to stand aside. Then there was another male that only achieved a mating now and then, but was still at hand at the end of the mating season, and remained with the females after all the other males had moved on. Will his residence in the group give him better chances in the next mating season, for example because the females will accept him more readily? The question is, which strategy was ultimately more successful? The balance of costs and benefits, along with other factors, determines whether a single-male or a multi-male system will prevail. Among the redtails, apparently, there is an equilibrium between the two alterntives.

The BLUE MONKEY *(Cercopithecus mitis)* and allies. This subgenus comprises three species, the blue monkey *(C. mitis)*, the white-throated guenon *(C. albogularis)*, and the greater white-nosed monkey *(C. nictitans)*. All told, there are 15 subspecies. These are comparatively large monkeys, contrasting in behavior with the lively representatives of

Perhaps the handsomest *Cercopithecus* species, the diana monkey, here pictured in two different color phases, is hardly ever seen by the visitor to Africa because it inhabits the upper levels of the forest, and has, moreover, become progressively rarer.

Mona monkey (*Cercopithecus mona*)
Campbell's monkey (*Cercopithecus campbelli*)
Dent's guenon (*Cercopithecus dentil*)
Wolf's monkey (*Cercopithecus wolfi*)
Crowned guenon (*Cercopithecus pogonias*)

the subgenera *Mona* and *Cephus*, with which they often travel in company. They are more reserved and relaxed in their comportment.

All species are decidedly arboreal, feeling most at home in the tall, dense forest trees. They are found in mountain forests up to elevations of about 9600 ft (3000 m), and also in swamp forests and even papyrus swamps.

The monkeys of the Mitis subgenus follow the general pattern of the genus, being primarily fruit eaters, while insects and other invertebrates are a minor component of the diet.

It has recently been reported that blue monkeys occasionally even look upon vertebrates as food. Once a wood owl was observed to be taken by surprise and devoured, and on two occasions a bush baby, a nocturnal prosimian. In certain regions where greater white-nosed monkeys live near human habitations, they will plunder crops and even sometimes steal chicks. To do this, they must come down to the ground, which they do not normally like to do.

Blue monkeys and their relatives live in groups of 10 to 40 individuals, including as a rule only one adult male. They are moreover territorial, that is, they defend their territory against intruders of their own species. The threat of a defensive foray discourages the groups from trespassing on the territory of neighbors. The warlike mood of the neighbor group is announced by signals transmitted especially by the leading male. Like most other territorial inhabitants of the primeval forest, these monkeys have a very loud call, whereby other groups can tell that they are near, and where.

Recent studies have shown how these calls are suited to their purpose – long-distance communication. In the deep forest, sounds are muffled even over short distances. Experiments have shown that low-frequency sounds at a pitch of about 125 to 200 Hz have the maximum penetrating power. The call of the blue monkey, by which it marks its territory, is just about 125 Hz. Their vocal apparatus has definite peculiarities enabling them to produce loud sounds of this low pitch.

Mitis monkeys are often observed in mixed groups with other guenons and with mangabeys or colobi. These mixed groups do not arise simply because the animals are attracted by the same food sources, for examle trees with a brief but abundant crop of fruit. The animals coordinate their excursions. There are two reasons for such accompaniment. One group may be led to a food source by another. The impression is gained that in the Kibale forest, blue monkeys join redtails,

Three representatives of the Mona subgroup: mona monkey (upper left), Campbell's monkey (upper right), and crowned guenon (bottom).

who utilize a much smaller territory and therefore have better knowledge of the state of varying food sources locally from time to time. The redtails doubtless tolerate the increased competition because they too reap an advantage. There is no doubt that larger groups are better protected against the depradations of the crested eagle.

An association with animals of another species, rather than an enlargement of the group itself, represents a lesser increase in competition for food, because there will be at least some difference in food habits between the two. Thus, redtails search for insects in the foliage, whereas blue monkeys catch them on trunks and branches. Both species will also attach themselves to red colobi, which prefer a different diet. Apparently the only advantage here is safety in numbers.

Such an alliance does not necessarily imply that friendly relations will arise, such as are expressed within one species by behavior such as grooming, sitting together, playing, and mating. The parties are only "using" each other. In our introduction to this genus, we mentioned that the great differences in coloration apparently serve to limit sexual intercourse to animals having the specific traits with which they have been imprinted from infancy. How effective this system is appears from the fact that crosses between species traveling together are indeed extremely infrequent.

L'HOEST'S MONKEY *(Cercopithecus l'hoesti)* and allies. The L'Hoesti subgenus comprises two species living far apart from each other. *C. l'hoesti* lives in the eastern Congo and western Uganda as well as in Rwanda and Burundi. About 1200 mi (2000 km) to the West lie the Cameroon highlands, where PREUSS'S MONKEY *(C. preussi)* is at home. The structure of both betrays their kinship to the subgenus *Mitis*.

These dark colored monkeys are inhabitants of high mountain forests. Dense fog envelopes the peaks, especially in the early morning. In this damp and often gloomy world, the monkeys seek chiefly berries and greens, which they find in the lush undergrowth of the trees. Among the forest-dwelling guenons, these are exceptional in that they must venture out on the forest floor in search of food.

They live in small groups including one adult male. Details of their mode of life in the wild are not yet known.

Even in captivity, they are rarely to be seen. Nevertheless, it has been learned in zoos that they have an extraordinary trait for Old World monkeys: the tail can be used as a grasping tool, though to a considerably more modest extent than by certain New World monkeys. Guenons of all species have this ability during early youth, but it is then lost almost completely. Adult L'Hoest's monkeys can bend the tip of the tail into a hook. This ability is useful to them when seeking a purchase while climbing lianas. It may also give them a sense of security, for when two individuals are resting side-by-side, they will curl their tails around each other.

Preuss's monkeys are reported to live in great spherical nests, hung in thick branches. This

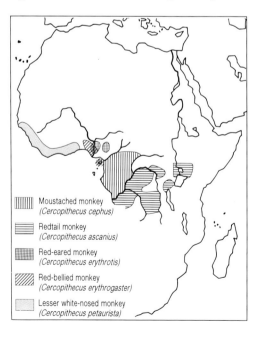

Monkeys of the subgenus *Cephus*: two lesser white-nosed monkeys of different colors (top and right), and a red-bellied monkey (left).

would be a remarkable habit for a primate. Scientific confirmation is still extant.

The OWL-FACED MONKEY *(Cercopithecus hamlyni)*. The monotony of the dark gray-green appearance of this species is relieved by a conspicuous white stripe, running from the eyebrows to the lip across the bridge of the nose.

The owl-faced monkey lives in a limited range on the right banks of the Upper Congo, in the mountain forests. An American museum has preserved the remains of an individual discovered at an elevation of about 12,800 ft (4000 m) in the mountain forests of the Virunga volcanoes. Thus the species inhabits the same range as L'Hoest's monkey. How these two species differ in their habits is as yet quite unknown. It is asserted that the owl-faced monkey is nocturnal. This would account

for the fact that the native population knows the animal by a name connoting resemblance to an owl. Pending confirmation of this alleged habit, one might consider the opposite explanation, namely that this animal's owl-like face has resulted in an attribution of owl-like behavior.

We have no certain knowledge of the social life of this species. The animals have been successfully kept in some zoos, but generally only in pairs. The coloring of the newborn differs strikingly from that of the adults.

DE BRAZZA'S MONKEY *(Cercopithecus neglectus)*. This powerfully built guenon's predominant green-gray color is relieved by the coloring of the face – a long white beard and moustache, and an orange-red forehead stripe.

The species has a very wide range of distribution in which its occurrence is remarkably uniform, compared to the multiplicity of other wild monkeys of the forest.

These monkeys inhabit the moist forests along rivers, and the swamp forests. They have been studied by Gautier-Hion in the Makokou forest of Gabon. There they shared their range with crested mangabeys and talapoin monkeys; all three species live near water. De Brazza's monkeys eat chiefly fruits. This diet is supplemented with flowers, insects, leaves, as well as mushrooms. In the search for food, the monkeys regularly come down to the ground.

De Brazza's monkeys live in small groups, presumably consisting as a rule of one pair with some descendants. The animals find what food they require within small territories in which they wander about at leisure. Because they live in rather small groups, and their behavior is not very temperamental, they do not readily attract the attention of

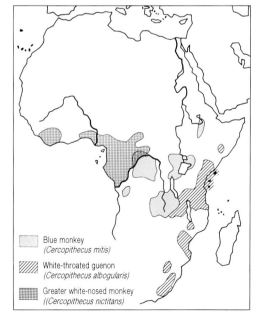

Blue monkey *(Cercopithecus mitis)*
White-throated guenon *(Cercopithecus albogularis)*
Greater white-nosed monkey *((Cercopithecus nictitans)*

The lesser white-nosed monkey is a forest dweller, ranging agilely through the lower and middle storeys.

▷ The moustached monkey, also belonging to the subgenus *Cephus*, looks as though its face had been dipped in a pot of blue paint – a curiosity among mammals in general, but common among the more colorful monkeys.

any enemies. When they sense danger, they wait motionless to see whether the threat will pass. Only if the danger becomes acute will they suddenly take flight with a loud cry. Remarkably enough, they run away on the ground.

The DIANA MONKEY *(Cercopithecus diana)* and allies. With its white goatee and conspicuous colors, this large but slender monkey makes an elegant appearance.

Besides the diana monkeys proper, which live in the primeval forest of West Africa, the subgenus includes two little-known species living in limited areas of the Congo Basin.

Diana monkeys are definitely arboreal. They live in the highest levels of the forest, where they subsist primarily on fruits. They travel about in rather large groups, up to about 30 individuals, in an extensive territory in search of trees bearing ripe fruit.

They are lively, energetic creatures, essentially more similar to the smaller monkeys of the subgenera *Mona* and *Cephus* than to the other, larger forest monkeys. Their expressive behavior is a good example of coordination of movement and traits of external conformation. Like all monkeys of the genus *Cercopithecus*, the diana monkey has a poorly developed mimetic repertory. On the other hand, head and body movements as components of expressive behavior are developed to a degree not found among the Old World monkeys of open country, such as the macaques and baboons. Movements and attitudes are moreover performed in a manner utilizing the contrasting color pattern to great effect.

The diana monkeys exhibit a threatening posture utilizing the handsome colors of the posterior; one will hang from a branch by all fours and gaze at another across the abdomen. This displays the orange-red hues on the backs of the thighs; in a V-shape, they frame the brilliant white of the neck and breast.

Two color phases of the blue monkey (top); owl-faced monkey (center left); L'Hoest's monkey (center right); De Brazza's monkey (bottom).

L'Hoest's monkey *(Cercopithecus l'hoesti)*
Preuss's monkey *(Cercopithecus preussi)*
De Brazza's monkey *(Cercopithecus neglectus)*

The GRIVET *(Cercopithecus aethiops)* and allies. When safari tourists relax on the veranda of their lodge in an East African national park, little yellow-green and gray-green monkeys with coal-black faces and long tails rush up across the lawns to see whether anything good to eat has been left over.

By this bold behavior, the grivet has become one of the most photographed of primate species. Its range of distribution is larger than that of any other *Cercopithecus* species. Unlike the others, all of which stay in the woods, the grivets live in open country. They are at home all over Africa south of the Sahara except in the rainforest and the dry desert areas. Within this extensive territory, several geographical forms may be distinguished which some scientists regard as subspecies of *Cercopithecus aethiops,* while others class them as independent species within a superspecies.

The GREEN MONKEY *(C. sabaeus)* is found in the extreme West, in Senegal. The TANTALUS MONKEY *C. aethiops tantalus,* likewise golden green but with a white headband, lives between the Sahara and the Congo forest. Well to the East, towards Ethiopia, the GRAY-GREEN GRIVET *(C. aethiops aethiops)* occurs, likewise wearing a headband and often a little white moustache on the upper lip.

The VERVET MONKEY *(C. pygerythrus)* is to be met with from North Kenya to the Cape Province. It has shorter sideburns than those aforementioned. The color of the back is gray-green, while the tip of the tail and the hands and feet are dark to black.

Michael Harrison, a student of the West African grivets or savanna monkeys in Senegal, describes their habitat as a mosaic terrain, in which acacia bush and treeless expanses of rock alternate, and the grass and bamboo country is intersected by gallery forests lining the banks of rivers. Although grivets like to travel on the ground, they never stray too far from the trees, into which they can take flight if danger threatens. In this respect, they differ from the patas monkey or red guenon, which will venture far out into the treeless savanna.

The group observed by Harrison had a territory of nearly 450 acres (180 hectares). Since the distances traversed by the animals range from about 0.3 to 1.5 mi (0.5 to 2.5 km), they can visit most of their territory only occasionally. Thus, the treeless heights, where short grass grows sparsely, are rarely explored. An exception was the period following the annual burn-off of the grassland during the dry season. Then the monkeys searched this area for baked and roasted invertebrates and locusts.

The size of a group's territory depends on how far they can roam before they must return to meet certain vital requirements. Water is a major requirement for the grivets, at least in the dry season. Monkeys rarely drink in the rainy season because the water contained in their food suffices; but in the dry season, they drink daily. A hot wind

Grivets are very curious and bold in their behavior in the African national parks, where they have largely lost their fear of humans. They will even invade hotel terraces and automobiles.

blows in Senegal at that time, with the result that the animals transpire more. Just at this time the water sources also gradually dry up, so that the groups must remain within a radius permitting them to return daily to the few active water sources.

There may be severe competition over this. In the Amboseli National Park of Kenya, Richard Wrangham observed three groups. In the territories of two, there was always plenty of water even in the dry season. In the territory of the third group, the single water hole dried up. Only a little dirty water remained in deep recesses at the forks of tree branches. The females alone used these holes. They would drink by dipping the hand and arm and then licking the water from the skin and hair. Females competed vigorously for access to such holes, while low-ranking animals were often left out and perished with their young.

The males of the disadvantaged group ventured daily far into the territory of the neighbor group, where there was still enough water. In so doing, they encountered stiff resistance from the males in possession. Some of these males died as well. All showed severe injuries, suggesting that they fell in combat.

The hostility to intruders, by the way, is not attributable to an attempt by the owners of the well-watered territory to bar access to the water. The trespassing males were regarded rather as a possible threat to the authority of the males in their own group.

Observations of grivets in various regions have shown that tolerance among groups varies as a function of living conditions. In certain places in East Africa, where the monkeys in wooded areas have comparatively small territories (about 50 acres or 20 hectares), these are stubbornly defended, and the overlap between neighboring territories is small. In other regions where the territories are larger, the overlap zones are comparatively large also, and there is greater tolerance.

In Senegal, it was found that a group's tolerance would also vary sharply with density and spatial distribution, especially of food sources necessary to survival and affording only a limited supply.

Serious conflicts between groups of guenons sometimes occur. More often, especially if a spatial partition has been achieved, the altercations are confined to threatening or imposing behavior, exhibited primarily by the adult males.

Grivets are the only representatives of their genus that are consistently found in multi-male groups. In this they resemble the macaques and baboons. In a group, numbering about 15 to 50 individuals a strict ranking prevails. High-ranking (dominant) males display their superiority by strolling past another with stiffly upright tail. The submissive one is thereby shown a red skin area under the tail, a snow-white posterior, and an azure scrotum beneath.

The striking colors of the genital region – a carmen penis contrasting with the blue scrotum – must have a signal function. Display of the penis in jerky erections is nearly always associated with aggressive manifestations towards other males, whether strangers approaching a group or inferiors within the group.

It is significant than the males are able to retract the blue scrotum between the hairs of the posterior. This is done, for example, when inferior males are in contention with the leading male. The conspicuous sign marking them as possible competitors is thus concealed. It is, as it were, a denial of their masculinity and of attendant claims. Very likely this permits the inferior males to live within a group in which they are tolerated by dominant males. It means also that the multi-male groups of grivets are different in kind from those of macaques and baboons. They are really single-male groups, much like those of other forest-dwelling

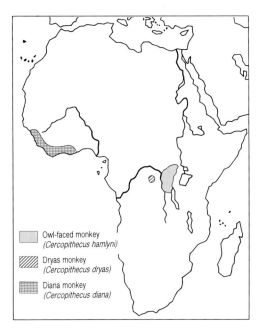
Owl-faced monkey (Cercopithecus hamlyni)
Dryas monkey (Cercopithecus dryas)
Diana monkey (Cercopithecus diana)

monkeys. But since in adaptation to life in open country the chances do not favor solitary males as much as in the forest, where they can more readily avoid predators' notice, the forbearance of male pretensions may be a strategy for participating in the protection of group life.

Grivets and their allies are small compared to other terrestrial monkeys, such as baboons and macaques. They do not seek safety from enemies in common defense, but rather in flight. Each of their enemies has its own methods of hunting and attacking, and the monkeys have their adapted responses. At the appearance of leopards, they take

flight into the trees, where, being light in weight, they are out of reach at the extreme tips of the branches. The native eagle is a bird of prey that attacks its prey either in the trees or on the ground by a power dive. The monkeys will leap down from the trees and escape in dense underbrush. But pythons lurk in the underbrush. All the monkeys have to do about them is see them before they can strike. Baboons are dangerous to young monkeys on the ground, separated from their group.

Grivets alert each other by means of warning cries, which the human ear can also differentiate. Scientists of Rockefeller University have determined that there are at least three types to which the monkeys react differently, and which are heard upon appearance of leopards, birds of prey, and snakes, respectively. To find out whether the animals "understand" these cries, that is, whether their response is selected according to the nature of the cry alone, and not because other perceptions have informed them of the nature of the threat, they recorded the cries on tape and played them back over loudspeakers hidden in the field on another occasion. It was found that the monkeys responded "correctly." This shows that these cries are more than mere utterances expressive of mood, as animal sounds are ordinarily interpreted.

It would seem that this understanding must be learned, for the young more frequently respond "incorrectly" than adults. Seemingly local "dialects" develop in this form of communication. Dogs generally evoke the same reaction as leopards. The monkeys utter loud calls and take flight in the trees. But in regions where the monkeys are hunted with dogs, as in Cameroon, they respond to dogs (and human beings) with a softer sound, of a timbre more difficult to distinguish against the background noise. Then they slip unobtrusively into the foliage of bushes. This is not unreasonable in the presence of humans and their weapons. Such adaptations, which must have developed rapidly in recent times, doubtless constitute true tradition.

Humans have extensively influenced the distribution of many species. Certain species have colonized new areas outside their original habitat after being relocated there by humans, often much against their will. Thus, there are populations of Old World monkeys on various islands in the West Indies, and some of them have led an independent existence for centuries by this time. Early in the present century, the occurrence of mona monkeys was reported in the Grand-Etang forest on Grenada. Grivets have settled on Nevis, Barbados, and St. Kitts. On the last-mentioned island, they were running wild as early as the seventeenth

Mother with young one, inviting an adult male to a grooming session. He ignores her.

Young grivet at safe observation post.

century, having been brought there by slave traders. They found an unoccupied niche, with no New World monkeys about.

On St. Kitts, for example, an islet measuring 18 by 6 mi (30 by 10 km), there was a luxuriant tropical forest on the mountainsides, and on a low peninsula, a drier area of bush, where food is harder to find. The animals settled into both terrains, including, that is, the forest, where grivets do not occur in Africa. The situation on St. Kitts is so interesting because differences can already be observed between the groups living in the hill forest and those on the peninsula. Among the mountain dwellers, the coal-black face color has given way to a slightly mottled appearance, and there are deviations in behavior as well. The peninsular inhabitants are terrestrial, the forest dwellers arboreal. The latter also lack the typical expressive behavior of the males in open country, such as threats with breast and penis display. Such behavior would not make sense in the forest. Here we are evidently witnessing the first beginnings of a further subspeciation. Whether we shall have opportunity to observe this process for any length of time, however, is questionable. The development of tourism on the island will presumably interfere increasingly with the undisturbed continuance of the populations living there.

Talapoin monkey (Genus *Miopithecus*)

To the hunters of monkeys in the forests of Gabon, the TALAPOIN MONKEY *(Miopithecus talapoin)* is not worth shooting. Hence this least of all simians may thank its minuteness for its safety from humans, the predator that represents the greatest threat to primates in large areas of Africa today. This attractive little monkey, with its small round head and comparatively large eyes, was long taken for a pygmy form of the genus *Cercopithecus*. Apart from its small size, however, it exhibits several features that justify a special position. The most conspicuous is the pronounced natural swelling of the female. The English primatologist Osman Hill reports that the talapoin, unlike members of the genus *Cercopithecus* proper, has been found to be immune to tuberculosis in captivity, indicating that there is a difference in blood serum as well.

During my comparative studies of facial expressions of Old World monkeys, I also noticed an important difference in behavior. Talapoins are often seen to grin in a manner that I have never observed in this type. The corners of the mouth are drawn back; nearby, the lips are retracted from the teeth, while remaining closed at the center. The lip outline thus resembles a figure-eight lying down, much as in mandrills and drills. As in the latter species, this grin is displayed by self-assured individuals to others of inferior rank. It appears to be friendly and reassuring in intent, and thus corresponds to a human smile. It is often seen during a remarkable embracing maneuver that I have observed in this species only. Two individuals stand erect face-to-face, look directly at each other, and throw their arms about each other's chests.

Anatomical, seological, and ethological (behavioral) traits thus justify a separate generic name, *Miopithecus*.

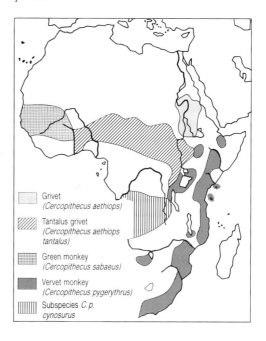

The talapoins do not make matters easy for the zoologist who would study their habits, for they are arboreal inhabitants of swamp forests along the rivers and mangrove forests on the coast. They are predominantly vegetarian, especially partial to palm nuts, but they will also hunt insects and other small prey.

Talapoins live in very large groups of about 70 to more than 150 individuals. There is a mating season, occupying not more than two months of the dry season of the year. Outside of this mating season, the males and the females with their young live apart in separate subgroups. Only in the mating season, when the female genital region swells, will the males join the female groups. Females ordinarily mate with several males in succession, and the males make no effort to prevent others from mating. In these large active groups, it may not even be possible for a male to claim a female for himself. His reproductive success is determined by number of pairings. This may account for the huge testes of these tiny primates.

It is remarkably also that among animals in captivity, it has been found that the females outrank the males. Where housing conditions do not enable the males to retire after the mating season, they may be severely persecuted.

This social form, in which the males and females live separately from each other outside of the mating season, is exceptional among Old World monkeys. Interestingly enough, there is a species with a similar organization among the New World monkeys also, namely the squirrel monkey *(Saimiri sciureus)*. They too are very small, live in very large groups, and have similar habits under similar environmental conditions. Since of course the two species are not at all closely related, their social organization must have been guided in the same adaptive direction by similar selective pressures. But just what the adaptive significance of this form of organization may be is still unclear.

Even though an individual in a given spot in the dense foliage of the swamp forest usually can see only its nearest neighbor, the groups, large as they are, nevertheless travel about in a coordinated manner. Since all individuals regularly utter short, explosive sounds, each can keep track of his changing position in the moving group.

Talapoins are good swimmers. Their sleeping trees are always on the bank of a river, and they rest in the boughs overhanging the water. If they are threatened by a predator, they simply drop in and swim away.

Allen's swamp monkey or guenon (Genus *Allenopithecus*)

Like the talapoin, the SWAMP GUENON *(Allenopithecus nigroviridis)* is *sui generis*, regarded as constituting a genus to itself, containing only this one species. In addition to *Cercopithecus*-like traits, it shows some resemblances to macaques as well as to mangabeys. The most striking, again, is the presence of a hairless, sometimes pink genital skin at the base of the tail in the males as well as in the females. The latter show menstrual swelling also.

Swamp monkeys have short, powerful limbs and quite a thick tail that is somewhat shorter than

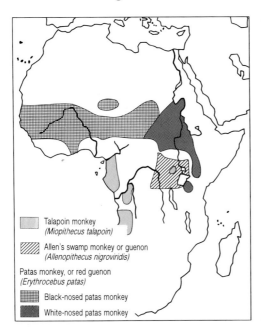

Talapoin monkey (*Miopithecus talapoin*)
Allen's swamp monkey or guenon (*Allenopithecus nigroviridis*)
Patas monkey, or red guenon (*Erythrocebus patas*)
Black-nosed patas monkey
White-nosed patas monkey

that of other long-tailed monkeys, being not much longer than the head and trunk.

The inaccessibility of their habitat accounts for our lack of information about this species. It occurs widely scattered in the region of the Upper Congo. There it is confined to swamp and flood forests along the many tributaries of the Congo. G. Pournelle has written about the behavior of some individuals kept at the San Diego Zoo. They took great pleasure in playing in the water. Perhaps they brought this predilection with them from their home, where probably they also like to come down from the branches and sport in swimming holes on the forest floor.

Consistent with this, the San

Diego animals liked shrimps, snails, and fishes, as well as vegetable fare.

Patas monkey or red guenon (Genus *Erythrocebus*)

Were one to write advertising copy for the RED GUENON *(Erythrocebus patas)*, one would note that it can attain a speed of 33 miles per hour (55 kilometers per hour) within three seconds. It is a real "sprinter" and its build is consistent with this. It has long but powerful arms and legs, and shortened hands and feet with somewhat rounded digits. All of these are adaptations to fast running.

The red guenon, or patas monkey, is definitely terrestrial. It is found on grassy savannas and in open acacia bush. Dense stands of trees and even islands of forest on the steppe are avoided. Thus, they have lost the option of taking refuge in trees when predators threaten. So they have no alternative but to run away as fast as possible and disappear into the tall grass of the savanna. Other behavior patterns also suggest that they rely primarily on their inconspicuousness for protection against enemies. They are especially silent when moving widely scattered and scarcely visible through high grass or between thorny acacia shrubs. To reconnoiter, they may stand on their hind legs, using the tail for additional support, like a third leg. In mixed groups, there is usually only one leading male, who from time to time will climb a tree, bush, termite hill or other elevation for a better view over the plain.

He is on the lookout not only for enemies but also for rival males and for other groups in the neighborhood. It is not clear which of these is most important.

If a human being approaches a group, the leading male exhibits very conspicuous behavior, in order to lead the trespasser away from the group. The other members meanwhile lie low until the threat has been averted, or rather diverted. Although the observation has not been made, it may be assumed that predators are dealt with similarly.

A day's march for the patas monkey may be anywhere between 0.3 and over 9 mi (0.5 and 15 km), depending on the food supply. This includes the tender parts of grasses, seeds, berries, pods, and shelled fruits. Animal fare, such as insects, especially locusts, eggs and even small birds and lizards, is not refused. In the search for food, the animals fan wide apart, over 0.3 mi (0.5 km) or more. In West Africa, a group of 31 individuals was found to use a territory of 19 mi² (52 km²). At night, the adults choose small trees or bushes for sleeping. The group is then even more widely scattered than during the day, often over an area of more than 68 acres (25 hectares). It may be supposed that this too is an adaptation to preda-

Curiosity, alertness and agility are traits of all the *Cercopithecus* allies.

tors, but we do not yet know in what the advantages of the strategy consist.

Patas monkeys can live even in areas that one would describe as semi-desert. In the dry season, when water holes are few and far between, they can apparently pass several days without drinking water. Unlike real desert dwellers, however, they are not able to produce metabolic water; or rather, they cannot utilize the water formed by their metabolism. So they must find water holes regularly. Yet these are the places where the greatest dangers lurk. When a group stops to drink, the male goes first, very cautiously. When he has actually decided to drink, the others will join him.

We owe our knowledge of the social life of the patas monkey chiefly to a pioneer in the field observation of primates – Hall, an Englishman who studied the simians in Uganda early in the 1960s. Supplemented by later observations of Thelma Rowell and her associates in Kenya, these findings present a fairly complete picture.

There are two types of groups: bisexual groups, in which, besides the females, infants, and juveniles, there is only one adult male, and male groups, comprising both deposed former group leaders and adolescents that have emigrated from their birth group. Within such a group of males, friendly and fairly peaceful relations prevail. But if a male assumes the leadership of a mixed group, he will drive out his predecessor, and show hostile behavior to other males also.

As terrestrial animals, patas monkeys need to be especially alert, because their open terrain provides but little cover. From time to time, the leader of the group will if possible climb a tree and keep watch for predators or rivals (above). Even while drinking at the water hole, the monkeys are always on their guard and ready for flight (below).

As in most simian species, relations between groups are not particularly friendly. Groups of patas monkeys avoid each other when possible. In the dry season, however, they may be forced to visit the same water hole. It has been found that male groups will generally yield to bisexual groups, which may even drive them away. The leading male of a mixed group will often succeed in putting all males to flight. Evidently it is not in the nature of the members of a male group to form an alliance against a leader.

Although the weight of a grown male is nearly double that of a female, the leader is no match for the females in his group. Quite often, they will attack the male and keep him more or less at a distance. The group is controlled by the highest-ranking female. The male is only a visitor, on sufferance so long as he has the strength to guard the group against other males and to protect the females and their young.

A patas monkey at his outlook post.

Long-tailed monkeys II *(Cercopithecus, Miopithecus, Allenopithecus, Erythrocebus)*

Nomenclature English common name Scientific name French German	Approximate Size Body length Tail length Weight	Distinguishing Features	Reproduction Gestation period Young per birth Weight at birth
L'Hoest's monkey *Cercopithecus l'hoesti* Cercopithèque à barbe en collier Vollbartmeerkatze	1.5–2 ft; 45–60 cm 1.6–2.5 ft; 48–75 cm 8.8–22 lb; 4–10 kg	Fur uniformly dark gray; posterior back reddish-brown; black face; white fringe beard extends up to the ears; no menstrual swelling	Not known 1 Not known
Preuss's monkey *Cercopithecus preussi* Cercopithèque de Preuss Preuß-Bartmeerkatze	Much like L'Hoest's monkey	Gray face with markedly contrasting chin beard; otherwise like L'Hoest's monkey	Much like L'Hoest's monkey
Owl-faced monkey *Cercopithecus hamlyni* Cercopithèque à tête de' hibou Eulenkopfmeerkatze	1.8 ft; 55 cm 1.8 ft; 55 cm Not known	Fur uniformly dark olive-gray; white stripe on bridge on nose; light blue scrotum; no menstrual swelling	Not known
De Brazza's monkey *Cercopithecus neglectus* Cercopithèque de Brazza Brazzameerkatze	Males 1.5–2 ft; 46–60 cm Females 1.3–1.6; 40–47 m. 2.1–2.8 ft; 63–85 cm f. 1.8–2.1; 54–63 8.8–17.6 lb; 4–8 kg	Fur predominantly dark olive-gray, white moustache and chin beard; light orange-red stripe on forehead; blue scrotum; no menstrual swelling	177–187 days 1 About 8.9 oz; 250 g
Diana monkey *Cercopithecus diana* Diane Dianameerkatze	1.3–1.8 ft; 40–55 cm 1.6–2.5 ft; 50–75 cm 8.8–15.4 lb; 4–7 kg	Strikingly variegated; general coloring black; pointed beard, throat and anterior side of arm white; posterior back and thighs red-brown to orange; no menstrual swelling	5 months 1 Not known
Dryas monkey *Cercopithecus dryas* Cercopithèque dryade Dryasmeerkatze	Much like diana monkey	Posterior back and thighs greenish-gray; otherwise like diana monkey	Like diana monkey
Grivet; savanna monkey *Cercopithecus aethiops* Grivet; singe vert Grüne Meerkatze	m. 1.3–2.2 ft; 40–66 cm f. 1.1–1.6; 35–50 m. 1.8–2.4 ft; 55–72 cm f. 1.6–2; 50–60 5.5–15.4 lb; 2.5–7 kg	Fur predominantly gray to yellow-green; black face; white stripe on forehead; blue scrotum; light red penis and buttocks area; no menstrual swelling	5.5 months 1 10.7–14.3 oz; 300–400 g
Green monkey *Cercopithecus sabaeus* Callitriche Gelbgrüne Meerkatze	Much like grivet	Side whiskers yellow; gold-green back; no white stripe on forehead; otherwise like grivet	Much like grivet
Vervet; vervet monkey *Cercopithecus pygerythrus* Vervet Vervetmeerkatze	Much like grivet	Short side whiskers; gray-green back; otherwise very similar to grivet	Much like grivet
Talapoin; talapoin monkey *Miopithecus talapoin* Talapoin Zwergmeerkatze	m. 1.1–1.5 ft; 35–45 cm f. 10–14.8 in.; 25–37 m. 1.2–1.7 ft; 36–52 cm f. 10.4–16 in.; 26–40 1.5–9 lb; 0.7–1.4 kg	Smallest old world simian species; upper parts gray-green to yellow-green; under parts gray-white; menstrual swelling	158–166 days 1 5.4–6.3 oz; 150–175 g
Allen's monkey; Allen's swamp monkey; swamp guenon *Allenopithecus nigroviridis* Cercopithèque noir et vert Sumpfmeerkatze	1.3–1.7 ft; 40–51 cm 1.1–1.7 ft; 35–52 cm 5.5–11 lb; 2.5–5 kg	Powerful build; upper parts dark yellowish-gray; under parts light gray; blue-white scrotum; menstrual swelling	Not known 1 About 7.1 oz; 200 g
Patas monkey; red guenon *Erythrocebus patas*, 2 subspecies Singe rouge; singe pleureux; patas Husarenaffe	1.6–2.9 ft; 50–88 cm 1.6–2.5 ft; 50–75 cm m. 15.4–28.6 lb; 7–13 kg f. 8.8–15.4; 4–7	Long, slender limbs; hands, feet and thumbs shorter than in *Cercopithecus*; rough fur; upper parts brown-red; under parts white; red penis; blue scrotum; no menstrual swelling	About 6 months 1 Not known

The strategies of patas monkeys are silence and inconspicuousness. This individual is cautiously peeking out from behind a thick tree trunk.

COMPARISON OF SPECIES

Life Cycle Weaning Sexual maturity Life span	Food	Enemies	Habit and Habitat	Occurrence
Age not known Age not known In captivity, over 16 years	Fruits, leaves, sprouts, invertebrates	Leopard, golden cat, python, crested eagle	Ground dweller in moist hill forests at elevations up to 8000 ft (2500 m); in small, single-male social groups	Seriously threatened
Like L'Hoest's monkey	Like L'Hoest's monkey	Same as L'Hoest's monkey	Much like L'Hoest's monkey	Threatened
Not known	Fruits, leaves, sprouts, invertebrates	Not known	Arboreal dweller in moist mountain forests up to elevations of about 12,800 ft (4000 m); presumably in small, single-male social groups	Not known
Not known	Fruits, leaves, sprouts, roots, invertebrates	Leopard, golden cat, crested eagle, python	Tree and ground dweller in swamp forests; in small family groups (parent pair and progeny)	Still fairly abundant
Age about 6 months Age about 3 years In captivity, 19 years	Fruits, flowers, young leaves, invertebrates	Not known	Dweller in upper levels of primeval forest trees; in single-male social groups (up to 30 individuals); territory size 0.19–0.38 mi^2 (0.5–1 km^2)	Seriously threatened by hunting and by destruction of forest
Much like diana monkey	Much like diana monkey	Same as diana monkey	Much like diana monkey	Seriously threatened
Age about 6 months Age 2–2.5 years Over 30 years	Grass, fruits, berries, young leaves, flowers, beans, invertebrates	Predator cats, baboon, jackal, hyena, eagle, python, crocodile	Terrestrial and arboreal; preference for forest margins and adjacent savanna; social groups include several males (10–60 individuals); territory size 0.03–0.38 mi^2 (0.1–1 km^2)	Abundant
Much like grivet	Much like grivet	Same as grivet	Much like grivet	Abundant
Much like grivet	Much like grivet	Same as grivet	Much like grivet	Abundant
Age 4–5 months Age 4.5 years Over 28 years	Fruits, flowers, sprouts, young leaves, insects	Leopard, golden cat, genet, large birds of prey, snakes, gavial	Arboreal in swamp forests; social groups include several males (40–150 individuals); territory size 0.03 acres (0.1 hectare)	Still fairly abundant
Age 2.5 months Not known Over 23 years	Probably fruits, leaves, nuts, invertebrates	Not known	Terrestrial and arboreal in swamp forests; habits not much studied	Not known
Age about 6 months Age about 2.5 years Over 21 years	Grass, fruits, beans, seeds, invertebrates, lizards, birds' eggs, young birds	Leopard, cheetah, hyena, jackal, crested eagle	Ground dweller in bush and grass savannas; in single-male groups (up to 30 individuals) and all-male groups; territory size up to 18 mi^2 (50 km^2)	Still fairly abundant

Head of a white-nosed patas monkey, with face, sideburns and shoulders lighter in color than in the black-nosed subspecies.

The Red-Faced Macaque "Culture"

by Kosei Izawa

The red-faced macaque *(Macaca fuscata)* is found in all parts of Japan except the Island of Hokkaido; it lives on the main island, on Shikoku, on the Kyushu Isles, and on some small islets. The northern limit ot its range – Oh-ma-cho (41°30′N, 140°55′E) on the Shimokita Peninsula to Aomori prefecture – is at the same time the northern limit of all living non-human primates.

The region inhabited by red-faced macaques extends from coastal to alpine terrain. Parts of this range are at elevations of more than 9600 ft (3000 m) above sea level. In their northern habitat, the snow may lie up to 13 ft (4 m) deep from December to March. The temperature occasionally drops to 5°F (-15 °C). The monkeys will then warm themselves by sitting in the trees on slopes protected from the wind and hugging each other.

Its naked red face, contrasting sharply with the gray-brown fur, gives the Japanese red-faced macaque its name.

The red-faced macaques in snow country are often called "snow monkeys."

Red-faced macaques from groups numbering from 15 to 200 individuals, but as a rule 20 to 100. These groups lead a nomadic life, having in a sense "no fixed dwelling place," and will cover 0.3 to 1.2 mi (0.5 to 2 km) per day in the search for food. The food supply in the forest affects territory size. Groups living in the laurel forest of the southern range of distribution have predominantly territories of only a couple of square miles. Particularly the groups on the Island of Yakushima, representing the southern limit of the habitat, often have territories of less than 0.4 mi^2 (1 km^2) compared to a typical territory of 7–11 mi^2 (20–30 km^2) in the deciduous forests of the north. To be sure, the range is contracting at the present time, because the forest is being cleared almost everywhere, and conifers such as the Japan cedar *(Cryptomeria japonica)* and the hinoki cypress *(Chamaecyparis obtusa)* are being planted instead. Groups of monkeys in areas so radically altered often have territories of 14 mi^2 (40 km^2) or more.

Although neighboring territories overlap to a considerable extent, the groups avoid each other for the most part. But when two groups do meet, they are far more likely to travel together for some hours than to fight.

The diversity of habitats is reflected in the diet. Red-faced macaques prefer fruits, berries, and nuts, but they will also eat young leaves of trees, grasses and their seeds, greens, stalks, root stocks, and fruit pits. Especially in Winter, monkeys in the North live chiefly on bark and on hard winter buds and sprouts. Besides vegetable fare, they will also eat various insects, their eggs and larvae, spiders, snails, beach crabs, and crayfish, as well as birds' eggs.

Monkeys in coastal regions often find food on the beach, and will eat various species of algae and kelp, including true and paper kelp, such as *Undaria pinnatifida*, *Laminaria japonica*, and *Scinaia*, which are easily collected from nearby rocks. They will also eat animals such as the gastropod *Cellana* and barnicles of the genus *Balanus*.

They also like mushrooms. From early June to late September, when there are no fruits, nuts, or young leaves to be found on the little island of Kinkazan, the macaques living there will eat more than 20 species of fungus.

When human beings supply a group of monkeys with food, the number of animals in the group will grow very rapidly. Groups of more than 100 animals are then no rarity; even 1000 individuals or more may form a single group, like one at present in Takasakiyama.

Most primates have no experience of snow whatsoever, but for many red-faced macaques snow is very much a matter of course. In the far north of their range, as here in the Shika mountains, they pass severe winters year after year.

▷ Red-faced macaques in their community bath. Fortunately for the "snow monkeys", hot springs abound in their otherwise inhospitable habitat. Submerged to the neck, they will often linger for hours in the agreeable warm water, but then they must go back into the cold Winter air to find food. Then their only protection against hypothermia is brisk movement and a coat of long, thick fur.

Two half-grown red-faced macaques warming each other by embracing and huddling together.

Red-faced macaques generally sleep in closely spaced trees on steep declivities or rock faces. Those in regions of heavy snow often sleep in pine trees such as *Thujopsis dolabrata* and the Japanese hemlock *(Tsuga diversifolia)*, which commonly occur mingled with deciduous trees.

Although it is said that bones of a young macaque were once found in the nest of a Japanese rock eagle *(Aquila chrysaetos)*, and a pack of feral domestic dogs is said to have given chase to a group of monkeys, such incidents are extremely rare. There are no predators in the macaques' natural habitat. Until about 1950, the monkeys were hunted for food and medical purposes and for fur. By this time, the species is protected throughout the empire. To be sure, it can hardly be said that this protection has led to any adequate measures of conservation. As a matter of fact, owing to the recent extraordinarily rapid destruction of its habitat, the red-faced macaque has now disappeared from many areas where it formerly existed. Its total numbers are currently estimated at around 35,000 to 50,000.

The mating season of the red-faced macaque is confined to a much smaller part of the year than that of its nearest relatives, the other macaque species and the baboons. In the south, mating occurs chiefly from November to February, whereas in the northern regions it extends over the months from October to December. Since the duration of pregnancies is about six months, births take place in Spring or early Summer. Usually a female will have one offspring; twins are very rare. The females bear young every two or three years in the wild, or at most twice in three years. Thus they are not especially fertile compared to other species.

The social organization of red-faced macaques has been under constant study since 1948. This thorough research on their community life was made possible primarily through a technique developed by K. Imanishi and J. Itani of Kyoto University. Wild monkeys are supplied with food and so become accustomed to the presence of human beings. Imanishi and Itani first succeeded with this technique in 1952 on the little island of Koshima. At present, more than 30 formerly wild groups of red-faced macaques are being maintained in this way in various parts of Japan.

Monkeys provisioned by human beings do not shun observers, and since the feeding places are located in open terrain, observation is not hindered by shrubs or trees. These methods not only serve to identify individual monkeys, but also provide an understanding of the "personality" of the species. It has also been an easy matter to identify rank, by offering peanuts or oranges at appropriate times. When this technique is supplemented by long-term observation, it is possible to recognize changes in the ranking of adult males and to decipher their social significance. As to the females, their genealogy can be recorded to discover the relationships between kinship group and social structure. Itani describes the social behavior of the red-faced macaques on the basis of the results he obtained up to the mid-1960s in the manner described.

On his compaigns, several younger males will lead the troupe. They are followed by a considerable number of females, juveniles, and half-grown individuals, accompanied by one ore more fully adult males. The rearguard again consists of young

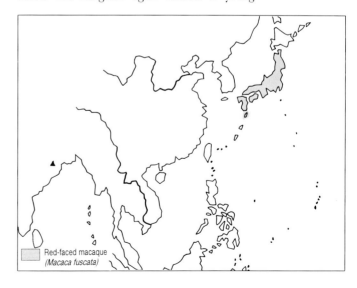

Red-faced macaque *(Macaca fuscata)*

adult males. At the feeding place, the animals form two circles, one inside the other. The inner circle consists of all the females, their young, and several larger, fully adult males. The outer circle consists of young adult males.

There is a clearly recognizable ranking among the adult males, but it is not invariable, and it changes frequently among the younger males especially. There are several "classes" of adult males, such as leaders, subordinate leaders, members of the outer circle. The leaders crouch with the females and their young in the center, while the subordinates are to be found surrounding this inner circle. Ranking within a class is far more pronounced than between members of different classes.

The leaders guide and watch over the troupe.

This gray head wears a chilly aspect, but at the same time bears witness to the adaptability of the snow monkeys. The animals put up well enough with a harsh climate, but not with the encroachments of humans, who used to hunt them in very recent decades and continues to constrict their habitat.

They settle disputes and keep watch for dangerous enemies, and will resist them if necessary. During the time when the females are caring for their newborn young, male leaders assume responsibility for the yearlings. The duties of subordinate leaders are much like those of principal leaders. Although they assist the leaders in their duties, seemingly they are not allowed a place in the inner circle. Males in the outer circle (peripheral males) act as announcers and expediters when the troupe is on the move. They will sense danger and attack enemies.

All of this is referred to as "troupe directed" behavior. The more zealously an individual participates in it, the more the animal consolidates its position within his or her own class. Some monkeys – males without exception – nevertheless become solitary. They appear to be unsociable by nature, and unable to adjust to community life.

The social structure described above pertains to "provisioned" groups; the narrow confines of the feeding place and the limited quantity of food create a state of perpetual tension on the part of each individual. Solitary individuals are unaccustomed both to human observers and to the prepared feeding place. Besides, virtually all provisioned groups hitherto studied have been groups living separately. The possible resulting distortion of the social structure has been a subject of debate since the early 1970s. Many scientists began to observe undisturbed groups of red-faced macaques, surrounded by several neighbor groups, in their natural surroundings, once the animals had got used to the observers' presence. As a result, the picture of their social behavior as outlined above has been corrected on some points.

Thus, we no longer regard the solitaries as alienated from their group, or as failures incapable of adjusting to the community. Rather, we believe today that for a male to leave his birth group after reaching sexual maturity and to migrate from one group to another are normal incidents of life. The females, on the other hand, tend to remain in their birth group for life. This difference would be susceptible to biological explanation, that is, it might contribute to the gene flow within a locally limited population (society), or to sociological explanation, that is, as avoiding incest. Furthermore, it is in a sense inappropriate to refer to single males as "solitary," since they do not necessarily live alone, but often in a group in common with other males; sometimes they will attach themselves to peripheral males. It is often not an easy matter to distinguish between solitaries and peripherals. Consequently, we today no longer regard the troupe as a closed community.

Group cohesion is not secured by the dominance of high-ranking males but by the strong ties between females. These ties are based on kinship as well as on the connecting mutual dependency of kinship groups. Still, in various situations of daily life these females are dependent on the support of a certain few grown males. Thus it is not unusual for a group of about 20 individuals, for example, to have only one male in the inner circle. Although this inner-circle male assuredly influences the conduct of the group members, leading males have no narrowly circumscribed function. We also no longer believe that there is a rigid class structure in the group.

Why and by what criteria the females select certain males as attendants (protectors) is not clear. So far as we know at present, they do not always choose the highest-ranking males. Besides, frequently a male on whom the females were dependent will suddenly leave the group, when he has for a time been the "alpha" male, as the highest-ranking male is called. Scientists are now trying to understand these facts not in the limited context of one single group but in terms of contacts among neighbor groups living in the same region under natural conditions.

Continuing observations of groups being supplied with food show that an individual may sometimes invent a new mode of behavior, which will then spread within the group and even be passed on from generation to generation. These forms of a monkey "culture" have been investigated by Masao Kawai and his associates since 1953.

One of the most celebrated examples of the monkey culture is the potato washing of the red-faced macaques of the little island of Koshima. Summarized, Kawai's account runs as follows.

One day in the Autumn of 1953, a one-and-a-half-year-old female named Imo picked up a sand-covered sweet potato at the feeding place. She dipped the potato in water and rinsed away the sand with her hands. By this act, Imo introduced a culture trait into her group on Koshima. A month later, one of Imo's playmates started washing potatoes and after four months, Imo's mother was doing the same. Through daily encounters between mothers and children, sisters and brothers, contemporaries and playmates, this behavior gradually spread. In 1957, 15 individuals were potato washers. Animals aged one to three years were the

(Left) The culture trait initiated by Imo in 1953 gradually spread over the island of Koshima. Ten years later, all the red-faced macaques of her group were washing sand-covered sweet potatoes before eating them, and their peculiar table manners were even transmitted from one generation to the next. (Right) Sea bathing has also become an established "tradition" for the macaques of Koshima.

most frequent learners. Three five- to seven-year-olds learned it also, as well as two adult females. But no male who was more than four years old at the time adopted the new procedure.

This difference between the sexes may have been the result of a difference in social position. From their fourth birthday on, males must live in the outer circle, where they rarely come into contact with the young females who, in this case, invented potato washing. The females of the inner circle had more opportunities to witness the practice.

At first, the new technique was transmitted between playmates and members of the same family, especially from the young to their mother and from younger siblings to older ones. Later, when potato washing had spread farther, mothers passed it on their young. Individuals born since 1957–58, who as infants spent the entire day with their mothers, wash their potatoes before eating them as a matter of course.

In 1962, 42 of the 59 members of the group would wash potatoes. Those who did not were all males or females who had been full-grown by 1953. Fully matured individuals are apparently too "conservative" to accept new patterns into their behavior, while the juveniles are still highly adaptable.

After ten years, potato washing was part of the normal feeding behavior of the troupe, and each generation was transmitting it to the next. Later, these monkeys began to wash potatoes in salt water, perhaps because a little salt made them taste better.

Strictly speaking, the invention of potato washing

The red-faced macaques made another important discovery. They learned so separate grains of wheat or peanuts from adhering sand by gathering them up and throwing them on the water. The sand will sink, but the edible grains float and are easily recovered. This form of feeding likewise became a tradition on Koshima.

is not equivalent to human cultural development, and might best be referred to as "simian" culture or "pro-culture."

Kawai observed the genesis of several other traditions. Whereas he sometimes deliberately influenced the monkeys in his studies, they would also come up with forms of behavior that he had not intended. Thus, Kawai incited a number of monkeys to enter the water by giving them peanuts. In three years, all juveniles and adolescents had formed the habit of bathing in the sea regularly. They would even swim and dive. They also began to wash the wheat that had been strewn on the sand for them. At first they would laboriously pick each grain out of the sand. Later they would simply throw a handful of sand and wheat into the water. The sand sank to the bottom, and the grains of wheat floated on the surface.

All the monkeys who were grown up when the monkey "culture" was founded have since died, and six generations have been born since Imo invented potato washing. This artificially evolved behavior, however, like the wheat washing, the bathing, and the diving, is a firmly naturalized pattern of behavior in the everyday life of the monkeys on Koshima, and has certainly been handed down from generation to generation.

The dissemination and transmission not only of the Koshima "culture" but also of other special modes of behavior invented by macaques in other provisioned groups – such as the bonbon-chewing of Takasakiyama, the medicinal-spring bath of Jigokudami, and the pebble game of Arahiyama – have been studied by numerous scholars. Their conclusions are much the same as those of Kawai. Experimental psychologists are now also investigating these developments in simian "culture."

The great question, however, is why the red-faced macaque, possessing these remarkable faculties, did not evolve any similar behavior in the wild in the course of its long history. Some students have identified regional or group differences in diet and in modes of feeding. When eating fruit, for example, some groups will swallow it whole, while members of other groups will squeeze a fruit and eat only the pulp. To be sure, regional differences in modes of taking nourishment are observed among nearly all other non-human primates, and there are similar reports concerning many other animals.

Other researchers are of the opinion that behavior observed in provisioned groups is to be compared with that of simian groups in the zoo; when new forms of behavior are developed there, they are often transmitted from group members to group members and across generations. They give the following argument: When the groups on Koshima and Jigokudani were divided, the members of the new subdivisions were no longer able to use the feeding places. They were once more obliged to make their entire living in the wild. From then on, they never again exhibited the "cultural" behavior that had been familiar to them.

A new behavior pattern may be discovered by one individual, disseminated among other individuals of its group, and even transmitted through generations. It is important to realize that this mode not only represents an adaptation to new surroundings; it improves the adaptive process itself, by a new procedure. It is also important that the manner of dissemination and transmission of new modes of behavior be understood in the context of the social structure of the group.

Refined methods of food preparation were learned primarily by females and juveniles. Older males live outside their circles, and are perhaps also too "conservative" to integrate new patterns into their behavior.

Langurs and Colobi

By Christian Vogel and Paul Winkler

This second subfamily (Colobinae) of the Cercopithecoidea comprises the langurs and the colobi. They are about as widely distributed as the other subfamily (Cercopithecinae). Out of the total of five genera, three are at home in Asia and two in Africa. They are of a slender build, the hind limbs are longer than the forelimbs, and the head comparatively round. The thumb is reduced, especially in the colobi of Africa. In the green colobus, there is only a slight prominence in the place of the thumb. Nevertheless, the langurs are extremely dexterous, and can pick up minute plant parts without difficulty and very quickly.

Despite many differences between species, the langurs have numerous traits in common that distinguish them clearly from other Old World monkeys. First, the highly specialized structure of the stomach should be mentioned. The langurs are generally considered to be leaf eaters. The digestion of leaves is no simple matter, either mechanically or chemically, because of their often quite hard surface and high cellulose content. The langurs have solved this problem by evolving a compartmented stomach, quite similar in principle to the complicated apparatus of ruminants. The German anatomist Hans-Jürg Kuhn has studied and described the structure and function of the langur stomach in detail. The stomach is divided into four sections. In the first two, which may be regarded as fermentation chambers, the contents are blended and decompose by cellulose-degrading bacteria. The stomach occupies much of the abdominal cavity, and its contents may account for up to 17 percent of an individual's total body weight.

The powerful chewing apparatus, with the relatively long molars, having cusps joined by transverse ridges ("bilophodontia"), may also be regarded as an adaptation to the diet. Unlike the macaques, the langurs have to cheek pouches for transporting food.

The description of the langurs as leaf eaters does not justice to the diets of most species. From leaf to root, all plant parts are consumed, and the proportion of leaves in the total diet is quite different for different species. The langurs might perhaps be more suitably described as vegetarian, for animal food is seldom taken, and even then is usually confined to insects, such as beetles or locusts, rarely spiders.

In addition to the physical peculiarities mentioned, the langurs are also distinguished by special behavioral characteristics. This applies to their mating behavior, for example. Unlike many cercopithecines, they have no conspicuous swellings of the genital region to indicate pairing readiness of the females; the females will take up a position in front of or near a male and turn their backs. In a slightly stooped posture, they then begin to shake the head to and fro frenetically, drooping the tail to the ground. Now and again, they will stop the head shaking and glance at the male over their shoulders. The red and the green colobi, which do have swellings of the female genitals, are an exception.

Other peculiarities are the far-carrying calls and associated menacing leaps of the males; they jump around in the trees stiff-legged, shaking the branches vigorously. This behavior is often observed when two groups meet, and the males – more especially the females – undertake to defend their group. Some species will hold regular concerts, especially in the early morning hours. Probably they can track the whereabouts of neighboring groups in this way. Private territory can thus be avoided, so that no confrontations need occur.

Every observer of the langurs immediately notices

The hanuman langur, also called the common or gray langur, is certainly the most familiar species. It is distributed throughout the Indian subcontinent, including the island of Sri Lanka (Ceylon).

the fur coloring of the newborn, which is altogether different from that of the adults. It ranges from snow-white among the northern black-and-white colobi, or guerezas, through brown-black among the hanuman langurs, to brilliant orange among the dusky langurs. When newly born, the babies are a great attraction to all group members except the adult males. They appear rather indifferent, while the others will carry and look after the infant.

Genus *Presbytis*

The classification of the langurs has not yet been satisfactorily settled. This is especially clear in the case of the genus *Presbytis*, in which four formerly separate genera have recently been merged. Still, all authors are agreed that there are four closely related groups to be distinguished, comprising an aggregate of 14 species and 84 subspecies. Probably the actual number of subspecies, perhaps even of species, is smaller. Presumably animals have sometimes been assigned to two different species merely on the basis of different fur coloration. In our sequence of presentation, we have observed the composition of the four groups of species as follows: The former subgenus *Semnopichtecus* comprises the species *Entellus;* the former subgenus *Kasi* comprises the species *Johnii* and *Senex;* the former subgenus *Trachypithecus* comprises the species *Obscura, Phayrei, Cristata, Pileata, Geei, Potenziani,* and *Françoisi;* and the former subgenus *Presbytis* comprises the species *Melalophos, Rubicunda, Frontata,* and *Aygula*.

The HANUMAN LANGUR *(Presbytis entellus)*. The hanuman langur (gray or common langur, formerly *Semnopithecus entellus*) is one of the best-researched of the Asian forms.

Seeing a hanuman langur sitting in a dignified human-like attitude (top), one can understand it being revered in its homeland as an incarnation of the Hindu ape-god Hanuman. These animals, even mothers with their young, spend much time in human communities, knowing from long experience that they will be tolerated and even fed (bottom right). Placid and even phlegmatic as the hanumans may seem, they can occasionally display astonishing temperament (bottom left).

OLD WORLD PRIMATES

Black faces, hands, and feet, contrasting with the handsome silver-gray fur, are characteristic of the human langurs. The babies are black or dark brown at first.

It is distributed throughout the subcontinent of India in 15 or 16 subspecies. The ecological adaptability of these animals is especially astonishing. They are found in the snow-covered Himalayas of Nepal at elevations above 13,000 ft (4000 m) as well as in the semi-desert region of Rajasthan or in the tropical rainforests of Sri Lanka (Ceylon). They are also a familiar sight in the densely settled provinces of northern India, in many cities and in temple precincts.

Hanuman langurs are vegetarians. All plant parts, especially leaves, flowers, and fruits, are used as food; beetles, locusts, or spiders are eaten only occasionally.

In many parts of India, the animals are fed more or less regularly, for religious reasons. Veneration of the hanuman langur by the Hindu population is rooted in the Indian national epic, the Ramayama, portraying the life story of the god Rama. It is told of the ape-god Hanuman that he marched on Sri Lanka at the head of his simian horde to liberate Rama's abducted consort Sita from the clutches of the demon prince Ravana. In the successful act of rescue, Hanuman set fire to the city of Lanka, suffering burns on face, hands, and feet. The black faces, hands, and feet of the langurs are seen as commemorating that exploit.

The hanuman langurs are quite flexible in their forms of community life in groups. As regards the relationship of adult males to adult females within a group, normally each species has its characteristic composition. There are four different forms of social organization. Most frequently encountered is the single-male group, in which one adult male lives with several adult females and their descendants. There are also groups in which there are two or three adult males, though of different ages. Presumably these males take rank according to age. Other hanuman langurs live in multi-male groups with a "promiscuous" system of reproduction; that is, at least theoretically, any male may mate with any female. In practice, the picture may be much the same as among the macaques and baboons; there are certain particular, high-ranking males who can mate with the females at any time, while the other males must await a moment of inadvertence, or else may copulate only at times when conception is unlikely.

As a fourth form of social organization, there are the all-male groups, in which – except for youngsters – only males of all ages are represented.

Group size varies widely on the whole. Between 8 and 125 animals have been counted per group. Territory size varies, correspondingly, being anywhere from 0.07 to 3.7 mi^2 (0.2 to 10.4 km^2). In some areas the hanumans are strictly arboreal, in others almost exclusively terrestrial. This will depend largely on the ecological context.

In many villages and towns of India, hanuman langurs climbing about on roofs and walls are a familiar sight.

Since the late 1950s, these animals have been studied in the wild. There are now systematic observations from more than twenty locations in Nepal, India and Sri Lanka. Especially important is a research project conducted by primatologists at the Anthropology Department of the University of Göttingen. They conducted thorough observations between 1977 and 1985 in the city of Jodhpur, in the Indian state of Rajasthan, that have revealed many details of behavior. In and around Jodhpur, more than 1200 hanuman langurs in a total of 29 single-male groups and about 12 all-male groups inhabit an area of some 54 mi^2 (150 km^2). Data on ecology, development of the young, reproductive behavior, and other topics

▷ A group of hanumans resting in the glow of the setting Sun. These social animals have different community forms, most commonly single-male groups, comprising one adult male and several females with their offspring.

were collected in a sample of two groups whose members have been known "personally" for years.

The city of Jodhpur is situated in a semidesert region on the eastern margin of the great Thar desert of India. Primarily the following four factors have enabled the langurs to survive under the prevailing extreme climatic conditions: (1) Throughout the year, there is an adequate food supply. There are wide fluctuations, but even in the dry seasons there are 6 to 10 evergreen plant species, which then become the main item in the diet. (2) Many of the groups are fed regularly by the Hindu population especially. (3) All groups have access in their home territories to reliable springs or to artificial lakes and ponds. (4) There are no predators except for stray dogs.

In Jodhpur, langur babies are born throughout the year. There is no season for childbearing, as among the hanuman langurs of Nepal. Quite soon after birth, not only the mother but also the other group members – except the adult males – will tend the babies; they are carried about, cared for, and not infrequently abused until the age of about five weeks. In that case, they may be kicked or struck so as to sometimes even cause them to tumble down over the rocks. They will cry piteously, and often sustain bloody wounds. The mother will rarely interfere; sometimes the youngster's own mother may be the abuser. As yet, little is known about the "sense or purpose" of this behavior. One should hesitate, however, to regard it as pathological. It appears rather to be part of the "normal" repertory. At all events, no severe injuries to the infants have been observed.

With the very first research activities in the wild, observers had noticed that infants are continually being handed about among the group members ("infant transfer"). Sometimes more than an hour will pass before a baby is brought back to its mother or the mother comes to fetch it. The aim and purpose of infant transfer is as yet quite obscure. Attempts have been made to draw up a sort of cost-benefit analysis by inquiring what advantages and disadvantages accrue to the several parties engaging in the transfer. The mother can consume a good deal more food while her baby is in the keeping of another group member than while she is carrying it herself. Her social contacts, especially mutual grooming, are also more extensive. For the baby, the direct benefit is slight. At best, an earlier familiarity with the members of the group might be an advantage. To be sure, the risks involved are not inconsiderable, when one recalls the abuses that have been described. What is quite unclear is the possible advantage to those group members who transport the young of others, apart from those cases, say, where a half or full

Hanuman langurs are terrestrial and arboreal. On the ground or, as here, on a fallen tree trunk, they move about on all fours.

sibling is being carried. Such cases might be described as kinship selection, a process in which the chances of survival and hence also the reproductive success of relatives who are not direct descendants are improved.

Among the adult males of the Jodhpur population, there is especially keen competition for the "headships" of the 29 single-male groups, for it is only in that capacity that a male can get descendants of his own. In a group, there will be a change of males every two years on the average. A change of males will usually occur whenever the head of a single-male group becomes unable to withstand the pressure exerted by an all-male group. The all-male groups, numbering up to 30 members, that roam between the home territories of the single-male groups, try to expel the leader of such a group. If they succeed, the highest-ranking male of the all-male group becomes the new head of the single-male group, and soon after begins to drive off this former allies. At the same time, all weaned male youngsters will leave their group and attach themselves to the male groups.

When a male has taken over a group, "infanticide" may occur. This practice of killing youngsters has given rise to a scientific controversy. The first thoughts were of an abnormal behavior pattern, limited to quite definite circumstances, for example overpopulation. Observations in Jodhpur have shown, however, that infanticide may quite well make biological sense. To understand this implausible suggestion, one must consider some other aspects of the reproductive behavior of the hanumans. The females become sexually mature at about 3 years of age. A pregnancy lasts about 200 days. The interval between two births averages 15.4 months. Now if a male takes over a group, he is under a sort of time pressure, for on the average he has only about two years in which to get as many direct descendants as he can. After that, he will be liable to displacement by another male in turn. But if the females happen to have young infants at the time of the takeover, the male will have to wait about nine months before the females will be pregnant again. A male can shorten this waiting period by killing the little ones, who after all are not his own but the predecessor's. Not long after, the females will be ready to mate again. In Jodhpur, it was clearly documented that there is a close correlation between the age of the infant killed and length of the interval before the next birth; the younger the victim, the shorter the interval. This represents a gain of almost four months of time for the new male, and signifies a reproductive advantage over a male who does not kill his predecessor's offspring.

(Above) Whether hanumans spend more time on the ground or in the boughs depends largely on the particular surroundings. In some areas they are strictly arboreal, in others almost exclusively terrestrial. (Left) Hanumans in typical attitudes of rest on a boulder. The largest individual, evidently the head of the group, sits at the summit.

▷ Hanumans like to live near bodies of whater, where they can quench their thirst regularly. But if necessary, puddles of rainwater and their juicy vegetable diet will serve to supply the needed moisture.

Hanuman langurs have been kept and bred in captivity for many years, as in the zoos of Frankfurt-am-Main and Hanover.

The NILGIRI or JOHN'S LANGUR *(Presbytis johnii)*. Nilgiri langurs are named for a mountain range in the South Indian state of Tamil Nadu, part of the Western Gap, along the southwestern coast. Their range, however, is not limited to the Nilgiri Mountains, but includes neighboring country as well. Nilgiri langurs inhabit not only evergreen mountain rainforests, in which they may be found up to an elevation of 7700 ft (2400 m), but also on lower terrain on the forest margin. They are strictly vegetarian, the diet consisting predominantly of leaves of trees and shrubs.

The American anthropologist Frank Poirier investigated the Nilgiri langurs in 1965 and 1966, and he made some astonishing observations. The animals live in single-male groups and in all-male groups. Multi-male groups may also sometimes occur. The groups are not very large, averaging 9 members or so. Within the group, small subgroups are formed, in which individuals of the same sex or the same age are associated. Social contacts were seemingly limited to the members of a subgroup. Remarkably enough, males and females have hardly any contact with each other. Hence, the adult male in a single-male group is rather isolated. He lives on the periphery of the group both socially and spatially.

A birth drastically affects a group's daily routine. Shortly before giving birth and for about a month thereafter, the mother will travel about less than the group as a whole. In fact, Poirier had the impression that the group itself did not range as far and moved more slowly.

Like hanumans, Nilgiri langurs are carried about soon after birth by other group members (infant transfer). However, only the adult females participate. At birth, the baby's fur is rusty-red, changing to the black of the adult at about four to five months.

Nilgiri langurs are predominantly arboreal. They spend between 87 percent and 100 percent of their day in the branches. Taking nourishment accounts for the major portion of the day's activity. The monkeys will spend as much as eight hours per day eating. They seem to drink very infrequently. In more than 1000 hours of observation, Poirier saw only five instances of an individual's drinking. Apparently the need for moisture is satisfied by the food.

Although Nilgiri langurs are quite abundant in some areas – in the Ashambu hills at the southern tip of India alone, their numbers have been estimated at no less than 5000 – the exact population is not known. They are increasingly endangered by the destruction of their natural habitat. To this day, the animals are hunted by the natives. Their

The Nilgiri langur, recognizable by its dark body fur and light head hair, inhabits chiefly the evergreen rainforests in southwestern India. It is distinctly arboreal, with excellent leaping abilities.

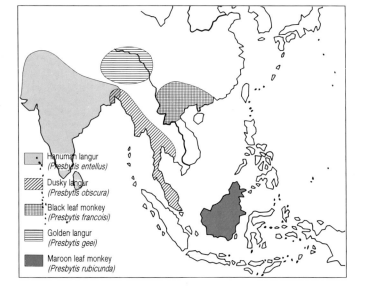

flesh is highly prized. It is also used to prepare medications for treating respiratory disorders. The hide is used for drumskins.

Nilgiri langurs are difficult to keep in zoological gardens because of their special dietary needs, and they are found in only a few zoos, for example the Erfurt Zoo, outside of India.

The PURPLE-FACED LANGUR *(Presbytis senex)*. The closest relatives of the Nilgiri langurs are the purple-faced langurs, formerly placed in a common genus, *Kasi*. Purple-faced langurs, limited in range to the island of Sri Lanka (Ceylon), live in small single-male groups numbering only five to eight individuals and in all-male groups. So-called "nomads" have also been observed. They are solitary individuals, or small groups of three, that roam about between home territories of single-male groups. Occasionally a juvenile or perhaps an adult female will join the nomads.

In Polonnaruwa, in the eastern province of Sri Lanka, the purple-faced langurs have been more closely researched. It was found that births are scattered throughout the year, however, more births occur in the months of May to August. This would mean that mating takes place chiefly in the rainy season, when food is most plentiful. A couple of French scientists, the Hladiks, have researched the menu of the purple-faced langurs and found that on an annual average, the diet consists of 60 percent leaves, 12 percent blossoms, and 28 percent fruits. It is remarkable that 70 percent of the diet derives from only three species of trees. The Hladiks calculated that one langur requires about 880 lb (400 kg) of vegetable food per year, but that ten times this quantity is available. Such a proportion is absolutely necessary, because otherwise the young leaves and shoots would be eaten off to such an extent that the trees would perish.

The Hladiks made another important observation. Comparing the diet with that of the hanuman langurs in the same area, they found that the purple-faced langurs are less amply supplied with nu-

The hanuman (below), often encountered on solid ground only, is a gymnastic artist, like the almost exclusively arboreal purple-faced langur (above). In the leap, langurs launch themselves from the treetop and land securely on a somewhat lower branch of a neighboring tree.

trients and calories, but are also less mobile than the hanuman langurs, and so require less energy.

Purple-faced langurs have comparatively small territories, seldom larger than 10 acres (4 hectares). However, these are vigorously defended against trespassers from other groups. About every three years, a change of males takes place in the group. In such a case, apparently all non-adults leave the group. Purple-faced langurs seemed to practice infanticide as well. Youngsters were found to disappear in several groups on the occasion of male displacement.

The DUSKY LANGUR *(Presbytis obscura).* The name spectacled langur derives from the white rings

(Left) A hanuman langur has fallen prey to a tiger. (Right) The drawing shows a Nilgiri langur (top) and a silvered leaf monkey (bottom).

around the dusky langur's eyes. The mouth is framed in white also.

The average group size is 17 individuals. Besides single-male groups, there are some in which two or three adult males live. All-male groups have not been observed. The single-male groups are rarely encountered as a closed unit. Immediately after leaving their sleeping place, smaller subgroups are formed, consisting as a rule of one or two females and their young. The youngsters form play groups, often sporting in the vicinity of an adult female who acts as supervisor while the other females rest or eat.

An adult male has three principal tasks. The most important is perhaps the detection of predators, and the males are often seen in the highest tree-

tops. At an alert, they first give warning by a soft call ("whoo"), then by loud cries and shrieks, to the other group members. Their second duty is to hold the group together. When individuals stray too far apart, or into the outlying parts of the home territory, calls serve as signals to assemble. The third function of the males is to patrol the boundaries of the territory. Imposing leaps through the branches and loud cries notify neighbors of the group's presence, especially in areas of overlap between group territories.

When the subgroups meet again after having been separated for any length of time, there are regular scenes of reunion, in which the animals first embrace and then groom each other.

Dusky langurs prefer the dense forests of the interior. They do not descend to ground level. In the trees, they generally maintain a height of about 112 ft (35 m). Progress through the branches is primarily by quadruped trotting and running, less often by leaping. Nourishment is provided by 87

different tree species. The diet consists of 58 percent leaves and 32 percent fruits.

At birth, the babies have bright orange fur, which changes at about 6 months to the gray of the adult coat. Presumably, only the adult females take part in infant transfer.

Dusky langurs are kept in many zoological gardens the world over, as in Wuppertal, Frankfurt-am-Main, and elsewhere.

PHAYRE'S LEAF MONKEYS *(Presbytis phayrei)*. Phayre's langur, regarded by some authors as a subspecies of the dusky langur *(Presbytis obscura)*, is widely distributed in Southeast Asia. Colored gray to gray-brown, these langurs live in the dense evergreen forests of Thailand, in the bamboo forests of Burma and in the primary and secondary forests of Bangladesh. Phayre's leaf monkeys may be recognized by white hairless rings around the eyes and lips. Males and females differ only in coloring beneath the ischial callosities; light gray in the males and yellowish in the females.

Our only close observations are due to R.P. Mukherjee, in the Indian state of Tripura, east of Bangladesh. Mukherjee counted 36 groups, including one all-male group. The other groups, numbering 4 to 38 individuals, included one to four males living with the females, but usually only one of the males was fully adult. The males seemed to be excluded from group social life.

Phayre's langurs eat chiefly leaves. As among most langurs, feeding begins immediately on waking. Towards noon, a siesta is observed. Feeding activity intensifies again in the evening. In the rain, the animals retire under the dense canopy of leaves and remain sitting huddled together until the rain passes. They seldom leave the trees. No drinking has been observed, but rainwater is licked off the foliage.

Little is known about the total numbers of Phayre's langur that are still living. The total for Bangladesh has been estimated at 1300 individuals; no estimates are available for other areas.

The SILVERED LEAF MONKEY *(Presbytis cristata)*. Represented by eight subspecies, the silvered leaf monkey is distributed over the southeast Asian mainland and the islands. The fur color of these monkeys is quite varied; all shadings from brownish-gray to dark brown have been recorded. These langurs live in the mangrove and coastal forests of Malaysia, the mountain forests of Java and Sumatra to an elevation of 5400 ft (1700 m), and the evergreen forests of Thailand. Though very shy, they are occasionally seen in the vicinity of human settlements.

At birth, the baby coat is orange. Soon after birth, however, the face, hands, and feet go from pale to black. The females are easily identified by irregular white patches on the inside of the thighs.

Silvered leaf monkeys live in single-male groups, comprising 12 to 51 individuals. The groups are strictly territorial, defending their home area against neighboring groups.

The silvered leaf monkeys doubtless also practice infanticide. The American primatologist Kathy Wolf noted that at about three months after a change of males, all the infants in the group had disappeared.

These langurs occasionally come down out of the trees, but will return in flight at the slightest sign of danger. They eat primarily leaves, preferring the fresh young leaves and tips of sprouts.

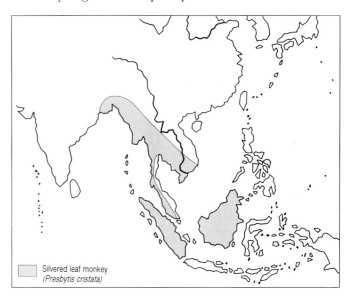

Their future in Malaysia is most uncertain, because their natural habitat is being destroyed by clearing for agriculture. Thus in a single preserve on the Selangor River, 78 percent of the mangrove forest was cleared in a period of only fourteen years.

The CAPPED LANGUR *(Presbytis pileata)*. The capped langur lives, separated only by the Manas River from the golden langur *(Presbytis geei)*, in the deciduous and evergreen forests of Assam, Burma, and Bangladesh. It is chiefly arboreal, leaving the trees only when two stand so far apart that the gap cannot be leaped.

Capped langurs live in small single-male groups of 7 to 13 members. The adult males spend much of the day on "sentry duty". They will sit in a tall tree and attentively observe the surroundings.

The capped langur is named for its long, erect shock of hair, which is laid towards the back. The fur color on the back is gray to black and brown-yellow to orange on the ventral side. The newborn are golden to orange-red. In 1982, the total population was estimated at 37,000.

The GOLDEN LANGUR *(Presbytis geei)*. The golden langur was not discovered until 1953, in Assam on the Bhutan border. It probably inhabits only a small area. Its discoverer, Gee, estimated the Assam stock at 540 individuals in 1961. Nothing is known of the numbers living in Bhutan. Golden langurs, counted by some authorities as a subspecies of the capped langur, live chiefly in a single-male groups in dry deciduous woods, where the trees are bare in Winter. The group size ranges from 3 to 40. Golden langurs are excellent leapers. In leaping, they assume a "seated" posture, hands and feet stretched forward, the long tail hanging down.

They prefer the dense branches of trees, and if possible will run and climb on the thick horizontal limbs. Occasionally they will be seen running along a branch on two legs. They will then hold on to the foliage on either side with their hands. In places where a leap is necessary, they will wait patiently one behind another, all using the same take-off and landing points. Analyses of stomach contents have shown that many flowers and fruits are eaten, as well as leaves.

Golden langurs rarely leave the shelter of the trees; they drink from streams or water holes, or ingest earth or sand. This behavior has been observed in many primate species, especially leaf eaters. Doubtless the earth contains minerals that are deficient or absent in the everyday diet.

Oboussier, a German zoologist, has observed how during a squabble two golden langurs fell from a tree into the water and then swam 25 or 30 ft (8 or 10 m) to reach the bank. This is one of the few observations of langurs' ability to swim; they are generally supposed to dislike the water.

The golden langur is hunted by the natives for its flesh. It is marked "rare" on the Red List. It has been kept successfully to date in only two zoos in India, at Kampur and Lucknow, as well as at the Kathmandu Zoo in Nepal.

The MENTAWAI LANGUR *(Presbytis potenziani)*. The few existing reports of the habits of the Mentawai langurs indicate that they are an extraordinarily interesting species. Their range is the Mentawai Islands, situated between 50 and 90 mi (85 and 145 km) west of Sumatra. The islands are all quite small, their total area being only 2500 mi^2 (7000 mi^2).

The Mentawai langurs live in family groups consisting only of a pair and their common progeny. Among Old World monkeys, this community form is most unusual. Except for gibbons, monogamy is found only among the Pageh snub-nosed monkeys, which also live on the Mentawai islands, and the Brazza monkeys of Africa. According to more recent observations, it seems that the Sunda Island leaf monkey may also be monogamous. Some authors surmise that the Mentawai langurs have developed this mode in reaction to severe endangerment by humans. There are no other predators on the islands. The natives hunt the monkeys with poisoned darts shot from blowpipes. The absence of sexual dimorphism (distinctive appearance of males and females) and the duet-singing of pairs (see below) are cited as evidence that contradicts this hypothesis. These traits presumably appeared long before the islands were settled by humans.

When the hunters draw near, the adult males execute elaborate diversionary maneuvers. They attract the attention of the hunters to themselves by vigorous leaps and cries. The female with the young will sit quite motionless and silent in the thicket. She may do this for as long as 45 minutes, while the male is luring the hunter away from the retreat. Of course the male runs a considerable risk, but the infant's chances of survival are improved.

Each family group inhabits a territory of about 38 to 55 acres (15 to 22 hectares). These home territories may overlap considerably. However, there are so-called "core" areas, used by one pair exclusively. Males and females both defend their core areas, which include sleeping trees, against intruders. Usually the male begins, with loud calls and leaps in the branches. As soon as he pauses, the female begins to call, and jumps around in the branches, stiff-legged like the male.

Among the Mentawai langurs, unlike all other *Presbytis* species, the females may utter long-distance calls as well. Despite the dense foliage, those of the males will travel more than 0.6 mi (1 km) and those of the females generally farther than 1600 ft (500 m).

A striking trait of the Mentawai langurs is the lack of any great

(Below) Purple-faced langur with typical sideburns. (Bottom) Two hanuman langurs in a dispute over rank.

difference in size between the sexes, such as we see in many others of the genus. However, the males are easily recognized by their conspicuous white fur in the genital region. Family groups comprise 2 to 5 individuals; presumably the pair bond is permanent. A female will bear offspring about every two years. The babies at birth have white fur, which gradually darkens. The adults have black coloration on the back, the ventral color being yellow to red-orange. The tail is also black. Mentawai langurs inhabit predominantly the evergreen rainforests. More rarely, they are found in mangrove forests. They are designated as threatened in the Red List. The number living in Siberut, the largest of the Mentawai islands, was estimated at 46,000 in 1980.

FRANÇOIS' BLACK LEAF MONKEY *(Presbytis francoisi).* There is no accurate information on the biology or habits of the Tonkin langurs. In their natural habitat – in Vietnam, China and Laos – they have not yet been studied. Individual specimens are found in some zoos.

The BANDED LEAF MONKEY *(Presbytis melalophos).* Misleadingly called the red langur in German, this species is quite variable in fur color. The back is gray or brown, the ventral side is light gray. Babies have white fur and a pale face at birth, and a conspicuous red-brown stripe, cross-shaped, from nape to tail and across the shoulder.

The banded leaf monkeys have been studied with exceptional thoroughness by the American anthropologist Sheila Curtin on the Malay Peninsula. Their diet is highly varied. It includes 137 distinct plant species. Fruits are preferred, accounting for 48 percent of the diet. Leaves represent only 35 percent. The preferred level in the trees is around 90 ft (28 m). The animals travel with great agility in the trees of the tropical rainforest and also on the margins of rivers. They are excellent leapers. Occasionally they will descend to the ground.

The members of a typical single-male group, numbering about 13, usually travel as a closed unit. Two tasks especially fall to the male; the diversion of possible predators is very important. When a member of the groups detects a predator, for example the snake eagle *(Spilornis cheela),* it will utter a cry. The male will hasten to draw the attention of the enemy by loud cries and imposing leaps, thereby leading him away from the group by as much as 240 ft (75 m), while the group members will remain motionless in the brush.

The male's other task is the acoustic identification of the territory. By loud, prolonged cries and imposing leaps, a system of communication is maintained between groups, informing them of each other's location at all times. Especially in the evening, as well as during the night and in the early morning hours, regular concerts of calls are held between the males of neighboring groups. These territorial calls, serving to identify a territory, are clearly distinguished from the alarm cries. In her observations, Sheila Curtin noticed that contacts between individuals are infrequent even within a group. Mutual grooming is rarely observed. Apparently there is no "infant transfer" either.

The MAROON LEAF MONKEY *(Presbytis rubicunda).* Reports concerning the maroon leaf monkey are merely anecdotal. It inhabits primary and secondary forests, and occasionally comes down to the ground to find another patch of woods. The ani-

LANGURS AND COLOBI

mals presumably live in single-male groups. Groups of 5 to 8 individuals have been seen on Borneo. The male sleeps away from the group at night, in a separate sleeping tree. Maroon leaf monkeys have also been encountered near cultivated areas.

The WHITE-FRONTED LEAF MONKEY *(Presbytis frontata).* This langur inhabits the tropical rainforests of Borneo. No studies in the wild have been reported.

The SUNDA ISLAND LEAF MONKEY *(Presbytis aygula).* The Sunda Island leaf monkey was scientifically observed only a few years ago, by the Indonesian primatologist Yayat Ruhiyat. He carried out his research in two different locations in the west of the Island of Java. Although the two places are only about 30 mi (50 km) apart, there are striking differences between the local leaf monkeys. In the region around Lake Patenggang, the animals reportedly live on monogamous family groups. In the Kamojang Nature Park, however, they form groups of one male and up to four females.

Whereas the Patenggang home territories overlap to a great extent, they do so only slightly at Kamojang. Ruhiyat surmises that the progressive destruction of the Patenggang forest has been a factor in the genesis of the family group.

The males watch over the members of their groups very carefully. Often they sit in the treetops and alertly observe the surroundings, and the group members themselves. At the least sign of danger, the males utter warning cries and engage in the aggressive leaping familiar among other langurs.

Sometimes infant tansfer has been observed, though only juvenile and adult females participated. Social contacts between individuals are otherwise quite rare.

Sunda Island langurs eat predominantly leaves, being 65 percent of their diet, of which 14 percent is fruit. A total of 74 different plant species are used for food.

Genus *Nasalis*

Assigned to this genus today are the proboscis monkey *(Nasalis larvatus)* and the pig-tailed snub-nosed monkey *(Nasalis concolor),* formerly accorded its own genus *Simias*. On the basis of numerous common traits in skull structure, and the prominent nose, a common genus for the two species seems justified.

The PROBOSCIS MONKEY *(Nasalis larvatus).* One of the most conspicuous representatives of the langurs is the proboscis monkey, native to Borneo, whose name refer to the long gherkin-shaped nose of the male. The nose of the female is by contrast

This male langur is intensely interested in the female, who is ready to mate.

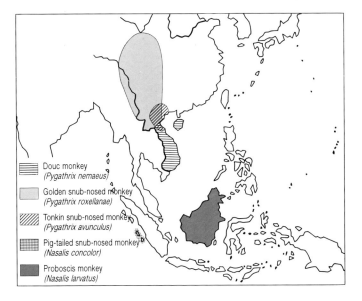

Douc monkey (Pygathrix nemaeus)
Golden snub-nosed monkey (Pygathrix roxellanae)
Tonkin snub-nosed monkey (Pygathrix avunculus)
Pig-tailed snub-nosed monkey (Nasalis concolor)
Proboscis monkey (Nasalis larvatus)

quite dainty. Also, the male proboscis monkeys are twice the size and weight of the females.

Little is known of the habits of these extremely shy monkeys. Observations in their natural habitat, the mangrove forests, must be based on watercraft. Proboscis monkeys live in multi-male groups numbering between 11 and 32 individuals. Single-male groups have also been observed. There do not appear to be any fixed ties between the members of a group.

Proboscis monkeys are good swimmers and divers. Sometimes entire groups will swim across the smaller rivers. When danger threatens, they will often drop many feet to the ground or into the water. Presumably these falls often cause injury. Examination of skeletons has revealed numerous healed fractures.

The proboscis monkey, an "endangered" species,

(Left) Proboscis monkeys are strictly vegetarian; their chief nourishment is the fresh foliage of inaccessible mangrove forests. (Right) Only the adult males, which are considerably larger and heavier than the females, wear such an enormous nose ornament.

is kept in many zoological gardens, including the Cologne Zoo. A baby proboscis monkey was born in a zoo (in Indonesia) in 1963, for the first time.

The PIG-TAILED SNUB-NOSED MONKEY *(Nasalis concolor)*. Like the proboscis monkey *(Nasalis larvatus)*, the pig-tails have a limited range. They live on the Mentawai Islands, west of Sumatra. Like all local species of primates, the pig-tailed snob-nosed monkeys are hunted by the human inhabitants. On some smaller islands of the group, therefore, they have already disappeared, and the Red List classifies them as "threatened with extinction."

Observations by the American Ronald Tilson and the Japanese Kunio Watanabe have shown that in regions where they are intensively hunted, the animals live in family groups composed of the parent pair and at most three offspring of different ages. Where hunting is less prevalent, on the other hand, small groups of one male and up to four females predominate.

Family groups and one-male groups differ distinctly in behavior. Within the family groups, there is close spatial cohesion; the animals are seldom more than 16 ft (5 m) apart. When they pass through the branches in the search for food, the male is in the lead, while the females and the young follow close behind. Insofar as possible, they avoid leaps from tree to tree, which would cause conspicuous movement in the branches. Instead, they often take long detours, and change trees at points where the tops meet. Communication by sound is limited; calls are sometimes uttered, serving to mark the home territory or to warn against predators.

If the animals are disturbed by humans, they will hide in the thick leaf canopy of the trees, and remain motionless for two hours or more. If the trees do not afford adequate cover, they will take

flight through the treetops or slide down the trunks. They will also seek safety by jumping several yards down to the ground.

Encounters between two families pass off almost noiselessly. Tilson observed them approaching within 80 ft (25 m) of each other. One of the males uttered a short call, answered by the male of the other group. Each then ran back to his group, and they withdraw in opposite directions.

The one-male groups, traveling noisily through the trees, behave very differently. In the morning hours, the males' calls are heard, carrying more than 0.3 mi (0.5 km), and showy displays of strength by leaping are regularly observed. Group movements are a good deal more noticeable, and take a longer time than those of family groups.

Pig-tailed snub-nosed monkeys occur in two color variants, a light yellow and a dark gray form. The stock of animals on Siberut, the largest Mentawai island, was estimated at 19,000 in 1980.

Genus *Pygathrix*

The genus *Pygathrix* of snub-nosed monkeys is regarded today as comprising the two subgenera *Pygathrix* and *Rhinopithecus,* formerly genera in their own right. Four species and four subspecies are distinguished.

The Douc monkey *(Pygathrix nemaeus).* Undoubtedly the douc monkey, which is sometimes called the costumed ape for its coat of many colors, is undoubtedly one of the gaudiest among primates. Years of war in Laos and Vietnam and intensive hunting by the natives have brought this species under the threat of extinction.

The American researcher Lois Lippold has done one of the few studies of the behavior of these animals in Vietnam. She found that one or two adult males will form groups of 5 to 11 individuals with several females and their offspring. Douc monkeys are very particular in their foraging; leaves and fruits are picked, and eaten only after close inspection. Old leaves and ripe or overripe fruits are discarded.

Observations in zoos have shown that baby douc monkeys are passed around from hand to hand within a group (infant transfer) to be cared for. The San Diego Zoo reports an observation, so far unique, of a child adoption by the father. After the mother's death, the head of a one-male group was particularly attentive to the baby. The three-month-old youngster spent about 70 percent of the day and the entire night in close contact with the male parent. The rest of the time was passed with the three females of the group, who also nursed it at the breast.

The douc monkey is bred successfully in many zoos around the world, at Stuttgart and Cologne in particular.

The habits and behavior of the snub-nosed monkeys are largely unknown. Besides the monkey *Pygathrix nemaeus* this group includes the golden snub-nosed *(Pygathrix roxellanae),* Brelich's snub-nosed monkey *(Pygathrix brelichi),* and the Tonkin snub-nosed *(Pygathrix avunculus).* From the few observations, made chiefly in China, we know that they live in evergreen subtropical leaf forests, and also in mountain needle forests up to an elevation of 9600 ft (3000 m). The group size of the golden

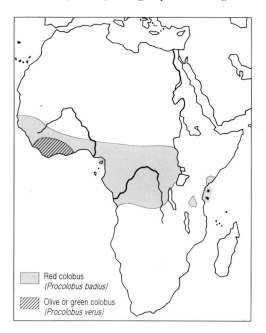

Red colobus (*Procolobus badius*)
Olive or green colobus (*Procolobus verus*)

snub-nose monkey, for example, is estimated at over 600 individuals. If these estimates should be correct, the habit would be unique among arboreal primates. Within the large groups, there are presumably subgroups of the one-male type.

The baby coat of the young, who are passed around to many members of the group, is said to persist until the age of two years. Snub-noses eat predominantly leaves. A peculiar feature is that the needles of conifers are also eaten. The diet is supplemented with birds' eggs, hatchlings, worms, and insects. The golden snub-nose is considered rare.

The African langurs are referred to as colobi (Greek *kolobos* "maimed") because of their much-reduced thumbs. Hence, too, the scientific name *Colobus*. Compared to Asian langurs, the African forms have a more compact build. Some of them have conspicuous head hair formations, and sometimes long body fringes and bushy tails. They inhabit tropical Africa from Senegal to Ethiopia, from the southern edge of the Sahara to Angola. Two genera are distinguished, *Procolobus* and *Colobus*.

The "costumed" douc monkey has a colorful coat that seems to be assembled of jacket, vest, knee breeches, stockings, and shoes of different colors. This species, which has been little studied and is threatened with extinction, lives primarily on leaves.

Genus *Procolobus*

The most striking feature of the genus *Procolobus*, distinguishing it from all other langurs, is the occurrence of genital swellings controlled by the ovarian cycle in the adult females, and the presence of a so-called "perineal organ" in the juvenile and adult males. This latter is deceptively similar in structure to the female genital region (exhibiting for example a pseudovagina and a pseudoclitoris), but of course is not liable to any cyclic variations in swelling. Males – especially those not yet fully adult – will "present" this conspicuous genital region to higher-ranking males, such as females do when ready to mate, and evidently derive the benefit of the aggression-inhibiting effect of the signal.

The difference in coloration between the newborn and the adults is not pronounced in *Procolobus* compared to other langurs, and correspondingly the "infant transfer" system, which is typical for langurs, is absent. At least the red colobus *(Procolobus badius)* is known to have a social structure deviating clearly from that of other langurs; there is no such information concerning the green colobus *(Procolobus verus)*. Lastly, the representatives of the genus *Procolobus* have no cry that is audible over long distances, as many langurs do.

The genus *Procolobus* is divided into only two species.

The GREEN OR OLIVE COLOBUS *(Procolobus verus)*. So called because of its olive-green to brownish-green dorsal coloration, the green or olive colobus is perhaps the smallest of the family, with an average body length of 20 in. (50 cm), not counting the tail, and a mean weight of about 9 lb (4 kg); the adult males do not surpass the females in stature, but they do so in the size of their canine teeth.

Green colobi occur scattered from Sierra Leone to Togo, and a population, today wholly isolated, in eastern Nigeria. To the North, they nowhere transgress the eighth parallel of latitude. They chiefly colonize dense swamps and rainforests, near waters and at low elevations. In the forest, they prefer the levels at middle trunk height of the trees.

Surprisingly little is known of their biology, perhaps because of the scattered, concealed, and unobtrusive mode of life of these rather plain creatures. Their diet is supposed to consist very largely of young unripe leaves, the dispersed occurrence of which perhaps accounts for the comparatively low social cohesion of groups in this species. Fixed mating and birth seasons are described, with considerable competition among males for access to willing females during the mating season. The newborn hardly differ from the adults in fur color, and there are persistent reports that mothers may carry their newborn with their jaws, a habit otherwise unknown among true simians.

Green colobi are considered rare, but not yet threatened; their status may change quickly, with the increasing destruction of their habitat. Zoological gardens have not yet been successful in keeping and breeding them.

The RED COLOBUS *(Procolobus badius)*. The red colobi, formerly divided into several species, are considered today to be a single species with 14 to 19 subspecies. The red fur color, for which they are named, varies greatly in intensity and distribution from population to population, from the deep chestnut-red of the subspecies *P. badius badius* to a rust-red cap, limited to the head, and black-brown back coloring of *P. b. tephrosceles*. The belly and limbs are mostly gray or gray-brown, and the face is dark, often framed by gray side whiskers. Adult males exceed females considerably in body size and weight, the head appearing especially massive because of powerful temporal muscles; they have larger canine teeth and a heavier beard.

The range of the red colobi extends over all equatorial Africa from Senegal to Kenya and the Island of Zanzibar. This zone of distribution, however, is much interrupted and fragmented, so that several populations are quite isolated, like the subspecies *P. b. preussi* in Cameroon, *P. b. rufomitratus* on the Tana River in Kenya, and *P. b. kirkii* on Zanzibar.

In West and Central Africa, the red colobus inhabits predominantly the continuous rainforest, but may also colonize gallery forests along the river courses and more open savanna woodlands. As a rule, it prefers the emergent layer of the tall trees, where it travels by running and leaping quadruped-fashion in the branches, but will occasionally swing overhand for short distances. Jumps from a falling height of 30 ft (10 m) have been described, the animals spreading their arms and legs laterally to catch in the leaf canopy. They are seldom to be met with on the ground at all.

Regional accounts of the composition of their diet vary widely. In the Kibale forest of Uganda, for example, Thomas Struhsaker found a proportion of 33 percent young unripe leaves and only 4–5 percent ripe leaves, as well as blossom and leaf buds (16–20 percent), sepals (13–19 percent), and also flowers, seeds, and certain fruits (5.7 percent). In the Gombe Reserve of Tanzania, however, according to Timothy Clutton-Brock, ripe leaves accounted for up to 44 percent of the diet, and in Senegal (according to Bernard Gatinot) again only 5.4 percent, young leaves 24 percent, and fruits no less than 35.9 percent. The chief activity of the day consists in eating (24–45 percent) and resting (about 35 percent), chiefly for digestion. Red colobi satisfy their fluid requirement mostly from their food. Danger threatens them from the air chiefly in the form of the crowned hawk eagle *(Stephanoaëtus coronatus)*, occasionally pythons and crocodiles, but above all humans, who in many places hunt them and esteems their flesh as a delicacy.

Red colobi are often found in association with other simian species, such as

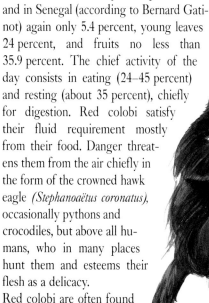

Brown snub-nose, a subspecies of the rare golden snub-nosed monkey.

various macaques, mangabeys, drills, and guerezas.

Young are born throughout the year, but in various regions there are one or two seasonal peaks. Pregnancy lasts 6–6.5 months, and a female's average interval between births is 20–24 months. The newborn are rather like the adults in fur coloring, compared to other langurs. Complete assimilation to adults takes place at the age of 3–4 months.

Red colobi live in sexually mixed groups of 12 to 80 individuals (groups in rainforest biotopes are approximately double the size of those in gallery or savanna forests). The sex ratio of adults in the groups is usually one male to one and one half to three females, each group counting at least two and usually more fully adult males. In striking contrast to other langurs, the males remain in their birth group more commonly than do the females, which leave their birth group as juveniles or young adults. Consequently the adult females of a group are rarely related, and seemingly for this reason do not tend to assist each other in rearing the young (no "infant transfer" system). The males, on the other hand, from cooperative subgroups with a linear ranking, the highest-ranking male in each mating most frequently. The associates of the American primatologist Thomas Struhsaker have observed infanticide by adult males among red colobi in the Kibale forest in Uganda, the infants killed being evidently unrelated to the males that killed them. Here also, the mothers of the victims were actively offering to mate with the perpetrator only two weeks after the event, with a distinctly shortened interval between births in consequence. The infanticides thus gained a reproductive advantage, as among the Asian langurs.

On the whole, the cohesion of the mixed groups of red colobi is looser than with most other langurs. Their territories overlap with those of neighboring groups. Encounters lead to threatening or menacing behavior and mock aggression on both sides, whereby one group may be driven away but injuries are hardly inflicted. Besides the mixed groups, there are solitary males who will not infrequently attach themselves to groups of other simian species (macaques, mangabeys, guerezas). All-male groups have not been observed to date.

Certain subspecies of the red colobus must today be regarded as threatened with extinction, such as *Procolobus badius preussi* in West Cameroon, *P. b. rufomitratus* in Kenya, and the Zanzibar colobus *P. b. kirkii*. The keeping and breeding of red colobi in zoological gardens has had no satisfactory success as yet.

Genus *Colobus*

The so-called black or black-and-white colobi are classified today in two to four species with 16 to 22 subspecies. We here recognize two species. The genus *Colobus* differs essentially from *Procolobus* by the complete absence of female genital swellings and of male "perineal organs"; by the occurrence of body fringes or bushy tails; by the black or black-and-white fur color, and the considerable greater deviation of the newborn coat (pure white or white-gray) from the adult color-

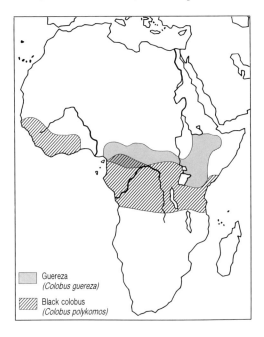

Guereza (*Colobus guereza*)
Black colobus (*Colobus polykomos*)

ing; by a pronounced system of "infant transfer" among the adult females of a group; and by the use of calls that are audible from afar. The species and subspecies are outwardly differentiated primarily by the coat of hair (body fringes, tail hair, head hair, whiskers) and the distribution of white color in the adult coat, unless white is completely absent, which is a trait of the black colobus *(Colobus polykomos satanas)*.

The GUEREZA or NORTHERN BLACK-AND-WHITE COLOBUS *(Colobus guereza)*. The term "mantle" has been used to refer to the white mane at the sides of the body that distinguishes the guereza. These fringes and the bushy tail (sometimes white) nearly lead to extermination of the guerezas about the turn of the century, especially the East African mountain forms, when their beautiful furs were in high fashion with European ladies. The Masai and other East African tribes had also used this "majestic" adornment since ancient times. Meanwhile, most stocks have recovered, insofar as increasing destruction of habitat permits.

The guereza is distributed from eastern Nigeria through Cameroon, eastern Gabon, northern Congo, the Central African Republic, northeastern Zaire, southern Sudan, Uganda, Ruanda, Ethiopia, Kenya, and northern Tanzania. It inhabits the continuous rainforests as well as gallery forest along the rivers and open savanna woodlands. On the whole, it appears to prefer somewhat drier habitats than the red colobus. Where the two species occur together, the guereza tends to occupy somewhat lower tree heights than the red colobus. Guerezas are outstanding leapers, and are able to great distances. Contrary to earlier reports in the literature, however, they virtually never swing through the branches hanging by their arms.

The guereza's dietary spectrum is narrower than that of the red colobus; according to the researches of the American primatologist John Oates in Uganda, it consists about 58 percent of young unripe leaves, 12.5 percent mature leaves, 13.5 percent fruits, 4 percent leaf buds, and 2 percent blossoms, but this distribution is highly varied seasonally and geographically; thus, mature leaves may at times account for up to 34 percent. Since mature leaves are especially difficult to digest because of their high cellulose content, guerezas require an especially large portion of their day for quiet digestion of what has been eaten. Up to 58 percent of the daytime is spent resting, 20 percent is devoted to eating, and 5.5 percent is used in traveling. Their comparatively one-sided diet may be connected with the fact that guerezas often obtain minerals, such as calcium, potassium, sodium, magnesium, and copper, by actually eating soil. Natural enemies of the guereza are the crowned hawk eagle, leopards, and sometimes

Their elegant black-and-white coat with its striking white fringes nearly sealed the fate of the Northern black-and-white colobus, or guereza, for about the turn of the century such monkey furs were a coveted item of "fashion." The slender, long-legged animals are excellent tree climbers and leapers, living primarily in rainforests and gallery forests.

Langurs and Colobi (Colobinae)

Nomenclature English common name Scientific name French German	Approximate Size Body length Tail length Weight	Distinguishing Features	Reproduction Gestation period Young per birth Weight at birth
Hanuman langur; common langur; gray langur *Presbytis entellus*, 15–16 suspecies Houleman; entelle Hanumanlangur; Hulman	1.7–3.6 ft; 51–108 cm 2.4–3.6 ft; 72.5–109 cm males 19.8–46.2 lb; 9–21 kg females 16.5–39.6; 7.5–18	Silver-gray fur; black face, feet, and hands eyebrows project; young dark brown to black at birth; subspecies distinguished partly by carriage of tail; on Sri Lanka, head hair erect, forming a crest	About 200 days 1, rarely 2 Not known
Nilgiri langur; John's langur *Presbytis johnii* Semnopithèque des Nilgiris Nilgirilangur	m. 2.6 ft; 78 cm f. 1.9; 58.5 2.3–3.2 ft; 68.5–96.5 cm m. 20–29 lb; 9.1–13.2 kg f. 23.9–24.9; 10.9–11.3	Fur shiny black or somewhat beige; yellowish-brown head; white tuft of fur on inside of female thigh; young reddish-brown at birth	Not known
Purple-faced langur *Presbytis senex*, 5 subspecies Semnopithèque blanchâtre Weißbartlangur	1.5–2.2 ft; 45–67 cm 1.9–2.8 ft; 59–85 cm m. 8.4–20 lb; 3.8–9.1 kg f. 8.4–20.1; 3.8–9.3	Brown-black fur; white to pale brown side whiskers; young gray at birth	About 200 days 1 Not known
Dusky leaf monkey, dusky langur *Presbytis obscura*, 7 subspecies Semnopithèque obscur Brillenlangur	1.6–2.3 ft; 48–68.5 cm 2.3–2.1 ft; 68.5–80.5 cm m. 16.3 lb; 7.4 kg f. 14.3; 6.5	Brown to black back; dark brown to dark gray ventral parts; mouth and eyes framed white; young light yellow, orange, or red at birth	Not known
Phayre's leaf monkey *Presbytis phayrei*, 4 subspecies Semnopithèque de Phayre Phayres Langur	m. 1.6–1.8 ft; 50–55 cm f. 1.5–1.8; 45–53 2.2–2.8 ft; 65–86.5 cm m. 17.4–19.1 lb; 7.9–8.7 kg f. 12.3–14.9; 5.6–6.8	Gray to gray-brown fur; white rings around eyes and mouth; males and females distinguishable by coloring below ischial callosities (light gray and yellowish, respectively)	Not known
Silvered leaf monkey *Presbytis cristata*, 8 subspecies Semnopithèque à coiffe Haubenlangur	males 1.7–1.8 ft; 52.5–56 cm females 1.5–1.6; 46.5–49.5 2.3 ft; 69 cm 11.4–18.9 lb; 5.2–8.6 kg	Diverse fur coloring; young orange at birth; white spot on inside of female thighs	Not known
Capped langur *Presbytis pileata*, 5 subspecies Semnopithèque à bonnet Schopflangur	males 2.2–2.3 ft; 68.5–70 cm females 1.9–2.2; 59–67 m. 3.1–3.5 ft; 94–104 cm f. 2.6–3; 78–90 m. 25.3–30.8 lb; 11.5–14 kg f. 20.9–24.8; 9.5–11.3	Gray back; brown-yellow to orange ventral parts; erect shock of hair on head; young golden to orange-red at birth	Not known
Golden langur *Presbytis geei* (regarded by some as *P. pileata* subspecies) — Goldlangur	m. 2.1–2.4 ft; 64–72 cm f. 1.6–2; 49–61 m. 2.6–3.1 ft; 78–94 cm f. 2.4–2.7; 71–80.5 m. 22–26.4 lb; 10–12 kg f. 20.9; 9.5	Cream-clored back; reddish to orange ventral parts; face nearly hairless with dark pigmentation	Not known
Mentawi leaf monkey *Presbytis potenziani*, 2 subspecies Semnopithèque de Mentawei Mentawailangur	1.5–1.9 ft; 44–58 cm 1.6–2.1; 50–64 cm m. 14.3 oz; 6.5 kg f. 14.1; 6.4	Black back; yellow to red-orange ventral parts; young white at birth	Not known
François' black leaf monkey *Presbytis francoisi*, 5 subspecies Semnopithèque de François Tonkinlangur	1.7–2.2 ft; 51–67 cm 2.7–3 ft; 81–90 cm 13.2 lb; 6 kg	Gleaming black fur; white from corners of mouth to ears; pronounced crest on head	Not known
Banded leaf monkey *Presbytis melalophos*, 19 subspecies Semnopithèque mélalophe Roter Langur	1.4–1.9 ft; 43.5–59 cm 2.3–2.8 ft; 68–85 cm m. 15.4 lb; 7 kg f. 14.5; 6.6	Regionally diverse fur coloration, from gray to brown; young white at birth, cross-shaped red-brown marking on back	Not known
Maroon leaf monkey *Presbytis rubicunda*, 5 subspecies Semnopithèque rubicond Maronenlangur	1.5–1.8 ft; 45–55 cm 2.1–2.6 ft; 64–78 cm 11.4–17.2 lb; 5.2–7.8 kg	Black-red or reddish-orange fur	Not known
White-fronted leaf monkey *Presbytis frontata*, 2 subspecies Semnopithèque à front blanc Weißstirnlangur	1.4–2 ft; 42–60 cm 2.1–2.6 ft; 63–79 cm 12.3–14.3 lb; 5.6–6.5 kg	Predominantly light, gray-brown fur; under parts yellow	Not brown
Sunda Island leaf monkey *Presbytis aygula*, 8 subspecies Semnopithèque des îles de la Sonde Mützenlangur	1.4–2 ft; 43–60 cm 1.8–2.8 ft; 55–83 cm 12.1–17.8 lb; 5.5–8.1 kg	Upper parts predominantly light gray, sprinkled black or brownish; under parts white, cap-like hood	Not known

COMPARISON OF SPECIES

Life Cycle **Weaning** **Sexual maturity** **Life span**	Food	Enemies	Habit and Habitat	Occurrence
Age 10–12 months Age 3 years In captivity, over 25 years	All plant parts, predominantly leaves, flowers, and fruits; occasionally insects	Tiger, leopard, snakes, dog, humans	Terrestrial and arboreal in nearly all climates and vegetation zones to elevations over 12,800 ft (4000 m); social form diverse; one to many males per group, all-male groups (8–125 individuals); territory size 0.07–3.9 mi² (0.2–10.4 km²)	Widely distributed
Age 10–12 months Not known Not known	Vegetable fare; primarily leaves	Humans, panther, jackal (?)	Predominantly arboreal in evergreen mountain rain forest to elevations of 7700 ft (2400 m), but also at perimeters near human settlements; single-male groups (at most 9 individuals) and all-male groups; territory size 0.2–0.9 mi² (0.6–2.6 km²)	Population not accurately known; hunted by natives; endangered chiefly by habitat destruction
Age 7–8 months Not known Not known	Vegetable fare; predominantly leaves, but also fruits and flowers	Occasionally, leopard	Arboreal in parks and mountain woods, rarely on ground; on plains and at elevations up to 6700 ft (2100 m); in small, single-male groups (5–8 individuals) and all-male groups; territory size, 2.3–25.2 acres (0.9–10.1 hectares)	Population threatened
Not known	Vegetable fare; predominantly leaves and fruits	Not known	Arboreal in woods, gardens, and bush terrain; social groups number 17 individuals, on average, with 1–3 adult males	Not known
Not known	Vegetable fare; predominantly leaves	Not known	Exclusively arboreal, in mixed or bamboo forests, sometimes along rivers; found in social groups of 4–38 individuals, usually with 1, occasionally up to 4, males; less frequently, all-male groups	Population not accurately known; about 1300 individuals remaining in Bangladesh
Not known	Vegetable fare; primarily young leaves and tips of shoots	Not known	Predominantly arboreal, occasionally on ground; in coastal and mountain forests and near human settlements; single-male social groups (12–51 individuals); defends home territory	Endangered by destruction of natural habitat
Not known	Vegetable fare	Not known	Predominantly arboreal, evergreen and deciduous forests; comes down to the ground for water; single-male social groups (7–13 individuals); territory size about 160 acres (64 hectares)	Total population estimated at 37,000 in 1982
Not known	Leaves, flowers, and fruits of various trees	Not known	Predominantly arboreal in tropical dry deciduous forests (bare in winter) and tropical evergreen rain forests; single-male social groups (3–40 individuals)	Rare and endangered
Not known	Vegetable fare	Humans	Arboreal in evergreen rain forests and mangrove forests; family groups (2–5 individuals); presumably permanent pairing; pair songs; territory size 37.5–55 acres (15–22 hectares)	Endangered by intensive hunting; only about 46,000 remaining on Siberut in 1980
Not known	Not known	Not known	In tropical monsoon forest and rock cliffs on river banks; habits have hardly been studied	Not known; believed to be endangered
Not known	Vegetable fare; primarily fruits and leaves	Not known	Predominantly arboreal, but will descend to the ground, in tropical rain and swamp forests, at elevations up to 6400 ft (2000 m); found near human settlements; single-male social groups (about 13 individuals)	Not known
Not known	Not known	Not known	Arboreal, occasionally on the ground, in rain forests and primary and secondary forests; also found near developed areas; single-male social groups (5–8 individuals)	Not known
Not known	Not known	Not known	In tropical rain forests; habits not known	Not known
Not known	Vegetable fare; predominantly leaves	Not known	In tropical rain forests; single-male social groups (6–10 individuals); possibly also monogamous pairing	Not known

LANGURS AND COLOBI

Nomenclature English common name Scientific name French German	Approximate Size Body length Tail length Weight	Distinguishing Features	Reproduction Gestation period Young per birth Weight at birth
Proboscis monkey *Nasalis larvatus* Nasique Nasenaffe	males 1.8–2.4 ft; 55.5–73 cm females 1.8–2; 54–60.5 m. 2.2–2.5 ft; 66–74.5 cm f. 1.9–2.1; 57–62 m. 25.7–51.9 lb; 11.7–23.6 kg f. 18–25.9; 8.2–11.8	Marked differences between sexes; males larger and heavier, with long gherkin-shaped nose	Not known
Pig-tailed snub-nosed monkey *Nasalis concolor*, 2 subspecies Rhinopithèque des Iles Pagai Pageh-Stumpfnasenaffe	m. 1.6–1.8 ft; 49–55 cm f. 1.5–1.8; 46–55 m. 5.2–7.6 in; 13–19 cm f. 4–6; 10–15 15.6 lb; 7.1 kg	Two color phases, dark gray and light yellow	Not known
Douc monkey; douc langur *Pygathrix (Pygathrix) nemaeus*, 2 subspecies Rhinopithèque douc Kleideraffe	1.8–2.7 ft; 55–82 cm 2–2.5 ft; 60–77 cm Not known	Extraordinarily vivid, high-contrast coloration	Not known
Golden snub-nosed monkey 2 subspecies (*Rhinopithecus*) *roxellanae*, Rhinopithèque de Roxellane Goldstumpfnase	1.6–2.7 ft; 50–83 cm 1.7–3.5 ft; 51–104 cm m. 39.6 lb; 18 kg f. 22; 10	Dark brown to black upper parts; whitish-orange under parts	Not known
Brelich's snub-nosed monkey *Pygathrix (Rhinopithecus) brelichi* Rhinopithèque jaune doré Weißmantelstumpfnase	2.4 ft; 73 cm 3.2 ft; 97 cm Not known	Gray-brown upper parts; light, yellowish-gray under parts	Not known
Tonkin snub-nosed monkey *Pygathrix (Rhinopithecus) avunculus* Rhinopithèque du Tonkin Tonkinstumpfnase	1.7–2.1 ft; 51–62 cm 2.2–3.1 ft; 66–92 cm Not known	Upper parts black; under parts yellow-white to orange	Not known
Olive colobus; green colobus *Procolobus verus* Colobe à huppe; colobe vrai Schopfstummelaffe; Grüner Stummelaffe	1.6 ft; 50 cm 2.2 ft; 65 cm 8.8 lb; 4 kg	Upper aprts olive to brown-green; under parts gray; male has perineal organ; female has genital swellings	Not known
Red colobus *Procolobus badius*, 14–19 subspecies Colobe ferrigineux, colobe bai Roter Stummelaffe	1.5–2.5 ft; 46–75 cm 1.3–3.2 ft; 41–95 cm m. 19.8–27.5 lb; 9–12.5 kg f. 15.4–19.8; 7–9	Male has perineal organ; female has genital swellings	6–6.5 months 1 Not known
Guereza; northern black and white colobus *Colobus guereza*, 10 subspecies Colobe de l'Abyssinie Mantelaffe; Guereza	1.5–2.3 ft; 45–70 cm 1.7–3 ft; 52–90 cm m. 19.8–31.9 lb; 9–14.5 kg f. 14.3–22; 6.5–10	Black body; frame of face, manes at side of body, and tail tassel white	About 6 months 1 Not known
Black colobus; southern black and white colobus *Colobus polykomos*, 12 subspecies Colobe à longs poils Bärenstummelaffe	1.6–2.3 ft; 49–68 cm 2.2–3.3 ft; 65–100 cm 13.4–25.7 lb; 6.1–11.7 kg	Black, with diverse white beards and manes; subspecies *C. p. satanas* completely black	147–178 days 1 About 16 oz; 450 g

The black colobus, or Southern black-and-white colobus, has brought forth variously marked forms in its extensive range. (Left) The black-and-white colobus *Colobus polykomos,* the nominate subspecies, (Right) a female black-and-white colobus of Angola with her baby.

even chimpanzees, which according to observations by Jane Goodall's group and the published reports of a Japanese study group, not infrequently prey on young colobi. Guerezas live in sexually mixed groups averaging 8 to 15 individuals, with usually only one fully adult male and three or four reproducing females. Sometimes several more or less adult males are observed in the mixed groups, but this would not seem to be a lasting condition. The fixed core of the association consists of the females, who seem to remain in their birth group

COMPARISON OF SPECIES

Life Cycle Weaning Sexual maturity Life span	Food	Enemies	Habit and Habitat	Occurrence
Not known	Vegetable fare; predominantly leaves	Not known	Predominantly arboreal in mangrove forests; social groups include one or several males (11–32 individuals); good swimmers	Endangered
Not known	Vegetable fare; predominantly leaves	Man	Arboreal in mangrove and rain forests; single-male and family groups; territory size 16.3–50 acres (6.5–20 hectares)	Threatened with extinction, especially by intensive hunting
Not known	Vegetable fare; predominantly leaves	Not known	In tropical rain and monsoon forests; single-male groups (5–11 individuals)	Facing extinction
Not known	Vegetable fare; occasionally worms, insects, birds' eggs and small birds	Not known	In subtropical evergreen and coniferous forests; presumably in groups including several males (over 600 individuals have been reported)	Rare
Not known	Vegetable fare; primarily leaves	Not known	In subtropical evergreen forests; groups include several males	Rare
Not known	Vegetable fare; predominantly leaves	Not known	In bamboo jungle; social groups include many males	Endangered; total population in China estimated at 500 individuals
Not known	Vegetable, with high proportion of young, immature leaves	Not known	Tree dweller in dense swamp and rain forests, near water; mothers transport newborn with the mouth; habits as yet hardly studied	Rare and endangered
Not known	Strictly vegetable	Crowned hawk eagle, python, crocodile	Predominantly arboreal in rain forests, but also in gallery forests along rivers, and in open, wooded savannas; social groups include several males (12–80 individuals)	Subspecies *P. b. rufomitratus*, *P. b. kirkii*, and *P. b. preussi* believed to be facing extinction
Age not known Males 6 years, females 3–4 In captivity, 24 years	Vegetable fare; predominantly leaves, less often flowers	Crowned hawk eagle, leopard, and chimpanzee	Arboreal in rain forests and gallery forests along rivers, and in open wooded savannas; single-male social groups (8–15 individuals); territory size 32.5–40 acres (13–16 hectares)	Still fairly abundant
Age about 15 months Males 4 years, females 2 In captivity, over 23 years	Vegetable fare	Eagles	Tree dweller in rain, gallery, and dry forests; also in savannas; single-male social groups	Still fairly abundant

The red colobus subspecies native to the Island of Zanzibar.

for life. It may therefore be surmised that the females of such groups are close relatives, consistently with the thoroughgoing mutual grooming among females and the well-developed "infant transfer" in these social units. The newborn, contrasting in their snow-white "baby dress" with the other group members, are highly attractive to the females (sometimes also to juvenile males), and are much carried about, which incidentally, when unskillfully done (especially in the care of juveniles), may involve dangerous falls of the helpless infants through the boughs. Interest in another's children fades noticeably when the color change has occurred at the age of 2–3 months. A female reaches sexual maturity at 3–4 years, a male perhaps not until 6 years of age.

Unlike the females, the adolescent males usually leave the group of their birth before they are fully mature. Apparently, however, they may remain members of their birth group into near or

323

(Opposite page) The Siamang is the largest and perhaps best-known representative of the small anthropoid apes, constituting the gibbon family.

young adulthood; mixed groups with one fully adult male and some nearly mature males have often been observed. However, it would seem that they will then leave the group, either "voluntarily" or under pressure from the fully adult male. They will then live solitary or associate themselves temporarily with other solitaries. Then they may again join groups of mixed sex (this may account for the occasional observation of groups with several full-grown males), perhaps eventually leading to a take-over of the "harem" by one of these younger males. In the case of the guereza also, there are definite indications that infanticide may occur in consequence of such male replacements, affording the new leader a reproductive advantage at the expense of his predecessor.

Mixed groups of guerezas live in well-defined territories of about 32 to 40 acres (13 to 16 hectares), which, though they may marginally overlap, are defended by the males with considerable "show effect" (leaps and cries) and sometimes also "hand-to-hand." Communication and "deterrence" are also served by a peculiar long-distance call, termed a "roar."

Taxonomists distinguish about ten subspecies of *Colobus guereza*, chiefly according to the white fur marking. Guerezas are no rarity today in zoological gardens, and several subspecies have been bred successfully in many zoos.

The black or Southern black-and-white colobus *Colobus polykomos*. About 12 subspecies of this species are recognized today, among them the all-black "satan" monkey *(C. p. satanas)* is perhaps the most well-known. It is distributed in Cameroon, Rio Muni, northwestern Gabon, and the western Congo. The white-epauletted subspecies *C. p. angolensis* lives in northwestern Angola, Zaire, southwestern Uganda, western Ruanda, Burundi, the southwest coast of Kenya, Tanzania, and probably down to Malawi and northwestern Zambia. Other subspecies inhabit West Africa from Guinea to Nigeria.

The black colobus also lives in continuous rainforests terrains as well as in gallery forests, savanna forests, and dry forests. It too forms mixed groups, predominantly including only one fully adult male. So far as yet known, all ecological, sociological, and ethological features resemble those described for the guereza.

The so-called Angolan black colobus on a postal stamp of the Republic of Burundi. The name is misleading, for this subspecies lives not in Angola only but in a much larger range, including Burundi.

GIBBONS

Category
FAMILY

Classification. A family of the order Primates, comprising only one genus (until some time ago, two genera were usually recognized, separating the siamangs from the gibbons in the narrower sense). The gibbons are referred to as lesser apes, or "small" anthropoid apes, in contradistinction to the "great" apes (the orangutan, gorilla, chimpanzee, and pygmy chimpanzee or bonobo). The terminology is appropriate because, zoologically, the lesser and great anthropoid apes together with humans all belong to the superfamily Hominoidea, representing the highest developmental level among the Old World or catarrhine primates and hence among the Primates in general.

Genus Hylobates (gibbons). Five species, or as many as nine species according to some. For good and sufficient reasons, the former view is adopted in this work; we list the species as follows:
– Siamang (*Hylobates syndactylus*; 2 (?) subspecies)
– Hoolock, or white-browed gibbon *(Hylobates hoolock)*
– Concolor, crested, or white-cheeked gibbon *(Hylobates concolor;* at least 5 suspecies)
– Beeloh, or Kloss gibbon *(Hylobates klossi)*
– Lar, or white-handed gibbon *(Hylobates lar;* 6 suspecies)

Body length: Up to 3 ft (90 cm; siamang); generally 18–26 in. (44–64 cm; other species)
Tail length: Absent
Weight: About 24 lb (11 kg; siamang); generally 11–15 lb (5–7 kg; other species)
Distinguishing features: Smaller and slenderer than the great anthropoids; rounded head; comparatively short muzzle; long canine teeth in both sexes; thick fur coat; fur coloring highly diverse from species to species, also color differences common between sexes and between juveniles and adults; in some species, throat sacs function as "resonance chambers" for loud calls; long slender limbs; arms especially long; power-

Hylobatidae
Gibbons FRENCH
Gibbons GERMAN

ful but slender and elongated fingers and toes; thumbs of grasping hands and great toe of grasping foot deeply separated; small ischial callosities (unlike the great anthropoid apes).
Reproduction: Times of mating not seasonally fixed; gestation 210–235 days; one young per birth, very rarely two; births every 2–3 years, hardly more than a total of 10 offspring per female; birth weight not yet known.
Life cycle: Weaning at 18–24 months; sexual maturity at 5–7 years; life span perhaps up to about 25 years in the wild, occasionally over 40 years in captivity.
Food: Diet not very specialized – predominantly fruits, sprouts, leaves, eggs, insects and nestlings, less often full-fledged birds or small mammals (depends on current supply); gibbons know when certain trees within their territories will bear fruit; siamang eats far higher proportion of leaves than the gibbons.
Habit and habitat: All species diurnal and arboreal; usually biped locomotion on the ground; acrobatic hand-over-hand swinging through branches; very quick and mobile; vocal, pairs often singing "duets"; formation of family groups; small territories, larger in the case of the white-handed gibbon (25–125 acres or 10–50 hectares); territorial defense; inhabitants of evergreen tropical rain forests and mountain forests, the capped, or pileated gibbon *(H. lar pileatus)* also in more open woodland; habits not yet thoroughly researched in some respects.

Swinging hand-over-hand. The gibbons is supreme in this exercise. For locomotion, it depends almost exclusively on the very long arms; the legs are quite short. The body is suspended below the arms, and skillful shifts of weight set it in motion like a pendulum, flinging it from branch to branch.

The gibbon hand (below) is very slender, with long, strong fingers, capable of curving to form hooks to hang and swing by, and short thumbs. The orangutan hand (right) is similar in structure but not so extremely adapted to swinging.

Arm movement. Most primates carry their arms below the body, like quadrupeds, in walking or running. The gibbons' shoulders are set laterally, and in swinging overhand, the immensely long arms are held above the body. The result is great freedom of movement in all directions.

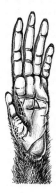

FAMILY Hylobatidae

Lesser Apes or Gibbons

by Holger Preuschoft

A Zoological Enigma

A hundred years ago the great German Zoologist Ernst Haeckel regarded the lesser apes, that is, the gibbons or Hylobatidae of Southeast Asia, as our nearest relatives. Just before the beginning of the twentieth century, that renewed support was given to this opinion with the discovery in Java of the remains of a human fossil, the so-called *Pithecanthus erectus*.

The gibbon seemed particularly human-like because of their rounded heads, with little protrusion of the jaws, and their tendency to walk erect on two legs. Certain other characteristics were overlooked.

Our view of the gibbons has changed radically since that time. While quite serviceable descriptions, some of them excellent, have existed since the eighteenth century, the "small anthropoids" remained the stepchildren of primatology until approximately 1970. It was only in the mid-1970s that any extensive field studies of the gibbons were published, from which we can begin to understand these remarkable animals. Despite our increasing knowledge, however, gibbons continue to present us with new enigmas, which we are only gradually beginning to solve.

There are three principal problem areas, all more or less clearly related to an understanding of humans as living creatures: that is, special body form, speciation, and the behavioral combination of song, monogamy, and territoriality.

In the adaptation of their behavior to life in the treetops and their highly eccentric mode of locomotion, as well as their body form, the gibbons differ quite distinctly from the large anthropoids (chimpanzees, gorillas, and orangutans), and still more so from the other Old World primates.

They are assigned to the Hominoidea (anthropoid apes and humans) chiefly because they will not fit in among the tailed monkeys of the Old World (Cercopithecoidea). This will become clear as we survey their traits in detail.

Outward Aspect: Head and Face

The fur is extremely thick compared to that of the other anthropoids; the hairs are set even closer than in most tailed monkeys. The individual hairs are of equal lenght; like the large anthropoids, the gibbons lack a soft, woolly undercoat. It may be that these smaller animals have longer and denser fur because, weighing 11–13 lb (5–6 kg; siamangs 22–26 lb or 10–12 kg), they are more liable to get chilled in their rainy home than their large rela-

All gibbons are exceedingly arboreal. If they osscasionally descend to the ground, they usually move about on two legs, where their long arms help them to keep their balance.

Still fairly abundant compared to the other gibbons, which are all threatened by increasing destruction of habitat, is the dark-handed, or agile gibbon. Today it is regarded as a subspecies of the white-handed gibbon *(Hylobates lar)*.

The agility of the gibbons is breathtaking. Their elongated, powerful arms are their principal organs of movement and of rest, in the boughs of the primeval forest. They spend much of their time suspended by both arms, or even one.

tives, whose ratio of surface area to mass is so much less. Unlike them, the gibbons have ischial callosities, though not so pronounced as those of the macaques and baboons.

Their short-muzzled faces are nearly hairless around the eyes, nose, and mouth; the face is often framed in a contrasting color, and clearly marked off from the longer fur advancing over the neck, scalp, and temples, including the ears.

The fur gives the gibbon head and even rounder look. The shortness of the muzzle and the moderate size of the jaw musculature emphasize the large brain case.

Very recently, it has been convincingly shown that the differences in mouth structure from the large anthropoids are indeed determined by the mechanical requirements of masticating food, and a function of size alone. The same study has revealed that the gibbons often adopt different approaches to the solution of mechanical problems than do the gorilla, chimpanzee, and orangutan. Thus, the superficial resemblance between the short-muzzled head of the gibbon and the similarly rounded human head does not reflect any closer relationship. If the two were of the same size, the similarity would vanish.

The mimetic facial gesture of gibbons is far less mobile and varied than that of their larger cousins. Facial expressions are not essentially different from those of the other apes, but far less striking. This is explained by the fact that the members of their small groups generally stay close together, and there are other methods of long-distance communication between groups (see below).

Body Structure and Locomotion

The paramount distinguishing feature of the gibbons is the length of their arms, justifying the alternative German name meaning "long-armed apes." This elongation affects every part – the upper arm, more expecially the forearm, the palm, and the fingers of the hand, and all are extremely slender and long-jointed. The arm length may attain 2.3 to 2.6 times the body length. (For comparison, the ratio of arm length to body length is 1.48 in humans, 1.7 in the chimpanzee and gorilla, and 2 in the orangutan.) The arms account for 20 percent of body mass in the siamang, 18 percent in the orangutan, 16–18 percent in the great anthropoids of Africa, and only 6–8 percent in humans. These arms provide the gibbon with a remarkably long reach; they are used for a mode of locomotion unknown to the rest of the animal kingdom, namely swinging hand-over-hand without food contact.

The elegance and seemingly effortless ease with which a gibbon will swing from branch to branch, change direction, and play its acrobatic game against gravity, is always breathtaking to watch. In a flash, it will change from a two-handed swing to a two-legged run, then jump, spin around tree

Gibbons demonstrate their innate acrobatic prowess on overhead bar.

trunk that is grasps, swing on in another direction, fly far through the air, and land on a higher branch. There it will sit quietly, huddled up, as though it had never been in motion.

When gibbons are not in a hurry, they will simply put one hand forward, let go with the other hand, and the body will swing ahead like a pendulum. Now it remains only to put forward the hand that has just let go, and locomotion with a very small expenditure of effort has been achieved; the fingers need only be kept crooked so as not to slip off. They actually travel at the physically determined speed of a pendulum bob; the longer the arms, the faster. A slight elevation of the center of gravity of the body on the upswing stores enough energy for the next, more rapid downswing.

A gibbon in a hurry will pull itself towards its new purchase with its forward-extended arm. The legs are drawn up close to the trunk, to avoid interfering movements. When the body has swung through under the pivot, the hand lets go before the other hand has found a grip. The body is projected freely through space. The phase in which a passive pendulum would have to slow down is thus eliminated. The trajectory may be longer than the strides in which the gibbon maintains a hold. Clear spaces between trees are negotiated in this way; flights of 25 to 32 ft (8 to 10 m) have been witnessed. The landing as always manual. Sometimes one arm will act as a brake, then let go, the other arm will brake again, before the animal lands feet first on a third branch, freezing into immobility.

According to Clarence Ray Carpenter, an American pioneer of behavior research in the wild, who did the first studies of gibbons in the 1930s, they spend about 90 percent of their traveling time hanging by one or both arms. Feeding, pausing for rest, reconnaissance, playing, mutual grooming, even mating, are accomplished in the hanging mode. It was often enough been asserted that this habit is related to arm length, but it is only recently that the relevant advantage of long arms has been worked out in detail. To understand this, one must know what exactly the animals do with their arms.

The most nourishing parts of trees are chiefly to be found in the outermost zone of the canopy – buds, young leaves, shoots, blossoms, and fruits. Access to these parts is difficult. From the outside, except for birds, only those herbivores that have long necks (many even-toed ungulates –

The efficiency of swinging hand-over-hand by the arms without using the legs (right) is an "invention" of the gibbons. The hands reach alternately beyond each other, while the body follows like a pendulum, facilitating the flowing progress. In mid-hanging flight, the animals can halt abruptly and take a break for resting or eating while hanging by one arm (above).

deer, antelope, and notably giraffes) or simply great size (elephant, rhinoceros) can reach them. From the inside, from the trunk, only small creatures can safely venture out on the weak terminal twigs. Under a great weight, the branches will bend until the animal slips off – unless it has grasping hands and feet so that it can hold on nonetheless. The only posture that can then be adopted without undue expenditure of energy is that of hanging by the hands or the feet. And in that posture, it it protected and safe from movements of the branches in the wind. An animal that can hang from the outer branches is independent of their limited bending strength; it utilizes their tensile strength. Provided it has the strength to keep its grip tight enough, its body is automatically in stable equilibrium. These conditions are confronted by many primates, and many accordingly are suspended during much of their feeding time. If they hold on by the hands and not the feet, the head maintains its customary attitude, and there are no added difficulties of orientation or of gauging distances.

The great masters of this art are the gibbons. Their physical structure offers a wide repertory of alternatives. Their long, slender hands will often grasp a number of twigs at once, utilizing their combined strength. The length of the arms enables them to distribute their body weight over several supports, thus enhancing safety. With their long reach, furthermore, they can sometimes harvest many fruits without any change of place, and lastly the same long reach affords them many

(Left) A juvenile capped gibbon, member of a seriously endangered subspecies of the white-handed gibbon, eating wild figs. (Below) This white-handed gibbon has a secure grip on the stump of a branch and is letting its body swing around over empty space.

▷ Clear spaces in the forest are traversed "in flight." A white-handed gibbon can project itself through the air like this for a distance of up to 32 feet (10 meters).

▷▷ This gray, or Müller's gibbon of Borneo is skillfully distributing and anchoring its weight among slim branches.

▷▷▷ Male and female crested gibbon differ in coloring. Only the males and the juveniles have the dark fur and the sideburns.

choices of a new firm hold from a given starting point.

Hanging by the arms alone will not only maintain normal head and body posture but also permit serial pendulum swings with little effort. The feet are left free to transport objects, and they are in fact used for that purpose. To be sure, this mode of locomotion is not of the safest. Missed holds and broken branches or twigs can be disastrous. Swinging hand-over-hand is made possible only by the fabulous coordinating and guiding faculties of the highly developed brain and by thoroughly familiar surroundings. Falls inevitably occur, and no less than one-third of the skeletons examined in museum collections by the great Swiss primatologist Adolph Hans Schultz show traces of fractures – that have knit by themselves, without surgical attention.

The long arms are of direct benefit in hand-over-hand progress over longer distances, and in emergencies; with increasing length of the pendulum, the swing of the suspended body gains velocity, without the need to expend additional energy. An 8-in. 20-cm) increment in arm length provides a velocity gain of about 25 percent. A gibbon surprised by an enemy will drop from its seat instantly, then take hold and proceed at once on the horizontal; the longer its arms, the higher its take-off velocity.

Then again, overhand swinging places a limit on body size. The centrifugal forces that a gibbon's bent fingers must sustain require a powerful forearm musculature. With greater body weight, muscle power becomes insufficient in maintain prolonged equilibrium between gravity and centrifugal force. In the weight range of gibbons and siamangs (between 11 and 29 lb or 5 and 13 kg); there is a great reserve of muscle power over the minimum required. The larger anthropoid apes cannot hang for long periods. On might imagine the arms to be equipped with even stronger muscles, but then the arms too would become significantly heavier, and their attachment to the trunk would have to be reinforced, so that their proportion of total body weight would be increased in turn. Such a gain in arm weight would result in body forms that will not swing like a simple pendulum. For the same reason, a still further lengthening of the arms would be problematical. A siamang's arms, at 20 percent of total body weight, are already heavier than those of any other anthropoid.

Given their long arms, it is easily overlooked that all hylobatids also have legs that are quite long enough – 130–150 percent of body length – to enable their owners to perform great leaps. (For comparison, the proportion of leg length to body length is 116–130 percent in the great anthropoid apes and 169 percent in humans.) When a siamang moves over the ground, it sometimes goes on all fours like a chimpanzee, the hands set down on their knuckles, but with elbows bent. Because of the length of the arms, the trunk will usually be erect enough to bring the feet under the center of gravity. Now the hands are not supporting any weight and may be lifted from the ground. Like a balancing pole, they are raised to the sides or over the head as the gibbon hurries along the ground

For the first three to four months of life, the baby gibbon constantly holds on to the mother's fur, even during her most daring acrobatics.

or along a branch, the trunk inclined slightly forward and the knees bent and spread outward. The heel takes much less of the stress than does the human heel.

In its biomechanical details, gibbon locomotion is more like a run or a series of hops than like the walk of a human being. So here again, the resemblance is superficial. Ten percent of gibbon locomotion is biped – definitely more than for any other non-human primate.

The long slender fingers and toes can develop a grip of impressive force. The bent form of each limb will transmit very high bending moments, such as occur in grasping. The fingers are ordinarily bent only slightly, for a "hooked" purchase.

When necessary, for example in hanging from lianas or ropes, the fingers will curl completely around the object, with the thumb clasped over them from the other side – a full-circle power grip. This is more often exhibited by the deliberately climbing siamang than by the faster moving, smaller gibbons. Small objects are also very dexterously grasped with the help of the thumb. Quite often too the thumb is used for grooming, or things are examined and palpated with it much as we do with our index finger.

Both the thumb of the hand and the great toe of the foot are separated from the second digital ray by a deep cleft in the soft part at the base of the hand or foot. This trait distinguishes the small anthropoids from all other primates. It makes for an exceptionally long span, and it perhaps enables the animals to climb the smooth trunks of giant forest trees that would not otherwise afford a firm hold. The thumbs and great toes are indeed shorter than the other digits, but not so much so as for the great anthropoids.

Here also the small anthropoids have taken different paths of development and adopted different solutions to problems of survival from their larger relatives'. Eighty percent of siamangs have their second and third toes joined by a bridge of flesh. Hence the specific name *Hylobates syndactylus*, that is to say, "gibbon with fused digits". However, the same feature is found in 16 percent of the lesser gibbons, occasionally in African anthropoid apes, and to a lesser degree among most human beings.

All details of body structure are recognizable adaptations to life in the trees. The perfection of adaptation of their method of travel has placed the gibbons in a dangerous predicament today. When connected forests are cleared in an area, the gibbons are trapped in the remaining "islands of trees".

The juvenile gibbon maintains close contact with the parents, in this case the mother, for about the first five years of life.

Phylogeny

Complete as this adaptation of physical structure to arboreal life may be, and remarkable as its peculiarities are, it would seem that they are phylogenetically of recent acquisition. A large number of finds of fossil anthropoids have been perceived as being ancestors of the gibbons. Their assignment to the superfamily Hominoidea is based almost throughout on tooth and jaw form, not on body form. Among the more familiar examples are *Aelopithecus* in the Egyptian Oligocene (about 33 million years ago), *Limnopithecus* (syn. *Dendropithecus*) in the East African Miocene, and *Plio-*

pithecus in a Middle Miocene fissure filling not far from Vienna (age about 15 million years). The two last mentioned forms have very slender, extended limbs, but arms and legs of equal length. On closer inspection, no such adaptations to swing-hanging are found as characterize the gibbons. Above all, there is no extraordinary lengthening of the arms. Doubts have accordingly been voiced recently as to the assignment of these finds to the Hylobatidae.

The short snouts and large eye sockets are taken to be rather the consequences of small size. Instead, new finds in the Early Miocene (*Micropithecus* in East Africa, *Dionysopithecus* in China, and *Krishnapithecus* in northern India) are being described as precursors of the gibbons, but there are no known fossil specimens of their limbs. From the Middle Pleistocene on, there is documentation of true gibbons in Southeast Asia, where early humans were also present.

Just lately, the findings of molecular biology have confirmed the position of the Hylobatidae between the Old World monkeys and the great anthropoid apes (Pongidae). The correct interpretation may be that they separated from the line of development of the anthropoid apes near branching on the way to the Pongidae slightly more than ten million years ago, that is, in the Late Miocene. That leaves scarcely five million years of less for differentiation within the familiy, about as long as the duration of the Pliocene (see also below). Considering this length of time, they have remained surprisingly uniform, doubtless because of similarity in mode of life.

The chromosome sets of the Hylobatidae differ from those of the great anthropoids even in the number of nuclear loops: 38 occur, as well as 44, 50, and 52. So far as known, the band pattern is the same among those with 44 chromosomes, but these bands are not found in the other species. Band patterns that can be identified in anthropoid apes and humans as well are detected only in exceptional cases.

(Top) The gibbon's whole life, inclusive of mating, birth, and rearing of the young, takes place high in the trees. (Bottom) Two white-handed gibbons conversing with eloquent gesture.

Brain and Intelligence

The gibbon brain is highly evolved, though we have less information about its internal structure than in the case of the great anthropoids. In the development of the several parts of the brain, the gibbons occupy a position between the Old World monkeys and the Pongidae. The evolutionary level of brain development is comparable to that of the most highly developed South Amer-

ican primates, the spider monkeys. In behavior, these animals appear to be not very adaptable, and quite highly specialized. In experiments, it is difficult to enlist their cooperation, and no outstanding feats of "intelligence" have been reported.

Lifestyle

The social life of the gibbons is quite different from that of other higher primates. In their natural habitat, they live together in pairs, being strictly monogamous. To be sure, this trait, comparable as it may be to our moral views, is based on intolerance of adult conspecifics of the same sex. Difficulties are consequently involved in the keeping of groups in zoos. In natural communities, the descendants live in company with the parent couple. The female gives birth every two or three years. As a rule, a pair is accompanied by two, sometimes three, very rarely as many as four, babies and adolescents.

Each offspring is born after a gestation period of 210 to 235 days. In the first three to four months of life, it hangs on to the mother's belly constantly, even through the most daring acrobatics; it never climbs onto her back. The fingers do not merely grip bunches of hair are literally braided into the fur. As the grow older, the children detach themselves from the mother more and more frequently. At first they climb about alone during intervals of rest and for short periods; little by little, the distance and duration of separation from the mother will increase. At the age of four or five months, babies begin to swing hand-over-hand, but by no means equaling the grace and seemingly effortless assurance of the adults. At six months, they will sometimes go on two legs. Contacts with the mother gradually diminish to cuddling at rest periods and at night, and weaning is begun at about a year and a half. At the age of seven months or more, a phase begins in which they play a great deal, especially with older siblings and later also with younger ones, and with the father as well. Play occupies about a third of the daytime. It consists of gymnastics, chasing, wrestling – all while hanging from the branches. Surprise attacks with rough slaps of the hand provoke hot pursuits with breakneck leaps. In the intervals, thorough grooming – usually mutual, occasionally solitary – is engaged in. This lively play activity gradually expands to include the younger siblings.

Group cohesion is close. Gibbons are not very possessive about food within groups. Even when several feed together, they rarely quarrel over the choicest morsels. Sometimes two will bite into the same fruit, or one will take fruit out of another's hand, even out of another's mouth. It can happen that one will find something especially good to eat and then pass it around until there is nothing left for itself. In most cases (86 percent), juveniles or older children will be given something by the adults, more rarely by females than by males. Cooperation in intelligence tests can be recruited in this manner.

At the age of 6 or 7 years, the young gibbons grow to sexual maturity. Now the parent of the same sex becomes increasingly aggressive in behavior towards the adolescent. The latter gives way until connection with the group loosens and at last breaks off entirely. This is the time when

Gibbons have been bred successfully in captivity for decades, especially the popular white-handed gibbon.

new pairs are formed, sometimes after a solitary interval.

Gibbons take their time in choosing partners. In the zoo, it is not to be taken for granted that two individuals of opposite sex placed together will proceed to enter into a pair bond. But once this has happened, the partners apparently stay together for life. "Remarriage" after the death of a partner does occur. The extreme age of gibbons is around 20–25 years; individuals in captivity have been known to live past 30.

Gibbons wake up at sunrise. After their morning concert, the group sets itself in motion; its members spread out moderately in search of food, they eat, they play, they roam, they eat again. The smaller forms take three to four hours, the larger siamang five hours, to eat their fill. This amounts to 22–50 percent of their active day, the time during which they are up and about. They roam about for 15–33 percent of the time. In this instance, the smaller figure applies to the siamang, which also makes the shortest day's march of only about 0.5 mi (0,75 km). These wanderings apparently serve primarily to search for food. Hours before twilight, after 8–10 hours activity, the group will move back to the sleeping place. They are especially active during the forenoon, and unlike most simians, they observe no siesta.

Gibbons are lively animals, constantly in motion. Here there will be something for them to see, there something to be examined closely, then a fur-grooming session with a fellow group member, a little playful biting or tussling, and ever and again the elegant, flowing progress hand-over-hand through branches, lianas, poles, ropes, or bars. Still, there is far less exchange of social messages between members of a group than among the baboons, macaques, langurs, or great anthropoids. This is attributable in part to a much simpler social structure, in which there is no occasion to arrive at many agreements and understandings among all present.

Choral and Duet Singing

One characteristic of the gibbons is unrivaled by any of the other anthropoids – their musical calls. There is nothing like this except among Old World langurs and South American howlers, and certain prosimians. Especially in the morning, when the Sun has dried the dew and dispelled the chill of night, and in the late afternoon, they lift up their voices and "fill wood and dale with their song" (Emil Selenka). In this performance, lasting 20 to 30 minutes, all elder members of a group take part; other groups will enter from a distance, prompting a reply. It usually begins with "dull, deep, bell-like tones," continues "with a shattering, high yell, followed by an overloud high-pitched laughter. This forest music, audible at a distance of an hour's walk, trails off in a prolonged diminuendo . . ." (Selenka).

In singing, the throat sacs of many gibbon species (siamang, concolor, beeloh) are inflated. Tonal pattern differs from one species to another, as do the numbers of individual sounds. Frequencies range between 2000 and 5000 Hz, so they are not

The gibbons are the vocal artists among the primates. Several species have an inflatable throat sac that serves as a "Sound box" to amplify their loud melodic vocalizations. This organ is most fully developed in the siamang.

The siamangs' duets and choruses evidently reinforce the bond between group members, at the same time advertising the group's station at the moment of neighboring conspecifics.

too difficult to mimic. Sound at these frequencies carry far in the moist forest.

Gibbons sit or stand in the highest branches of trees, throw the head back with mouth wide open, or from side to side, when calling. Towards the end of the song they will swing through the branches in tight circles. The most striking feature of this entire ritual is that male and female partners sing in tune with each other, in some species forming a regular duet, the two taking different "parts." The siamangs and the lar gibbons are outstanding in the duet. Accurate tuning is characteristic of older couples accustomed to each other. When fathers and adolescent daughters, for example, sing together, the ensemble is not nearly so good. Apparently, community singing creates or contributes to a bond among the group members. It is at this point, too, that the parents first clash with their adolescent progeny of the same sex.

We know today that these melodic songs are inherited in all details. Different gibbon populations have special calls of their own. There is no deviation between animals living in captivity and in the wild. There are borrowed elements from the songs of neighboring gibbons. Hereditary determination is especially clear in hybrids between species. In the various parameters, their musical utterances are precisely intermediate between those of the parents. In the case of a cross back to one parent species, sound patterns move away from the midpoint towards that form which represents 15 percent of the parentage. There are same known cases in which young gibbons grew up with no opportunity to hear the parent of the same sex. Since male and female animals sing different tunes, such an individual could not have learned the typical notes from the parent of the other sex.

Territory and Habitat

Choral singing conveys information of neighboring groups concerning the current position, perhaps even the state of health and temper of their neighbor. Certain distances are maintained as a rule between singing groups, and all stay within a clearly bounded home territory. When two groups meet at the boundary of their territories, and excited outcry is raised by both parties. Here, conspicuous swinging and leaping during the song is recognizable as display behavior. If neither group yields, biting and swift pursuit will follow. Battles are begun without warning by an individual that has attained a more elevated point of departure. It will approach the adversary at very high speed, inflicting a slap of the hand or a quick bite in passing. After a brief exchange, the opponents pursue each other at great speed through the branches.

Gibbons will defend a circumscribed territory; they are the only "territorial" Old World catarrhines. Solitary individuals, not calling, can apparently hold their own in between territories, but it

Altercations between gibbons may be very heated. Adversaries attack each other with bites and blows and at lightning speed.

is uncertain whether these may not be elder descendants of the pair in possession, who as yet barely tolerate them. Both sexes participate in confrontations over boundaries, although the main responsibility seems to lie with the male. In one instance, a gibbon "widow" was able to hold her territory for a considerable length of time after the loss of her mate. Consistent with this, and with the monogamous habit generally, the males are only insignificantly larger than the females, sometimes indeed smaller. In this again they contrast with the great anthropoid apes, among whom the males outweigh the females by 18–48 percent.

In any event, the males tend to confront an intruder, while the females tend to retire from sight. One result of this has been that hunters always bag more males than females. There are always more males in the large museum collections.

The gibbons are at home in the evergreen rainforests of the lowlands, as well as in the somewhat drier forests of more nothern latitudes and higher elevations, where trees begin to shed their leaves with the changing seasons. The territories of four family groups (northern Thailand, Assam, Bangladesh), or three (Sunda Islands) can be accomodated about 0.4 mi^2 (1 km^2). In Central Thailand, the population density may reach 6.5 groups per square kilometer ($1 \text{ km}^2 = 0.4 \text{ mi}^2$); on the Malay Peninsula, it may be only 2 groups. In this latter region, Lar and Syndactylus gibbons occur side by side ("sympatry"), and compete with each other to some degree. Gibbon territories are large where other primates abound, and smaller where they do not.

Each group has its established range, of 125 acres (50 hectares) in Malacca, 55 acres (22 hectares) in Assam, and 39 acres (15.5 hectares) in the Khao-Yai Park in Thailand. The average area is 75–100 acres (30–40 hectares), of which some portion, 28–95 acres (11–38 hectares), is defended against conspecifics as home territory. Within this area, groups will traverse 2100–5400 ft (850–1700 m) each day. Territory size evidently depends on food supply. The prerequisite is that fruit should be available throughout the year.

Besides flowers, young leaves and buds, gibbons prefer fleshy fruits of high sugar content, ripening in limited quantities and available over widely scattered areas. These minor crops are of interest neither to the much larger orangutans nor to the populous troupes of langurs or macaques, which would drive the gibbons away from the more productive fruit trees. Their sole competitors are birds and squirrels. In some regions, notably on the Mentawai Islands, as well as in Thailand and Malacca, a good many insects are ingested. Wild gibbons will rob bird's nests and catch lizards. Eggs can be offered in the zoo as well. Once in a while a gibbon may even snatch a flying bird out of the air, in mid-leap. The prey is dismembered and probably also consumed completely. In the Berlin-Friedrichsfelde Zoo, all singing birds gradually disappeared from the trees in which the white-cheeked gibbons were free to climb. They would also go after Guinea fowl or pea-hens – although a Guinea fowl was caught only once. When she began to scream, the gibbon let her go. A white-handed gibbon on an island in the Cologne Zoo pond would catch not only sparrows an magpies but also, on one occasion, a duck. He was prevented from further exercise of his abilities by being put back in his cage, because the zoo was not inclined to sacrifice its celebrated collection of aquatic birds to the gibbons. When drinking, gibbons dip the back of their hand in the water, raise the hand above their head, and lick off the drops.

It is in their feeding habits that the large siamangs diverge most from the smaller gibbons. The sia-

In zoos, different gibbon forms are not infrequently kept together. (Left) A male capped gibbon. (Right) A dark white-handed gibbon.

mangs' diet consists 43–48 percent of leaves, as against 32 percent for the smaller gibbons, along with 66 percent fruits. The Mentawai gibbons eat hardly any leaves. Young leaves contain more protein but less energy-supplying carbohydrates. Smaller forms, as the hoolock, white cheeked, and white-handed gibbons, move faster and travel longer distances than siamangs. By virtue of their small size alone, they would need more calories for basal metabolism. Their preference for high-energy fruits supplies this need. The energy required to find scattered fruit sources is not excessive just because they are so small and light. To meet the protein requirement, they must resort to animal fare or to leaves.

Siamangs move about less and have smaller home territories, and they are about a third slower in their wanderings. Their mode of climbing is more deliberate, and they have a greater tendency to anchor themselves by several limbs at once, something like orangutans. They stay closer together when feeding; lar groups scatter more. Larger animals have a more extensive digestive tract, better able to process chemically resistant leaves. It has been calculated that siamangs, given their size, are efficient in not wandering far in search of fruit and filling up instead on lower-energy leaves that can be obtained everywhere. To do this, they need more time than their smaller cousins (5 hours instead of 3 or 4). A siamang would have to be active for 12.5 hours in collecting an equivalent amount of fruit, instead of slightly more than 10 hours, like a lesser gibbon.

On the Malay Peninsula, where, as on Sumatra, both large and small gibbon species occur, their dietary interests overlap, especially in fruits. They are in competition. Usually the lars give way when siamangs approach. Sometimes the stronger siamangs will chase their smaller relatives out of a fruit-bearing tree, but they are too slow to evict

Three natural enemies of the gibbons are not a serious threat under natural conditions: clouded leopard, king python, and panther.

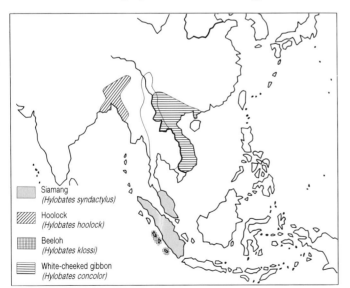

them entirely. The territories of both species are larger where they occur together.

Because of their smaller surface area relative to mass, siamangs give off less body heat. This may explain why their fur is sparser. In any case, the circumstance enables them to colonize forests at higher elevations, where it is colder and where there is a less abundant supply of fruits. It may be that they first appeared and evolved their characteristics in higher mountain forests. By virtue of their size, they are also less vulnerable to enemies.

The chief natural enemies of the gibbons are doubtless panthers, clouded leopards, giant snakes, and the larger birds of prey. They are threatened with a deadly peril, however, by the progressive destruction of their habitat, chiefly because of advancing human occupation. This destruction is multiplied by such practices as slash-and-burn clearing of areas for farming, to be abandoned after only a few years, when they have become useless to humans and beast. Fresh wounds must then be inflicted upon nature elsewhere. Most devastating of all is logging, which

concentrates upon the giant forest trees that are indispensable to the gibbons. Occasional reforestation programs are hardly adequate to improve the situation. In the first place, the gibbons have meanwhile disappeared, never to be seen again, and in the second place, trees are planted that may yield valuable timber fur us but no sustenance for gibbons. Recent distribution maps are painful to see. All gibbon populations have been fragmented and reduced to tiny island patches of woods. Since the specialization of these animals prevents them from crossing open country, they are endangered even within these minute habitats, by the effects of inbreeding.

Life in Captivity

Gibbons have traditionally been kept as household pets in their native countries. In most cases, the little fellows, chained and neglected, are a picture of misery. Nevertheless, the young ones display a friendly, trusting nature, and are happy to be hugged, petted, and carried about. They are extremely clean animals and will painstakingly remove any dirt from their fur and from the hair of their human friends. They will not infrequently form friendships with other animals. They become so attached to people that some may be allowed to climb about unconfined. They will then sport in the trees without straying far, and return voluntarily.

Experience in zoos has been equally favorable. Good relations with human beings may last for life, even beyond the attainment of reproductive age. At that stage, however, they will accept only the persons with whom they are familiar; they will instantly attack visiting strangers.

Under favorable conditions, they can be left at large in zoos as well, as at Berlin-Friedrichsfelde. They cannot, however, be housebroken.

They should be in the hands of experienced keepers only. There are several reasons for this. Because of their long, sharp canine teeth and their lightning-quick movements, the adults are not to be trifled with. Their requirements are stringent in terms of living space and diet. They were long considered difficult to keep in zoos, but since the 1930s, they have often enjoyed good health and long life. Today they can be quite reliably bred in captivity; this was accomplished first in 1936 at Aarhus in Denmark, then in Philadelphia. The Zurich Zoo also has had early and repeated successes. Siamangs are successfully bred at Frankfurt. It has been found, contrary to earlier expec-

An "unnatural" enemy of the gibbons, and their most dangerous one, are humans, whose chain saws and burn-offs are clearing the forests and thereby increasingly constricting and fragmenting the habitat.

tations, that the requirements of all species are much the same. That includes – point often overlooked – separate accommodations for couples.

Gibbons have surprisingly little difficulty in contending with the climate in our latitudes. At the Clères Zoo in Normandy, they live out-of-doors all year; at Kronberg near Frankfurt, as well as in Hannover, they are free to go out for many hours, even at temperatures of 5° F (–15 °C). They only need a dry, warm interior in which they can take refuge from rain or cold. At Berlin-Friedrichsfelde, they are kept on two islands in the pond from early May to November.

Difficulties in Distinguishing Species

Today all forms of the family Hylobatidae are assigned to the one genus *Hylobates*. The SIAMANG *(Hylobates syndacrylus)* occurs in the same areas as lesser gibbons in Malacca and Sumatra. This proves that it is genetically well distinguished from the others. Accordingly, it must be recognized as a species; formerly, it was given a genus *(Symphalangus)* to itself. Siamangs have 50 chromosomes.

The situation is clear also as to the HOOLOCK *(Hylobates hoolock)* of Burma, Assam, and Bangladesh. It is separated from other forms by the Salween River. By reference to accurate measurements of skull shape and to its size, it was recognized as a species even before it was known to have only 38 chromosomes, thus being distinct from all relatives.

In the eastern part of the range, that is, in Vietnam and China, to the east of the Mekong River, we find the crested or WHITE-CHEEKED GIBBON *(Hylobates concolor)*. It too deviates in outward appearance, cranial form, and number of chromosomes (52) from the other gibbons.

The BEELOH *(Hylobates klossi)* of the Mentawai Islands was formerly called the "pygmy siamang." In actual fact, it is not close to the siamang; it is more closely related to the white-handed gibbons, but it is distinguishable from them in coloration, behavior, and skull measurements.

Difficulties arise in Thailand, on the Malay Peninsula, and on the Sunda Isles. The fairly small gibbons in this large region all have 44 chromosomes. As might be expected from so wide a geographical distribution, there are differences between inhabitants of different areas. They can indeed be separated by skull form using variance statistics, but the differences are so slight and the ranges of variation overlap so far that they should be regarded as populations of the species, *Hylobates lar*. There are, however, other opinions about this. Apart from confusing and elusive deviations in fur coloring, these gibbons sing in quite distinct modes. There are researchers who would divide them into five species according to their songs: the SILVERY GIBBON *H. moloch* of Java; the GRAY GIBBON *H. muelleri* of Borneo and north of the Kapuas and east of the Barito rivers; the AGILE GIBBON *H. agilis* south of both rivers, which sings like those of central and southern Sumatra; the CAPPED GIBBON *H. pileatus* of eastern Thailand and

Northern and southern white-handed gibbon
(*Hylobates lar carpenteri* and *H. l. entelloides*)

Agile gibbon
(*Hylobates lar agilis*)

Gray gibbon
(*Hylobates lar muelleri*)

Silvery gibbon
(*Hylobates lar moloch*)

Capped gibbon
(*Hylobates lar pileatus*)

Kampuchea (Cambodia) as far as the Mekong, which is very different in coloration and in song from its immediate western neighbors; and the Lar or White-handed gibbon *H. lar* scattered over west Thailand, the Malay Peninsula, and northern Sumatra. Between the Tepha and Perak Rivers, there is a narrow belt of *H. agilis* across the peninsula. However, there is good reasons to separate the populations on either side as well into a northern *H. carpenteri* and a southern *H. entelloides*.

There is not indeed any dispute concerning the existence of all these forms. The disagreement is merely over their status. Both sides have argued at length in the book *The Lesser Apes* by H. Preuschoft, D.J. Chivers, W.Y. Brockelman and N. Creel (1984). According to one opinion, the forms are geographic variants of one species, that is, according to the rules of zoology, a reproductive entity; according to the other opinion, they are distinct species, which should not exchange any genes under normal conditions.

The single-species view is supported by the fact that the forms are not sharply distinguished, but exhibit transitions and overlaps. Besides, the animals interbreed at three out of five locations where their ranges adjoin, and have fertile offspring. Thies applies to animals in the wild; in captivity, they will cross without difficulty in any event. There are even known hybrids of Lar and Syndactylus, forms that are much less closely related. At the present time, attempts are being made to let the many hybrids in zoos die out gradually by attrition and replace them by authentic pure lines.

The mixed populations in the wild (one each in Thailand, Malacca, and Borneo) are being carefully observed, and in some instances have been so for a long time. They exhibit peculiarities that again support the view that the differences between the forms reflect a reproductive barrier after all. Hybrids occur only in a narrowly limited region; they do not spread out over larger areas. In the Khao-Yai Park of Thailand, they occupy only a strip about 5 mi (8 km) in width. This seems to suggest that the reproductive chances of the hybrids are indeed limited. As a matter of fact, they are not taken quite seriously by the main populations. Thus, in Thailand there are groups consisting of one male capped gibbon, one female white-handed gibbon, and one additional female, either a hybrid or a capped gibbon. Within the pure populations, interfemale aggressiveness suffices to prevent the formation of such triads. These females do not seem to feel challenged by the presence of a female not belonging to the same population.

Practical reasons are also urged for the subdivision into five or six species. Especially under field conditions, song is a convenient criterion. As has been emphasized by J. Marshall, W. Brockelman, and D. Schilling, the confusing variety of fur colorings according to sex and age can be reduced to order if the songs are accepted as the paramount trait.

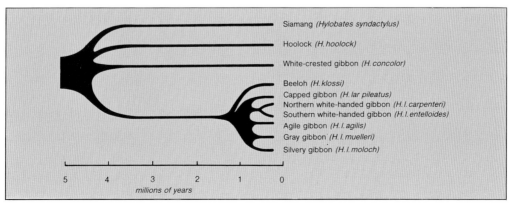

Evolution of the gibbons. The lar group has split into six subspecies in geologically more recent times.

Variety of shapes and colors in the gibbon family. (Opposite page) White-cheeked gibbon male and male hoolock or white-browed gibbon (top); dark white-handed gibbon and silvery gibbon (bottom).

(This page) Capped gibbon male (top); lighter white-handed gibbon and white-cheeked gibbon female (middle); white-cheeked gibbon baby and male gray or Borneo gibbon (bottom); hoolock female (above).

GIBBONS

Gibbons (Hylobatidae)

Nomenclature English common name Scientific name French German	Approximate Size Body length Tail length Weight	Distinguishing Features	Reproduction Gestation period Young per birth Weight at birth
Siamang; great gibbon *Hylobates syndactylus*, 2 (?) subspecies Siamang Siamang	Males 1.5–1.9 ft; 47–59.5 cm Females 1.4–2.1; 44–63 0 24.2 lb; 11 kg	Black fur; throat sac; long anterior limbs; thumb of grasping hand and great toe of grasping foot deeply detached	230–235 days 1, very rarely 2 Not known
Hoolock gibbon; white-browed gibbon *Hylobates hoolock* Hoolock Hulock; Weißbrauengibbon	1.4–2.1 ft; 44–64 cm 0 13.2–15.4 lb; 6–7 kg	Distinct sex differences in coloration; adolescents of both sexes and adult males black with white, separated eyebrows; adult female fur becomes light yellow-brown	Much like great gibbon
Concolor gibbon; crested gibbon; white-cheeked gibbon *Hylobates concolor*, at least 5 subspecies Gibbon noir Schopfgibbon; Weißwangengibbon	Much like hoolock gibbon	Distinct sex differences in coloration; adolescents of both sexes and males black, with or without light cheeks, depending on subspecies; adult females light ocher-yellow, at times with gleaming red-gold tinge	Much like great gibbon
Kloss gibbon; beeloh *Hylobates klossi* Siamang de Kloss Biloh; Mentawai-Gibbon; Zwergsiamang	1.4–2.1 ft; 44–64 cm 0 11–13.2 lb; 5–6 kg	Slender body with uniformly brown-black to black fur; comparatively short hair	Probably like great gibbon
Lar gibbon; white-handed gibbon *Hylobates lar carpenteri* and *H. l. entelloides* Gibbon lar Lar; Weißhandgibbon	1.4–2.1 ft; 44–64 cm 0 11–13.2 lb; 5–6 kg	Coloration varies from light yellow to brown, dark brown, black; a more or less broad face ring; white areas of fur on backs of hands and feet; gray shades rare	210 days 1, very rarely 2 Not known
Moloch gibbon; silvery gibbon *Hylobates lar moloch* Gibbon cendré Silbergibbon	Same as lar gibbon	Silver-gray, with darker under parts and scalp; hair comparatively long, lying close at top of head	Same as lar gibbon
Gray gibbon; Müller's gibbon *Hylobates lar muelleri* Gibbon de Müller Grauer Gibbon; Borneo-Gibbon	Same as lar gibbon	Gray to brown-black (gray shades predominate); eyebrows lighter deep-lying	Same as lar gibbon
Agile gibbon; dark-handed gibbon *Hylobates lar agilis* Gibbon agile Ungka; Schlankgibbon	Same as lar gibbon	Slender, short-haired, mostly brown-black to black; pale eyes; either no contrasting light areas, or two white eyebrows, usually separated from each other	Same as lar gibbon
Capped gibbon; pileated gibbon *Hylobates lar pileatus* Gibbon lar à bonnet Kappengibbon	Same as lar gibbon	Scalp black; sexes distinctly different; juveniles brown-yellow; females with comparable light coloring after maturity on back, arms, and legs; proportions of yellow and gray vary sharply; males almost completely black	Same as lar gibbon

One of the silvery gibbons, now rare, in his natural habitat on Java.

COMPARISON OF SPECIES

Life Cycle Weaning Sexual maturity Life span	Food	Enemies	Habit and Habitat	Occurrence
Age about 18–24 months Age about 6–7 years Probably about 25 years	Leaves, fruits; incidentally, insects, birds' eggs, small vertebrates	Leopard, big snakes, large birds of prey	Diurnally active tree dwellers in evergreen tropical rain and hill forests; family groups (3–5 individuals); defends territory; 1.9–3.3 individuals per km² (Malaysia), 10 (Sumatra); duet singing by pairs (1 km² = 0.38 mi²)	Still fairly widespread, but endangered; only 4% of natural habitat protected
Much like great gibbon	Fruits, leaves, flowers, insects	Same as great gibbon	Also found outside of tropics; much like other gibbons, but as yet poorly studied	Endangered by rapid destruction of habitat
Much like great gibbon	Probably like hoolock	Same as great gibbon	Probably much like other gibbons; habits as yet poorly studied	Much endangered; surviving only in small isolated enclaves, for example, on Hainan and in Vietnam
Probably like great gibbon	Fruits, almost exclusively	Probably same as great gibbon	Probably much like great gibbon, but more mobile; very small territories; no duet singing	Much endangered by hunting and destruction of habitat
Much like great gibbon	More fruits, less leaves, than great gibbon	Same as great gibbon	Faster migration over longer distances than great gibbon; also, larger territory (25–125 acres; 10–50 hectares); duet singing by pairs	Still quite widely distributed, but endangered
Same as great gibbon	Same as lar gibbon	Same as great gibbon	Much like lar gibbon	Facing extinction
Same as great gibbon	Same as lar gibbon	Same as great gibbon	Much like lar gibbon	Still fairly abundant, but endangered
Same as great gibbon	Same as lar gibbon	Same as great gibbon	Much like lar gibbon	Still fairly abundant, but endangered
Same as great gibbon	Same as lar gibbon	Same as great gibbon	Much like lar gibbon; also in more open, partly deciduous forest	Seriously endangered

In an earlier time, the boundaries of distribution were contour lines over 4800 ft (1500 m), interruptions of forest growth, and above all rivers or arms of the sea. Today, human occupancy – or mere exploitative clearing of the forest – is fragmenting the gibbon populations. The subdivision of the Lar group definitely occurred not more than a million years ago, perhaps much less. It dates from the time of fluctuating sea levels due to glaciation. Just when the gibbons had retained their adaptation to the rainforest, their habitat was repeatedly restricted by new limitations in the course of the Pleistocene. The advances and retreats of the sea on the Sunda shelf, and the drying of more elevated terrain with fluctuations in climate, were hazards that the gibbon population had to avoid. In their separated ranges, distinguishing features developed (especially of coloring and of song), and these survived geographical displacement. The present situation can be explained to some extent in terms of at least two marked changes in climate. Direct causes of change, to all appearances, were not so much environmental factors as the characteristic hylobatid tendency to chromosome transpositions. These may have occurred independently in the several populations, and brought with them the outward changes in fur color and in melody. Specialization was favored also by the habit of dwelling in a fixed location. Not only do the old couples cling to their home territory; the offspring try to settle in the neighborhood, and do not wander very far.

(Opposite page) The orangutan is called "man of the forest" by the natives. It embodies one of the four species that together form the family of great anthropoid apes.

The differences between populations, however, have not evolved so far that a renewed mingling and exchange of genes can not occur upon a subsequent meeting at the common boundary between ranges.

Now that the behavior of gibbons in their home is accurately known, it is apparent how greatly they differ in behaviour from most other primates, in any event from all other anthropoid apes. This applies to their permanent monogamy, their defense of a fixed home area, and their community singing.

This same combination of traits is widespread among birds, but confined among primates to South American forms and a few prosimians. We can certainly recognize it among the tarsiers. Why is this so? A list of reasons has been worked out by the zoologists John and Kathy MacKinnon. Strict monogamy affords the utmost security and provision for all offspring. The gibbons – like most human societies – thus continue the tendency of all primates to bring forth few progeny, but to ensure the best possible chances of survival for them. Monogamy likewise reduces the expenditure of effort in the wooing of partners and in altercations over partners to a minimum – but only if there is also a spatial separation from neighbor pairs.

Thanks to separation of territories, there is little trouble or competition between members of one species. Also, the groups or individuals know their own territories very intimately, and can tell what fruits will ripen when and where – an important consideration in view of the specialization in fruits available only in small quantities at any one time.

By virtue of their long lifespan, individuals can accumulate very extensive knowledge and experience. Demarcation of territories is usually by calls, that is, with comparatively low outlay of energy, and no risk of injuries. The securing of small territories by patrolling the borders is as it were a side effect of wanderings in search of food; pitched battles, with all their hazards, rarely need occur.

With the adaptation of their behavior to their surroundings, they once again are entering a phylogenetic blind alley: their commitment to a fixed territory renders it very difficult to discover food supplies outside. Nor can they augment the probability of propagation of the species by cohabitation of one male with several females. The small group of gibbons have no prospect of prevailing over large bands of macaques or langurs, particularly by reason of their small stature. Gibbons that do not possess a territory have no hope of a partner and so cannot reproduce. In short, gibbons are creatures of perfection in a stable environment, but are threatened with extinction when their surroundings are altered.

All recent gibbon species and subspecies are as still living, but their future is precarious.

On the whole, one has the impression that the confinement to inherited behavior patterns, the atypical conformation, and the rigidity of environmental relationships, are signs of a phylogenetically primitive position. A group of animals separated for a long time past from the other Hominoidea has survived by managing even better, in certain respects, with the "old-fashioned" tools of specialization, than their more flexible and changeable relatives.

GREAT APES

Category
FAMILY

Classification: A family of the order Primates, comprising four species in three genera. Termed "great" anthropoid apes in contradistinction to the "lesser" apes, or gibbons (Hylobatidae), they represent the highest stage of development among the primates in general and the Old World of catarrhine simians in particular; together with the gibbons and humans, they constitute the superfamily Hominoidea. This hierarchy emphasizes the close kinship of the great anthropoids to humans, based on common phylogenetic origins. Of all anthropoid apes, the chimpanzee is closest to us.

Genus Pongo (orangutans). One species, *Pongo pygmaeus* (two subspecies)

Genus Gorilla (gorillas). One species, *Gorilla gorilla* (two or three subspecies)
Genus Pan (chimpanzees). Two species, the chimpanzee *(Pan troglodytes)* and the pygmy chimpanzee, or bonobo *(Pan paniscus)*

Body length 2.3 ft (70 cm; pygmy chimpanzee) to 6 ft (185 cm; gorilla)
Tail length: Absent
Weight: 66–440 lb (30–200 kg), male gorillas up to 605 lb (275 kg)
Distinguishing features: In stature, conformation and physiology, the most human-like simians; protruding jaw parts (muzzle); powerful dentition with large incisors projecting forward and great canine teeth (in males especially); flat nose; face more or less naked; rounded, hairless external ears; eyes directed forward, capable of depth perception and gauging distances, with full color vision; very mobile lips; lively facial gestures; barrel-shaped trunk; arms consistently longer than legs; strong, long-fingered hands with "flat" nails and short, laterally extensible and opposable thumbs; short, bowed legs; feet with laterally extended great toes; no ischial callosities; genital swellings in females chimpanzees; use of tools and toolmaking known among chimpanzees only.
Reproduction: Gestation period, much in an humans, averaging a scant nine months; only one young per birth (twins far rarer than in humans); birth weight 3.3–4.4 lb (1.5–2 kg).

Pongidae
Singes anthropoïdes; pongides FRENCH
Große Menschenaffen GERMAN

Life cycle: Weaning at 2–4 years; sexual maturity after 7 years, "social" maturity not until about 12 years; lifettspan in the wild 30–40 years, longer in captivity.
Food: Almost exclusively vegetable; leaves, fruits, sprouts, pith, bark, tubers, and roots; rarely insects, worms or birds' eggs; very rarely (among chimpanzees) smaller mammals.

Habit and habitat: All anthropoid apes are diurnal; arboreal (orangutan), terrestrial (gorilla), or arboreal and terrestrial (chimpanzees); in tropical rainforests, mountain forests, and wooded savannas, occasionally in secondary forests; highly differentiated social behavior, usually with distinct ranking and a variety of community life according to species (more or less permanent family associations, or looser, unstable groups of greater numbers); extent of territory or range depending on species, population density and habitat, varying widely up to about 18 mi^2 (50 km^2); constructs sleeping nests; generally peaceable among themselves and towards other species; chimpanzees an exception, often quarrelsome and occasionally in deliberate pursuit of larger prey.

Physical proportions of great apes, unlike those of most simians, resemble the hanging pattern of the gibbons (2). The trunk is comparatively short, arms are long, and legs are short. Whether this structure can have been derived from early hand-over-hand locomotion is not certain. In humans (6), only the proportion of arms to trunk conforms to the hanging pattern; the legs, by adaptation to going on two feet, are much longer. 1. Howler monkey. 3. Orangutan. 4. Chimpanzee. 5. Gorilla.

Facial gesture (left). Development of the facial musculature, well advanced in all primates, attains its peak among the great anthropoids. Reactions as well as mood are expressed with precision, as exemplified by the four different expressions of a chimpanzee face in this illustration. From the top down: aggression, begging, placating one of higher rank, extreme fright.

The hands and feet of the great anthropoids and of humans (right) hark back to the simian hand as a grasping organ. The widest deviations from this pattern are the orangutan's hand (1), adapted to swinging hand-over-hand, and the gorilla's hand (3), modified for travel on the ground. The hand of the chimpanzee (2), like tat of humans (4), is less specialized and better suited to universal use. The human hand especially, freed from locomotive function, has developed into a true precision instrument. Foot structure (bottom) exhibits an increasing adaptation to walking and running on the ground in the following order: orangutan, chimpanzee, gorilla, human.

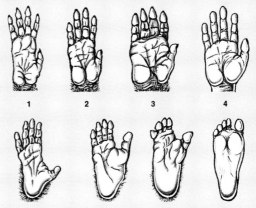

Great Apes

Introduction
by Holger Preuschoft

To so vain a species as humans, the great anthropoid apes – namely the Eastern and Western gorillas, chimpanzees, and pygmy chimpanzees of Africa, and the orangutans of Borneo and Sumatra – must be the most interesting group of animals of all. Why is that?

Human and Anthropoid

The explanation commonly offered, that the anthropoid apes stand on an especially high or indeed the highest level of development in the animal kingdom, really tells us only to what degree the particular author's frame of reference is human-oriented. After all, there are a great many other animals on a very high evolutionary level – only not in a direction pointing towards humans. It is in fact difficult to talk about the anthropoid apes without saying something about humans at the same time. The resemblance extends not only to physical conformation; as Thomas Henry Huxley put it in classic phrase in 1863, the gorilla is closer to humans than to any other animal. Consequently, the great anthropoids together with the gibbons and humans are rightly placed in their own superfamily, the Hominoidea.

As we know today, the resemblance extends to similarity in the structure of the chromosomes, those stainable cell constituents in which genetic traits are encoded; the composition of the endogenous proteins; and the vast realm of behavior.

These matters are to be discussed in detail in the following pages. In the spirit of the doctrine of evolution, the situation is interpreted to the effect that humans and anthropoids represent end branchings of a long line of common development. If the anthropoid apes offer the most pertinent and instructive clues to the human form of existence by virtue simply of their comprehensive affinity to our species, that model gains additional value by reason of common descent. In terms of the situation as it applies to the anthropoids, we might be able to guess at the circumstances of our

Our nearest relatives look at us - the four species of the anthropoid apes, or Pongidae - from left to right: orangutan, gorilla, chimpanzee, and pygmy chimpanzee, or bonobo. With the exception of the orangutan, which lives on Borneo and Sumatra, the great anthropoids are all inhabitants of Africa.

own antecedents. Comparisons between humans and animals, much deprecated but nevertheless continually drawn, in sociology, in psychology, in genealogy, in advertising, and as a matter of day-to-day routine in the pharmaceutical industry (very profitably so), carry conviction here because the forms are closely akin and have solved numerous problems of existence in quite similar ways.

Of course, caution is in order when inferences are drawn from human-animal comparisons. Such caution can be well documented here, for not only the phylogenetic path that led to humans but also the evolutionary path to the anthropoid apes can be well traced on the basic of thorough researches in just the past few decades. In these pages, we shall deal again and again with both the great value and the hazards of comparison.

The Head and Sense Perception

Looking and anthropoid in the face, we find alongside the conspicuous human-like qualities, certain alien features: hairiness, wrinkled skin, flat nose, the deep oral cleft set in the protruding muzzle, with incredibly mobile lips (lacking any red color). In gorillas and chimpanzees, add to this the bony shelf over the eyes, rendering the forehead more receding in appearance that it actually is. The orangutan lacks these ridges, but in older males the soft parts form check pads that give the whole face a broad, flat aspect.

These peculiarities in no way alter the fact that all the skin muscles of the face are organized in very much the same way as ours. They equip these animals for very lively facial gestures. Such gesture is very important to the apes; it is a carrier of messages addressed to conspecifics. The interplay of movements of facial muscles is so similar between the anthropoid apes and humans that we quite often understand their expressions directly.

The shape of the head in the great anthropoid apes is further determined by a very large cranium, accommodating a large and highly developed, very human-like brain. We shall have more to say later about its capabilities.

As in all simians, the eyes are set in bony sockets, making closed circles. They are directed forward, enabling depth perception and range-finding. It is generally thought that such vision is a prerequisite for the evolution of a grasping hand. In acuity, anthropoid vision is much like that of humans and other primates; the resolution is about 1 minute of arc. Anthropoid apes, like humans, have full color vision. This distinguishes them from the other

Old World simians, whose vision is weak in the red band.

The external ears, almost incapable of active movement, are set at the sides of the head, like ours. In the chimpanzee they are large and prominent; in the gorilla and orangutan, they are small and hidden in fur. The sense of hearing is quite similar to that of humans, although at least the chimpanzees have a somewhat more acute sense of hearing on the whole.

The external nose is quite variable in conformation from species to species, but in none does it project from the line of the profile. The bony nasal cavity is not very spacious and its muscular subdivision is much like that of humans in simplicity. There is no precise information on olfactory powers, but the sense of smell appears to be rather keener than in human beings.

The sense of taste also has been tested for so few substances that we cannot generalize. A "bitter" taste seems to be far less disagreeable to them than to us.

The jaw and dentition of the anthropoid apes are powerfully developed, and the "muzzle" protrudes but is not much wider than the human jaws. Compared to nonsimian animals of like size, however, the jaws are quite short. Furthermore, they are not anterior to the cranium but below it, giving the anthropoid head its characteristic short, high shape. Typical of today's anthropoid apes are the large, broad, projecting incisors, the impressive conical canine teeth of the males especially, and (in the lower jaw of the gorilla and orangutan) somewhat canine-like premolars. The molars have the same cusp anatomy as in all the Hominoidea.

The incisors appear to serve as tools for such tasks as spooning out or squeezing fruits, while being assisted or contained by those extraordinarily mobile lips. The canines are no doubt effective weapons, but more often, they are used to open hollow stalks and to husk fruits. They are much exposed at the corners of the jaw, where the rows of front teeth and side teeth meet. The latter, forming a straight line, are an effective grinding and masticating apparatus, adequate for hard kernels as well as for the strong cellulose cell walls of leaves.

The more closely spaced the rows of teeth in the lower jaw, the more the strength of the jaw muscles on either side can be utilized for masticating food. The same is true of the increase in chewing surface required as body mass increases from the bonobo to the gorilla, especially in jaw length, not width. The lever arms of the jaws in the biting function are thus lengthened. Now if the pressure of the bite is to be maintained between the teeth, and entire chewing apparatus must be "remodeled" accordingly. It has been shown recently that the differences in form of upper and lower jaw among the anthropoid species, as well as in the point of origin of the chewing muscles including the bony crest of the skull, may be attributed entirely to differences of scale ("allometric" morphology). This interpretation, by the way, does not apply to the impressive bony superorbital ridges of the chimpanzees and gorillas.

Anthropoid skull shape can be described by measurement. On this basis, similarities can be compared objectively, without preconception, and with quantitative results. They fully confirm earlier conceptions of kinship among the hominoids as well as the findings, to be descirbed later on, of molecular biology.

The baby gorilla (above) has a quite expressive face, but the facial gestures of an old male gorilla (right) are far more marked. The mobile lips especially - a trait of all large anthropoids and of humans - permit a richly varied mimicry, of great importance in intraspecific communication.

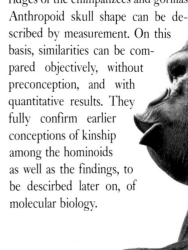

GREAT APE – INTRODUCTION

Body Structure and Locomotion

The "great" anthropoid apes (unlike the "lesser" apes, or gibbons) surpass most other primates in body mass, and approach human proportions in this respect also. Adults weigh from about 66 lb (30 kg; bonobo females) to over 40 lb (200 kg; gorilla males). Their size preludes endangerment by any but a few predators, and gives them an advantage over competitors for food. For arboreal animals in particular, however, a heavy body presents difficulties, for obvious reasons.

Like nearly all other primates, the anthropoid apes have hands with five digits, four outer fingers in parallel arrangement and a thumb that can be spread or opposed. All digits are armed with "flat" nails. On the ventral side of the fingers, and on the palm, a hairless, horny skin is formed that is extremely tough but serviceable as a tactile organ. In experiments, a chimpanzee was able to distinguish a ten-sided disc from a twelve-sided one. Both were presented to him in a sack into which he had to put his hand without looking.

From the tailed Old World monkeys of the superfamily Cercopithecoidea, the anthropoid apes (and the gibbons, but not humans) are distinguished by a substantial lengthening of the second to fifth fingers, far exceeding the length of the thumb. In fine manipulation, using the misleadingly so called "precision" grip, the tip of the thumb does not reach the tip of the index finger, but opposes its middle joint. Despite this digital proportion, which seems strange to us, the dexterity of the anthropoid hand seems if anything to surpass that of other Old World simians. A more highly differentiated, and in that sense more human-like musculature, is consistent with this condition.

One thing that the anthropoids are still less able to do than most humans and other primates is to curve their fingers beyond full extension, that is, concave on the dorsal side of the hand. On the contrary, in most individual animals the contracting muscles are so short that they cannot fully extend the fingers and the wrist at the same time. As a rule, the middle and end finger joints, except the proximal joint, remain bent, resulting in the characteristic "hooked" attitude of the hand. The hook shape is further emphasized by the curvature of the several finger joints. As a result of the shortness of the muscles, they function in a "prestressed" condition, in which they can exert exceptional force. The hooked hand is doubtless to be understood as an adaptation to climbing, especially with hands lifted overhead. It permits grasping and holding with great force. When such an animal walks the ground on all fours, it of course cannot set down the palms, let alone the fingertip, as all other quadrupeds do. Instead, the orangutans rest the entire first on the ground, and gorillas or chimpanzees only the backs of the middle joints. Hence the "knuckled" gait that has given rise to much speculation in recent years. There is some reason to believe that it is in fact only an escape from the predicament of an adapted climber walking on the ground. Juveniles repeatedly attempt to solve the problem in other ways, but any other placement of the hand in walking is difficult for anatomical reasons.

On the stressed dorsal side of the fingers, imperfect callosities are formed during adolescence, and sometimes there are early indications of a pattern of skin ridges. Thus, body form is not thoroughly

The hands of the great anthropoids seem quite human, and so does the way they use them. The external ears set at the side of the head also resemble ours, at least in the chimpanzee (above); in the orangutan (left) and the gorilla, they are smaller and often hidden in the fur. The paramount distinction of the anthropoid head, however, is the protruding "muzzle."

363

adapted to mechanical requirements. The knuckles support only a comparatively small share of the body weight. The mobility of the hand itself is supplemented by that of the wrist and by the ability to rotate the forearm about its longitudinal axis. The elbow joint is fully extensible, and the shoulder has extraordinary mobility. The arms appear very long and powerful relative to trunk length; their mass is about 16 to 18 percent of that of the trunk (8 percent for humans). Here again, mobility conforms to structure: the muscles of the pectoral girdle do not extend as far on the upper arm in the anthropoid apes as in the Old World monkeys, and therefore do not tether the arm to the wall of the trunk as much as in these and in all quadrupeds. Thus the hands of the anthropoids have more scope for movement than those of other Old World simians. In this, we human beings resemble the anthropoid apes, even though our shoulders and wrists are not capable of such sweeping motions.

The structure, tactile sensitivity, and mobility of the hands have an important consequence. Through their inclination and ability to pick up objects in the hand, to examine them with the hand, and to handle them, these animals approach typical human behavior very closely, and all that has been said about the significance of hand use upon depth perception, brain development, and powers of "comprehension" applies in like manner to the other higher primatess. Most importantly, this also means that all "intelligence" tests, all tasks designed to explore intellecutal faculties – which in this case must emanate from and refer to human beings – will yield fairly "good" scores, because of the readiness of the anthropoid apes to pick things up with their hands and move them about.

If human similarities predominate where the hands are concerned, it would seem at first glance that differences take over when it comes to the feet. Like the hands, they have four long, well-formed outer toes opposed by a great toe of thumb-like mobility. This great toe is well separated from the four parallel outer toes, and it is shorter (not longer) than the other toes. It can be actively spread out and drawn in, it can be brought more or less into one plane with the other toes, and it can even be opposed to them. The ventral side of all toes and the entire sole of the foot are covered with the thick, horny skin, that is richly innervated; this was previously described. The pattern of ridges resembles that of the hands. In their form, the anthropoid feet are likewise grasping tools, which inspired the old scientific name "quadrumana".

Upon closer consideration, to be sure, differences in the use of the hand and feet emerge, as well as structural similarities to the human foot. The feet are quite predominantly employed for purposes of locomotion. Objects are only occasionally picked up and held with the feet; fine "manipulations" are virtually never seen, and very rarely indeed are the feet used to carry off objects such as a towel, or fruits. But when an anthropoid ape is traveling on an irregular substrate, such as branches or the bars of its cage, or if the underfoot is especially

Hand (left) and foot (right) of a mountain gorilla. Note the flat, "humanlike" nails and the opposable thumb and great toe.

narrow or precarious, like a branch or rung of a ladder, or much inclined, like vertical holes or trunks, the toes will take hold with great skill and strength to secure a purchase. On the level, as measurements have shown, two-thirds or more of the total body weight rests on the hind limbs. This is true even of the macaques and baboons, in contrast to the "true" quadrupeds. Furthermore since the trunk is short and the center of gravity, therefore, is never far in advance of the feet, they can easily move all the way under the center of gravity, so that a biped posture is easily and frequently assumed.

Then the feet support the entire weight of the body – in the case of chimpanzees, over distances of several hundred yards, or for more than ten minutes. The preponderant body load is transmitted to the broad foot soles, which are thickly cushioned with fatty tissue. In this sense, they do just as human beings do, and as other Old World simians do not; their feet are narrow, not much cushioned, and often incompletely ridged. In running, the heels are not set on the ground at all; at rest, this seldom happens, and only under moderate load. Like all true quadrupeds, the Old World monkeys are on the way to a mode of locomotion in which only the distal half of the sole makes contact ("semiplantigrade" locomotion).

The action of the great toe, as well as the action of the heel, is a trait that is common to both the African great apes and humans. The gorilla's great toe is spread far inward, and the four outer toes outward, created a "hollow" on the inner side of the foot, that is, an arch, much as in the human foot, bridged only by the great toe. Chimpanzees often – orangutans always – set the foot down with its outer edge on the ground and curve the toes in slightly, in which case the great toe can rest again the second toe and not assume any load. The orangutan's great toe is not only very short compared to the long, massive fingers, but it is also small compared to the base of the foot. Frequently, it lacks one joint and the nail, as well as the retracting tendon, but this does not alter the fact that the great toes are regularly used for grasping, and can do so with considerable force.

The chimpanzee's "knuckle" gait; only the knuckles of its hands make contact.

The grasping foot of the anthropoid ape is the extremity of a very short leg; the legs represent only 18–24 percent of the mass of the trunk (30 percent in humans). Once more, the high mobility of the joints is remarkable; the ankle and knee joints in particularly will flex farther than we can expect of our own. At the hip joint, the great thing is the ability to move the thigh and with it the entire leg in front of the abdomen or along the trunk, with muscular force, towards the head. This depends not so much on the structure of the hip joint itself, or the arrangement of the muscles, as on the form of the pelvis. The broad pelvic bones are parallel to the dorsal plane, so that they do not present a hindrance to the thigh. They are very long, and the length of the muscle fibers (of which only the great muscle of the buttock is much differently arranged from that of humans) makes possible considerable mobility, allowed the leg to be placed in any position at will, even quite close to the trunk, and to be held or moved there with great force. In the shank, the calf muscles so typical of humans take second place to the powerful retractors of the digits; in the thigh, the extensors of the knee joint are not very greatly developed.

Whereas in humans, as in all highly developed walkers, the leg movements are in a "sagittal" plane, oriented from front to rear and parallel to the direction of progress, the anthropoids are able in addition to move their legs freely in a "frontal" plane, oriented to left and right. The freedom of motion of the thigh in this plane, laterally to the trunk, is especially great, being favored by the placement of the pelvis.

Now since each foot must be capable of stepping under the center of gravity of the body, to support it alone in certain phases of locomotion, and since the center of gravity is located more or less in the center of the trunk, the knees must be bowed. Actually, the bow legs and inturned toes are characteristic of the anthropoids. It is attended by a reinforcement of the inner bone of the joint at the thigh.

The physical image of the great ape is completed by a markedly short but deep and broad, that is, barrel-shaped trunk. Whereas the pelvis is very long, the several vertebrae of the spine are flattened. The number of vertebrae in the thoracic region is often greater (thirteen) than in most higher primates, but in the limbar region it is diminished to four, and not infrequently to three. The barrel-shaped trunk distinguishes the great anthropoid apes both from the Old World monkeys and from the gibbons and humans. One gets the impression that the point is to accommodate as much volume as possible in a minimum of length. This is of decided mechanical advantage to forms that very often are half erect, without having carried that specialization as far as human beings have. Humans and Old World monkeys have a fairly long trunk, supported by arms and legs in quadruped posture. To prevent sagging

Gorillas are distinctly terrestrial, traveling chiefly on all fours. Only at nightfall do the lighter-weight youngsters like to climb trees, in which they build their sleeping nests. The large individual in the center is an old male, called a "silverback."

GREAT APE – INTRODUCTION

Unlike the African anthropoids, which have become more or less terrestrial, the Asian orangutan has remained a tree dweller, seldom descending to the ground.

under the weight with a minimum of muscular exertion, the distance between the straight abdominal muscles and the spinal column must be maximal. In other words, in cross section, the shape of the trunk should be the familiar narrow "keeled" conformation of all specialized quadrupeds. Because of the flat curvature of the ribs, the shoulder blades take up a position to the sides of the chest. In humans, the adaptation to erect posture is possible because the spinal column as a supporting member is shifted to the extent possible towards the center of the trunk in cross-section. The result is the well-known lumbar lordosis (curvature of the spine, concave to the rear) and a chest with ribs first curving out dorsally and then sharply inward towards the breastbone. Its distance from the spine is small, while the breadth of the rib cage is enlarged. The shoulder blades are at the back. In the anthropoid apes, the shoulder blades are so placed between the dorsal and lateral aspects of the trunk that it is possible to place the weight on the arms just as advantageously as to move them freely to the sides of the trunk or over the head.

The large anthropoids have no ischial callosities whatever. The hair covering of the body is sparse – there are few hairs per unit area – and follows the same pattern as in humans, in that it lacks and undercoat of fine woolly hair; only stiff, rough guard hairs are developed.

Of course there has often been a question as to what mode of life, or more precisely speaking, to what form of locomotion, the anthropoid physique is properly adapted. With our present knowledge, this question is much harder to answer than it was twenty years ago. Today we do not recognize a special climbing technique as the one to which the animals are adapted. One point seems certain. The most nourishing parts of trees – buds, blossoms, young shoots, usually fruit as well – are found in the outer canopy. But here the branches are too flimsy to remain in any position near horizontal under the weight of an animal. An animal running along the outer branches will slip off, and only if it can hang or travel anchored beneath the branches by its fingers and toes will it be able to deal with the deflection of the branches. It then utilizes their tensile strength. The preferred opinion was based on the assumption that the anthropoids were originally at home in the emergent zone of the tropical rainforest, where they became adapted to hand-over-hand climbing. This fits in with the strong, hooked hands, the long muscular arms, the short, weak legs, and the short, barrel-shaped trunk.

But what we have since learned about the habits of the animals does not support that conception. Only the orangutan is clearly at home in the boughs of the rainforest. It spends the greater part of its life climbing, often hanging by its arms raised above its head. Seldom does it set foot on the ground. At the same time, it is not much of a swinger nor does it hang by the arms like the gibbons. The gorillas have been found to be distinctly terrestrial. Probably this is not simply because of their size; witness the example of the orangutan, which is also very heavy. Gorillas rarely climb trees higher than 16 ft (5 m). About 40 percent of the range of the chimpanzees is located in

Orangutans, which will tip the scales at up to 198 pounds (90 kilograms) when full grown, can get a secure hold even in light branches and lianas with their powerful grasping hands and feet. However, they are not good at leaping.

dry forests, wooded savannas, and regions that were cleared and have since become reforested, but not in the original rainforest where their population is less dense. In forest areas, they spend 50 to 70 percent of their time in trees. When alarmed, they will often take flight by coming down from the trees. Once on the ground, they will roam for remarkably long distances, even where there are no stands of trees. The pygmy chimpanzees spend rather more of their time in trees, will climb higher, and take flight in the branches as a rule. But they too will travel on the ground, without any apparent compulsion to do so.

The most frequent mode of locomotion among all African anthropoids is the four-footed gait, the hands being set down on the knuckles, but supporting less weight than the feet. The form of the hands, however, is not very convincingly adapted to this "knuckle" gait. Chimpanzees especially often stand up on their feet, for example to cross water courses (jumping and also landing on two feet). Like the gorillas, they will then stand stooping slightly forward, knees bent and turned outward. In open country, they will also walk on two feet, if they have anything to carry or if they are wading in shallow water.

Arm length does not have the function once ascribed to it either. Measured by body mass, the arms of gorillas and chimpanzees are no longer than would be expected for Old World simians of their size. Thus they have merely continued a widespread development. Still, many reasons have been found that render long arms advantageous for heavy tree dwellers. Ultimately, the recurrent crucial factor is the reach from a point of vantage once found secure. The short, apparently rigid trunk is by no means a condition for successful hanging by the hands. By well-timed muscle action, gibbons, as well as leaping prosimians, suppress interfering oscillations without anatomical shortening of the trunk. Furthermore, the mobile arms render large trunk movements superfluous. The proportion between the strength required in the digital retractors for swinging hand-over-hand and the required increment of mass in the arms becomes too unfavorable for animals over 65 lb (30 kg) body weight to permit swinging hand-over-hand for any long distance. These animals would need radically different physical proportions in order to adopt such a hanging form of locomotion.

The short bowed legs with grasping feet are perhaps easiest to explain. If the arms take over the seizure of remote branches in locomotion or to gather fruits, leaves, or other food, nothing remains for the legs but to support to body weight. In climbing up or down, each "stride" passes through a critical instant when the center of gravi-

(Left) Female chimpanzees display readiness to conceive by the conspicuous swelling of their genital skin. (Right) A pregnant orangutan.

ty of the massive body must be leveraged away from the tree trunk or bunch of branches on which the animal is climbing by a thigh's length. The shorter the thigh, the smaller the dangerous turning moments due to body weight. This accounts for the ability of the anthropoids to spread their legs so wide at the hip joint; the knee can be moved laterally along the trunk, so that the thigh will not get between the tree and the body.

Shortness of the legs, of course, more or less sacrifices the ability of leap; the animals lack the acceleration displacement that their great mass would require to keep the necessary force within reasonable limits. Evidently they can do without that ability. Orangutans and gorillas do not leap at all; chimpanzees are indifferent leapers. The bonobos leap quite well, and their legs are longer and

A gorilla who became famous for his unusual coloring: a true albino.

stronger, being 24 percent of trunk mass. Gorillas, with the lengthening of their heel, have lost the ability to move the forepart of the foot quickly for a jump. However, the longer lever arm saves effort in extending the ankle. Their great size alone renders their life fairly safe; few enemies can endanger them, even without the power of swift locomotion.

In recent decades, a number of speculations have been offered to explain the physique of the great anthropoids, usually in the context of the origins of humans. Unfortunately, they hardly go beyond the considerations already mentioned. Only the Netherlands chimpanzee researcher Adriaan Kortlandt has emphasized the possibility that after the lines of descent diverged, a separate development may have removed the anthropoids farther from humans. On the whole, we must confess that although we can reliably assess the individual traits in their mechanical function, we cannot completely explain the biological role of these striking characteristics in combination.

Is it possible that the supposed adaptations to swinging hand-over-hand are in truth nothing more than means to perfect climbing, permitting even large, heavy animals to exploit the "niche" of the treetops, even the outer canopy?

Life in the trees is not without its dangers to the anthropoids. Several observers have seen falls from heights of more than 30 ft (10 m). According to a study of very large numbers of skeletons, about a fifth of all adults show traces of severe bone fractures from which they completely recovered.

Phylogeny

At this point, of course, we ask for fossil documentation of the history of body form, that is, on the early stages of phylogeny.

Fossil anthropoid remains are not exactly rare. Most of them consist of jaw parts. They are known from Africa east of the Rift Valley, Asia, and Europe. The oldest date from the Upper Oligocene and are about 33 million years old: *Aegyptopithecus* and *Propliopithecus* in the Egyptian Fayum, southwestward from Cairo. Of the former, there is an almost complete skull, very archaic in appearance, with a dentition leaving no doubt of membership in the Hominoidea. The skeletal remains that can be associated with the skull and dentition in terms of size may be interpreted as those of a fairly slow-climbing quadruped, with movements perhaps similar to those of a modern howler monkey. Through the Miocene and Early

Pliocene (25 to 7 million years ago), anthropoid remains are the commonest of all primate fossils. Jaw bones are found, in the vicinity of water and forest when the terrain is identifiable. Whereas formerly their owners were thought of as living in gallery forests, wooded savannas, and dry forests, indications have recently come to light that, at least in Africa, extensive forests were their home.

The skulls as well as the dentitions, as presently known, exhibit all the features of modern anthropoids, though not in the same combination. *Dryopithecus (Proconsul) maior* resembles the modern gorilla quite closely, though it does not attain quite the same size. *Gigantopithecus* of southern China and northern India, on the other hand, is larger, more compact in build, and far more massive than any species now living. Furthermore, these remains are Recent: they pertain to the Late Pliocene or the Pleistocene, and so are less than 3 million years old. Tooth form deviates somewhat from that of modern great apes.

Finds in Spain, France, Turkey, and the Sivaliks on the southern slopes of the Himalayas, as well as in China, range in size from the slender, delicate *Pliopithecus vindobonensis* of the Vienna Basin, which is about the size of a gibbon, to *Dryopithecus,* easily as big as a chimpanzee but reminiscent rather of gorillas in dental anatomy. A well-preserved, large facial skull from Pakistan, named *Sivapithecus,* is strikingly evocative of the orangutan, but lacks its characteristic dental features. Smaller forms from the Sivaliks and later also from Africa have been regarded as very early pre-hominids *(Ramapithecus),* chiefly by reason of their short rows of teeth and small canines. For some time now, however, these same finds have been referred to *Sivapithecus* once more.

Tooth and jaw remains from Africa vary in size from the above-mentioned *Dryopithecus (Proconsul) maior* to the gibbon-sized *D. (Limnopithecus) legetet,* both of which are the despair of taxonomists. Skeletal remains of the smallest form, which has been described as *Limnopithecus* (now *Dendropithecus) macinnesi,* are so slender and delicate that its resemblance to gibbons was at first emphasized. That connection is not supported by further studies. The same applies to the *Pliopithecus* of Europe. Neither form appears to have left any living descendants. Meanwhile, nearly the entire skeleton of the somewhat larger *D.(P.) africanus* has become known through several finds. Known parts of the chimpanzee-sized *D.(P.) nyanzae* are very much like it. The clearly identifiable hands and feet are long and slender, befitting none but tree dwellers. The animals probably lived as highly mobile, active climbers, in trees, though it cannot be ruled out that they may have come down to the ground. They do not clearly show traits of the quadruped monkeys that run on the pads and toes only, nor are there any other specializations that we can interpret by analogy with living forms.

A baby gorilla trying to cling to the mother's fur. In the first few days, it needs additional support from her.

There are no signs of special leaping abilities, or of hand-over-hand climbing, or of a "knuckle" gait. Such forms as mangabeys are quite comparable. But it is by no means certain that the archaic hominoids must have evolved modes of locomotion that occur among New World monkeys now living, and that were (perhaps) inferior to the forms of movement of today's Old World monkeys.

From the Early Pliocene, we have *Oreopithecus,* which has a different dentition but the body dimensions of a modern anthropoid.

A mother orangutan with her baby of about 2 month old. Waking or sleeping, the baby can hold tight to her fur, so that her hands and feet are always free for climbing.

The very earliest prehominids (*Australopithecus afarensis*, of Hada, Ethiopia; "Lucy"), relative to their body mass, estimated to be about that of a bonobo or chimpanzee, possessed powerful, long arms and short legs. The flat rib curvature and the shape of the pelvis suggest a barrel-shaped trunk. From the leg skeleton as well as the foot shape of later finds in East Africa, in addition to footprints that are also 3.5 million years old, one would expect two-legged locomotion, but the fingers and toes are long and curved like those of climbers.

The fossil record gives no compelling answer then to the question of the adaptation of the modern anthropoid body shape. We shall probably have to content ourselves with the assumption that it was a phylogenetically late adaptation that may have similarly occurred among the ancestors of humans. Beginning with a stage perhaps 5 million years ago, developments have continued in quite different directions.

Modern Affinity Research

With improvements in microscopy and tissue culture techniques, it has become possible to examine anthropoid chromosomes (from Greek *chroma* "color" and *soma* "body", because these loop-like configurations in the cell nucleus can be made visible by certain staining methods). The resemblance to human beings as illustrated above extends to the chromosomes. Chromosomes are identified in the first instance by size, location of the centromere (to which the mitotic spindles attach), and relative length of the arms. For some years, the "band pattern", that is, the serveral transverse bands of the loops, most plainly visible just before cell division, has been determined with great reliability. These band pattern are so distinctive that even segments of chromosomes can be recognized. For humans and the anthropoid apes, virtually all readily stainable (euchromatic) segments of all chromosomes can be matched up. Only their numbers is different. The great anthropoid apes have 48 chromosomes. Apparently the union (fusion) of two formerly separated chromosomes to form the human "chromosome No. 2" resulted in the typical human complement of 46.

Apart from the fusion, the only differences lie in the frequent interchange and insertion (translocation) of individual chromosome segments. This type of modification seems especially typical of the anthropoid apes and humans; among the other Old World primates, translocation and fission occur more frequently. In gibbons, the attempt to identify chromosome segments that are present also in the great anthropoids and humans has been successful in only a few cases. Apparently, translocations of chromosome segments are especially numerous in this group of animals. It may be that the variety of fur coloring or song with no apparent difference in habits is quite simply a consequence of this "chromosomal" evolution.

The number of differences in the band patterns between humans and the orangutan is again greater than that between humans and the gorilla, or indeed the chimpanzee or bonobo; the three species mentioned last are especially similar to each other. The sex-determining X-chromosome (XX is female; XY is male), as in all mammals, does not affect the band pattern.

Advances in the field of biochemistry, which have been so significant to the whole of present-day bi-

A young "black back" playing with a two-year-old. During their long childhood, young male apes at play practice may modes of behavior characteristic of their species.

ology, have opened up new possibilities also in taxonomy (as the science of affinities). Human blood groups are found in almost the same form in the anthropoids, although the antibodies (blood defense against foreign proteins) are deployed differently, and not all blood groups occur in all species. The bonobo, in fact, has no antibodies against human blood, so that transfusions of human blood of the same group would be possible. In the other anthropoids, transfusion could be risked only with blood corpuscles isolated from serum.

It has been possible, using various methods, to make accurate determinations and comparisons of the structure of endogenous protein compounds and of the genetic substance DNA (deoxyribonucleic acid). These methods permit quantitative definition of the similarity between forms. In terms of the probability that changes occur in the sequence of the DNA building blocks, attempts have also been made to estimate the time since various lines of evolution separated (the "molecular clock"). Objections have indeed been raised against these experiments because of the different rates of change that have actually been proven. Still, these considerations have had a lasting influence on interpretations of affintiy and phylogeny.

The results of specific scientific studies largely agree with the morphological results that have been known for some time. Among the primates, the group of the Hominoidea is clearly distinguished from tailed Old World simians, the Cercopithecoidea. Within the hominoid group, the gibbons (Hylobatidae) stand farthest apart, then the orangutan. Clearly separate from the latter, the African anthropoid apes are found in close proximity to humans. In this respect, the results of molecular biology deviate from those of morphology, which places the orangutan and the African anthropoid in the family Pongidae, in contradistinction to the family Hominidae, comprising humans and their pre-human ancestors. The least distance is between chimpanzee and bonobo, though eastern and western gorillas have not even yet been compared by the methods of molecular biology. These findings are interpreted rather straightforwardly as reflecting a very diverse line of descent. Any very close agreement is interpreted as indicating that the lines of development did not diverge until comparatively recently.

The Brain and Its Capabilities

The connecting link between physical structure and behavior is the controlling organ, the brain. Anthropoid cranial capacity – 340 cc. (orangutan average) to 530 cc. (male gorilla average) – is much less than that of humans (average about 1400 cc.). There are no essential differences among the species in the internal structure and general form of the brain. In all cases, especially the cerebral cortex and also the mass of basal ganglia (a ganglion is a nerve junction), as well as the cerebellum and the centers of vision, are highly developed, while the olfactory brain is reduced. The most highly developed parts of the cerebral cortex are the phylogenetically recent ones, the "neocortex." In the Old World simians, its volume is between 40 and 76 times the brain volume to be expected at the lower end of the mammalian tree in an insectivore of comparable size; the volume of the chimpanzee neocortex is 84 times that of a lower mammal of like size, and the human neocortex is 214 times that volume.

A young chimpanzee challenging its mother to a chasing game.

As in all primate brains, the "primary" regions are separated by the progressive spread of large areas of "integration" cortex. The cell structure of the cerebral cortex exhibits the same differences in humans and orangutans, but the extent and arrangement of the cortical types do not quite agree, and the anthropoid apes lack those regions that accommodate the sensory and motor speech centers in humans.

Reproduction and Juvenile Development

The great anthropoid apes have no season for reproduction. The males are sexually active at all times, and the females have a hormone-controlled cycle like the human menstrual cycle, usually longer than thirty days. In mid-cycle, the females are willing to mate and frequently offer themselves to the males. Sometimes they prefer certain partners; usually they will accept any male present. After a gestation period of seven to nine months, a rather small baby, weighing 3.3 to 4.8 lb (1.5 to 2.2 kg) is born. Twin births are extremely rare. The young are quite helpless in their first months, and are dependent on the mother in virtually every respect. Newborn orangutans and chimpanzees will cling to the mother's fur by themselves; baby gorilla cannot do even this, and must be supported. Nurslings of this age level expand their relations with their surroundings in stages much like those of human children. By the age of four to five months, the specific modes of locomotion have been mastered, and the milk teeth emerge. Children are generally nursed for several years, during which time the mother is not ready to conceive. Until the birth of a sibling, youngsters seek refuge from danger with the mother. Solid food is of course taken early, and prospects of survival are fairly good even without the mother. Sexual maturity, at about 7 to 8 years, is preceded by a long "juvenile" phase. The permanent dentition develops during this time. The survival chances of these animals depend very largely on their ability to learn during this protracted adolescence. The readiness to confront novelty is far greater in the juveniles than in the adults. They gather independent experience and learn most of the behavior patterns characteristic of the species. So far as we know, surprisingly few anthropoid modes of behavior are innate. Continual "play" activity provides experience with objects in the sourroundings and especially with conspecific relationship. The "social maturity" of the males is preceded by a period of further growth, as well as the appearance of jowls in the orangutan and of the silver back and mane in the gorilla. It is not until the age of 12 years that an anthropoid male assumes a role of importance among his peers. The females appear to reach reproductive age earlier. The lifespan of the chimpanzee is more than 40 years, that of the orangutan and gorilla is about 35 years. In first approxi-

(Top) This little chimpanzee, who lost his mother a year ago, looks as unhappy as a neglected child. As a manifestation of consequent disturbed behavior, he has pulles off almost all the hair of this lower legs. (Bottom) Physical contact is as important in the social life of chimpanzees as in ours. Touching hands gives a sense of security and safety.

mation, then, it may be said that the developmental process after birth take about half as long as in humans.

Community Life

All great apes need to relate to conspecifics. The strongest incentive to effort is the striving for conspecific company and for a rank within it. This means that apes can and should be kept in communities. The most effective reward is give-and-take or social play. "Solitary confinement" causes the same despondency as in humans. This is why so-called "Kaspar Hauser" experiments in which individuals grow up with no parents or conspecific companions, and which are intended to identify innate behavior patterns ("instincts"), do not work well for primates. If anthropoid babies are raised in isolation, they will indeed grow faster, but their behavior will be seriously disturbed, so that they will deviate considerably from the norm in all respects. This applies even to such seemingly biological matters as sexual behavior. In human company, if offered sufficient companionship, anthropoid babies will thrive.

Under natural conditions of life, which is to say in a conspecific social context, most actions are addressed to other members of the group. The existence of any large associations requires some sort of order. The form this order takes is ranking. It is instituted and maintained by the exchange, usually of visual, less often auditory, "signals." Such signals are continually being transmitted by each individual and received by the others, with the result that all group members are at all times informed of their fellows' state of mind.

The signals for this "emotional communication" are of the same kind as ours: condition of the facial musculature, gesture, and the placement of the limbs, the head, and the trunk – matters that have for some time received attention under the name of "body language". An animal's entire posture, indeed location, as at the same time an expression of the inner state. Minute details are perceived and answered as messages by the others. The transmission of such signals is linked to certain emotional stimuli, and will not occur without them. For the reception of signals serving the exchange moods and feelings, there is a remarkable attentiveness and a quite astonishing acuity of perception. As surely as each human being transmits these signals, that is, as a person's feelings are written in countenance and movements, so can the person accurately interpret and response to these clues in others. Our "good manners" are often nothing more than modes of social behavior that

A typical gesture of conciliation: a old male chimpanzee reassuring a younger member of his group.

existed from the first, with a retroactive rationalization. Both the transmission and the reception of emotional signals are quite regularly unconscious.

Who realizes that the face of an acquaintance is studied from afar to verify whether there has been any recognition, and whether a greeting will be welcome or not? Who is aware that the set of the eyes and the facial expression are the controlling signals? Only actors and actresses plan such signals deliberately, but even they are not always conscious in detail of what they are doing, or why they contrive a nuance in one way or another.

It is possible to make direct comparisons of those modes of behavior that are widespread among anthropoids and which govern their social behavior as well as human behavior patterns. In many specific cases, corresponding gestures of different species differ in subtle detail and the rapidity with which they are executed; they may simply seem strange to us because of differences in facial configuration. There are also gestures and expressions that are very important in one species and absent in another.

That visual signals are the most meaningful is consistent with the acuity of the sense of vision. The role of hearing in social communication is limited. Language, which is so important to humans, is most effective in terms of objective content. Whenever one tries to express feelings in language, one must confess all too soon: Words fail me. The audible utterances of anthropoids are highly varied and variable. Accordingly, the content of messages transmitted by sound is vaguely defined.

The relationship of individuals in a group to each other differs from species to species. Of course, truly reliable conclusions concerning conspecific groups are possible only with animals in their natural habitat, which must not be unduly disturbed.

Gorillas live in "harem" groups. These are composed of one adult silverback male, usually four or five females, and their as yet dependent offspring. These may include sexually mature but not yet fully adult males. Not infrequently, another silverback male is more or less loosely attached to the group. The focus of the group is always the highest ranking (dominant) male; all moves emanate from him, or are prevented by his simple failure to move. His dominance manifests itself primarily in the attention paid to him by all members of the group; all make room for him, often quite unnoticeably.

The group leader does not necessarily bar all sexual relations between the females of his group and the other male members. The bond among the female individuals does not seem to be especially strong. Harems move about independently of each other in overlapping territories. Upon meeting, the groups remain apart, and retire from each other after some prolonged contests of display, looks, or threats, between the male leaders. Serious combat between harem leaders is extraordinarily rare. The males exhibit very little aggressiveness within the group, and infant members often approach them during rest periods.

Group ties are not as close among orangutan. Adult males move about alone in very large territories. They are usually far away from females with young, who seem to limit their movements to smaller areas. Males will come closer only when

Long-distance communication among chimpanzees: a mixed group makes a loud reply to calls from conspecifics far away.

they are ready to mate. The females for their part have small young ones with them, while the older ones stay nearby. Several females with young may remain together for longer periods and form "mother groups." Males, also occasionally form small bands.

Chimpanzees are often together in larger numbers, so that several males and several females are observed, the latter with their several offspring of various ages. The composition of the chimpanzee groups in a locality changes frequently. The males, several of whom often remain together for a considerable length of time, travel fast. They are usually the first to find the trees on which fruit is just becoming ripe. Since they make a great deal

Family life of the "gentle giants". Gorillas live in harem groups composed of a few females and their offspring and led by an adult silverback male (left). Silverbacks (right) are peaceable family men, patiently allowing their little ones to climb all over them.

of noise on such occasions, females will follow from all directions. Now a large assembly of mixed sexes has been gathered, embracing all age levels. After some time, the males will first leave the spot, attended by younger females, until at last only the oldest females, or those with the youngest offspring, are left behind. Conversely, several males will gather around any female that happens to be ready to mate. Since there may be older or even grown-up young in company with this female, the result is again a sexually mixed assembly. Chimpanzees gathered in any one place will react in concert to any special events, as for example presentation of foreign objects, or attack. All this applies only to members of good-sized societies inhabiting a particular territory. These animals know each other and will greet each other, often boisterously, when they meet. This will not prevent them, hours or days later, from separating and moving on in other parties. Extremely violent altercations can occur between members of different societies. At the boundaries of territories of neighboring societies, small detachments behave very quietly and unobtrusively.

Conditions among the bonobos are much the same. However, the assemblies are larger, or the animals in one place are more numerous. The tendency to aggression seems to be less, despite the greater excitability of the animals. This is consistent with the more friendly behavior of partners before and during mating. Very recent observations in the wild suggest that female bonobos are ready to mate for longer periods than female chimpanzees, which are rather roughly induced to pair by the males only during the height of the estrus. Bonobo pairings are quite often belly-to-belly instead of belly-to-back, as with most anthropoids except orangutans. Homosexual relation-

ships are also common. Bonobos are less inclined to form all-male bands than chimpanzees or orangutans. Also, the bands remain together longer.

All great apes build sleeping nests, mostly in trees; the big male gorillas make their nests on the ground. In this, and in the changing of territories used by several groups as well as their larger social organizations, the great anthropoids differ clearly from the gibbons, which are monogamous and claim a fixed home territory.

The ranking of the males is evidently not based exclusively on physical fighting strength. A certain sort of calm, deliberate behavior, suggestive of assurance, is often observed on the part of those of very high rank. One male chimpanzee owed his peak position in the ranking to his idea of accompanying a dance of display with a infernal noise on some sheet-metal drums. Another gained his leading position through cooperation with his brother: the two would support each other in challenging and eventually attacking the predecessor. Within the group, the males generally dominate the females provided they belong to the same age class, and the older dominate the younger. Very close relationships may exist between individuals. Long-term observations have shown that these are attributable as a rule to kinship (mother-daughter; siblings).

Whereas relations between gorillas are generally peaceful, in fact gentle, aggressive outbursts continually occur among chimpanzees: male against females unwilling to mate, or males among themselves. In the best-known chimpanzee society on the Gombe River (see the article by Jane Goodall), there were several killings of group members – 12 in five years, among about 50 individuals! Among the dead were four adults, and eight juveniles. These results are in gross contradiction with all conceptions held from earlier observations of anthropoid social life until the early 1970s. First it had been noticed in what measured channels, set about with rules ("ritualization"), community life usually flowed. The indulgence of formidable male gorillas towards juveniles, or the solicitude of adult male gorillas or chimpanzees for motherless infants, were the yardstick, as well as the survival of crippled animals in some groups; all this was thought to represent some care by other group members. Recently, it has turned out that members of a group may be severely abused or killed in confrontations, or that individuals may be gradually forced out of the group and killed after months or years of estrangement. Conspicuously friendly relations subsist primarily within close kinship circles.

These observations remind one of frequent reports of infanticide among other simians (hanu-

Chimpanzees greet each other with a "kiss" after any long separation.

man langurs) by males who have just taken over the leadership of a harem. Something similar has been observed among mountain gorillas. Adult males of neighboring groups, or even of the same group, direct their attacks at infants, with the result that these die and the mother promptly becomes ready to mate again. The next issue will be that of the male who killed the infant, not begotten by himself.

The killing of babes in arms by males improves the reproductive prospects of the latter. If it is assumed that all behaviour ultimately is aimed at replication of their own genes (genetic code), then these infanticides are accounted for in the same way as self-sacrificing assistance to others – provided the others are blood relatives. There

▷ All great anthropoid apes build sleeping nests nightly, but only the tree-dwelling orangutan do it exclusively in airy heights. For this purpose, they bend twigs and branches so as to form a springy platform.

are also instances of such altruism among gorillas.

We must perhaps leave the question open as to whether such events are observed only when an ape population is kept under observation for an extended period, or whether external influences, such as overpopulation pressure in the face of constantly constricted range boundaries, lead to such calamities.

Long-term studies have revealed another deviation from what was at first established rule. In principle, the anthropoids are vegetarians, with an emphasis in the diet on leaves and bark (gorilla) or fruits (orangutan). The chimpanzee and bonobo eat chiefly fruit, but also worms or insects, and usually eggs. Both species will repeatedly catch smaller, generally young mammals (duikers, pigs, monkeys) and devour them. As amoung the baboons, only the males are predators. Regular and seemingly planned hunts have been reported, staged by several animals in cooperation.

A great deal has been written about these hunts, especially with an eye to the supposed prehistory of humans. Of course, one should not forget how insignificant the prey is in the sustenance of the hunters, and that these are after all exceptions to the rule that the anthropoids are vegetarians. They evidently lack any regular taste or knack for killing. Victims are seized somehow, flung about with immoderate expenditure of effort, or beaten until they go down. Such quarry is usually shared among all present (including females, that is) and the flesh is consumed together with leaves "on the side."

Here again, it may be noted that the observations in question all represent recent long-term studies. From earlier years, there are only undocumented and implausible horror stories about "beasts" abducting women, or lurking in trees while companies of natives pass beneath them and then snatching the last one, pulling him up, and afterwards dropping the mangled corpse from the treetops at the feet of the searchers.

Insightful Behavior

The behavior of many animals, especially other than mammals, is wholly or largely composed of ready "programmed" routines, innate to each particular species. These are not learned, but rather make their appearance, like physical changes, at certain stages of development. Such behavior patterns, referred to as "instincts" or "hereditary co-ordinations", are of minor importance in simians, and the anthropoid apes in particular. Outside of social behavior, namely in relations between the individual and the environment, the search for innate behavior patterns has had little success – apart from the movements themselves, which in turn are closely linked to physical conformation. Instead, their behavior appears to be extraordinarily malleable and suited to the particular situation.

This quite general statement – aside from an outward resemblance to human beings – may well account for the fact that systematic studies of anthropoid "intelligence" were undertaken quite early (1912 and 1914).

One of the first in the field was Nadja Kohts in Moscow, who tested the powers of discrimination of her young chimpanzee Joni. She introduced the method of "choice by prototype," in which the subject chooses an object, that is to say, a card of a certain color, or a twelve-sided disc, or the like. The task is to pick out the like object from among a number of similar cards or blocks spread out at hand. After half a dozen repetitions, it had become clear to Joni that he really was to be rewarded – by play – only if he chose the like object, and then he made few mistakes, which simply revealed his limitations of perception. Thus, he had trouble with fine gradations of red and orange; his vision was not as good in the red band. But he was quite able, for example, to distinguish dodecagons from decagons. He succeeded also when the blocks for selection were handed to him in a sack, while Ms. Kohts held the prototype before his eyes. The change from using one sense to using another caused him no difficulty whatever. The key to his success, according to Kohts, was his ability to make "practical generalizations." She did not credit him with abstraction.

The method of "choice by prototype" has been much used since, and the tasks have been made more difficult by increasing the number of objects offered or by complex combinations. However, no more conclusive results has emerged.

The psychologist Wolfgang Köhler replaced the shadowy concept of "intelligence" with capacity for behavior controlled by insight into the existing situation. In 1912, he began his now famous work at the anthropoid ape station on Teneriffe. His experiments are designed so that the ape cannot accomplish his objective directly. In other words, to reach the goal, for example to eat a banana or an orange segment, the ape must make a detour. The "detour" may consist in using a tool. Use of tools has been deemed highly significant.

The experiments and the performances of the chimpanzees are widely known. They built piles

These two photographs illustrate an important step in the evolution of the anthropoid apes and of humans: the ability to travel on two legs and the consequent freeing of the grasping hands. It is astonishing how skillfully the two chimpanzees will pick fruit and hold it with their hands. In order to carry off as much of the coveted food as possible, they will use their mouth to help out.

of boxes to reach something hanging from overhead and developed proficiency at climbing freestanding poles of up to 12 ft (4 m) in length with "ape-like rapidity" before they had time to topple. They used tools, and they even made tools: they broke a bolt from the door when they saw a use for it, gnawed sticks until they could insert the end in a pipe to lengthen it, and even more.

W. Köhler dwells on the manner in which the solutions were arrived at. The chimpanzee Sultan would pause for a time, as after a vain attempt to grab the banana hanging beyond his reach by jumping. His eyes begin to move to and fro between the "goal" and the objects available as "tools" – a rod 28 in. (70 cm) in length lying casually near his cage and a box. Then he stands up, purposefully fetches the rod, drags the box under the goal, climbs up, lifts the rod, and knocks the banana down. The whole thing is done in one motion, without interruption. This comprehension of a connection in a flash has made its way into the literature as an "A-ha!" experience. The prospective actions were evidently anticipated inwardly by the subject.

The chimpanzees were less successful when it came to clearing obstacles from their path. An empty box standing in the way of a goal was pushed aside. When it had been weighted with sand or stone, the subjects made strenuous efforts to push the box aside. But it was only when the box had been tipped over that it occurred to them to remove some of the contents to reduce the weight – a move the animals repeated in subsequent experiments.

When they made mistakes it was often because they failed to consider circumstances not in plain sight. Boxes, for example, were not supported but simply held at a suitable height against the wall (where of course they would not stay); or a rope wound around a horizontal beam was not unwound when a longer end was needed, but was pulled at ineffectually, and the attempt was abandoned unless a loop happened to drop off. On the other hand, they would pull the correct string to which a reward was attached even when several strands intersected.

There were great differences in achievement among the individual chimpanzees worked with by Köhler. Sultan, two or three years old, was the first to build towers out of boxes, and afterwards he lengthened sticks by thrusting two or three tubes one into another; he even gnawed down a board so that it would fit into a tube. He would also fetch objects from other rooms that he might need for solving a problem. Tschego, an adult, hardly ever solved anything difficult. Chica distinguished herself by special acrobatic abilities. It was she who, to the researcher's surprise, devised

Intelligence tests of anthropoid apes have been performed for a long time, and often yield informative results. (Above) This orangutan at the Osnabrück Zoo is reflecting how he might obtain his lofty reward. (Right) A solution is afforded by the boxes of different sizes standing nearby. The ape stacks them one upon another and then climbs cautiously to the pinnacle.

"chimpanzee" solutions that would never have occurred to him: swinging by a rope and letting herself fly many yards through the air to a target, or setting up a ladder with one stringer parallel to the wall. She was also the first to set up a pole vertically and get to the top before it had time to fall over.

These gradations between original "inventors" and "imitators," who take over the discoveries of others by mere observation, and finally "dullards," who cannot either one thing or the other, have since been repeatedly verified.

It is by no means the case that only the lively chimpanzees show insights and are especially human-like in this respect. Some years ago, the zoologist Jürgen Lethmate repeated the experiments of Köhler and others at the Osnabrück Zoo using orangutans. Phlegmatic and unreceptive as they are to attempts at assistance or instruction by the researcher, they performed quite similar feats by dint of thoughtfulness, and one some points did even better. Thus, they were able to assemble as many as five rods (not merely three). Their box structures were often very stable and comprised as many as four stacked units. Orangutans have also proved expert at tying and untying knots. In zoos, they will twist ropes out of wood chip, straw, or waste and tie them up in the cage to swing by.

Some of the Köhler experiments have also been successfully repeated with gorillas. However, gorillas are less easily moved to cooperate in such projects.

Zookeepers have tales to tell of anthropoid escapes from cages. Ordinary locks are no problem at all for apes. Tests of their powers of discrimination and association and of memory were based on this foundation from early times. Thus Jürgen Döhl of Münster presented a chimpanzee, Julia, with whole series of problems for solution. A receptacle containing in orange segment as a tiny reward can be opened only with a key. The key, however, is hidden in another little box, the lid of which is screwed on. The screwdriver is inside of a pipe, whence it must be knocked out with a bar; the bar is in a container the lid of which is fastened with a wire. To cut it, nippers are needed ... Only the tool for opening the first container was acces-

sible. Once Julia was acquainted with the several opening mechanisms – having discovered them herself or been shown how – she could run through the entire chain of operations in any given sequence or spatial arrangement. She would often run at first to the goal box and satisfy herself of the sequence, in which she would then proceed purposefully and rapidly.

Against the supposition that anthropoid apes can act from insight, the same objections were raised very early that are still voiced today: It's a matter of training, or self-training, or imitation of human examples. However, circumstances better known to us today than in the time of Wolfgang Köhler diminish the significance of these objections.

Very few mammals can be enticed into operations such as anthropoid apes perform in tests. Prosimians, for example, show no inclination to carry out complex procedures of the kind described. If one attempts to "teach" solutions of problems to a New World monkey, for example, in some cases numerous repetitions (several hundred) are required. This is in clear antithesis to the rapidity with which anthropoids will make such procedures their own. For example, many opening mechanisms had to be shown to Julia only once before she mastered them. In other cases a few successes or failures experienced by an individual in dealing with an apparatus will suffice for realizing how it works. Obviously such rapid adoption of procedures must be judged otherwise than of course of training that finally "takes" after 237 repetitions.

Spontaneous imitation also occurs, but surely it presupposes that the course of action has been recognized as such and is associated with its result. As Amélie Koehler reports, for example, "Julia snatched up the empty scrub pail" from the feed pantry in the morning, "carried it over to the drain, placed it under the spout, put a rag in it, and turned on the tap."

Again, an orangutan who had been raised in captivity and was to be rehabilitated in the wild comes into his keeper's tent one morning and demands coffee, but she is unwilling to give him any. He then prepares a cup for himself, with hot water, powdered coffee, milk, and sugar, and tries to cool it by stirring.

One may easily imagine that the chimpanzees at the primate center in Covington, Louisiana, got

This orangutan is trying to solve a problem otherwise then intended: he fetches a branch with which to knock down his reward.

certain ideas from observation of workers, keepers, or scientists, but what they did had no exemplar either in the combination of action or in the result. A number of them live in a large outdoor enclosure, bounded by a high fence continued at the top by a sheet-metal barrier, several yards high. An observation tower with large windows, in smooth concrete, seemed beyond their reach. But one day when there were visitors in the observation room, one of the chimpanzees, Rock, picked up a long pole, rested it against the wall, and then knocked at the window. To put a stop to this sort of thing, the wall below the window was secured with electric fencing. Nevertheless, one morning a whole band of chimpanzees climbed up on poles unobserved, broke the pane with stones, and wrought a frightful devastation. Trees standing within the enclosure had indeed been stripped up to a considerable height, but the chimpanzees climbed up all the same and damaged the leafy canopy. Attempts were made to prevent this by wrapping the trunks in electric wire mesh. Once again, the ingenuity of the chimpanzees had been underestimated. They simply bridged the electric fence with branches resting on the planking. When the branches were taken away, they took the planking apart for the construction material they needed. Short pieces of wood were also inserted between the plates of the fence to make a convenient ladder.

It must not be considered essential to insightful behavior that the steps should be devised all at once. Lethmate rightly emphasizes that the inward anticipation of future actions may include learned routines, so to speak, as components. The chimpanzee Sultan on Teneriffe as well as the orangutans Mano and Suma at Osnabrück had first assembled two lengths into one while playing. Then, when one proved too short in an experiment, they availed themselves of their experience and recalled that part of the solution. This type of linking of component actions to meet the given situation is characteristic of anthropoid behavior.

Anthropoids have also been induced to use tools in contrary fashion. The "goal" is inside a drawer so placed outside the cage that its opening faces away from the animal. With a hooked stick, the subject must now push the desired object away first, and then retrieve it – a problem that presents greater difficulties than a roundabout walk or climb. A variant of these experiments is the maze, a system of pathways more or less hidden from view with only one route to the exit and which is covered with a sheet of Plexiglas. An iron ring inside the maze can be moved about from the outside with a magnet. Julia would study the arrangement of pathways (changed from one test to another) at leisure, then pick up the magnet and drag the ring to the exit.

At the Yerkes Laboratory (when it was in Florida), labyrinths were presented totally enclosed, so that the subject, when moving a button to and fro in a

When this orangutan found that he could not reach the goal with the hooked stick held in his hand, he slipped the hollow end over his index finger to lengthen the "fishing pole."

slot, while the labyrinth was in motion behind the panel, had to memorize distances, speeds, and lengths of time.

Duane Rumbaugh drew mazes as lines on a sheet of paper and tested the animals for their ability to recognize minute differences in the patterns – a performance at which the orangutans were especially brilliant.

"Intelligence" or "insight" includes memory, which permits interruption of the chain of mere stimulus and response – perception, processing, controlled action – so that situations and the potentialities of things can be stored. The experiments of J. B. Wolfe at the Yerkes Laboratory, then still in Florida, were addressed to memory as well as to the formation of value concepts.

His six young chimpanzees were familiar with colored tokens as playthings. They were shown how a white token could be dropped in a slot machine to obtain a grape. One of them, Moos, needed to see this done only once in order to imitate it; another had to be shown 237 times. Blue tokens were introduced, worth two grapes. It took 31 to 91 repetitions before the animals positively preferred the blue tokens. They also learned to distinguish tokens for drinking water, and others for "open door" or "play with keeper."

An apparatus was presented that contained a token but could be opened only by lifting a weight. Again, Moos saw the point immediately; the others needed a demonstration.

Now if the machine for changing tokens into grapes was locked up temporarily, at the end of the test period or for a few hours or even next day, individuals would accumulate different numbers of tokens before quitting; Moos averaged 18.5, Bimbo only 4. This "nest egg" was then saved and cashed in later. When the chimpanzees were given a few tokens to begin with, they earned fewer, bringing them to a total of 20–30. But when the token was withheld at the dispenser, they lost interest much sooner.

Clues to intellectual abilities are perhaps also to be found in simian painting. Many anthropoids – capuchins also – like to use pigments, crayons, or ball points and draw lines. If suitable media are offered with paints and brush, they soon discover

Problem box experiments in manipulation and tool behavior. In this serial test, the orangutan is required to open several different boxes with different implements. After passing the test, he can enjoy his reward (right).

the possibilities, and will be kept busy for 15 minutes or half an hour. They do not aimlessly scribble over the sheet; after a phase of concentrated work, they leave the sheed aside or destroy it. It has been possible to smuggle their works into exhibitions of contemporary painting. Mostly, the images are colored stripes in a fan shape, converging towards the originator. The choice of colors reveals a sort of balance. Simple drawings are easier to analyze for components and composition. Apes have a tendency to mark the center or the corners of the sheet. If a shape is placed on the blank sheet beforehand, marks will be added to "counterbalance" it. Some drawings are reminiscent of the patterns (circles, crosses, triangles) drawn by very young human children. A combination of these "patterns" brings the child to the next stage, drawing "faces," of which there have been only occasional signs among apes.

Tool Use by Anthropoid Apes in the Wild

A common feature of all laboratory experiments is that they try to find a typical human faculty, namely the free use and combination of abstractions, in animals. Information about their abilities

Art, or just making a mess? Anyway, anthropoids enjoy working with brush and paints, and they display some sense of form and design.

is sought by presenting them with objects that have no significance for them, and actions are prompted that have no value for the animals. Probably anthropoid priorities concern quite different matters from ours.

This shortcoming does not apply to certain observations in which the animals were able to act freely and uninfluenced; W. Köhler got some of his best insights by observing his apes on their own time, outside of testing. In the meantime, observations have accumulated that give us a good picture of the abilities of anthropoid apes living in the wild, and to a large extent confirm the experimental results.

Gorilla and chimpanzee youngsters in captivity, as well as adult chimpanzees living in the wild, would put on "adornments" in the form of lianas, and orangutans would deck head and shoulders with leafy branches in rainy weather or intense sunshine.

Kortlandt, and others since, have presented wild chimpanzees and bonobos with foreign objects, such as luminously colored surfaces, stuffed toy animals, a lighted lamp, tools, strange fruits, pictures, snails, spiders, snakes, chameleons, chicks, monkeys, or placed dead creatures or pictures of conspecifics in the cage. In general, not much interest was taken; the adults often seemed to pay no attention at all. Juveniles and infants would touch the objects, look at them, study their properties, and then tussle over them, that is, use them as objects of common play.

Chimpanzees living in the wild would obtain water from knot-holes that were no directly accessible by chewing leaves to a fibrous, sponge-like mass and stuffing this into the recess. The "sponge" was pulled out, sucked, stuffed back into the aperture, and so on.

To fish out termites from their subterranean galleries, chimpanzees on the Gombe River would insert twigs or straws, and lick off the insects clinging thereto. W. Köhler observed similar actions by his chimpanzees on Teneriffe.

Tool use definitely involves planning. The Gombe chimpanzees would go hundreds of yards out of their way, gathering a handful of straws of various lengths and thicknesses, and then use the best-fitting ones on a termite hill; or they would strip twigs for the same purpose. Orangutans have been observed to do the same.

Young orangutans who had group up in captivity and were to be released in the wild would beat or pierce coconuts with sticks to open them. They would use paper or leaves to cover the spines on the surface of durians before opening them. They would also contrive to use boats, rafts, or drift-

▷ Astonishment at a great idea: In rainy weather or glaring sunshine, orangutans place leafy boughs on their head and shoulders.

wood to return to a feeding place. They did not paddle, but would pole with boards or sticks, having first undone the knot in the mooring line. A young male actually held onto the line with his hands or feet while gathering food, so that the boat would not drift away. Wild mother orangutans soiled with feces of the young in the sleeping nest would wipe themselves clean with leaves. A wild orangutan male broke off a dead branch for scratching his back.

In the Tai forest on the Ivory Coast, Hedwige and Christophe Boesch from Geneva have been observing for some years how chimpanzees open several species of hard-shelled nuts (especially cola nuts) by means of a tool. The males find a root, the females usually a forked stick. Then, sometimes by a long route, they bring as many nuts as they can carry in one load, and with them a heavy stone or club of wood for a "hammer." The nuts are set up and cracked one after the other with several blows. In a chimpanzee community not far away in Bossou, Guinea, the apes crack oil palm kernels. However, they carry neither the kernels nor the (stone) tools to the point of use. Oil palms occur in the Tai forest also, but the fruit is not eaten there. In still other places, cola nuts are opened by striking them against a tree trunk with the hand. According to close calculations, the energy gained by consuming cola nuts is 16 times the energy that goes into all the labor required to collect and open the nuts, procure the hammer, and find a suitable anvil. This shows that the Tai chimpanzees, at least, have good cause for their use of tools.

We have an instance here – perhaps much as in "hunting" – where even such vital matters as eating behavior are evidently not innate and hence typical of a species, but depend on "traditions," which, once "invented," are learned within the ape community by one individual after another. Among the Japanese red-faced macaques, it has been seen how new behavior patterns spread. Another example of such "traditions" is the differential tendency to use clubs as "weapons" or as threat "tools." The behavior researcher Adriaan Kortlandt of Amsterdam concealed a stuffed

Chimpanzees are the only anthropoid apes that regularly use tools, and even make tools for specific purposes in the wild. (Left) This male chimpanzee is brandishing a stick at his mirror image. (Right) A female chimpanzee "fishing" for termites with a straw. Her little sister watches attentively, learning how to obtain this delicacy by observation.

leopard in places where wild chimpanzees were living so that the dummy could be rolled out into view suddenly. An incredible uproar always ensued. Some chimpanzees seized clubs and sticks, swung them about in a display dance, shook them at the leopard, or in exceptional cases even struck it vigorously. This use of sticks as weapons has been found only among chimpanzees living outside of the dense forests.

Among apes in the wild, it is of course as difficult to get hard information about the limits of insight as to ascertain whether modes of behavior that seem "impressive," such as use of tools or weapons, have been learned through occasional contact with human beings. According to what we know today, however, the likely answer is learning within the natural social context. But suppose that

some tool use had in fact been borrowed from human beings at work, as has been suggested, would it be less remarkable on that account?

By insightful behavior, then, we mean use of objects, neutral in themselves, as tools; willingness to reach a goal by roundabout routes – which may themselves consist in tool use or pursuit of several intermediate goals; and the making of tools. The effect is of course especially convincing if the manner in which a problem is solved crops up for the very first time in a particular situation, or in other words is an "invention."

Tool use is clearly definable, and so has tended to be recognized as evidence of "higher" faculties. The behavior trait of "tool use" has also been of service in discussions of the origin of humans, because prehistoric stone tools are the most common evidence of the presence of prehistoric humans. They are more often preserved than skeletal remains. We have no difficulty in imagining that before tools were made out of hard stone, the occasional use and eventually the production of tools and weapons out of materials more easily shaped, such as wood, were practiced. Unfortunately the intelligence of anthropoid apes as gauged by toolmaking cannot be directly compared with the phases of human prehistory documented by fossil. The "tools" made by apes are all of compliant material. It would seem that making a stone tool calls for far more planning of action than has ever been demonstrated by an anthropoid ape.

Utterances of Sound and "Speech"

Neither the tool use by anthropoid apes living in the wild nor their ability to learn certain procedures from having seen only a single instance necessarily presupposes insight into cause and effect relationships. Conversely, however, it must be assumed that insightful action presupposes notions of objects and their properties. Manifestly, their ability to conceptualize and to abstract is limited in some way. For human beings, it seems a natural step to judge the presence of higher mental faculties by the occurrence of messages about complex states of affairs and connections between them. Human beings communicate such messages by means of language.

For the French philosopher René Descartes (1596–1650), language was a sure sign of thinking. He accordingly separated human language from the utterances of all other living creatures. Descartes ascribes this human uniqueness to a soul standing above the brain, characterized by thought, and in his opinion lacking in the beasts. This notion of the fundamental difference be-

This female orangutan, experienced in taking tests, has learned to make a tool. Her intense expression suggests that she does not find it a very easy task to assemble the two parts.

tween humans and all other creatures has stubbornly persisted to this day, and still dominates opinion.

But on that course one runs the risk, like the Nobel laureate John C. Eccles, of arriving at the not very novel conclusion that anthropoid apes do not speak and hence are not human. Such a conclusion is not fruitful, either in insights into the nature and abilities of the anthropoids or in understanding human origins and development. Improvements are not made by replacing the word "soul" with the word "consciousness." This term is not a good substitute for something we don't know about or don't like to mention because it's not "in." "Consciousness" becomes really intelligible only in contradistinction to "unconsciousness," say in sleep or under anesthesia, both being conditions observable in all vertebrates and induced in them in the same way as in humans.

In view of the similarity illustrated above between humans and anthropoid apes, the question is whether the gap between absence of language and highly evolved powers of thought is one of kind or only of degree.

So far as vocal utterances are concerned, in the particular case of the anthropoid apes there is not much activity. In anthropoid nurseries, the first thing one notices is the silence. In the lively and persistent play of animal children, sounds are of subordinate importance. Their calls, in the first place, are hardly separable from the mood, or emotional attitude, of the "transmitter." In the second place, the phonetic patterns, as in all Old World simians, are variable and not very firmly tied to particular circumstances.

The strong tendency to imitate is apparently prompted only by things that are seen. Things that are heard interest the apes only in terms of their mood content. One of course wonders whether animals with no inclination to imitate sounds and little inclination to utter any can be expected to speak at all.

Two American psychologist couples, in the 1930s and in the 1950s, raised chimpanzee infants in their homes, as or together with children of their own, and tried among other things to get them to say English words. Both attempts, in the face of the fact that the apes were not ready to utter any sounds, yielded scant success. Elsewhere, a young orangutan was rewarded for each sound he produced outside of his usual mood repertory. After months, he connected four distinct sounds,

The educated chimpanzee Nim gets language lessons. The instructor touches the puppet's ear with his left hand, whereat Nim grasps his own ear between thumb and forefinger of his right hand, mirror-image fashion.

70 percent of the time, with beverages, solid food, being picked up, and brushings. This is pretty much the same as what had been achieved with the chimpanzees.

Now the fact is that the larynx and upper air passages of the anthropoid apes are not especially well suited, to all appearances, for the production of vowels. These happen to be indispensable for the multitude of verbal configurations.

If language is viewed more generally as any means of sending messages from a transmitter to a receiver with the aid of some medium, then many animals must be granted the power of speech, including the anthropoid apes, which do after all exchange messages about their moods. Of course, that still leaves two questions to be discussed: (1) Whether the transmitter has any intention of informing the conspecifics, or whether the expressions of mood are simply understood by the others and provoke suitable responses; and (2) what purposes these communications serve. Human language serves primarily as a medium for thoughts and for articulating intelligible associations.

A cogent reason for the inability of the apes to produce sounds of arbitrary form and use language as an aid to thinking has seemed to be that the animals simply do not need a highly developed system of communication of vehicle for objective information.

The American psychologist E.W. Menzel has investigated the performance of the chimpanzees' "system of emotional communication" in a test situation where human beings would doubtless have adopted communication by means of language. Early in the day, before the time at which the chimpanzees were usually released into their large outdoor enclosure, he would take individuals of the group out of the sleeping quarters and, once outside, show them things of interest to a chimpanzee: fruits singly or in quantity, or scorpions, or whole boxfuls of spiders. These things were then concealed nearby. The animals were put back, and released hours later with the rest of the crowd. Not long after, numerous chimpanzees were always to be met with near the place where attractive things had been hidden – and more of them, the larger the quantities of tasty fruits shown to the individual subjects before. The places where less agreeable things lay concealed were avoided. The more offensive the concealed item would be to a chimpanzee the greater the radius within which there were no chimpanzees. Evidently the directness and haste with which the pre-indoctrinated individuals directed themselves to certain sites, or the shyness, if not apprehension, with which they avoided others, were accepted by the rest as a signal and message concerning the nature and importance of what was to be expected out there. They certainly did not require any briefings about the objects themselves.

A number of psychologists, on the basis of such intelligence tests as described above, have hypothesized that the incapacity of anthropoid apes to learn language lies not in their "limitations", but merely in their inability to produce fine gradations of sounds and noises in a controlled manner. Experiments have therefore been undertaken to find out whether and to what extent anthropoid apes can deal with a vehicle system for objective content if they are armed with such a medium.

In 1966, Beatrice and Alan Gardner, of Reno, Nevada, began the endeavor to communicate with the female chimpanzee Washoe, then about one year of age, by means of the American Sign Language for deaf mutes (ASL). The chimpanzee child Washoe was a laboratory inmate, but kept in the utmost freedom and in loving association with quite a number of human companions. All concerned had first to learn ASL and then use it actively in their dealings with Washoe. Washoe was exposed to numerous events and objects, which on each occasion were named by the appropriate sign. This was done in the hope that she would learn little by little to associate the signs with the corresponding objects or terms, and later on to make the signs herself.

With good reason, the Gardners believe that Washoe learned to understand a large number of signs at an early stage. The first that she herself made, almost from the first day, was a gesture of request, widespread also among apes living in the wild and equivalent to the sign for "please", or "give me."

The first strictly objective name, for "tooth brush," was observed in the tenth month, and in the fifteenth, "flower" was added. Washoe at first used the "flower" sign for all other strong odors as well. Later she was taught the "smell" sign. After she knew this, she distinguished of her own accord between merely fragrant objects – for example a tobacco tin – and actual flowers.

In a state of relaxation, she would recite names of objects, things, or concepts without apparent stimulus. Of course, it is typical of human children also at a certain age to name all possible things, accompany all actions with talk, and thus practice speech.

Washoe's tendency to prefer the expression of wishes rather than of objective situations, and to express wishes often with much emphasis, likewise resembles the behavior of small children. It has been known for some time that chimpanzees at 12 to 16 months have about as much understanding as a human child of about the same age. Children from one-and-a-half to four years old, however, can understand much more than Washoe learned.

Deaf-mute visitors were able to understand about 70 percent of Washoe's sign language immediately, and about 95 percent on the second visit. The Gardners explained that the chimpanzee spoke with a "moderately strong accent."

Washoe's vocabulary grew with increasing speed. Whereas the first four signs took seven months, nine new ones came in the next seven months, and 21 more in the ensuing seven months. Washoe eventually had command of about 130 different signs. Questions would elicit simple conversations, such as: "Who tickles?" "You." "Who is tickled?" "Me."

The teacher makes the "drink" sign with her fingers for her pupil Nim, and he tries to imitate it.

Other young chimpanzees have also learned sign language. As one would certainly expect, there are wide differences in the rapidity with which individuals will understand and reproduce new signs. The time required to learn a new sign varies between ten minutes and six hours – unless, as in Washoe's case, one waits for the signs to be reproduced spontaneously.

Even simple generic terms are formed, such as "fruit", or "beverage." For a watermelon, the chimpanzee Lucy invented the term "taste-drink" or "drink-fruit." Lemons, oranges, and pineapples are "smell-fruit." Unfamiliar objects are described; a swan on the water as "water-bird," hard-shelled nuts as "stone-berry".

Spoken language is understood by the animals to some extent, and they are also able to translate words that they know into sign language.

Among themselves, the chimpanzees use the sign language to only a very limited extent. Some of them have addressed conspecifics by ASL signing again and again, and in a large number of cases signs that one of the individuals already knew were adopted and used correctly by others. However, no extensive exchange of signs ensued; in particular, no objective information was passed on among the apes, not even concerning things they had mastered completely in exchanges with human beings.

"Speech" or "Writing"?

Likewise, beginning in 1966, David Premack, a psychologist in Santa Barbara, California, provided Sarah, a chimpanzee then aged six, with a kind of "writing." Plastic chips of different shapes and colors were to count as words. By their metal backs, they could readily be placed on a magnetic surface. A bit of fruit for Sarah to eat was placed next to the chip representing that fruit. In the next step, she was required first to mount the chip on the board before receiving the designated fruit.

Then other symbols for other fruits were introduced.

To these symbols, new ones were added as names for the keepers who worked with her; now, reading down, she was required to set up "Mary banana" when Mary was sitting opposite her and Sarah wanted some banana. By placing a symbol for "give" between the keeper's name and the fruit name, a simple sentence was formed, to be extended by specifying the recipient: "Mary give apple Sarah." Here difficulties arose, for instead of Sarah, Jim or some other member of the staff might be put in as recipient. Sara was unwilling either to write anything like this herself or to comply with it when – in writing – it was suggested. It was only with the help of an extra piece of chocolate that the apple could be duly routed to Jim. It was also easy enough to replace "give" with "cut," "place in ...," or the like.

More abstractly, "like" was inserted between two objects, for example cups, or "different" between a cup and a spoon. Sarah mastered this very quickly. Now if only one object is presented, and a question mark instead of the second, a question can be put: What is like (or different from) spoon? Or Sarah could be given a choice: Are two objects like or different? Understanding of the names of properties is easy to test, with such instructions as "Sarah place brown in red dish," for which purpose there must of course be several colors and several dishes at hand. Chains of such sentences presented little difficulty: "Sarah place banana in pail, apple in dish" would be performed 75–80 percent correctly.

Generic concepts worked too. She distinguished shapes and colors with assurance. "Yellow is not shape." Sarah knows the difference between "all," "several" and "none" out of a supply of five biscuits. So she might also be told, "Sarah place several (no, all) biscuits in dish." When conditions were prescribed, "If red is on green, (then) Sarah take apple," or "If green is on red (then) Sarah take banana," more learning time was needed than in other tests. Eventually, objects were successfully described by four characteristics, each offering two choices. These descriptions would be given for the symbol "apple," not only for the real apple. When completely separated from her teacher or companion, by the way, Sarah would have nothing at all to do with her experimental tasks.

Since only a limited number of symbols out of her "vocabulary" of about 130 were at hand on any one occasion, the test results were nice and clear, but Sarah could not perform any feats but those the investigator expected of her. This chimpanzee was allowed no opportunity to help in solving the question whether chimpanzees can understand speech or writing.

Since 1970, the animal psychologist Duane Rumbaugh of Atlanta, Georgia, has given the chimpanzee Lana, at first two years old, access to a computer console. On large keys, shapes and colors from different signs each designating a word of the artificial language "Yerkish." (It is so called after the Yerkes Primate Research Center.) The signs on the imprinted keys appear on a luminescent strip, so that both Lana herself and people outside her room can read what she has keyboarded. Outside of her room, there is another console at which the investigator or attendant can write something, which is then displayed to Lana on the tape.

First, simple sentences were regularly addressed to Lana and their correct application monitored by the computer. Rewards were Being Picked Up,

GREAT APES

Great anthropoid apes (Pongidae)

Nomenclature English common name Scientific name French German	Approximate Size Body length Tail length Weight	Distinguishing Features	Reproduction Gestation period Young per birth Weight at birth
Orangutan *Pongo pygmaeus*, 2 subspecies Orangoutan Orangutan	males 3.2 ft; 97 cm females 2.6; 78 0 m. 176–198 lb; 80–90 kg f. 88–110; 40–50	Long reddish coat; juveniles bright orange, changes to brwon in adults; Borneo subspecies darker; short legs; adult males with cheek ridges	260–270 days 1, twins very rare About 2.6 lb; 1.5 kg
Gorilla *Gorilla gorilla*, 2 or 3 subspecies Gorille Gorilla	4.6–6.2 ft; 140–185 cm 0 m. 308–605 lb; 140–275 kg f. 132–220; 60–100	Massive build; black, often sparse, coat; old males have silver-gray back ("silverbacks") and nape crest; knuckled gait; subspecies distinguishable by face and nose region	251–289 days 1, twins very rare About 4.4 lb; 2 kg
Chimpanzee *Pan troglodytes*, 2 or 3 subspecies Chimpanzé Schimpanse	2.3–5.6 ft; 70–170 cm (subspecies widely diverse) 0 m. 94.6–132 lb; 43–60 kg f. 73–103; 33–47	Black, sparingly hairy body; subspecies show diverse face coloring; genital swelling in female; knuckled gait, also biped walk; highly mobile	225 days 1, twins very rare About 4.2 lb; 1.9 kg
Pygmy chimpanzee; bonobo *Pan paniscus* Chimpanzé nain Zwergschimpanse; Bonobo	2.3–3 ft; 70–90 cm 0 m. 99 lb; 45 kg f. 72.5; 33	Much like chimpanzee, but more slender, with relatively longer limbs, rounder skull, and less protruding jaws; face usually black; knuckled gait; very lively; large, pink genital swelling	8 months 1 Not known

Play, Looking Out of the Window, as well as projected slides or short films. Lana was expected to make all her day-to-day needs known by request such as "Please, machine, give Lana kibble" or "Please, machine, give Lana drink." More elaborate requests were also honored, such as "Please, machine, show Lana film" or "Please, Tim (her favorite staff member), take Lana along outside."

This procedure safely rules out influence exerted by the investigator. Furthermore, all utterances of the chimpanzee and the staff were recorded unaltered.

Lana also learned the use of Yes and No. "No, no," she wrote for the first time when a staff member drank a cola before her eyes without giving her one.

She began of her own accord to carry on conversations. One of the first occurred when she had received something to drink but wanted something else: "Lana drink this outside." Attendant: "Drink what?" Lana: "Lana drink cola outside."

Later there were longer conversations. Tim showed her an orange and asked, "What color is that?" Lana's answer: "Color is orange." Tim: "Right." Now Lana wanted the fruit, but as yet she had no name for it. She tried: "Please, Tim, give cup that's red." Tim then gave her a red cup – which of course was not at all what she wanted just then. After some random attempts, she asked: "Please, Tim, give ... that's orange." Tim pretended not to understand, and asked again, "What that's orange?" Lana: "Please, Tim, give apple that's orange." She evidently knew that this description was pretty nearly right, for she sprang to the door before she got the confirmatory "Yes."

Here, then, we have fairly coherent and objective handling of language. But this works only in exchange with people. At the same Yerkes Laboratory, another experiment concerned the passing

Chimpanzees Sherman and Austin at the "console" of their computer, with which one can ask the other for treats, or for tools to reach them with.

COMPARISON OF SPECIES

Life Cycle Weaning Sexual maturity Life span	Food	Enemies	Habit and Habitat	Occurrence
Age 2–3 years Age 7–10 years 35 years; in captivity, 50 years	Predominantly fruits; leaves, birds' eggs, bark	Tigers, red dog (Sumatra only), clouded leopard, humans	Diurnally active tree dweller of lowland forests (rain, mangrove, marsh and mountainside); rarely at elevations over 3200 ft (1000 m); sleeping nests; loose family groups, but rather solitary; foraging area 0.7 to over 3.8 mi² (2 to over 10 km²; males), 0.5–1.9 mi² (1.5–5 km²; females)	Entire population seriously endangered by destruction of habitat
Age 2–4 years About 7 years 25–40 years; in captivity, over 47 years	Strictly vegetable; roots, sprouts, leaves, bark, pith, tubers	Leopard, humans	Acitve by day; 90% terrestrial; in rain and mist forest; in bamboo forest to elevations of 12,800 ft (4000 m); family groups each led by a "silverback" male; foraging area 3.8–9.5 mi² (10–25 km²)	Endangered by destruction of habitat and by hunting; mountain gorilla population now only 200–300 individuals
Age 2–4 years Males 7–8 years, females 6–10 years 30–40 years; in captivity, over 50 years	Fruits, leaves, nuts, bark, seeds; termites, ants; occasionally, smaller monkeys, pigs and duikers	Leopard, humans	Active by day; arboreal and terrestrial, in tropical rain forests and treed savanna at elevations up to 9600 ft (3000 m); sleeping nests; unstable groups which may be composed of both sexes; foraging area about 5.1–19 mi² (15–50 km²) on the Gombe, less elsewhere	Endangered by destruction of habitat and hunting
Not known Age 8–13 years Over 20 years	Predominantly fruits; also leaves, buds, flowers, insects, earthworms, sometimes small mammals	Possibly leopard; humans	More arboreal than chimpanzee; in primary and secondary forests of lowlands; also in marsh forest; communities of 50–120 individuals, in loose groupings of 2–50	Threatened with extinction; very scattered occurrence; much endangered by destruction of habitat and hunting

on of objective information between apes. A reward was placed in closed receptacles. The tools for opening the receptacles were accessible only at first. Once the use of the particular "opener" was known, the chimpanzees were next merely shown the tools, writing: "This is a ... (screwdriver etc.)." The animals would then ask for it right away: "Give (me) screwdriver." This step was also reversed. The apes were required to pick out and hand over the requested tool out of a considerable collection of tools. Now the receptacle with the reward was placed in the room of one of two equally trained chimpanzees, and the tools in the neighboring room of the other. After only five or six not quite successful attempts on the first day, the first chimpanzee asked for and received the right tool from the others. At first the investigator rewarded the bearer of the tools, but eventually the ape who had taken the reward out of the receptable did that himself.

If the impression has been given that studies like this produce such remarkable results only when the subjects are chimpanzees, that is chiefly because they are easier to keep and will enter into experiments of all sorts with considerable enthusiasm. Bonobos have also participated in similar experiments successfully. Gorillas seem to be especially good in the subject of language.

The female gorilla Koko, born in zoo and since grown up, reportedly learned over 600 ASL signs from her teacher Francine Patterson in California, 375 of which are regularly used. The first ("drink" and "more") Koko acquired in the very first week of instruction. In the first year and a half, she learned about one new sign a month; after that she learned faster. Her intelligence quotient has been measured several times and is around 85.

Koko is credited with long conversations. On the day after she had bitten her teacher Penny, Penny asked her, "What did you do to Penny?" Koko: "Bite." The day before she had claimed it was only "scratch." Penny: "You admit it?" Koko: "Excuse bite, scratch." After a pause: "Bite wrong." Penny: "Why bite?" Koko: "Because crazy." Penny: "Why crazy?" Koko: "Not know." On many occasions, unpleasant matters are postponed. Thus when Penny says to Koko, "Tell me what you did", Koko: "Later, me drink." Or she is evasive, when asked for example about a broken-down sink: "Kate was her. Bad!" sometimes she is insulting. Concerning a poster showing herself, her keeper Cathy (a deaf mute) asks her: "What is that?" Koko: "Gorilla." Cathy wants to know more: "What gorilla?" Koko: "Crap" – an expletive of hers. Cathy: "You crap?" Koko: "You!" Cathy: "Not me, you crap!" Koko: "Me gorilla." Cathy: "Who crap?" Koko: "You crazy." Koko can explain things. Penny asks her, "What is a stove?" Koko points to the range. Penny: "What is it for?" Koko: "Cook." Or in answer to the question, "What is an orange?" Koko: "Eat, drink."

Koko is said to understand some hundred spoken words. The signals she transmits, however, are always signs only, never vocal. To a limited extent, she also exchanges signs with her gorilla friend Michael.

These experiment results make clear firstly that speech in fact plays a very substantial role in what is broadly called thinking. In the second place, all attempts to describe the peculiarities of humans in contradistinction to those of the anthropoid ape, that is to find a qualitative distinction between the two, become increasingly blurred. It is quite clear that the scope, the quantity of insight-directed behavior and use of abstract ideas by apes fall far short of what human beings achieve even in early childhood.

In closing, it should be observed that anthropoid apes know perfectly well who they are. Chimpanzees recognize their image in a mirror and make use of it in grooming. If their image is altered by cutting the hair or staining the ears under anesthesia, the result is considerable confusion, lasting for hours. In this respect, by the way, chimpanzees are quite different from macaques, who just don't care. Animals using deaf-mute language have no trouble using the word "me" – despite the substantial theoretical significance with which the ego has been freighted. The gorilla Koko, in answer to the question whether she was animal or human: "Fine animal gorilla."

The teacher shows Nim his shirt. He evidently gets the point, and tells her in sign language: "Me - shirt."

Orangutans

by Kathy MacKinnon

Myth and Reality

Literally translated, the Malay name "orangutan" means "forest man." This is an apt designation, for the largest anthropoid ape of Asia is astonishingly humanlike in appearance and countenance. So it is no wonder that its origin and its solitary habit are the subject of many native legends.

One tells how an orangutan stole a human bride and dragged her into the woods. She bore him a son who was half human and half ape. One day the young woman escaped from the lofty nest in which she was being held captive. With the child on her arm, she fled to the river bank, where hunters had moored her boat. But the ape was at her heels, and she dropped the baby to delay him. While the ape paused to pick up his child, the woman escaped. When the ape saw that his bride was gone, he flew into such a rage that he tore the child in two, threw the human half after the receding boat, and the ape half back into the jungle. There the orangutan has been roaming solitary ever since.

Nowadays, orangutans are found only in the forests of Borneo and northern Sumatra. The two island races, which are distinct in fur color and facial features, are considered separate subspecies.

Orangutans are found in many different forest habitats, such as mangrove and coastal swamp forests, lowland fruit trees, woods on a limestone substrate, and mountain forests. They have been observed at an elevation of more than 6400 ft (2000 m) on Mount Kinabalu in Sabah, but generally they are seldom to be met with above 3200 ft (1000 m). The densest populations are found in the luxuriant mixed rainforest of the lowlands, especially on the flood plains of the larger rivers, but they are nowhere abundant.

The earliest account of an orangutan is ascribed to the physician Bontius in the seventeenth century, but it is not quite certain whether his observations refer to an anthropoid ape or to a half-witted native girl. In 1712, the Englishman Captain Daniel Beekman visited South Borneo, and he described orangutans in his memoirs. In 1776, the first orangutans reached Europe. They had been brought to Amsterdam for the natural science collection of Prince William V, and soon more followed.

During the nineteenth century, orangutans were hunted and collected by such naturalists as Alfred Russel Wallace and Hornaday, who shot many specimens for their museums. Still, although orangutans were esteemed as trophies and as household animals, and captured in large number, their habits and their behavior remained a secret. As late as the early 1960s, when Tom Harrisson,

Portrait of an orangutan. The facial expression, astonishingly humanlike despite the bulging mouth area, of this forest dweller moved the Malay natives to call him the "man of the forest."

curator of the Sarawak Museum, first pointed out the sorry predicament of the orangutans, next to nothing was known about their private life. It is only since the great field studies initiated on Borneo and Sumatra by such inspired biologists as John MacKinnon, Peter Rodman, David Horr, and Herman Rijksen that we have learned a good deal more about the environmental conditions and behavior of the orangutan. The longest study has been carried out by Birute Galdikas since 1971 in the swamp forests of southern Kalimantan on Indonesia. At Camp Leakey, named after Birutes' patron Louis Leakey, scientists are now studying the day-to-day activities of juveniles who are grandchildren of the animals observed at the beginning of the study.

Results of Studies in the Wild

To study orangutans is not an easy thing to do. One of the first to succeed was my husband John MacKinnon. John's procedure was to search the forest until he met an orangutan. Then he stayed with this animal wherever it went, even sleeping in the wild when the ape slept and following it the next day. This method was arduous and self-sacrificing, but soon he began to make some sense of the daily campaigns of the animals. Unlike the other great anthropoid apes – the chimpanzee and gorilla – orangutans are by nature solitary. They generally travel about alone or in small groups, a mother with her young offspring or a cohabiting pair. Because of this unsocial and quiet behavior, orangutans are far more difficult to find than a troupe of noisy macaques squabbling over food and storming through the treetops. Besides, the red coat of the orangutan blends completely with the merging green and brown shades of the forest. The coloring of the sparse fur ranges from brilliant orange in juveniles to dark chestnut or chocolate in adults, the animals of Borneo being consistently darker in color than those on Sumatra. The hair is never very thick; length varies individually, and the hair of some males looks like an imposing tassled curtain. The skin of the face is light in young individuals, but darkens with age and ultimately turns black, as in the chimpanzee. Juveniles are rosy about the mouth and have light-colored "spectacles" around their eyes.

Orangutan males in the wild weigh about 176 lb (80 kg), being twice the size of the adult females; fattened zoo animals may attain double the weight of their wild counterparts. With advancing age, the males acquire broad jowls and long beards, presenting quite an awesome countenance in combination with the long hair of the head. Orangutans have a throat sac, which is most prominent in the adult males. This lends resonance to the calls of the animal, especially the loud cry.

Orangutans are the largest true arboreal animals. Their long and powerful arms, and their hands and feet, with great toes that are mobile like thumbs, are excellent grasping tools, enabling them to hold onto branches and swing through the treetops. A young orangutan is able to hang from a limb by the feet, or even only one foot, so that its hands are free to seize any fruit within reach.

Orangutans live predominantly in the trees. Most of them travel effortlessly through the middle and lower storeys of the treetops, swinging from branch to branch, and bending branches to bridge gaps between neighboring trees. Such large animals need much energy for locomotion through

This young orangutan judiciously distributes its body weight with all its limbs.

This baby orangutan peers curiously over its mother's shoulder while clinging tight to her fur.

the treetops, and orangutans traverse only short distances each day. The large heavy males cannot always find a safe route over branches that will support their weight. Often they are obliged to come down to the ground in order to move on. They can stand and walk on two legs, but they rarely do it; on the ground, they generally travel on all fours. In a secure environment, such as a rehabilitation station, females and young will also venture down to the ground. But in the wild this seldom happens; locomotion on the ground requires less energy, but is more dangerous because of terrestrial enemies.

The orangutan's chief enemies include several large predators, for example, the lurking tiger and the red dogs (to be found on Sumatra but not on Borneo). Clouded leopards are agile climbers, and pursue young orangutans. Large pythons may also catch young orangutans if they venture out on the forest floor, where they are also liable to attack by blood-sucking leeches.

Nest Building

The threat of predators is probably the reason why all orangutans sleep in the trees at night. At dusk, the orangutan will climb to a suitable forked branch, and use its powerful arms to pull in other branches and build a springy nest. Such nests are often build in places affording a good look-out over the surrounding forest. Nestbuilding behavior seems to be innate rather than acquired. The orangutan Jacob who escaped from the London Zoo built himself a perfect sleeping nest in a tree in Regent's Park, although he had never witnessed nest building by conspecifics and had had no previous experience in dealing with live branches. The nest building of orangutans is similar to that of the chimpanzee, but differs in one important point: orangutans often build a roof out of branches, presumably to shelter themselves from the heavy downpours in Southeast Asian rainforests.

Occasionally, wild orangutans will return to their sleeping nests to rest up after breakfast. However, they seldom use a nest for more than one night, and prefer to make a fresh bed each evening. The apes will seldom nest in a fruit tree where they have been eating; more often they choose a tree at some distance, rendering it more difficult for predators attracted by the fruit and reducing the

The orangutan's appearance changes considerably from one time of life to another. The babies (top) have big, dark "baby" eyes and a brilliant orang-red fur. More brownish shades of the coat and especially the protruding jowls characterize the grown male (bottom).

probability of finding animals feeding. Sometimes orangutans will also build daytime nests, where they will nap not far from a useful fruit tree after eating or wait out a heavy shower.

Feeding

The orangutan is a late riser, becoming active at about 7 a.m. – a good hour after the more lively gibbons have hastily adjourned for their morning rounds. Each orangutan has a large home territory, sometimes covering several square miles. However, it will traverse only a small part of it each day, wandering no farther than a couple of hundred yards. It will spend more than half of its day feeding, often in a single fruit tree, where occasionally it will rest between meals in a nearby nest.

The orangutan's diet is varied, but about 60 percent of the food consists of fruits. The teeth and jaws are comparatively massive, for opening spiny husks and cracking hard nuts, and for grinding coarse-fibered plant parts or bark. The jagged teeth scrape the clinging pulp off the kernel, which is then spit out.

With their long, strong arms, the orangutans can reach fruits far distant from a support that is sturdy enough to support so heavy an animal. Orangutans will reach far out and bend branches towards themselves so that they can pick fruit directly with their lips; sometimes they will hang completely upside down to get at morsels far below their position.

When fruits are scarce, orangutans forage farther each day in search of fruit trees, and must resort increasingly to other sources of nourishment. In every season, they supplement their diet with young leaves and sprouts, ants and termites, tree bark and woody lianas; occasionally they will steal eggs and young from the nests of birds or squirrels. They chew the nutritious cambium off tree bark and spit out the fibrous lumps, like a person chewing tobacco. They will also come down to the ground to eat of mineral-rich earth dug up by elephants or other forest animals in search of trace elements. The orangutan gets much of its water supply by eating plants or by licking rainwater from wet leaves or its own fur; however, it will also drink water collected in the hollows of trees. It will dip its hand into the water and then lick the water from its hairy wrist.

The trees of the rainforest that provide most of the orangutan's food supply are far apart, and do not all bear fruit at the same time, but the orangutans have an uncanny knack of locating these scattered sources of food. When John MacKennon decided to survey all the durian trees growing in his research area – their fruit, the zibet fruit, is

Orangutan (*Pongo pygmaeus*)

The orangutan feels safe and sheltered in its sleeping nest, which is built in a different place each day. Here the animal is fairly well protected from predators and also has a good view of its entire surroundings.

▷ Adolescents continue to maintain close contact with their mothers.

the orangutans' favorite – he discovered that one of his subjects used the ery same route, the shortest path, to inspect all fruit trees in the vicinity. This shows that they are thoroughly familiar with the fruit trees in their area and that they check the trees regularly, like any good fruit farmer.

The home territories of orangutans overlap, and sometimes several animals will gather at a tree that bears a great deal of fruit. Adult orangutans seem to avoid these encounters. This is in sharp contrast to the excitement with which the more sociable chimpanzees meet. But when two juveniles arrive at the same time with their mothers, they will seize the opportunity to chase around together until the mothers have had enough and leave the tree, moving off separately. The young and the half-grown are the most sociable of orangutans, and when two adolescents meet in a fruit tree, they will often join company happily for several days or even weeks.

In the rainforest, orang-utans must compete with many other species – gibbons, siamangs, monkeys, fruit-eating birds and squirrels – for the fruit supply. The orangutan has two advantages; it can deal with hard-shelled or prickly fruits that other fruit eaters cannot open, and it is able to eat many fruits unripe, despite the unpallatable substances rendering them repellant to animals until the seed is mature.

It is instructive to consider the eating habits of the orangutan and compare them with those of their near relatives and their neighbors. On North Sumatra, besides the big orangutan, there are the medium-sized siamang and the smaller Lar gibbon. At first glance, they all seem to be doing the same thing; all are fruit eaters, using the same forest areas. There are some differences, however, in their feeding habits.

The orangutan has a larger home territory than the other two, but because of its great mass and its

Orangutans live almost exclusively on a vegetable diet, and the varied fruits of primeval forest trees provide their principal diet. These intelligent animals know quite well which trees in their territory are bearing fruit at a given time. Occasionally, however, they will also eat leaves, bark, or birds' eggs.

adaptation to arboreal life, it can search for food over only a short distance each day. However, it is able to utilize a good crop by eating large quantities and storing the excess as fat. Gibbons and siamangs live in smaller districts, which they defend by singing duets and even by hand-to-hand fighting. If a group of siamangs comes upon a feeding orangutan in a fruit tree, they will try by their combined efforts to drive it away from this source of food. Siamangs as well as orangutans traverse only comparatively short distances daily, prefer sources affording larger quantities at a time, and visit but few trees in meeting their requirements. The smaller gibbons travel at least twice as far, start earlier in the morning, and visit more trees; thus, they not only take the ripest fruits (the first choice) before the siamangs and the orangutans arrive, but will also visit trees with much smaller supplies of fruit, which for the siamang and the orangutan would not be practical. The orangutan moves only one quarter as far in a day as a gibbon, and needs six times as much nourishment and even so the fruit component in its diet is greater than that of the gibbon, which specializes in eating fruits. How does the orangutan manage this? The answer is, by superior intelligence and an excellent memory.

Reproduction and Raising the Young

In intelligence tests administered by the American psychologist Duane Rumbaugh, a two-year-old orangutan was more successful than any of the gorillas and chimpanzees similarly tested. True, intelligence is normally associated with a more complex social life. Yet the high intelligence of the unsociable orangutan and its problem-solving ability might have something to do with its unusual knack of locating fruit in an environment where fruit trees are generally scattered and bear at irregular intervals.

Orangutans can also be remarkably inventive in the use of twigs as tools which they use to bring fruit within reach. Such behavior is comparable to the tool use of chimpanzees. Herman Rijksen tells how a young orangutan floated from one tree to another on a liana. Arriving at its goal, it carefully tied the liana up, instead of letting it swing back, thereby frustrating the efforts of another animal to follow it.

Attempts have been made to teach American Sign Language (ASL) to orangutans for the purpose of communicating with them. The American biologist Gary Shapiro taught ASL to two young orangutans being prepared for rehabilitation in the wild, for still another reason. At his station, the orangutans that are to be rehabilitated will communicate unhindered with their conspecifics living in the wild. What could be more exciting than to "debrief" such informants concerning the activities of the orangutans at large? Unfortunately, the experiment so far has been disappointing to the scientists.

Like the other apes and humans as well, orangutans have a few – but well-cared-for descendants – rather than many offspring with a high mortality. Females are sexually mature at about eight years, and may then bear a single infant about every three years for the next twenty years. In ac-

(Above and left) These cheerfully munching orangutans seem almost human.

▷ Among the solitary orangutans, the mother-child relationship is especially intimate, as witness the photographs on the following four pages. The mother orangutan will roam through the forest alone with her offspring - usually an only child, sometimes several of different ages - for years. Even an older youngster will not stray far from her side.

tuality, the average interval between births is longer (about six years) for most animals in the wild and a successful female will probably raise not more than four or five young in her lifetime. Females are willing to mate throughout the entire month, and show no such particular signs of readiness as the swollen genital parts of chimpanzees for example. When a female orangutan is ready to mate, she herself will look for an adult male and find him, guided by his long-distance call. The pair will remain together for several days, even months; as soon as the female is pregnant, she resumes her solitary life and roams about either alone or with her older brood.

During the first year of life, the young orangutan is in physical contact with the mother almost uninterruptedly. It clings to her side as she moves about, or scrambles around over her while she stops to eat or to rest. Young orangutans are not completely weaned until they are three years old, or even later if there is no new baby, but they will start trying out other sources of nourishment at about four months. The infant will take a morsel from its mother's lips and savor it. If the infant has an older sibling, or meets another youngster in a fruit tree, it sometimes gets a chance to play, but mostly his mother is the infant's only companion.

At two years old, the young orangutan will leave its mother temporarily to build play nests or play with other toys, such as a piece of bark, but will never stray far from her side and will always share her nest at night. The youngster seldom needs to be carried, but its mother will help it to cross gaps between trees by extending her body as a bridge. The young orangutan becomes increasingly independent, and will no longer share the mother's

(Left) When a young orangutan has an elder sibling, or happens to meet a contemporary, the youngsters become temporarily independent and will romp about in the trees together. (Right) This orangutan in the water is not an everyday sight.

nest after a new baby has been born. At three to seven years of age, the half-grown animals will gradually separate from the family and spend longer periods of time alone. By the time they are sexually mature (7–10 years), they will have separated from their mothers completely. Young females are slightly less enterprising than young males. They stay near the mother and their younger siblings longer, thereby probably learning the elements of "baby care," for when they themselves have offspring there will be no older females at hand to guide them.

Grown males as a rule have large territories, often extending over several square miles and overlapping the home territories of several females. The male associates with these females when they are willing to mate and until they become pregnant, and behave aggressively towards other males who violate his boundaries. Once the females are pregnant or occupied with dependent young, they are without sexual interest for several years, and an adult male can then move to a new territory and try to impregnate other females.

Young adult males (10 to 15 years of age) are mostly solitary; they do not yet utter the far-reaching call, but they often maintain relationships with females. Although they travel about with these females for several months and mate with them, it is usually the larger adult males who actually fertilize them, by associating with them only in those few days when their readiness to mate is at its peak. Even for the large male, it would in fact be difficult to remain with the female longer. Because of his bigger body, he needs more nourishment daily and must forage farther than the female. The smaller, young adult male is not subject to these limitations. On Sumatra, it is common to find a young adult male in the company of a female with infant; the male is probably not the father of the infant, but his presence effectively protects the family group against intruders and enemies.

Young adults and even high-ranking males occasionally attack females an abuse them sexually. Unknown females passing through alien territory, perhaps because the trees in their ranges have been felled, seem to be especially exposed to such attacks. In taking such females by force, a male gains a chance, however slight, to impregnate them, thereby enhancing the probability of passing on his genes. Such encounters also give young adult males an opportunity to prevail over grown females, even though they dare not yet challenge the high-ranking males.

Although orangutans roam about singly and seem to take little interest in their conspecifics, they are personally acquainted with the individuals whose territories overlap their own. Grown males are very observant of each other's movements, and utter their far-carrying cries to announce their location. When a male calls within another's territory, the owner will call back, and hasten towards the challenge. As a rule, the intruder will then withdraw quietly. Sometimes, however, the two males will meet and threaten each other; they stare, inflate their throat sacs, shake their tasseled coats, and even break off branches and rage about in the trees. Now usually one of the two males will retreat, running off on the ground. But sometimes such display leads to regular fights, in which the two rivals will grab and bite each other. Most adult males have facial scars and broken fingers due to such conflicts.

Proberbial "monkey love" is well demonstrated here. "Mouth to mouth" contact is an especially tender expression of the affection between mother and child.

Vocal Utterances

The far-reaching cry of the male is a peculiar, unmistakable call. It consists of a long sequence of loud roaring tones, at first increasing to maximum volume, then gradually subsiding into repeated gurgling, groaning sounds. The whole phrase may last one or two minutes. Many jungle animals stop what they are doing to listen. Sometimes the males reinforce the effect of the cry with an introduction of tossing a dead tree or limb around. The resulting crash puts all animals within hearing on the alert.

Scientists do not agree as to the function of the call. Is it supposed to attract a sex partner, or frighten away other males; does it express irritation on the part of the utterer, or is it simply a message telling the orangutans living in the neighborhood where the highest-ranking male is to be found? Field observations show that males call more frequently when the weather is bad, when they hear the crash of a fallen tree, and when other males are calling.

How other orangutans will behave after the call of a male depends on who they are. Most show no visible reaction but male intruders as well as young adult males generally withdraw discreetly. Other adult males may call back or even approach the calling individual to champion their own territorial rights.

Females with nurslings hide in the treetops when the calling male is near, but females willing to mate will approach and themselves invite mating. Biologically speaking, this decision makes sense, since the fact that the male is calling at all announces that he is one of the highest ranking in the area and therefore a "good match."

The long-distance call probably has other functions. It seems at least to distribute the adult males over the available space. Some announce their presence by their cry, and the others can follow their movements. The cry perhaps serves also to decide seasonal migrations and the distribution of orangutans in the forest. A remarkable fact in this connection is that male orangutans will call more frequently if the orangutan population is crowded, too numerous, or otherwise socially disturbed. Clearing of timber and destruction of the forest region in which these anthropoid apes live lead to overpopulation, increasing aggression among

(Left) The mother orangutan attentively follows the first exploits of her offspring (Right) For secure anchorage and better weight distribution, as a rule all four limbs are used in climbing.

males, and a diminished number of successful matings and of births.

Humans and Orangutans

Humans and orangutans have a long common history, which has not been a very happy one for the orangutan. For at least 35,000 years, the apes have been hunted for food, and it is likely that overhunting by early humans is responsible for the erradication of the orangutan on Java. But not only primitive humans have played an important part in diminishing the orangutan stocks; modern descendants have continued the slaughter. The natural science collectors of the nineteenth century, including many distinguished biologists, bagged orangutans with no less zeal than they mounted butterflies, and gave hardly any thought to the effects of their depredations on the ape population.

In more recent times the species has become locally extinct in many parts of Borneo because, in the traditional headhunting rites of some Dyak tribes, especially the bloodthirsty Ibans, the human victims have been replaced with orangutans. Outraged by such barbaric customs, the British colonial masters of Sarawak had prohibited the taking of human heads as trophies. The Ibans turned instead to the poor orangutans, which they hunted with poisoned darts and blow guns. Headhunting was considered a trial of courage. Another such trial required the Dyak warrior to measure himself against a big orangutan in single combat. The fighter would attack his adversary on the ground, and stretch out a hand in a feint. Almost invariably, the orangutan would try to grab the offered part and bite it. Then the Iban could cut open the orangutan's grasping arm with his sharp jungle knife and soon disable his far stronger opponent. Today, such trials of strength are prohibited by law, the orangutan being fully protected both in Malaysia and in Indonesia.

The natural behavior of the orangutan when encountered in the forest by a human being perhaps reflects this long history of enmity. Usually the orangutan will hide and remain perfectly still when it sees its archenemy. But as soon as the ape is aware that if has been discovered, it tries another tactic. It resorts to displays, grunts, utters shrill, sucking squeaks, and breaks off branches

Orangutans can travel swinging hand-over-hand, though by no means so swiftly and elegantly as their smaller and lighter relatives, the gibbons, some of which dwell in the same habitat.

and throws them to the ground. Females generally climb higher and higher into the treetops, to scream and threaten from a safe height. Once in a while, large males will come down to the ground in order actually to put human intruders to flight. Unfortunately this behavior has seldom saved the apes.

During the 1960s and early 1970s, stocks living in the wild were further threatened by the thriving trade in orangutans for zoos and as exhibition animals. The demand was for infants, but several died for every orangutan sold. To capture the baby, the mother was usually killed, and many nurslings died in the process of capture and transit.

Effective measures to abolish this traffic were not taken until the late 1960s, although officially the orangutans had been protected previously in Malaysia and Indonesia.

The trade in baby orangutans has meanwhile been suppressed but not completely abolished. The red anthropoid ape has become a symbol of the threatened lowland rainforests. Today the orangutan is found only on Borneo and northern Sumatra, and its range is vanishing year by year, as the lowland rainforests fall to the chain saw. A glance at the scientific stations at which most orangutan research has been carried on illustrates how rapidly these forests are disappearing.

To reach his observation post at Ulu Segama in Sabah, MacKinnon had to journey several days by boat from the nearest town; today there is a tarred road, and clear-cutting extends to the banks of the Bole River. The trees in Horr's observation site in Lokan have been felled, as has happened in more than one-third of the forests of Kutai in the Indonesian part of Borneo, where Rodman did his research. On Sumatra the situation is no better. Cultivation has advanced to far beyond the Ranun River, swallowing up the area in which MacKinnon made his observations on Sumatra.

Illegal farm clearing along the Alas Valley is gnawing at the borders of the Gunung-Leuser National Park on Sumatra, the last stronghold of Sumatra's race of orangutans. The horrible fact is that all this has been destroyed in only fifteen years.

The sorry plight of the orangutan has been a matter of concern to conservationists for more than a quarter-century. In 1965, the World Wildlife Fund stated that only 5000 orangutans remained in the entire world. Meanwhile, the lowland rainforests, with frightening rapidity, had diminished by one-third. Since the early estimates, we have improved our census techniques, and studies have shown that orangutans, with a population density varying between one and five individuals per square kilometer ($1 \text{ km}^2 = 0.36 \text{ mi}^2$), depending on the environment, are not so rare as originally believed. Today we believe that there are still more than 150,000 of these magnificent apes living in the wild. This figure sounds reassuring, but the orangutan happens to be especially sensitive to the destruction of its habitat and especially endangered by hunting. The only safe orangutans are the roughly 20,000 individuals living in national parks and protected areas.

Eighty-five percent of the orangutan range is located within Indonesia. That country has an excellent network of parks and preserves, but they should be effectively protected. The orangutan stocks in Sabah and Sarawak are much smaller, but both Malaysian provinces are at present taking steps to ensure that the apes will be preserved within the existing protected areas. Recently, Sarawak established a new national park, Lanjak Entimau on its border with Indonesia, chiefly to protect the orangutan and other rainforest animals.

This female orangutan is stretched out comfortably, taking a siesta in the fork of a tree.

Nature conservation measures in favor of these great red apes should now be addressed to achieving the protection and administration of these existing reserves: Gunung Leuser (Sumatra), Kutai, Tanjung Puting, Gunung Palung, Gunung Karimun dan Bentuang, Bukit Raya and Bukit Baka (Kalimantan), Lanjak Entimau (Sarawak), and Danum (Sabah).

Most people have enjoyed the spectacle of orangutan antics in zoos. Yet these animals are most admirable when seen in their natural surroundings, how they swing through the trees with such astounding grace and speed for such large animals. To weigh the value of an orangutan against the commercial yield of timber is impossible, "but it is definitely time for us to stop and think before we annihilate another species, a fellow ape, our cousin the orangutan." These words of John MacKinnon were written in the late 1970s; since then, the situation for the orangutan has worsened. Let us hope that the orangutan will have be apportioned a secure future in well-protected preserves throughout its entire range.

Orangutan Rehabilitation

by Monica Borner

Orangutans have been protected by law in their countries of origin, Indonesia and Malaysia, for almost fifty years. Nevertheless, young animals especially are traded illegally in large numbers. To catch a baby orangutan, its mother, to whose fur it clings, would be killed. Many of the infants then died in transit, or because they were poorly kept, so that for each youngster that came to the West alive, there would be at least five or six dead orangutans. The responsible zoos of Europe and the United States therefore decided not to procure any more orangutans from their countries of origin. But the traffic did not stop. Private persons thought it chic to raise young orangutans as child substitutes.

There was only one way to put an end to the illegal orangutan trade. The orangutan had to be taken away from the captors, dealers, and owners. The authorities in the Malaysian provinces of Borneo, Sarawak, and Sabah started to confiscate orangutan youngsters back in the 1950s. They brought them to Barbara Harrisson, then living in Sarawak with her husband. She kept the orangutan orphans for some years, but then faced the difficult question of what to do with her half-grown charges. She sent the first orangutans to good zoos in Europe and the United States. But she found that this could not be a solution in the long run, for more and more infant orangutans were being seized, and needed a refuge.

In 1965, she initiated the founding of a Rehabilita-

At first, orangutans to be rehabilitated at the Bohorok Station in northern Sumatra must be kept in a roomy cage to accustom them to the unfamiliar surroundings.

tion Station for orangutans. The station, Sepilok, is located in Sabah, in a protected area where wild orangutans live. The station was directed by the warden G.S. de Silva. In 1967, the oldest grown female, after an absence of some length, brought her firstborn back to the station. This showed that orangutans released in the wild can survive and reproduce.

Other centers for rehabilitating rescued orang-

As soon as the little orangutans, at about the age of two, are sufficiently sure of themselves, they can spend the day frolicking in the woods around the station. That is the first step towards eventual return to the wilderness.

utans, therefore, seemed to be needed in other parts of Borneo as well as Sumatra, and in the early 1970s three such centers were started in quick succession. Birute Galdikas, who was studying the behavior of wild orangutans in Kalimantan, the Indonesian province on Borneo, admitted sequestered babies to her station, Tanjung Puting. In Ketambe, in the Sumatran province of Aceh, Herman and Ans Rijksen, likewise as a by-product of a field study, cared for rescued orangutans. I came to northern Sumatra myself in 1972, and my colleague Regina Frey half a year later. On behalf of the Frankfurt Zoological Society and the World Wildlife Fund (WWF), we founded the Bohorok Rehabilitation Center in the Langkat preserve.

Both of us are zoologists, and we had had experience with young anthropoid apes in the Frankfurt Zoo. We now had to learn to solve many practical problems by ourselves. In time, a usable method of successful placement of orangutans in the wild was developed, enabling us to reintroduce more than a hundred individuals in seven years.

On arrival, we placed all orangutans in quarantine for at least one month. During that time, blood and feces were checked for causative agents of disease, we innoculated the animals against infantile paralysis, and we wormed them. The purpose of the quarantine is to prevent rehabilitated animals from afterwards infecting wild orangutans in the forest with diseases communicated by humans. We also utilized this period of time to get acquainted with the new arrivals and their characteristics. After quarantine, if an orangutan proved physically and mentally sound, it was put into the forest. It turned out that reintroduction of orangutans is simplest from about four years of age on. They were big enough to get along without a mother, and enterprising enough to want to explore their new environment.

Although orangutans tend to be loaners, the Station, with its ten to fifteen half-wild orangutans at a given time, was a great help to novices. They learned by imitation, and they had company to play with and in exploring the forest. This went far to make up for the connection, lost at an early age, with mothers and siblings. Older individuals, almost or quite grown up, went their ways alone. Females were sometimes "abducted" by wild males.

Small orangutan children, aged up to about three years, require the close bond with the mother for healthy development. If they lack any such close relationship, they show disturbed behavior. For this reason we would carry such infants around with us a good deal during the day, for example on excursions into the woods, where they would have contact and opportunities to play with older orangutans. We gave adult females the opportunity to adopt these waifs, to prepare them for their future role as mothers. The little ones would spend the night in a sleeping box in our house.

With the passage of time, they became more assured and venturesome, and when they were about two years old, we were able to leave them in the forest during the day in company with other orangutans, and we visited them only at feeding times.

Twice a day, we fed milk and bananas to the apes on a platform in the forest. We deliberately kept the selection monotonous in order to encourage them to discover greater interest and variety for themselves.

The skills of climbing and finding suitable food, like nestbuilding behavior, are innate. Climbing paths through the trees are discovered by trial and error, or copied after other orangutans. Everything that seemed edible was placed in the mouth and tasted, and some things were promptly spit out. The very littlest ones would start building nests in play. When they were older, they would build sleeping nests for themselves every night. There was no occasion for us to teach our charges anything important to the survival of a wild orangutan. We had only to give them the opportunity to discover and practice for themselves.

When orangutans stopped showing up regularly at the feeding station, we stopped feeding them. However, we weighed them on occasion, to make sure that they were indeed feeding themselves adequately. Some few wandered off during this time, and were not seen again unless by chance. Others found the separation from feeding place and friends difficult, even though they no longer got anything to eat. We would round up these in-

The half-wild apes are fed on a platform in the forest two times. They are given only milk and bananas, to encourage them to obtain a more varied diet in the forest.

dividuals about once a year and fly them by helicopter into the interior of the preserve, where they were released.

The natives wondered at the two women living with apes in the forest, and they would often ask us whether we wanted them to catch some orangutans for our station. It became clear to us that we would have to open our ape facility to visitors if people would understand what we were trying to do. Since the farmers were burning off large areas of forest and thereby destroying orangutan habitat, we wanted to talk with them about the utility of the forest to human beings as well as to animals. Visitors, admitted only once a day, did cause some problems. Orangutans and people were not to be allowed to come too close to each other, and so supervision was constantly required. But we realized that educating the populace was far more important even than the successful rehabilitation of a few individual animals, both for the protection of the orangutans and also for the preservation of their habitat.

Hardly any market for the sale of orangutans remains today. But enormous forest ranges are still being cleared, and without rain forests the orangutans are doomed. Several times we had to capture wild orangutans with nets when they had taken refuge in the last trees of a clearing and could not escape. Thus, the function of rehabilitation stations has changed. The orangutans are no longer threatened by illegal trade, but by loss of habitat. Stations must therefore concentrate on education in environmental protection.

John MacKennon found in Borneo in 1974 that steady immigration will render an orangutan population too dense. As a result, the number of young will diminish, each wild female having a baby only every seven or eight years instead of every four. Care must therefore be taken not to augment the total population by rehabilitation. For real benefit to the orangutan species, a rational rehabilitation facility today should be located in a suitable, well-protected tract of primeval forest in which no wild orangutans occur.

Breeding and Life span of Orangutans in Captivity
by Marvin Jones

Between 1776 and 1925, some hundreds of orangutans were born in captivity. Most did not live very long, and there was no continuing reproduction. A change in methods of capture in 1925 made possible the taking of whole family groups, which came into zoological collections. Most originated on the Island of Sumatra. Then on the 12th of January 1928, a female Sumatra orangutan was born at the Berlin Zoo to parents who had arrived as adults in May 1927. More births occurred in 1928 at the Zoological Gardens of Nuremberg and Philadelphia, and from then until 1960, 60 orangutans were born into more than 20 animal collections in Europa and North America.

With great advances in the proper keeping of anthropoid apes and the many births since 1946, reproduction finally became a yardstick for success; 14 orangutans were born in 1961 alone. Four female orangutans who came into the world in that year are still living and have offspring of their own. From 1961 to 1971, more than 235 orangutans were born, and 60 collections reported their first orangutan birth ever. In the period from January 1928 to December 1984, no fewer than 880 orangutans were born in captivity, in about 127 collections; 530 of them were still living at the close of 1984. Their numbers exceed those of wild-born individuals to be found in captivity.

Ten collections accounted for 32 percent of the births, as the table below shows.

The first second-generation birth – that is, where both parents had themselves been born in captivity – occurred at the Rotterdam Zoo in 1963, and more than 45 second-generation births have been reported since. Some of these youngsters in turn have produced young by this time.

A female Sumatra orangutan born in the wild in about 1919 gave birth in a private collection in Cuba in 1929. The mother later came to the Phil-

The ten most successful orangutan breeding centers

Place	Beginning	Births
Yerkes Primate Center, Atlanta, Georgia	1966	60
San Diego Zoo, California	1943	30
Cheyenne Mountain Zoo, Colorado Springs	1961	29
Frankfurt-am-Main Zoo, West Germany	1958	29
St. Louis Zoo, Missouri	1933	26
Rotterdam Zoo, Netherlands	1951	25
London Zoo	1961	22
Philadelphia Zoo,	1928	22
West Berlin Zoo	1928	21
Duisburg Zoo, West Germany	1971	21

A female cross between Borneo and Sumatra orangutans, born in Australia at the Sydney Zoo in October 1957, had her first baby – which in turn reproduced in 1981 – in December 1967. The hybrid female had five more babies, and was still living in December 1984. A female orangutan, Sungei, born in the wild on Sumatra in about 1956, had her first child at the Yerkes Center in 1967, gave birth eight times all-told, and died in 1984. Bulu, whose parents were from Borneo and who was born in the London Zoo in March 1961, had her first baby in January 1970, and a second in June 1971, which later herself gave birth in March 1982. Bulu is still having healthy babies at the London Zoo.

A few captive-born orangutans have reproduced at a much earlier age. The most familiar example is that of Schorsch and Uta at the Hanover Zoo. The male was born at the Frankfurt Zoo in March 1971, and Uta in Hanover in July 1971. Both were raised by hand in Frankfurt and afterwards sent to Hanover, where they had their first baby in August 1978. This shows that orangutans brought up together from an early age may nevertheless mate and reproduce. Incidentally, Schorsch was a hybrid.

The first twin birth occurred in 1968 at the Seattle Zoo; both of the twins have since reproduced. There have been a number of more recent twin births. Many of the twins survived and reproduced, for example Bruno and Hella, born at the Hellabrunn Zoo in Munich in 1969.

The orangutan population in captivity is continuing to grow, with about 40 born each year, more than 75 percent of whom survive their first year.

So undoubtedly the future of the species in captivity is assured. The gene pool is large, that is to say, ancestry is varied, and there would not seem to be any fertility problems like those of other anthropoid species. The prohibition of traffic in wild-born orangutans is helping to preserve this species in the wild.

adelphia Zoo, where she had eight more offspring. She died in 1976, at the age of 57. Her mate, probably born in 1919 also, begot his last offspring when he was 40 years old. At his death in 1977, he held the longevity record for orangutans in captivity. Another male born in the wild on Sumatra, Moritz, came to the Frankfurt Zoo in 1956 at about age 22, similarly begot his last offspring at 40, and died when he was about 44.

At the Rotterdam Zoo, a female orangutan from Sumatra, born about 1940, was still living in December 1984.

Zoo keeping and breeding of anthropoid apes, especially the difficult orangutans, has made great advances in recent decades. Nowadays - as here at the Zurich Zoo - animals are kept in large, hygienically outfitted quarters, affording them enough freedom of movement and climbing opportunity. Thus they can show zoo visitors far more of their natural modes of locomotion and behavior than in the cramped ape cages of the past.

The Gorilla

by Bernhard Grzimek

Suddenly the black male gorilla charges, screaming. He comes closer and closer. When almost upon me – 5 ft (1.5 m) away – he turns aside, makes an arc, retreats about 30 ft (10 m), sits down, and looks away.

I can hear my heart pounding; I am gasping for breath. But I didn't budge! For twenty years, I have wondered whether I could face the attack of such a raging monster, weighing half a ton, outstretched arms spanning 8 ft (2.5 m), and not take flight at the last moment. Running away is in fact the stupidest thing one can do, for then the giant will certainly grab you from behind and bite you in the shoulders, the hip, or the legs. But you never know whether you won't lose your nerve at the critical instant. This same gorilla, Kasimir, who was taking off after me without warning, had killed the brother of the pygmy chieftain Mushebere in 1966, almost on the very spot where we were now standing. On that occasion, the gorilla would not allow the corpse to be removed; he held his position, with his entire family, numbering eighteen. Adrien Deschryver had to use teargas to disperse the mighty apes. There was a similar case some years ago near Dian Fossey's camp, when a mountain gorilla ripped up to visitor's entire face and shoulder.

It is nearly 2500 years ago that a fleet of 60 large vessels set sail from the Phoenician merchant city of Carthage on the Mediterranean Sea – about where the city of Tunis stands today – to explore the western coasts of Africa. On board were some 30,000 sailors, oresmen, and soldiers; their commander was Hanno, one of the principal statemen of the Carthaginian realm. The fleet sailed through the Straits of Gibraltar and on southward down the coast of Africa until it came to what is now Cameroon.

Full-grown gorillas are formidable beasts. Yet they are hardly a danger to humans; rather, humans are the danger to them. Through human encroachments, their stocks have been much reduced. This is true of both subspecies, the lowland gorilla *Gorilla gorilla gorilla* (above) and the mountain gorilla *G. g. beringei* (right).

That country – which for a time was a German colony – was named for the crabs, the *camerões*, that Portuguese sailors found there, many centuries after Hanno.

Hanno's fleet, which had penetrated farther south than any Mediterranean seafarers before them, tried to proceed up the Niger River, but were turned back by rapids. To the south of Cameroon, perhaps in present-day Gabon, the party of explorers came upon giant, hairy black "people". These bared their teeth, snorted, threatened, and when the Carthaginians, like the practiced slave hunters they were, tried to catch a few, even the females bit so fiercely that they had to be killed.

Hanno had these hairy animals skinned and brought the hides to Carthage, where they were hung up in the temple. After another half-millenium, the Roman conquerors saw them there, as is reported by the Roman historian Pliny the Elder, who perished when Vesuvius erupted.

Interpreters had told Hanno that these strange wild creatures were called Gorilla. Whether they really were members of that great anthropoid ape

GORILLA

Moist evergreen mountain forests up to elevations of 13,000 feet (4000 meters) are the home of the mountain gorilla. The luxuriant vegetation affords these mighty animals all they require for life, notably an abundance of vegetable food.

The lowland gorillas, among which a gray-brown western form (here shown) and a black eastern form can be distinguished, is somewhat shorter-haired, especially on the head and arms, than the mountain gorillas, who make their home at cooler elevations. But all gorillas spend their time almost exclusively on the ground, even when, like the lowland subspecies, they live among the trees deep in the rainforest.

species will never be known; some scholars surmise that they were merely baboons. But the Phoenicians would probably have become familiar long since with baboons, on their extensive trading voyages in North Africa. In that case, they would hardly have brought baboon skins home from Gabon as a rarity.

Not until nearly 2000 years later did reports concerning these giant beasts reach the North again. Late in the sixteenth century, the English sailor Andrew Battel was held captive by the Portuguese in West Africa. He tells of two species of anthropoid apes, which we can easily recognize as gorillas and chimpanzees. The gorillas he calls Pongo, and he asserts that after the humans have moved on, they will come to the campfire to warm themselves, but have not the wit to put fresh wood on the fire. Much else that he has to say about the black giants has been confirmed by more recent studies.

As early as 1640, the first live chimpanzees came to Europe, but little was learned about the other anthropoid species. In scientific books of the time, gorillas, orangutans, and chimpanzees were constantly confused with pygmy folk and races of "wild" humans. It was not until 1860 that the missionary Savage and the adventurer Du Chaillu brought back more precise descriptions of the gorilla to Europe, as well as skulls and skins. Du Chaillu portrays the gorilla as a bloodthirsty forest monster, ready to attack any human beings and tear them to pieces. He describes the fear-inspiring visage, the giant stature, the terrifying roar, and the beating with the fists upon the chest before the charge. His tales have placed the gorilla in the consciousness of Europeans as a merciless demon of the forest for half a century, even though Alfred Brehm exposed Du Chaillu's account as a lie in an 1876 book *Thierleben* [Animal life]. Only a few years ago, a sport journalist claimed that he had had to "execute" a male gorilla in West Africa because he had "raped all the female gorillas," broken into natives' huts, killed a mother gorilla, and seized her baby. I was able to prove later that that "murdering" gorilla was a female. At all events, until just recently very little was known about how these creatures, so closely related to us, lead their lives in Africa, and what they are like. The thorough researches of George B. Schaller and Dian Fossey (see separate articles by these authors) have enlightened us.

Living gorillas were first seen in Europe towards the end of the last century. A colonial officer, Major Dominik, with the aid of nearly a thousand natives, managed to catch three gorillas in nets. Two of them lived to reach Hagenbeck in Hamburg,

The first of the "gentle giants" to reach sexual maturity in captivity is commemorated by a monument in the Berlin Zoo (right). It is an anachronism that gorillas are still occasionally pictured as bloodthirsty monsters (left).

but there they died after 13 and 17 days – old man Carl Hagenbeck thought, of homesickness. For older animals, this perhaps should not be ruled out altogether. But the then "animal houses" contained no such facilities as exist today as a matter of course in an ape exhibit.

At the old Berlin Aquarium directed by Alfred Brehm, where a baby gorilla was brought at a later time, they would feed this strict herbivore a couple of frankfurters for breakfast, and sometimes a sandwich of hamburger steak, cheese or the like. This would be washed down with beer. At midday, first a cup of bouillon, then rice or potatoes and vegetables, beets or cabbage cooked with meat. Today we know more about how to keep wild animals in zoos, and, accordingly, the zoos are becoming more and more successful at keeping them.

Anyone seeing a full-grown male gorilla in a zoo for the first time will need a little time to digest the experience. If the giant were to straighten his knees, he might stand 7.5 ft (2.4 m) tall, although its legs are short; some individuals measure over 40 in. (100 cm) across the shoulders. Male gorillas look enormous and superhuman, and one can well imaging the horror of the first whites who, inadequately armed, found themselves face-to-face with a giant beast so human in appearance. In the first decades of the twentieth century, adult gorillas were hardly ever to be seen in zoos. The male gorilla Bamboo died at the age of 34 in 1961 at the Philadelphia Zoo, and so did the equally enormous Massa, at nearly 54 in 1984. Gorillas in the wild would hardly live to such a great age.

To keep a monkey or an ape alone is very cruel. Gorillas in captivity should live in families so that

Two gorillas in the zoo - but what a difference! One (top) stares moodily into space in a bare, old-fashioned cage, while "silverback" (bottom) is cheerful in his open, planted setting.

adolescent females will have an opportunity to see births and babies. At the Frankfurt Zoo, even adult male gorillas live without bars behind panes of glass measuring 6 by 10 ft (2 × 3 m) and about 1.5 in. (40–42 mm) in thickness. The possibility newly introduced in Frankfurt, of letting the gorillas go out-of-doors at any time, Summer and Winter, through heavy plastic drops, has proved beneficial here as elsewhere. In some zoos, gorillas and the other anthropoid apes now live in line-of-sight facilities with no bars or glass at all, behind "dry moats" that are almost unnoticeable but deep and wide enough to prevent escapes. So that disease germs will not be transmitted to the apes by coughing and sneezing zoo visitors, an artificial breeze moves the air – and with it, of course, any odors – from the enclosure into the visitor area, rather than the other way. At Frankfurt, the hydraulic sliding doors to the sleeping quarters can be set at any height. Thus it can be arranged that only infants, or only females and infants, can slip through, and not the large or even middle-sized males. This makes it much easier to keep whole families. Even in the wild, members of a social group must occasionally avoid each other.

The number of gorillas that are still living in the wild is much disputed. Only a few years ago, some researchers estimated 15,000 to 20,000, others even 50,000 to 60,000. These figures include the mountain gorillas in the region of the Virunga volcanoes on the borders between Zaire, Ruanda, and Uganda, then estimated at about 4000 or 5000. It is quite certain that these figures are greatly overstated. Gorillas are not reluctant to enter areas being used by humans, near villages and plantations or roads and mines. That is where the giant forest trees have been felled; a second growth of shrubs and young trees affords the gorillas far more sustenance than the forest interior, not to mention human crops. For this reason, the numbers of the gorillas are consistently overestimated; they are assumed to be as numerous in the trackless jungle as near roads and settlements.

From May to October 1981, a research team tried to take an accurate census of the mountain gorillas in the Virunga territory. A total of 242 gorillas were counted; allowing for an additional 10 percent that might have been missed, we get 266. This is about as many as in the counts of the early 1970s and 1978. Conrad Aveling and Alexander H. Harcourt, two of the counters, estimate the number of mountain gorillas in the gorilla preserve in the Bwindi Forest, Uganda, of about the same extent, at 130, and hence the entire population of this subspecies at less than 400! To be sure, this does not include the mountain gorillas living in the Bukavu preserve founded by Adrien Deschryver, and doubtless also farther to the West, because they may belong to a different subspecies.

In any event, the mountain gorilla is one of the most endangered animals, and the lowland gorilla is not much better off.

Seldom has a creature on Earth been so shamefully misrepresented as the gorilla. Such must be the conclusion arrived at upon reading what George Schaller and afterwards Dian Fossey wrote in their journals during many weeks and months of observation.

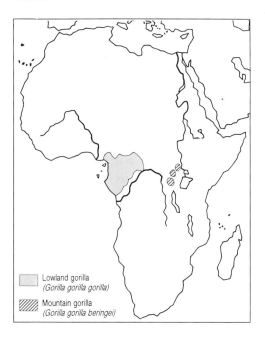

A great many of the higher animals, including human beings, will take possession of certain territories and defend them ferociously. Humans, in fact, will slaughter each other indiscriminately on this account. A gorilla group too has a home, a territory of 9 to 14 mi^2 (25 to 40 km^2). Within it, the gorillas wander about continually, seeking food; sometimes traveling only 300 ft (100 m) in one day, sometimes up to 3 mi (5 km). But in the same area, there are other gorilla groups. When they meet, usually there is no fighting; the two leaders may simply stare at each other menacingly, but usually not even that. George Schaller never saw any disputes, let alone fights, between gorillas, and Dian Fossey in only a few exceptional cases.

There is no quarreling over the females either. In general, sexual activity among the gorillas, unlike many other apes and monkeys, plays a rather subordinate role. Sometimes a leader will tolerate a second male's mating with one of the females of the group, only a few yards away.

Schaller observed gorillas mating on two occasions, and in both cases the male partner was not the highest-ranking "silverback." In one case, the female knelt, elbows on the ground, while D.J., the second-ranking male of the troupe, stood behind her and embraced her hips. Because there was a steep incline, the animals slipped slowly downward.

The other time, the seated male held the female on his lap. Usually so silent, gorillas are often quite noisy when mating; Schaller heard a characteristic mating cry uttered by the males especially. The other members of the group would pay no attention.

Female gorillas have a baby every three and a half to four and a half years. Nearly half of the offspring die in infancy or early youth.

There are twice as many females as males in the groups. Counting male gorillas living solitary or in twos, there are perhaps three females to two males. Probably mortality is higher among the males.

Gorilla pregnancy is not very noticeable; their figures are rather stout anyway. According to zoo experience to date, the term of pregnancy is nearly eight and a half months, compared to barely eight months for chimpanzees, eight months for orangutans, and nine months for humans. During pregnancy, some mother gorillas have temporarily swollen knuckles. Birth occurs supine within a few minutes, whereupon the mother severs the umbilical cord and carries the nursling pressed to her breast. Newborn anthropoid apes – unlike monkeys – cannot hold on unaided, although their "grasping reflex," like that of human babies, is fully developed. Births occur throughout the year; no special season is preferred. This is in accord with the habitats of all anthropoid apes, and perhaps also of our own ancestors – the moist or semi-moist tropics, where there are hardly any "seasons" that could affect the reproductive cycle.

Breeding gorillas in captivity is no simple matter. There is joy in any zoo when a female gorilla becomes pregnant.

When we acquired our little girl gorilla Makula, who lived at the Frankfurt Zoo when she grew up and became a mother, she was so small that she could not even lift her own head. We had to keep her at home like a human baby and tend her in a crib beside the bed. And although we now know that mother gorillas will occasionally "park" their newborn, parking babies in cribs or carriages is not "natural" for either human or gorilla babies. Monkeys, apes, and humans are not born to be nestlings, but to be carried about, spending most of the day and night in close physical contact with their mothers.

Young gorillas develop about twice as fast as human babies. Within six months, they become lively, playful creatures. In the Kabara region, they begin to eat some plants at two and a half months, and at six to seven months they live on these in large part. Even so, they will still nurse now and then at a year and a half. It is much the same among human beings, allowing twice as much time. Human children growing up naturally as among the Bushmen in southern Africa or – before we indoctrinated them with the customs of white people – among the Kikuyu in the East African highlands, are or were nursed by their mothers up to the age of three to four years, their mothers usually not having had another child in the meantime. Gorilla babies can crawl at three months, go securely on all fours at four and a half, and climb at six to seven months.

Our first gorilla, Max, to be born in Frankfurt, who came into the world on the 22nd of June 1965, weighed 4.6 lb (2.1 kg) at birth, and 35.6 lb (16.6 kg) at one year. His first milk incisors came in the sixth week of life, and the first molars in the seventeenth. As early as the second week, he would follow moving objects with his eyes, laugh when he was tickled, and lift his head when lying on his stomach. When he was ten weeks old, he knew his foster parents from other people and would turn over from his belly to his back without assistance. At nineteen weeks he would run on all fours, at twenty-six weeks he would pause to stand up on two legs, banging on the wall, and at thirty-four weeks he was able to walk a few steps erect.

On the 3rd of May 1967 – nearly two years after the birth of Max – the same mother had female twins, though not identical, and quite different in weight and in their natures (identical twins originate from a single fertilized ovum, and so have the same inheritance; fraternal twins develop from two different ova that matured at the same time, having been fertilized by two different spermcells of the father or fathers).

My Frankfurt Zoo was the first zoo in the world to succeed in breeding all four great anthropoid species – orangutan, gorilla, chimpanzee, bonobo. By 1987, 26 orangutans, 12 gorillas, 21 chimpanzees, and 11 bonobos had been bred in Frankfurt.

An old hand at living with gorillas in the wild is my Belgian friend Adrien Deschryver. As a ten-year-old, he came with his parents to the then Belgian Congo in 1949. When the country became independent in 1960, Adrien remained, while his parents, like nearly all of the white farmers, returned to Europe. At first, Deschryver was an enthusiastic hunter, but like so many true hunters, he lost his taste for shooting down defenseless animals, and became a conservationist. His first experience with gorillas, like my own, goes back nearly 30 years. In 1957, a male gorilla in his neighborhood killed two people and, on the same day, injured two others so seriously that they were in the

(Above) A gorilla baby not accepted by its mother is raised on the bottle like a human infant. (Below) A place for gorillas. Since 1970, the Kahuzi-Biega forest hills in Zaire, which still accommodate a fair number of mountain gorillas, have been a fully protected national park. This model facility goes back to an initiative by the Belgian animal lover Adrien Deschryver.

hospital for a month. Deschryver took ten days to drive the animal back into the nature park. At each new charge, he would shoot into the ground in front of the gorilla or intimidated him with blanks.

A couple of years later, another male gorilla who had been considered quite peaceable suddenly grabbed an old woman who was working in the field, pulled her about 30 ft (10 m) into the nearby bushes, and detained her there. He tore her clothes, the woman screamed, the gorilla screamed also. The other workers threw sticks and stones into the bush, until finally the gorilla ran away. The woman's clothes were in tatters, her skin scratched, and she was quite beside herself. However, the gorilla had made no attempt whatever to assault her sexually. Perhaps he was lonely and wanted company.

Deschryver was hunting in the vicinity at the time, and was aware of all this. About then, Mushebere, the pygmy chief, together with his brother, killed the gorilla Kasimir's father. "He tasted good," Mushebere observed, laughing, many years later. He seemed to have forgotten the dreadful revenge of Kasimir, who later killed the pygmy's brother.

In 1965, Deschryver began to make friends with the gorillas, and a year later founded the Kahuzi-Biega preserve, an hour by car from the town of Bukavu on Lake Kivu, but 1600 ft (500 m) higher. For all this, the young Belgian, who by the way is married to a black African woman, accepted no money. In fact, he bought a small plane with his own money to patrol the preserve. Ever since the civil war, there have been many firearms in the possession of the population, and they are used to kill not only elephants (for their ivory) but also gorillas.

Kahuzi-Biega is a mighty forest range, whence rise the summits of Kahuzi and Biega, about 11,000 ft (3500 m) high. This private nature preserve of Mr. Deschryver's was officially declared a national park in 1970, and enlarged tenfold. According to international rules, which hold universally in the poor developing countries and also in the United States, a national park is an area of complete protection of nature, protected by law for all time for any human use – whether agriculture, stockraising, forestry, or even the mere construction of huts or houses.

About 200 mountain gorillas live in the Kahuzi-Biega national park of Zaire. Their body hair is somewhat shorter than that of the gorillas in the uplands and on the volcanic slopes at the three-country point between Zaire (Virunga National Park, formerly Albert Park), Ruanda (Volcano National Park), and Uganda (Kigezi Mountain Gorilla Preserve).

The man who was appointed director of the new Kahuzi-Biega National Park is Mushenzi Lusenge, a black African trained at the Garua rangers' school in Cameroon. Deschryver became the other director. There were very difficult negotiations with the pygmies, since these were their traditional hunting grounds. Deschryver agreed with them that they might continue to hunt small animals and gather plants, but must kill no gorillas, elephants, or other large animals. Nevertheless, in 1972 four more gorillas were slaughtered by pygmies; two poachers were shot down by rangers. One of Deschryver's best moves was to engage the pygmies themselves, as the most experienced hunters and poachers, to sign on as guides and wardens. They thus gained by the new national park, and tried to maintain the gorillas and keep them as gentle as possible.

In 1973, the directors Lusenge and Deschryver heard that vacationers on the Kisangani (formerly Stanleyville) road had bought a baby gorilla for the equivalent of about five dollars. Such anthropoid babies never come into human hands except when the mother, and usually the entire family, has been slaughtered in order to sell the baby. This gorilla nursling was taken over, picked up by plane, and raised in the Deschryver family with a bottle. Four months later, Adrien took the young gorilla with him into the national park to see what Kasimir's family would do.

First the male gorilla, Hannibal, came to see exactly what was going on. He came within 5 ft (1.5 m), then stepped back and pushed aside the branches so that Kasimir could see the baby too. Presently he advanced upon Deschryver, crying out, but slowly. The baby screamed in fright, and so did Deschryver himself. Kasimir passed close by him twice, almost touching him. There was nothing for Deschryver to do but put the baby on the ground. Kasimir rolled it towards himself and held it carefully against his breast. He took it over to the female gorillas, who had meanwhile come out of the bushes too. The silverback male put the infant on the ground in front of them. One of the females took it to her breast, and the baby stopped crying right away.

Eight days later, there was a severe hailstorm and a rainfall of 2 or 3 in. (5 or 1.5 cm). The hailstones lay inches deep, still covering the ground in places next day. The temperature went down to about freezing. All gorillas caught cold and coughed, and so did the pygmies. They were still coughing when we arrived three weeks later. Probably the baby perished in this unexpected bad weather. In any case, it has not been seen since then.

Mother gorillas with babies of their own are not unkind to others on that account. It may happen that a strange youngster will come to them, sit on their lap with their own, and press against the breast without being rejected. Even the big males will tolerate infants playing around them or climbing over them. Today all this can be seen in zoos, at Frankfurt and in other places. A female gorilla living in the wild had an eight-month-old with a great wound in the flank. Unlike other young ones, it never rode on her back, being apparently too weak to hold on. The mother always carried it carefully on her arm, with no part of the wound touching her body. She always held it ventral side down.

Sitting down, small baby gorillas are often held in both arms. A mother gorilla living in the wild carried her dead baby around for four days. Until the age of three years, the young remain with their mother and sleep in her nest. At four or five years, the bond is gradually loosed. Not that the mother will simply drive off her child. Often an older child will continue to sleep with the mother when she already has a new infant. Sometimes she will quietly detach the older one's hand if he should hold on by her body hair on the march.

Female gorillas become sexually mature at six to seven years, males at nine or ten years. At about the age of ten years, the midportion of the back of the males begins to turn silver-gray. Even without their silver saddles, one could hardly mistake them, with their imposing size and calm dignity. Nevertheless, their supremacy – in contradistinction to baboons, chimpanzees and rude people – is exercised with hardly any loud strife, blows, or bites. There is no squabbling over food and hardly any over females.

Gorillas seldom practice mutual grooming; most often this is done by mothers with children, occasionally silverbacks, but never the younger black males. The lowland gorillas at the Frankfurt Zoo do not groom each other at all, as B. Schirmeyer found. However, they would often pluck their own hair, scratch themselves, clean their nose, eyes, or ears with the index and middle finger. With the mouth, they clean only the fingernails, not the fur, as chimpanzees will do, for example. Coarser dirt is shaken off, or the fur is rubbed against stationary objects. Whereas baboon and chimpanzee young can hardly get enough of being groomed by their mothers, and will demand it insistently, small gorillas seem rather to be irritat-

(Bottom) Professor Grzimek in Arusha, Tanzania, congratulating native rangers for services in defense of wild life in the national parks.

▷ A mother mountain gorilla holds her nursling to her breast. Babies usually hold on to the mother's fur, but at first they need the support of her hand, because the innate grasping reflex is not yet strong enough.

▷▷ Mountain gorillas at home. These social animals, with no fixed territories, roam the woods in family groups under the leadership of a silverback, eating almost without interruption.

ed by it. They struggle like little children against being bathed.

Gorillas yawn, sniffle, cough, belch, get the hiccups, and scratch themselves, much as we do. They utter at least 22 distinct sounds in communicating among themselves, but there is no "language" such as all human beings have. Gorillas do not seem to have marked personal friendships between members of a group, as chimpanzees do. Perhaps that is why they get along so well. Personal friendship, after all, entail a distinction from the non-friends.

The silverback male holds his group in check, and meets outside aggressors, human or not, with imposing intimidative behavior. He begins with a series of cries, sometimes interrupted by taking a leaf between the lips. Then the cries come in rapid succession, he stands up on his hind legs, throwing leaves and twigs into the air. At the climax, the giant will beat his breast with his hands several times. Then he runs sidewise on all fours, tearing up vegetation. Lastly, he beats the earth with the flat of his hand.

Youngsters four months old will beat their chests and tear up plants, but only large males accompany this intimidating display with cries. The same purpose is served by male dances of Indian, African, and other aborigines, as well as by our own marches, parades, demonstrations, and maneuvers: the idea is to make an impression, to intimidate, without immediate need to kill or maim, but among us this is typically group behavior.

Gorillas behave otherwise when they really mean to attack, and again not so differently from humans. In this context, a sharp gaze may be understood as a threat. We too perceive a hard stare as a challenge. Only decades ago, members of student associations felt insulted when gazed at intently by someone in a restaurant. They would ask the offender to step outside, faces were slapped, visiting cards exchanged, seconds dispatched, duels fought.

After all, we retain much hereditary behavior that we have in common with the great apes. The fatalities in such academic encounters may have been few, but hardly so few as among the "wild" gorillas.

Occasionally the threatening male gorilla will knit his brows, or compress his lips. In doing so, he utters short, clipped grunting sounds. Finally he will throw back his head towards the opponent, and often his body as well, as though he were about to charge. And when at last he does charge, he nearly always stops a short distance away or runs past on one side. If a person is carrying a rifle, the animal will be shot will shoot down "in self defense." Reports of such killings have been commonplace in past decades.

There are safeguards that nearly always prevent gorillas from killing or crippling each other. A female or lower-ranking male will generally avert his or her gaze, usually turning the head aside, when exposed to the stern regard of a "superior," suggesting no intention of starting anything. On the other hand, at the Zoo in Columbus, Ohio, the male gorilla courting a female would always at first look away from her, by way of reassurance. At Frankfurt, our male gorilla Abraham and I had an understanding whereby each would observe the other out of the corner of his eye at meeting in the morning. If one happened to catch the other's eye, he was expected to glance away at once, as if

Gorillas are not swimmers. Hence caution is indicated in dealing with the wet and cold element.

There is less than meets the eye in this impressive display of intimidating behavior on the part of gorillas, but usually it is enough to chasten rebellious group members as well as aggressive strangers, including humans. At the climax of the imposing ceremony, the gorilla may rise to his full height and beat his chest in a loud tattoo.

unconcerned. When a pair of adult gorillas at the New York Zoo who had been getting acquainted through the cage bars for some time were put together, both would avert the face on approach.

Apparently, however, gorillas, like human beings, make a distinction between the stare of a possible adversary and intimate eye contact between old friends meeting unexpectedly. So at least I interpret Dian Fossey's account in this volume.

Jakob Schmitt, a veterinarian, carried out elaborate blood tests of monkeys, apes, and humans at the Frankfurt Zoo years ago, using electrophoresis and other techniques. Electrophoresis involves a weak electric current for separating protein constituents to determine their composition in the blood serum of living creatures. According to Schmitt's studies, the gorilla is second-closest to humans, after the chimpanzee, and these results agree with findings in other scientific fields. The Asian orangutan is far more remote. A close relationship to humans seems to be manifested also in the habits and disposition of gorillas, and although chimpanzees are even closer to us, most people find gorillas more attractive – perhaps because some chimpanzee behavior is too reminiscent of some less pleasant human traits.

Gorillas will ford streams 1 or 2 ft (0.3 or 0.6 m) in depth. They will not cross deeper or wider waters, unless a log or the like has made a bridge. Like all hominoids including humans, gorillas are not natural swimmers, as are monkeys and nearly all other mammals. The New York male gorilla, Makobo, drowned in the moat of his newly built line-of-sight enclosure in 1951. He had stooped over the water, fallen in, and sunk like a stone. He made no swimming movements, and did not use the ropes and ladders provided on the bank. The Frankfurt Zoo's adult male, Abraham, similarly perished in 1967. We then replaced the moat with a glass wall around the outdoor enclosure.

Unlike chimpanzees, gorillas in the wild have never been seen to use sticks or other objects as tools. They show hardly any curiosity about strange objects. Once Schaller's backpack was lying in full view of a blackback male, only 15 ft (5 m) away. The gorilla took one look and then paid no attention. Gorillas are introverted by nature, in the wild and in the zoo. Perhaps we human beings could live together in greater amity and peace if we were more closely related to the gorillas than to the chimpanzees.

Like the orangutan, the gorilla – and especially the mountain gorilla – is most threatened today, not so much by hunting and poaching or the prohibited traffic in the offspring of slain mothers, though that is still a problem, as by advancing destruction of the habitat of these peaceful giants, to accommodate ever-growing human numbers.

Because we are increasingly destroying our own vital resources also – famine in the overpopulated and so denuded "sahel" zone between the Sahara and the rainforest is an obvious example – we shall have to limit population growth in any case, sooner or later. If we do it sooner rather than later, it may be possible to save at least some of the wild animals, including, I hope, the gorillas.

For many species, certainly the gorilla, that is feasible within national parks in the long term. Some figures compiled by C. Aveling and A. H. Harcourt's research group are instructive here. In the region of the Virunga volcanoes, scientists determine not only the total number of mountain gorillas but also the proportion of juveniles (infants and adolescents). This is a measure of the birth

Eyeball to eyeball with a maligned forest giant: The gorilla acts quite peaceful, evidently because he does not feel threatened by Carlo Dani, the Italian photographer.

rate and hence of survival prospects. They found that the number of young individuals was greatest – 54 percent – in those gorilla communities that were regularly visited by small groups of tourists. Gorilla bands not visited by tourists but observed by scientists continuously or during certain periods, had as many as 49 percent juveniles. In unsurveyed gorilla troupes, however, neither visited by tourists nor under scientific observation, the proportion of juveniles was only 39 percent. Moderate tourism is even more beneficial than research activities in the preservation of gorillas. The least favorable condition is for nothing of the kind to be going on. However, Aveling and Harcourt also point out that large groups of vacationers that were not properly supervised would not be helpful. This applies to wild animals in national parks everywhere.

Raising gorillas in zoos can hardly help directly to counteract the present endangerment of populations living in the wild. There have hardly been any serious attempts – scientifically planned and monitored, that is – to "rehabilitate" gorillas born in captivity, as with orangutan and chimpanzee programs. For the most endangered subspecies, the mountain gorilla, it could not now be done at all, since to date only three babies have been born in captivity. At this moment, there are believed to be fewer than ten mountain gorillas living in captivity. The situation of the lowland gorilla is somewhat better; in late 1976, 647 individuals were living in captivity around the world, 139 of them born there.

Field studies, so important to the survival of the gorillas, were begun in 1959 by the German-born American researcher George B. Schaller in the Virunga volcano area and continued by the American zoologist Dian Fossey in a part of the same territory until she was murdered in 1985.

In the next two sections, these two field investigators will themselves describe the salient results of their work and their fascinating experiences with these giants of the misty mountain forests.

Our First Field Studies
by George B. Schaller

If a threatened species is to be preserved, careful scientific field studies in the wild are indispensable. In February 1959, my wife Kay and I made a journey to the then Belgian Congo. Our mission was to observe the free-living mountain gorillas of the Virunga volcanoes – their behavior, their group formation, their diet, their adaptation to climatic conditions, their reproduction, etc.

Our headquarters was a hut in the Kabara district, deep in gorilla habitat. Kabara is located in an enchanting park terrain with good distance visibility, at 11,000 ft (3500 m) on the saddle between the two volcanoes Mikeno and Karisimbi. There is not much undergrowth, and the trees are widely spaced. They consist largely of the kusso tree *(Hagenia abyssinica)*, whose large plumed leaves form umbrella-shaped crowns.

It is not easy to watch these apes on their daily rounds; they keep to the dense bush. The Kabara terrain was favorable to our research because the gorillas here have had little or no contact with human beings. When an observer approaches them singly, they rarely take flight.

On the first two days in Kabara, we found no gorillas, only abandoned sleeping nests. On the third day, I went reconnoitering again with our African assistant, N'sekanabo. We were passing through

The Schallers - Kay Schaller, in the picture - for the first time made thorough studies of the habits of the mountain gorilla in 1959, dispelling many legends that had grown up around these "savage" dwellers in the forest.

thick growths of lobelia; the bare trunks rose nearly 7 ft (2 m), crowned by bunches of big leaves. Up ahead, we suddenly heard a shrill cry. I signaled N'sekanabo to wait, stole carefully forward, and reached a depression which I observed from the cover of a tree trunk. A female gorilla stepped out of the thicket and slowly climbed a stump; she had a stalk of wild celery in her mouth, like a cigar. She sat down, grasped the stalk with both hands, tore off the tough outer layer, and ate the juicy part inside. A second female with a young one on her back came out, took hold of a celery stalk near the root, and pulled it out of the ground with a jerk. Then she patted the place on the ground with one hand, crouched down, and began to eat. Wild celery is the gorillas' second-most important food plant in the Kabara region. To get a better look at the other members of the group, I ventured to advance a little farther, and one of the females saw me. She uttered a short cry and disappeared in the underbrush. A large juvenile, weighing some 80 lb (36 kg), ran up an oblique log, gazed sharply in my direction, and came right down again. All at once, hardly a 100 ft (30 m) away, seven gorillas passed in single file, a large male with silver-gray back fur bringing up the rear. He stopped briefly and watched me, hidden behind leaves, so that I could see only the top of his head. Then he uttered grunting sounds in rapid sequence, evidently intending to warn both me and his band, and disappeared.

Soon three females followed with four young, two of them being carried pickaback.

Next day, Kay and I saw gorillas only from a distance, but on the third day I succeeded in approaching the troupe. I was crouching, under partial cover of some branches, in the lower fork of a kusso tree, and observed the unsuspecting animals at leisure, finding food and resting between tall bushes. The male with silver-gray back advanced slowly upon a *Veronia lasiopus* shrub, gripped it at a height of about 6 ft (2 m), and with a sudden jerk pulled up the entire plant. He deliberately selected a branch, tore it apart with his teeth, and nibbled the tender white flesh as though he were eating corn on the cob. A female was leaning against a slanting trunk. She had drawn the top of a nearby veronia within reach and was eating the deep-red blossoms, which she picked one by one with thumb and forefinger and then put in her mouth.

I identified these first gorillas that I saw in the Kabara region as Troupe I; it was not the only band in the vicinity. On the 22nd of August, about 300 ft (100 m) from the troupe I had under observation, I heard a male beating his chest, and next morning when I surveyed the woods more closely, I found that no fewer than three troupes had built their sleeping nests close together. One of these troupes was very large; nineteen individuals were later counted. The third troupe, however, numbered only five – one male, two females, and two

Gorillas have every reason to be apprehensive at the approach of humans (left). When they feel they will not be disturbed, they give free rein to their quiet playfulness (right).

young. These three troupes (I, II and III) remained in the same part of the forest for five days. Later I became acquainted with many other troupes.

When following a troupe, usually I would first find the place where I had last seen them the day before. I would carefully follow their trail, never knowing whether they had moved on 300 ft (100 m) or a mile or so, or whether they had doubled back and were now behind me. Sometimes there was a musty odor in the air, like a barnyard, and then I knew I was near the place where the animals had spent the night. The nests and the immediate vicinity were covered with droppings, and many little brown flies were swarming about and laying tiny white eggs on the fresh feces. Often it took me half an hour to find all the nests, because sometimes one individual might have slept 60 ft (20 m) or more from his neighbor.

Gorillas will make their bed wherever night overtakes them; all they need is the presence of suitable plants for nesting materials. Nests are built either on level ground or in trees. In the kusso woods around Kabara, 97 percent of the nests were on the ground, because the brittle kusso branches are unsuitable. In the Utu region, however, only 20 percent of the nests are on the ground. Juveniles will build their nests in trees twice as often as will adult females or males of immature color. Mature males always sleep on the ground in Kabara, but in nearby Kisoro they would sometimes build nests 8 ft (2.5 m) up in the undergrowth. It is widely believed that females and the young sleep in trees for safety, while the males remain below to keep watch. But this is a myth. Immature males will sleep 60 ft (20 m) from the troupe, juveniles often close to females, sometimes even in the same nest with them. A small gorilla will stay with its mother; only now and then an older youngster will build itself a little nest of its own next to her sleeping place.

In building a nest, the gorillas on the ground bend all the saplings and shrubs towards the center in one place to create a springy platform. They will do much the same in the tops or forks of trees. Of course such a nest will support a human being without difficulty. Chimpanzee nests look similar but are usually found much higher in the trees. A gorilla will build a fresh nest each night, and nearly always select a new location for it as well. The nests are only a few yards apart; there are no fixed rules about where the females, the young, or the chief will sleep within the sleeping place as a whole. In the districts of Utu and Kisoro, the gorillas sometimes slept only 90 to 225 ft (30 to

Gorillas in the wild are strictly vegetarian, feeding on many plants. Wild celery is one of their favorites.

75 m) away from human dwellings. Gorillas, incidentally, do not snore.

It is sometimes stated that the ability of young gorillas to build nests is acquired entirely by learning, but I doubt it. Of course, the youngster does learn how to build a sound nest in given circumstances and how best to use the various plants, but I think the gorilla builds instinctively, to have something around and beneath him, before he retires for the night.

As a rule, the gorillas would build nests at dusk and get up in the morning after about thirteen hours sleep, when the Sun was up, not later than about an hour after sunrise.

In the morning, gorillas must first eat for two hours or so until they are satisfied. In the Kabara region, their basic diet consists of the trailing herb galium, wild celery, thistles, and nettles. In season, bamboo shoots and blue Pygeum berries may be eaten as well. Principal foods are leaves, shoots, and the pith of various plants. In eating, they use their hands throughout, rarely biting off a piece directly with the mouth. In this also they differ from the chimpanzees, which eat primarily fruits, not so much leaves and sprouts. I identified about a hundred different species of plants picked by the gorillas. Following the trail of a gorilla group in the Kisoro region, Grzimek tasted all the plants they ate. For the most part they tasted bitter. Since these leaves and plants are not very nourishing, they need to consume large quantities. The dry mass of one day's droppings of an individual will weigh 2 or 3 lb (1 or 1.5 kg). Probably the vegetable diet also meets the water requirement of gorillas. They have not been observed drinking in the wild.

Wild gorillas have never been found eating any animal matter. I inspected their droppings several thousand times without ever finding any hairs, insect chitin, bones, skin parts, or other signs of animal fare. In captivity, gorillas will eat flesh rather readily. This, though, may be attributable to the complete reorganization of their diet and a lack of vegetable proteins.

In the wild, they disregard freshly killed animals passed on the way. Once a group of gorillas rested 10 ft (3 m) away from a brooding dove, impossible to overlook. They left the bird undisturbed. Natives have told me that gorillas often rob the nests of wild bees, but I never observed this around Kabara.

Between 9 and 10 a.m. they generally stop eating, and then it's time to rest until the afternoon. Gorillas present a picture of utter contentment when lying on the ground, gathered about the silverbacked leader, especially with the Sun shining on their bodies. They do not avoid the Sun at all; sometimes they will lie supine for more than two hours in full sunlight. Often they will even get up from their resting place in the shade and go to places on which the Sun is shining. Many stretch out on the ground, lying not only on the back but also on the side of the belly, with their arms and legs stretched out from them; others may lean against a tree. Occasionally a gorilla will build a nest to nap in. But during the noon interval, there is only sleeping, drowsing and sitting around. Some gorillas will groom and scratch and make their toilet. Often they may clean each other, especially the younger ones. But for adult gorillas, mutual grooming is by no means so important as for other apes and monkeys. Gorilla youngsters spend the noon rest period playing and wandering about. It is as difficult for them as for human children to sit still, especially when mother wants to rest undisturbed. They dash about between their rest-

Old male gorillas have an especially prominent crest on the vertex and occiput. This crest is required as an attachment for the powerful jaw and neck muscles of this herbivore.

ing elders, and play alone or with other young ones. But young gorillas are not apt to play continuously; sometimes I saw no one playing for days at a time, especially when the clouds hung low and the vegetation was wet. Gorillas in the wild stop taking pleasure in play at the age of about six years, that is, at sexual maturity. Favorite games among the children are tag, or defending a stump or hillock against assault by the others, some of the same games engaged in under various names by human children.

The gorillas will rest for from one to three hours at noontime. Sometimes, for mysterious reasons, the leader of the troupe will decide to walk on 150 to 300 ft (50 to 100 m) then lie down again and go back to sleep. Of course the entire troupe always follows him. The actions of the highest-ranking individual determine the outline of the day's activities – when and where to rest, how far to travel and in what direction. He guides the behavior of the troupe with simple gestures and vocal utterances. If he suddenly rises and walks rather stiff-legged in a given direction, the others know not only that he is adjourning but also in what direction he intends to go.

Until afternoon, the gorillas pass the time alternately sleeping, eating and sleeping, and then duly starting to eat again, until twilight overtakes them. They search for food in a leisurely manner, sit about a good deal, and from time to time travel on at a rate of 2 or 3 miles per hour (3 or 5 kilometers per hour). When it grows dark in the woods, they slow their movements, and gather around the highest-ranking male. They sit around indecisively, as though each were waiting for somebody to make a start on nestbuilding. At about 6 p.m. or sometimes even 5 p.m. if the sky is overcast, the highest-ranking male will start breaking branches to build a sleeping nest, and the rest follow his example.

In the Virunga region, it rains frequently and abundantly. The gorillas are used to it but not by any means enthusiastic. At the beginning of a shower, they will get up, move under trees, or sit down close enough to a tree trunk to be more or less sheltered from the wet. They are just as likely simply to remain sitting in the rain. In a gentle rain, usually they will merely stop looking for food. If they are already resting, they will turn over from back to belly, or sit up. At the beginning of a heavy downpour, gorillas sitting in the trees will come down to the ground; babies go back to their mothers, who often take them in their arms and bend forward to keep the little one dry. If they are sitting unprotected in the open, they will bow the head so that the chin touches

A mountain gorilla yawning. High time for a noonday nap!

the chest, cross their arms over the chest, and often place a hand on the opposite shoulder. Thus they remain sitting motionless and without a sound, letting the water run down from their shoulders and eyebrows. They make quite a pitiful impression. Hardly anything will disturb them. Once I walked straight through such a sitting group unannounced, and only one individual lifted his head. Another time I arrived in full sight of the gorillas and sat down under an overhanging bough some 10–30 ft (3–10 m) away. They looked at me but did not go away. It must have been raining for more than two hours before they decided to look for food during the heavy rain. They will also leave their sleeping nests because of rain. Their behavior is the same in a hailstorm.

In my first encounters with gorillas, I had tried to approach the animals unnoticed. However, I soon found that I was barred from valuable observations if I hid myself too thoroughly, for then as soon as I tried to work forward into a more favorable position, the animals would catch sight of me and become excited. So I would usually approach the gorillas slowly and openly, climb a tree and make myself as comfortable as possible, without seeming to trouble them. If I took up an elevated observation post, I could not only see them clearly in the woods, but also be seen by them, so that they could keep an eye on me.

I was convinced that if I came into their vicinity alone and made clear that I had no hostile intentions, the gorillas would soon recognize that I was not a threat. With some creatures – such as the dog, the rhesus monkey, the gorilla, and humans – it may constitute a sort of aggression to look to other straight in the eye. Even if I was observing gorillas from a distance, I always had to remember to turn my head aside from time to time, and not stare at them too steadily, lest they become restless. Similarly, they felt threatened if the lenses of binoculars or the camera were pointed at them, and I had to use both sparingly.

When observing Troupe IV, I had several times seen the young male "Junior," still black-backed, or sometimes another gorilla, come within 60 ft (20 m) of me and then shake his head. This remarkable gesture apparently meant, "I mean no harm." To see whether the gorillas would react if I shook my head in turn, I waited one day until Junior was only ten paces away and watching me closely, while I was reloading my camera. When I began shaking my head, he immediately looked away; possibly he felt that I had mistaken his inspection for a threat. When I looked straight at him, he shook his head. We kept this up for ten minutes. After that, when I came upon gorillas unawares I would always shake my head to announce my peaceful intentions, and they always seemed to understand the gesture.

Some years later, Dian Fossey found out that when a gorilla looks "deep into another's eyes," the intent may be quite friendly. Anyway, gorillas are sometimes swayed by curiosity, and this without apparent cause. One day, a group that I had met and observed 76 times before, came gradually closer and closer, to a distance of 30 ft (10 m), and looked me over with the greatest curiosity. A female with a three-month-old baby at her breast came closer still, reached up, and sharply rapped the branch on which I was sitting. Then she looked up to see what I was going to do about it. Presently a half-grown individual did the same, and one female even climbed up the branch for a couple of seconds. More often, the males, especially the leader, would make mock attacks, perhaps to see what I would do, or to drive me away. Once a large male made a seven- or eight-yard run at me in this manner, but stopped short while still 15 to 90 ft (25 to 30 m) away.

Gorillas can be recognized individually, like human beings, not only by physical conformation but also by their whole manner, and especially by their faces. I had named the members of the troupes I most frequently observed; this greatly assisted me in my observations and in reading my notes. I found out that the troupes are of fixed composition and closely coherent, quite unlike the small groups of orangutans and the large communities of chimpanzees, in which a constant coming and going seems to prevail.

Troupe IV, with which I was especially familiar and which was led by "Big Daddy," was attended by another full-grown male gorilla. He would range about the group unconcernedly, paying no attention to the others, so I named him "Outsider." He was of giant size and in full possession of his powers, definitely bigger than Big Daddy and by far the heaviest individual in the entire area. During my first weeks in Kabara, Outsider left the group and returned at least twice. Some male gorillas live solitary lives but keep joining some troupe for a time. In the Kabara region, I found seven such loners – four silverbacks and three males of the less mature color – but there may have been a few more. These loners attached themselves to Troupes IV and VI exclusively. Apparently only certain troupes are willing to accept loners, and the latter find out in the course of time where they are welcome and where they are not.

The most familiar of the loners to me was Lonesome Stranger, a silverback in his prime with a pensive glance and a faint look of disdain about the mouth. He didn't care for me, and usually hurried off yelling when he caught sight of me. I observed him for the first time on the 18th of November near Troupe VI. "Mr. Dillon," the highest-ranking male of this troupe, was resting, surrounded by the females and young; they seemed to take no notice at all of Lonesome Stranger, only 30 ft (10 m) off. A quarter of an hour later, Mr. Dillon stood up. The members of his troupe passed by him, but he remained behind and stared at the visitor. Seemingly he was giving him to understand that it was time he took himself off. Lonesome Stranger disappeared into the woods, but in the coming weeks I observed him following the others several times. When, in May, Troupe IV took to the slopes of Mikeno, Lonesome Stranger joined them for at least a week. But soon he got the wanderlust again, and departed, to roam the forest by himself once more.

When more than one male of mature color belongs to a group, there is always a fixed social ranking. The higher-ranking will take precedence on a narrow trail, and may at any time displace the lower-ranking. Such rankings prevail among many social vertebrates. Contrary to a widespread misconception, this does not lead to conflict and resentment, but makes a peaceful coexistence possible, because each has a definite position in the social structure. Each individual accurately knows its rights and duties in relation to all other members of the group. The whole grop is guided by the silverback chief, though he does not seem to give any orders. He is the most alert and excitable of all; he will roar and threaten long after the others have become calm. But by the same token, he is the most shy, and tends to blend into the background.

The males with silver-gray backs are superior to all other members of the gorilla troupe. To some extent, an individual's social rank seems to depend on size and strength. Similarly, the females are superior to the adolescents, who in turn rank higher than the juveniles no longer with their mothers. Once when it was starting to rain, an adolescent sought shelter under a leaning tree trunk. He pressed close to the trunk and looked out into the heavy rain coming down around his dry spot. But

Little gorillas grow up in the shelter of their family group. During their long childhood, their elders prepare them for adult life. The pictures show a four-month-old lowland gorilla with its mother (above) and a two-year-old (below) with a young adult male.

when a female made for the tree, the young fellow got up at once and ran out into the rain. Hardly had the female taken shelter than a fully adult male emerged from the bushes. He sat down beside the female and thrust her away gently but firmly with one hand until she was sitting in the rain and he had taken her place.

Among the females of a troupe, however, there does not seem to be any permanently fixed ranking. It is perhaps significant that it is chiefly the females who quarrel, whereas the adult males take no active part in such altercations. I am not quite sure, but I believe that the females have a "sliding scale" of rank; mothers with newborn or very small infants rank above females with older children. Of course, the temperaments of individuals are a factor also; the more excitable members of the troupe are generally deferred to.

The highest-ranking males, of mature color, are absolute monarchs, who always get their way because of their size and social position. On the other hand, they are considerate and peaceable, as is especially apparent during rest periods. The females and the young of a troupe really seem to feel affection for the highest-ranking male. Sometimes a female would rest her head against the silver-gray fur of his back or lean her body against him. Occasionally as many as five infants would come to him, sit by his feet or in his lap, climb all over him, and generally behave rather impertinently. He paid no attention to them unless they began to overdo it. Then one glance sufficed to restore order.

Although the government of the new independent State has been protecting the mountain gorillas even more strictly than did the Belgians – and in spite of the civil war – they are still threatened by farmers who want to clear more of the forest and bring it under the plow, by hunters, and by the pastoral Watusi, who try to graze their big herds of cattle, of which they make no economic use, in the gorilla country.

Young mountain gorilla riding on the shoulders of a black-backed male, who treats the little one with indulgence.

New Observations of Gorillas in the Wild

by Dian Fossey

Two Memorable Experiences

"What was your most memorable experience during your fifteen years with the mountain gorillas?" I have often been asked this question, and when answering it, I always call two events to mind. One occurred when I had been doing "field work" with gorillas – studying them in the wild, that is – for only ten months. I felt that they were still reluctant to accept my invasion of their private world. One day I saw a young male gorilla about 25 ft (8 m) away, eating. To my surprise, he put down his leafy branch, turned, and looked me attentively in the face. The depth of the expression of his eyes held me fascinated. His calm glance combined curiosity with good will. When our eyes met, I felt as though an invisible barrier between humans and apes had been lifted. Some moments later, with a deep sigh, the young "black back" slowly resumed his meal, as though he has just solved a puzzle. I withdrew, at once honored and chastened.

The second extraordinary encounter occurred fifteen years later. It may perhaps be more remarkable in terms of behavior research, but not less significant to me on that account. From 1980 to 1983, I had been absent from the research camp and from "my" gorillas. Mixed feelings, varying between anticipation and apprehension, were mine when in July 1983 I prepared to renew acquaintance with one of my favorite subject groups. Would they remember me? To be honest, I doubted it when I considered how many other human observers had met them during my absence – not to mention the day-to-day events of their life among themselves. I approached the group as they were feeding one rare sunshiny day, and uttered my usual greeting sounds – a soft train of inarticulate murmurs imitating the gorillas' own sounds of satisfaction. When I was still about 20 ft (6 m) away from the nearest member of the group,

Peace and concord between humans and apes: Dian Fossey surrounded by her friends. The American zoologist and dedicated friend of animals spent may years among mountain gorillas, until her cruel assassination in December 1985. In the course of time, she had come to be completely accepted by the animals, and thus gained direct experience of their life. She tells in these pages of what she learned.

I sat down. She was a nice old lady, a mother of six, one of them born during my absence. She glanced at me briefly as she chewed a long stalk of celery. She looked away again, then did a double take, as though she couldn't believe her eyes. She stared at me questioningly, intently. Then she threw her celery stalk down and marched over to me very fast, to take a careful look at my face. Her eyes were a couple of inches from nine. She held that pose for about 45 seconds. Then she put her arms around my shoulders, pressed herself against me and sat down close beside me. In so doing, she communicated her discovery to the rest of the group, scattered through the underbrush, with loud cooing sounds.

Her plaintive calls brought one after another of the older gorillas out of the bushes. Each repeated the eye-to-eye examination before settling down with the rest, on or around me, with long arms embracing us, making a black, hairy heap. I wept; for the second time, I felt chastened and honored, by the fact that the gorillas had recognized and acknowledged me, after a three-year absence.

Significantly, this sort of eye-to-eye examination, attended by a close gathering of the group, has never been repeated in later encounters. From that day on, I was treated as an "old acquaintance," and I continued my studies as though I had never been away.

These two incidents are impressed upon my memory largely because they were so different from the comparatively few occasions on which I met with any aggressive or threatening behavior. If only because of the great excitement attendant upon them, such confrontations would necessarily be the exception, as indeed they are in my notes on meetings with gorillas in the early period. Happily, I seldom observed behavior of this kind, and I must confess that it was usually due to my unintentionally approaching a group before I had noticed their near presence. The gorillas were startled, especially when there had recently been an encounter with other groups or with a solitary silverback male – events involving considerable tension.

Even if I had remained convinced that attacks by

Siesta in a sunny forest clearing.

silverbacks were simply a defensive operation, with no intention of harming the opponent, their threatening display can only be described as fairly terrifying. The cries accompanying this display are frightful, and seem to come from all directions at once – especially when more than one individual is coming straight at you like a bulldozer. At some point, the spell is broken, and eye contact between the menacing animal and the observer closes the circuit. The danger ist past. That may be the moment allowing a bilateral, even a courteous, withdrawal, ending the incident.

What tragic irony that the erect posture of a disturbed breast-beating silverback turns the most vulnerable side of his body towards the cause of his unease, whether it be a poacher's weapon or another gorilla.

Confrontations Between Male Gorillas

Silverbacks, experienced in encounters with other adult males, seem to be aware of their enormous strength, and this must be why adults have never been known to be killed in these hostile episodes. Male gorillas have developed remarkable resources for various equivalent threatening behavior patterns, enabling them to avoid open combat while nevertheless announcing it or at least placing their physical strength in a proper light. They may parade up and down side-by-side, stubbornly stiff-legged, their heads just perceptibly averted from each other. Willingness to make peace is suggested by the luxuriant black hair of their arms, the increasing visibility of the silver saddles, and the bristling hair on their high bony crests. If tension between gorillas threatening each other exceeds the limit of tolerance, it can be released by a maneuver of running apart, whereupon both may beat their chests or break off branches. Such ritualized conduct is a highly effective means of intimidation, preventing injury – "agreeing to disagree."

Altercations between adult male gorillas lead to serious injuries only if one or both lack experience in group behavior; then caution is thrown to the winds. I felt quite helpless when I had to watch young, impatient silverbacks clumsily interfering with older, experienced males in their courtship of females. The lives of three such overzealous individuals were permanently altered by the severity of the bites inflicted upon them by more experienced males during brief quarrels over a female. One of the three young silverbacks was presumably killed by poachers. The second is still a loner with a festering bite wound in the chest – eleven months after he recklessly made his way into a strange group to get himself a female. The third is severely handicapped by a fourteen-year-old bite in the head, incurred when he was half grown. He foolishly interfered in a fight his father had started with another group, over two females. This young male is still living, but he merely wanders about with other young males, and is barred from reproduction by his disabilities.

Infanticide in the Interests of Reproduction

Encounters between groups can result not only in injuries but also in deliberate killing of infants.

In the past eighteen years, 12 infanticides have been reported within gorilla groups under study, but only eight of them are fully documented. In seven cases, the mothers of the victims left their group and formed a new connection with the killer. In eight cases, the mother entered into a new sexual relationship within her own group. At first the causes of infanticide were hard to understand. But as this behavior – mothers leaving the group when their young had been killed was repeated from time to time, it became clear that this process favors exogamy and hence the spread of genes throughout the population. Among gorillas, infanticide is a reproductive measure prematurely restoring the readiness of the mother to conceive by eliminating the alien offspring, and enabling the perpetrator to father an offspring of his own. George Schaller, by the way, has observed quite similar infanticides by male strangers among lions

▷ The picture of peace: mountain gorilla having a midday nap.

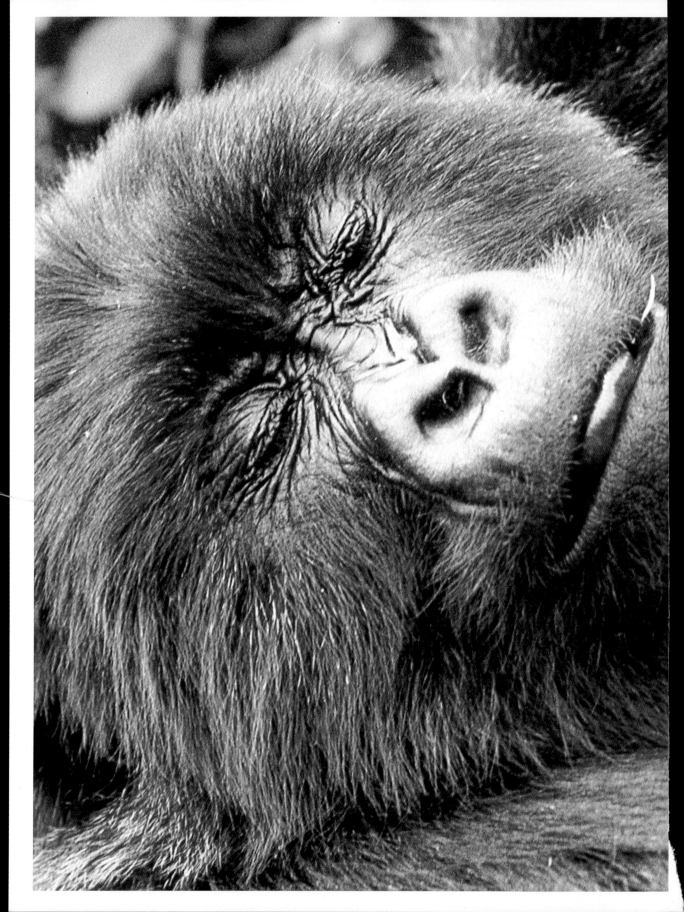

in the Serengeti, except that it is not the female but the male lions that change groups.

The imposing image of a silverback male as defender of his family, and this merciless elimination of an infant not related to him seen as an obstacle to reproduction with its mother, is in marked contrast with the benevolence so characteristic of the male gorilla's paternal role. The polygamous organization of gorilla society occasions frequent and very close relationships of fathers as well as grandfathers with their mates and descendants.

A background mood of contented togetherness is constantly observed during lengthy rest periods in the course of the day, when gorilla females and young attach themselves closely to the highest-ranking silverback. That he good-naturedly tolerates their presence is one of the hallmarks of his success as leader of the group. He exhibits a certain preference for the higher ranking females of his harem.

Many years ago, when I was beginning my work and before the gorillas had gotten used to me, I once spent a few delightful hours in the dense foliage on the banks of a ravine. Far beneath me, in a little clearing in the bush, a group of gorillas were Sun bathing, in peaceful contentment and with not the slightest suspicion of my presence. The only silverback in the group, a magnificent old male, lay stretched out near one of his consorts, the highest-ranking female, while their young played idly about them in the hot Sun. Playfully, the old male reached over to lift the youngest infant, a six-month-old female, from her mother's lap. Smiling, he held and swung the baby overhead, absently groomed its coat, and put it back with its mother. During the rest of the midday recess, the other youngsters and the consorts took turns grooming the old male, while exchanging soft murmurs of utmost contentment. This observation is typical of the usual close cohesion in a gorilla group as well as of the paternal amiability of the silverback.

Now, seventeen years later, I returned to the same place, following the old male's group. He had died recently, and I was saddened, although the group had remained intact under the leadership of one of his grown sons. His consort, now very old, has survived him, and four of her six offspring were still living in the group. The once six-month-old female who had been played with by her father is now the mother of two children by her half-brother. The passage of time and its consequences to these magnificent endangered animals is disquieting. I am often astonished by their relationship to time, and by their faculty of memory.

Behavior towards Sickness and Death

Their reactions to a death are quite noteworthy. The old male's sole surviving spouse, who had always been quiet and good-natured, became irritable after his death, and violent towards other

The family group, led and defended by a "silverback," is the basic unit of gorilla community life. There are close if not always conflict-free relationships between the individual members of such a small band.

gorillas and the observers. Perhaps she felt insecure without the protection of her mate of many years, or – more likely – she was suffering from old age. I was surprised to find that none of the other individuals showed any changes in behavior following the death of the patriarch. After the natural deaths of four other silverback leaders, the members of their groups had seemed unwilling to leave the body. In one case, the females and the young followed for several hundred feet while it was being removed, and then returned to the place where the dead had lain for several days. Such behavior raises the question whether gorillas comprehend the notion or the finality of death. When infants lose their mother, they stop playing, lose their appetite, and fall into a profound depression that may last for as long as a year. The ability of some infants to deal with the loss of a mother depends largely on whether there are willing siblings to console the youngster, pay attention to it, groom it thoroughly, and hug and kiss it. The silverback also willingly adopts motherless infants in his group, sheltering them from the unruliness of older ones, taking them into his nest at night, and grooming their fur.

A mother gorilla who has lost her baby shows manifest signs of grief, especially after an infanticide. After the loss of a firstborn through infanticide, young mothers take refuge in childish play behavior. Perhaps such unforeseeable reactions reflect the young mother's natural relief at suddenly finding herself no longer burdened with a mother's duties. However, it is also possible that this play behavior on the part of the mother represents an attempt to restore the close bond with the group members after the horrible experience of the infanticide.

In four cases, the behavior of gorillas towards fatally ill or severely weakened companions may be considered very strange, indeed from a human point of view cruel or murderous. All four cases had one feature in common; there seemed to be a deep compulsion to elicit a response from the fallen individual. The longer the "victim" was incapable of responding, the more violent the efforts of the "tormenters."

The most striking of these incidents concerned an old female gorilla who was seen from a distance, lying either dead or in deep unconsciousness and in any case fortunately in no condition to realize what was happening to her. For about 24 hours, the all but hysterical gorillas beat her up, pulled her about, jumped around, jumped on her, examined her, and mounted her. Up to a certain point, this excitement of the gorillas is reminiscent of human reactions when seeing someone unconscious, consisting in attempts to rouse the person by slapping and shaking. This marginal resemblance may suggest that such behavior need not be perceived as cruel.

Mating, Birth, and Childrearing

Another human-like emotion of the gorillas is a tendency to anger. Most friction within a group occurs between the matrilineal "clans," the descendants of the several females, when one of the females is in heat. At these times, the members of the fertile female's clan have preferred access to

This mountain gorilla has taken the trouble to climb up high for food.

▷ Mountain gorillas in Ruanda have climbed a kusso tree.

▷▷ The facial expressions of gorillas always seem slightly melancholy, as though they were aware of the gloomy prospects for the future of their species.

the highest-ranking silverback, independently of their rank at other times. Jealousy between clans leads to increasing tension and outbursts of shouting between different clan groups.

The silverback decides such disputes rationally in favor of the female who is fertile at the time, either restoring order by scolding or simply rushing into the midst of the restive females. After such a call to order, the disputants reluctantly separate, while the silverback gives his undivided attention to the female in heat. Fortunately for the silverback, it does not happen often that more than one female is in heat at the same time.

The presence in the group of a female in estrus causes animals of the same sex or adults and adolescents not closely associated to mount each other more frequently than usual. Females on the verge of sexual maturity begin to flirt with young males in the group. These young gorillas preen themselves and strut about in a comical imitation of their elders. Infants are attentive observers of displaced sexual activity as well as of real sexual intercourse between group members, but despite their curiosity they very rarely interfere.

As in the case of human children, a long period of dependency on the mother provides young gorillas with many learning opportunities and prepares them to deal not only with their environment but also, no less important, with their conspecifics. Maternal status favors very striking differences in behavior between individuals. The indulgence of one mother may be in marked contrast to the strictness of another. I suppose that the basic causes of these differences may have to do with the young female's past opportunities to devote maternal care to younger relatives. At the same time, these differences reflect differences in the amount of nurturing that the female experienced during her own formative years.

Low-ranking females bearing their young in large, tightly structured groups apparently do not become very successful mothers. Their lack of self confidence is easily transmitted to the infant, which then lacks assurance in its relations with older individuals. For the mother, to be sure, the situation is not irremediable. She may get the chance to transfer into a smaller group or to join a solitary silverback – events that would automatically increase her rank and influence. Of course, if she emigrated while the baby was still dependent on her, infanticide would certainly result. Weaned children are usually left behind in the group of their birth if their mother emigrates, and the highest-ranking silverback in the group will make himself responsible for them.

As an example of these situations, I may mention a female gorilla who was orphaned at the age of three years when both parents died natural deaths. The silverback who took over the leadership of her birth group was probably her uncle. He adopted her willingly and looked after her until she was sexually mature. She then mated with one of the males of the group, probably her cousin, and had her first baby when she was ten. But shortly before the baby was born, both the uncle and her mate were killed by poachers, leaving her with no aid or protection for her firstborn. When eight months old, the baby was killed by the silverback of a large group into which the unfortunate female had been admitted. For six years she remained the lowest-ranking of the six adult females in the group, and never became a full-fledged member, although she had a second child three years after she joined the group. When this youngster was three years old, the mother changed again, this time to a solitary silverback. She left her three-year-old behind in her birth group, although its ranking there was very low. She and the lone silverback stayed together for 104 days; they were often seen mating. Just as was beginning to see a secure and fruitful future for them, they were separated by a violent counter with the silverback leader of another large, close-knit group. As an ape adult female, her ranking will remain very low indeed unless has another chance to join some small formed group or a lone silverback. The eve this one individual's life show what a train

This juvenile mountain gorilla is looking up at the photographer with curiosity rather than with alarm.

Gorillas are more adapted to life on the ground than the other three species of great apes, and heavy adults even have some difficulty in climbing a tree.

happy events may follow from a single poaching incident. How different this female's life might have been if her mate had not been killed.

Protection from Poachers and from Destruction of Habitat

During our eighteen years of research at the Karisoke Institute, six gorillas of the groups under

Despite all protective measures, poachers are still the greatest threat to the diminished stock of gorillas. The ruthlessnes and brutality of the hunters is exemplified by this picture: the foot of a mountain gorilla, caught in the wire snare of a poacher.

study were killed outright by poachers; six others died of immediate consequences of poaching (for example, infanticide following the death of the group leader). During the same time, 23 died of natural causes – including infanticide unrelated to poaching. In the entire Virunga volcano territory, 49 killings by poachers were reported. But since news of more remote groups not under regular observation is scanty, probably the figure is an understatement. Also, one cannot tell how many other gorillas perished in consequence of these kills.

In 1960, the total population of Virunga mountain gorillas was estimated at 400 to 500; in 1981, on a closer count, at 230 (hardly more in the early 1970s). This reduction by half is essentially a result of poaching, although loss of habitat through agricultural development (36 mi^2 or 100 km^2 in the Ruanda sector of the park alone) will have contributed to the decline in population.

The Karisoke Research Center was founded chiefly with the thought that research and active protection are equally important goals. Observers should not have the right to intrude upon the gorillas' habitat or jeopardize the openness of these magnificent animals towards humans without becoming completely subordinate to the preservation of their species. Some of our tasks have been to cut wire snares, conduct compaigns against poachers, confiscate poachers' weapons, train park rangers, and bring pressure to bear on the authorities to obtain long prison sentences for any wrongful despoilers of the park. In the past year (1984) alone, we devoted 5 percent more time to poacher patrols than in all the previous years, with the result that we had 75 percent fewer traps to destroy.

It is my passionate hope that we shall not be too late in our efforts to hold the line for the future of these splendid animals.

Note to reader:
In December 1985, a few weeks after we received this article by Dian Fossey about her research on gorillas in the wild, she was assassinated in Ruanda at the Karisoke Research Institute of which she was the director.

Chimpanzees

The Chimpanzee
by Jane Goodall

The chimpanzee *(Pan troglodytes)* is the closest living relative of humans. There are striking resemblances – biochemical, physiological, and anatomical as well as behavioral – between the two species. Even the life cycle of a chimpanzee is not much different from ours. A seven-year childhood is followed by a period of maturity. The female in the wild becomes capable of reproduction at the age of 11 to 13. She is then able to carry a pregnancy to term. The male becomes sexually mature at about the same age, but it may be another two to three years before he is fully integrated into the society of adult males. His maximum weight (103–121 lb or 47–55 kg) is not attained until he is 20 years old. In captivity, chimpanzees reach sexual maturity much earlier, and weight gain sets in correspondingly earlier in both sexes. Old age begins at 35 to 40 years of age; the greatest age of a chimpanzee in the zoo is 55. Because of more severe infestation with parasites and absence of medical care, chimpanzees living in the wild probably die somewhat younger. Laboratory and zoo data tell us that chimpanzees can contract virtually any infectious disease of humans. Thus, the population in the Gombe National Park, Tanzania, which I have been studying since 1960, was once decimated by an outbreak of poliomyelitis (infantile paralysis), and on another occasion there was a suspicion of infectious pneumonia in a series of fatalities.

The chimpanzee's brain closely resembles ours in structure. As is becoming clear, their intellectual capacity is on a far higher level than was formerly supposed. All who have worked with these anthropoid apes unanimously report that their expressions of feeling – joy, grief, curiosity, fear, rage, etc. – are very similar to ours.

The better we understand the behavior of chimpanzees in the wild, the more readily we should be able to interpret certain aspects of human phylogeny. Such, at least, was the view of the paleontologist and anthropologist Louis Leakey. That chimpanzees and humans proceeded from a common ancestor has now become the scientific consensus. If such was indeed the case, modes of behavior common to modern humans and the modern chimpanzees should have belonged to their common ancestor as well, and hence to humans of the Stone Age. Some environmental changes faced today by chimpanzees in the wild may resemble those that early humans had to contend with.

Community Life

The structure of chimpanzee society is complicated. On the Gombe River, the strength of the group or community chiefly observed varied over the years between 38 and 60 individuals, including babies and juveniles. Usually there are about twice as many adult females as males. The area ranged by one community is 3 to 5.5 mi^2 (8 to 15 km^2), within which the chimpanzees have a nomadic habit; the exact size will depend largely on the number of sexually mature males, compared to neighboring communities. Adult males traverse larger portions of the common range than do females, which once they have borne young are more settled in location.

The members of a community do not travel as a unit. They often move about and feed in four to six small, looser associations. The composition of these bands changes continually, as individuals or groups separate from or join with others. Some members meet only when they have a common interest, for example when attracted by an especially abundant food source or a charming female. Others are together often, and strong bonds develop

Unlike the terrestrial gorillas, the chimpanzees, which also spend much time on the ground, are very good climbers. This is here demonstrated by a half-grown female, climbing effortlessly up a pole.

between some of them. But only a mother and her dependent offspring, under eight years of age, will be found invariably together. As a rule there are strong bonds of care and assistance among all family members, so that they will spend a good deal of time together.

Reproduction and Child Development

The female sexual cycle is characterized by tumescence and detumescence of the genital parts. The complete cycle averages 38 days, of which the period of detumescence includes two to four days of menstruation. Full swelling continues for ten days, coinciding with the estrus, in which the female is sexually attractive and receptive. Mating procedure varies. At the time of maximum tumescence, a female may be followed and finally also mounted by most or all of the males in the community. Sometimes the highest-ranking male in the group will claim possession and try to prevent rivals from mating with his chosen female; such behavior is exhibited especially during the last three or four days of the female cycle, when the probability of conception is highest. Then again, it may happen that a male leads a female away into an outer area of the range. Even a male of very low rank has opportunities to reproduce. He need only contrive to seduce or abduct a female in an early stage of the cycle in which she is not yet attractive to high-ranking males. Thus all males have a real chance to pass on their genes, and doubtless this is an important factor in the very noticeable individuality of chimpanzees, in appearance, personality and behavior.

In the wild, a female will have a baby every five or six years, unless it dies, in which case she will be pregnant again within a few months. So far as I know, there has been only one twin birth on the Gombe. It was very difficult for the mother to take care of both infants, and one died when ten months of age.

Upon the birth of a baby, the mother's older child gradually becomes more independent; it is weaned, forages for itself instead of riding on its mother's back, and builds its own sleeping nest, in which it sleeps alone. However, the strong bond with the mother is maintained. Since the older

Unusual family idyll in the African rainforest. Chimpanzees live in communities of considerable size, in which primarily mother and child constitute fixed units; in this case, however, father has put in an appearance for the family photograph.

child remains completely integrated into family life, usually loving, caring relationships are formed between siblings.

Until the age of nine years, the young male spends most of his time with the family. Then he begins to associate more frequently with other conspecifics and to travel about with them. As a rule, he will choose one or more male members of his community. Fascinated by their behavior, he watches his models attentively. But the male chimpanzees are less indulgent towards him than they were in his infancy, so that he becomes increasingly nervous in their presence. In the course of time, the young male becomes larger and more aggressive. He begins the long struggle for his place in the community by intimidating the females, before venturing upon confrontations with the lower-ranking adult males. During all these quite exciting years, he maintains the tie with is mother and his family, and finds fresh strength among them in his struggle for rank.

Before the young female is about ten years old, she rarely leaves her mother for more than a couple of hours. This changes at the onset of the first estrus. Towards the end of her adolescence, she begins to withdraw farther from the group. During estrus, she will even sometimes mingle with members of a neighbor group and mate with the males. After several such visits, she may transfer to a new group for good, become integrated into it, and thus lose touch with her family. Instead, she may form a family of her own in her home community. Exchange of young females between communities and their visits help enlarge the gene pool and prevent excessive inbreeding.

At evening, each chimpanzee builds a sleeping nest of branches and twigs bent together, wherever his group happens to halt. Only the little ones are excused from this task; they creep into bed with their mothers. The older youngsters must make their own beds (above). Sometimes the animals require an opportunity for sleep during the day. An adult chimpanzee (below) is stretched out comfortably in such a "day nest."

Feeding and Foraging

Chimpanzees, like human beings, are omnivorous; their chief source of nourishment is fruits, but they will also eat leaves, buds, seeds, pith, bark, insects, birds' eggs, and meat. They will hunt small to medium sized mammals. On the Gombe, these may be young bushbucks, bush pigs, baboons, and young or old monkeys of various species, their preferred and most frequent prey being the red colobus *(Colobus badius)*. Occasionally they will catch small animals such as squirrels or rats. On the Gombe, both males and females hunt, but the males do so more often. A community of some thirty adults and juveniles may kill more than one hundred animals a year, sometimes cooperating very cleverly. When they have made a kill, there is great excitement and loud shrieking. All chimpanzees already present and all within hearing gather in haste about the possessor or possessors to demand or snatch a morsel of meat.

Chimpanzees sometimes practice cannibalism. There have been four known cases on the Gombe in which male chimpanzees killed and at least partly devoured infants from neighbor communities. One female together with her nearly adult daughter would seize newborn young of other females in her own community, kill them, and eat them. Only when the female had her own baby and the daughter became pregnant did their baby-killing phase come to an end. Why they had developed this behavior at all remains a riddle.

Infanticides have also been reported from Mahale, and a case of cannibalism in Uganda is known. Chimpanzees will also occasionally eat the remains of animals killed by other predators, for example by leopards, as in Mahale. And on the Gombe, they sometimes steal freshly killed prey from under the nose of baboons *(Papio anubis)*.

This is another respect in which the chimpanzees are closest to us of all anthropoid species; they will sometimes eat meat, and will kill even fair-sized mammals for that purpose. Relatives and acquaintances will promptly gather around the might hunter and demand their share. Here, a male chimpanzee has bagged a colobus and is dividing the prey more or less cheerfully with his comrades.

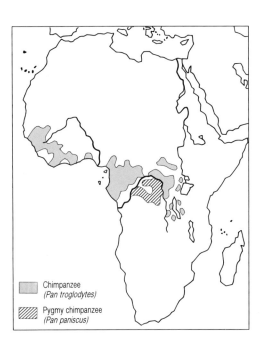

Use of Tools

There is no other beast that has the ability to use so many different kinds of objects as tools as does the chimpanzee. On the Gombe, they use straws and thin twigs to extract termites from their hills, and long sticks when they wish to enjoy the delicacy of driver ants. They also use branches to investigate objects that they cannot reach. When a chimpanzee finds a piece of rotted wood, it inserts a twig into a hole, pokes around, pulls it out, and sniffs at it. The chimpanzee then either breaks the wood open to look for edible caterpillars or else loses interest. A young chimpanzee who was not allowed to touch his newborn sister broke off a twig, touched the baby with that, and then sniffed it. Chimpanzees use leaves as wash cloths to remove dirt or blood from their bodies. Leaves are also used like sponges to pick up rainwater out of holes in tree trunks when the chimpanzees cannot reach it with their lips. They use stones and sticks as missiles or clubs, especially when fighting with baboons over food.

Chimpanzees often prepare objects that are to be used as tools. They will strip the leaves from a branch or remove the bark from a stick. To render leaves more absorbent, they will first be chewed and kneaded. Chimpanzees are also capable of making simple tools, a performance originally credited to humans alone. When Leakey first heard of these abilities of chimpanzees, he quipped, "Now we must either redefine 'man' and 'tool,' or else reckon the chimpanzees as men."

It is an especially important fact that the chimpanzee populations in different parts of Africa have developed different traditions of tool use. In West Africa, for example, they use stones as hammers to open hard-shelled nuts, behavior that has never been observed among chimpanzees in East Africa. Even between chimpanzees on the Gombe and those in Mahale, living only about 95 mi (160 m) apart, there are differences in tool use. These different traditions represent "precultures" passed on from generation to generation, the young watching and imitating the old, and mastering what they have seen by constant practice.

Communication and Aggression

Chimpanzees communicate with each other by a multitude of cries. Some can be heard over long distances and serve to maintain contact among scattered members of the community. If the chimpanzees are close together, they will chiefly employ gesture and posture as means of communication. These may appear very human-like. When chimpanzees meet after a period of separation they can be seen to embrace, kiss, touch or stroke each other, or hold hands. When a chimpanzee is begging, it will often stretch out its cupped hand. During a fight, the aggressor may strike his opponent with the flat hand, hit him, or kick him. Scratching and hair-pulling are favorite tactics of

There is no other species that has developed tool use and simple tool-making to so high a level as the chimpanzees, and this in the natural state, not under the influence of a trainer. (Left) This chimpanzee is using stones and lumps of clay to help. (Right) Another individual uses a smooth pointed stick, probably to poke some dainty morsel out of the ground.

▷ Relaxed sociability on the forest floor.

▷▷ A group of chimpanzees have climbed a leafless tree to reconnoiter "enemy territory."

▷▷▷ These animals find their chief nourishment in the trees: fruit, leaves, seeds and bark.

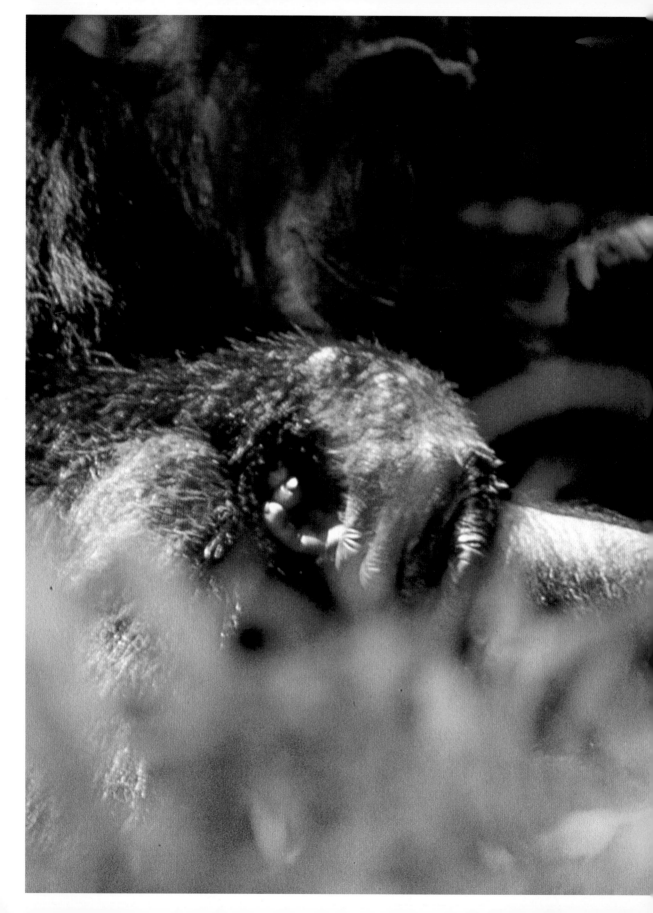

(Top) Chimpanzees, noisy and communicative as they are, habitually carry on "local" and "long distance" conversations. This individual is uttering a penetrating long-distance call because he has a message for his consorts, who have drawn away from the group. It is considered certain that chimpanzees can recognize one another by voice.
(Bottom) Ranking is an important feature of chimpanzee society. The outstretched hand of the high-ranking male is a gesture of conciliation vis-à-vis the lower ranking. It will have the effect of gradually calming and relaxing the obviously excited individual.

females especially. But biting, stomping and dragging the vanquished about are also common aggressive behavior patterns.

After a quarrel, the vanquished will often approach the attacker, crouch humbly, weep, and sometimes hold out his hand. The superior individual responds to this behavior in a quite typical way. He will touch, stroke, embrace, or "groom" the supplicant, who then feels visibly relieved. In this way, social harmony is restored. Friendly body contacts are of utmost importance in the chimpanzees' social life. Undoubtedly this goes back to the long period of childhood in which the hurt or frightened youngster seeks support in the mother's arms. The less time it spends with its mother, the more it seeks similar shelter by contact with other chimpanzees.

Mutual grooming, giving grown chimpanzees occasion to spend many friendly hours together, is of vital importance in chimpanzee society. In the first place, it calms and tranquilizes the one being groomed, whether it be a cranky child or a frightened individual of low rank; in the second place, it is also relaxing for the groomer. Mutual grooming helps preserve close relationships and consolidate looser ones.

Like nearly all social animals, including humans, chimpanzees act out their disputes by threatening sounds and gesture. Direct conflict is thereby avoided, with its danger of injury even to the aggressor. Especially effective in this context is an imposing display practiced by male chimpanzees and occasionally exhibited also by females, in the following manner. The chimpanzee will charge over the ground or through the trees, pulling down branches and swinging them to and fro, or dragging them on the ground, throw sticks and stones, and stamp and trample about on the earth. In this way it gives the impression of being larger and more dangerous than it actually is. Most male chimpanzees are busy from early youth dealing with the ranking order of their community. The more violent and impressive their display, the higher their position in the male ranking will be.

Often the end of an argument between two individuals will be brought about by the intervention of a third, taking the part of one of the adversaries, either by threats or by violence. Family members especially will support each other in this way. A male maintaining a guardian relationship with a brother will gain a higher place in the male ranking than one having no such special ally.

In the struggle for social position, there certainly may sometimes be confrontations among the males that are quite violent. Other causes of aggression in the community are defense of family members, squabbling over meat, and also frustration, meaning a feeling of being disadvantaged or overlooked. Frustration appears when an individual has been attacked or warded off by one of higher rank against whom he dare not take revenge; he will then turn upon some lower-ranking individual in the area and vent his rage upon him. Sometimes, though not often, strife will erupt between males over a female, and sometimes males will attack females unwilling to follow them in the early phase of courtship. Sometimes also, females may be attacked violently for reasons not clear to us. It almost seems as though the males need to beat up on the females occasionally because the society is a patriarchal one, given shape, that is, by the fathers.

Confrontations between Communities

Although very rough conflicts may arise between members of a community, they seldom last more than a minute or so, and usually there are no casualties. A fight between members of different communities, on the other hand, is carried on without restraint, and nearly always results in serious injuries and sometimes even deaths.

The males of a community will travel regularly, for ex-

Chimpanzees are no strangers to "intimate" forms of communication. This male greets his girl friend with a tender kiss.

ample about every four days, in groups of not less than two, into the margins of their range, where it overlaps with those of neighbor communities. They then move about with extreme caution and without a sound, always staying close together. Any sudden noise, for example the cracking of a twig, will startle them, and with anxious, grinning expressions, they will stretch out a hand to make sure of their companions. Often they will stop to sniff the ground where it is scented with discarded rinds of fruits, droppings, or fallen twigs. Sometimes they will climb a tall tree to survey "enemy" terrain. They may remain quietly at the top for an hour or more, searching the surroundings for significant clues.

If they see or hear a neighbor group comprising a larger number of adult males, they will silently withdraw to safer and more central parts of their own range. If they encounter a group of like size and composition, both sides will perform a wild dance; rocks are thrown, tree trunks are beaten, savage cries are uttered. After this impressive display, usually both parties will retreat into the security of their own ranges.

Behavior of this sort serves to intimidate, and thereby diminishes the risk of a regular battle. But when the patrolling males see or hear a solitary stranger or a mother with her offspring, they will give chase and attack if they are able. Between 1970 and 1985, we recorded 17 severe assaults upon strange females on the Gombe by two or more males, and in ten cases we could tell afterwards that the females had been seriously wounded. In three cases, these females' babies were killed and partly eaten during the encounter. Remains of another juvenile were found in the feces of the highest-ranking male (no juvenile was missing from his own community at the time). A fifth young chimpanzee died of injuries.

Why were these females so cruelly attacked? In some mammalian social systems, for example among lions and among langurs, the young are regularly killed by a male who has displaced the former leader and taken over the females. The mothers robbed of their infants are then very soon available for conception, assuring him of fathering the next generation. From the point of view of sociobiology, infanticide is a strategy ensuring transmission of the male's genes to a maximum number fo descendants. But in chimpanzee society, the females are the primary targets of aggression. That their young also perish seems rather coincidental. The impression is gained that the males of a community will tolerate no alien chimpanzees, regardless of sex, within their range; the only exceptions are females in heat, either close to sexual maturity or sometimes a good deal older.

In 1970, the community I had been studying since

(Left) The profound obeisance before the conspecific superior is an unmistakable gesture of submission, as in human society.
(Right) This young female chimpanzee has been successful in begging and receives a dainty morsel.

1960 began to subdivide. The reasons for this are still obscure, but possibly the number of adult males – 14 individuals – was too large. Seven males and three females with their young were spending more and more time in the southern part of the community range. Relations between the two subgroups became increasingly hostile, and late in 1972 the separation became complete. When males from the northern Kasakela community met males from the southern Kahama community where their ranges overlapped, they would perform their display, set up their imposing howls, and eventually withdraw.

Two years later, Tanzanian rangers observed the first rude assault of a band of Kasakela males upon a Kahama chimpanzee. The victim was attacked by five males. They beat, kicked and bit for 20 minutes, and then left it bleeding badly and scarcely able to move. The Kahama chimpanzee was never seen afterwards. During the next four years, at least four more attacks of this kind took place. In three cases, the attacked males were left so severely injured that they must have died. On another occasion, five Kasakela males pursued the oldest Kahama female, who died five days later. The body of the highest-ranking male of the Kahama community was found near a stream; it was covered with signs of a similarly savage attack. Late in 1977, when the last Kahama chimpanzee disappeared, the Kahama community was finally eliminated.

Why such outbreaks of violence take place is not clear. All of these attacks except one occurred in the middle of the range claimed by the Kahama community. Possibly the Kasakela males were attempting to reconquer territory to which they had had free access before the partition. Alternatively, the aggressive behavior may have been triggered in part by a lack of sexually stimulating females in the Kasakela community at that time. Thus, we know that Kasakela males three times returned, having penetrated deep into Kahama territory, with Little Bee, a nearly mature female, when she was in heat. The first time, we observed this operation after an attack upon a Kahama male. The aggressors released their victim to go in pursuit of Little Bee, who was accompanying the Kahama male. The Kasakela males forced Little Bee to follow them northward, and would threaten her whenever she tried to escape. After estrus, she would return to her family in the south every time. However, there must have been other reasons besides the repeated abduction of Little Bee for the violent acts of the Kasakela males, since Little Bee had become a regular member of the Kasakela community long before the last killings of Kahama chimpanzees.

After the Kahama chimpanzees had been wiped

Who's bigger? Quarrels over rank are carried on by male chimpanzees in the form of intimidating behavior. The individual to the right lifts his shoulders, thrusts his head forward, crooks his arms, and fluffs up the fur on his arms, shoulders, and back.

▷ A mother chimpanzee caressing her baby.

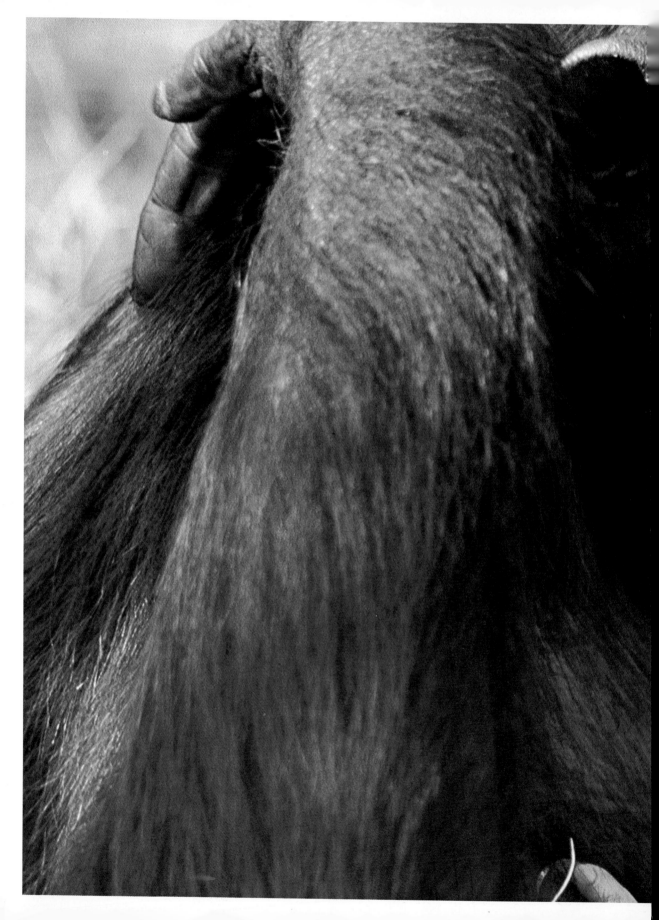

out, the victors with their females and their young roamed for a time in their newly acquired territory, where they would forage and sleep. But they were not to enjoy possession for long. In 1979, the large and powerful Kalande community living to the South began to extend its range northward. By 1980, the Kalande chimpanzees had gained control of the whole territory that had been seized from the Kahama community, and were still advancing into the precincts of the Kasakela community. Late in 1980 the territory of the Kasakela chimpanzees, which had covered 5 mi^2 (13 km^2) in 1978, had shrunk to a scant 3 mi^2 (8 km^2). Three of the adult males, last seen in good condition, had disappeared. The children of three females who had been frequenting southern boundary areas were no longer seen. Early in 1981, another Kasakela female suffered severe injuries while in the southern part of her range. Her two-year-old had been so mauled that it can hardly have survived.

For a while, the prospects for the Kasakela community were gloomy. Only four male chimpanzees were still living. They hardly ventured to visit the southern district of their territory, and when they did, they were sometimes driven back by Kalande males. At the same time, the large Mitumba community was penetrating the Kasakela range from the north.

However, the Kasakela community did not give way, and in 1983 two more males had grown up, bringing the group to an effective strength of six. These six were often attended upon their rounds by four nearly adult males. These young chimpanzees were unable to be of much help in a direct confrontation with Kalande or Mitumba males, because they lacked strength and experience; however, their voices and their dance may have helped to render the group's intimidation display more impressive to the adversaries. For after a time, the Kalande community retreated more or less to their 1978 borders, while the Mitumba chimpanzees in the north drew back. Thus, a state of affairs in which all three communities recognized the territories of the others was restored, and violence ceased.

Seemingly, violence between neighboring groups is an innate behavior trait of chimpanzees, suddenly making its appearance after comparatively peaceful periods. The immediate causes of violence are as yet unknown to us. However, it may be that these aggressive behavior patterns of chimpanzees resemble certain preliminary stages of warfare within our own species. Attacks upon strange chimpanzees are carried out with uncommon savagery, and often pursued to the point of incapacitating the victims. Certain modes of behavior such as twisting of limbs, pulling off strips of flesh, and drinking blood have been observed only in attacks upon strange chimpanzees, never in aggression between members of a community, although they often figure in the killing of large prey.

Because of our close phylogenetic relationship to the chimpanzees, these observations are at once instructive and alarming to us. If the aggressive tendencies of human beings have their roots deep in the phylogenetic past, and are not a consequence of unhappy childhood experiences or undesirable cultural influences, what hope can we have for the future of the human race?

For some years, a chimpanzee is cared for by the mother, until little by little the youngster becomes independent.

It should be noted firstly that aggressive behavior is only one aspect of chimpanzee character. One may observe a group of chimpanzees for hours without ever witnessing so much as a single misunderstanding, or follow them for days without seeing anything more than a faintly threatening gesture. Besides, chimpanzees exhibit a large measure of care, sympathy, and helpfulness towards conspecifics. Usually such behavior pertains to family members, but it will also happen that adults not related to each other will risk their lives for those in danger.

If we too have perhaps inherited our aggressive tendencies from our remote ancestors, as very likely we have, the same would apply to our capacity for love, loyalty, and self sacrifice.

The study of chimpanzees can help us realize what place we occupy in nature. We do certainly distinguish ourselves from the beasts in many respects, but we know today that this difference is not one of kind but one of degree. All the same, the difference in degree is very great. Speech, morality, the ability to discuss ideas, and to lay plans for the future on the basis of experience, are only some of the traits that define humanity. Above all, we should never forget about the human brain. Owing to its capacity, we are in a position to search and inquire after the biological roots of our own behavior; at the same time, it also furnishes us with a refined mechanism for controlling our behavior, and enables us to act in accordance with a fixed plan even if such action is contrary to certain inborn inclinations. The chimpanzees stand at the beginning of an evolutionary path upon which humans have proceeded much farther.

This baby chimpanzee is a year old but is still always carried in its mother's arms.

Chimpanzee Rehabilitation

by Bernhard Grzimek

I was perfectly aware that the enterprise before us was not without its dangers.

Most people have had contact only with cute, lovable young chimpanzees, trained to do stunts or at play in zoos. But when such an ape has reached sexual maturity at the age of eight years, or becomes a fully adult personality with social perquisites within his group at twelve years, one must have a care. Although a male chimpanzee fully erect is only three-quarters the size of a human, because of his short legs, he has nearly double the muscular strength, and teeth not much inferior to a leopard's. A raging wolf or leopard may bite into a broom held out to him; an infuriated chimpanzee bites only to draw blood.

My right middle finger is stiff from the bite of a male chimpanzee. I carry numerous scars due to another young male chimpanzee, barely adult, whom I kept in my Berlin home together with other apes during the bombing attacks on the zoo. A zoo director of my acquaintance had his knee cap half torn out by a male chimpanzee, whose cage he had been in the habit of entering almost daily; another friend of mine had both thumbs bitten clean off.

Our plan was to release such individuals on Rubondo.

I had learned about Rubondo a year and a half before through Peter Achard, wildlife warden in the town of Mwanza at the South end of Lake Victoria. He and I boarded a small plane and flew 85 mi (140 km) to the island of Rubondo. I fell in love with the island immediately. It is 23 mi (38 km) long and averages 5 mi (8 km) wide; three-quarters of its 125 mi^2 (350 km^2) of dry land are primeval forest, the rest grassy hummocks. Best of all, Rubondo is not inhabited by humans.

Peter is an active enthusiastic fellow, and enthusiasm is contagious. So that is why the Zoological Society of 1858 in Frankfurt sent out circulars to the zoos of Europe, and on the 16th of May 1966 ten chimpanzees were dispatched from Antwerp on the steamer "Eibe Oldendorff." They had now been three weeks at sea. They are housed in roomy, easily cleaned boxes. They must travel singly or in pairs, for they are not acquainted with each other, and in such close quarters during a journey lasting weeks, there might be trouble.

I myself had flown to Africa three weeks later and far more comfortably by overnight jet, and was now in Arusha, Tanzania, preparing for arrival and onward journey at the port city of Mombasa. Mombasa to Mwanza by rail would have taken eight days at best, so I had sent two trucks and two cross-country vehicles to the coast. I had also obtained permission for entry and passage through Kenya; veterinary police authority to import chimpanzees into Africa – that's news!

When the "Eibe Oldendorff" arrived in Mombasa, there were already 42 vessels in port waiting to be unloaded. They were going to have to wait at least eight days, was the word. So Captain Wehlitz restarted the engines and sailed on to Tanga, the first port in Tanzania. On bumpy country roads, our motor convoy followed. But Tanga too was overcrowded, so the vessel continued to Dar-es-Salaam, and once more the receiving committee pushed on behind through dust and dirt. Since the caravan was now not going to reach Arusha, I got into my Volkswagen bus and quickly drove – in two days – across the Ngorongoro, through the Serengeti, and along the shore of Lake Victoria to Mwanza. The chimpanzee convoy pulled in late at night, on the same day.

The trip by water to Rubondo Island was not child's play. Lake Victoria is really a sea rather than a lake. With an area of nearly 25,000 mi^2 (69,000 km^2), it is next in size to the Caspian Sea among bodies of water on the globe, and nearly 200 times the extent of Lake Constance.

With provisions for ourselves, tents, cots, and five heavy crates of chimpanzees, Dr. Fritz Walther and a colleague sailed from Mwanza across the south corner of the lake to Rubondo in seven hours. Dr. Walther, considered to be the world's leading antelope specialist, who will report on his experiences in Volume 5 of this Encyclopedia, had been living for two years in the Serengeti to research the life of the Thomson's gazelles. During the early weeks, he was to observe the European anthropoids as they settled in on Rubondo.

The rest of us drove round about with the others to the lakeside village of Bukindo, whence we crossed to Rubondo in two hours. We were kept busy entertaining and reassuring our hairy passengers in their boxes.

If these had not been zoo chimpanzees, but newly-captured somewhere in the jungle, we need only have taken the boxes ashore next morning and opened them. They would have run away from us and vanished into the bush, never to be seen again. Chimpanzees are shy; they never attack human beings in the wild if they are unmolested. But zoo chimpanzees have no fear of humans, even though they are not necessarily altogether friendly and affectionate. They would probably have remained here near these shelters, instituted a reign of terror, taken away people's food and clothing, and possibly done worse harm.

So in the gray dawn, we carried the animals in two transports for twenty minutes farther along the island to an opening in the shoreline. The heavy boxes were carried ashore and set up in a row close to the water. For the last time, the animals received Resoquine syrup against malaria.

First, we released those chimpanzees that were considered quite friendly and good-natured. One very slender female and one other simply strode across the clearing, making for the woods in a straight line and vanishing as though they had always lived here. Others visited neighboring crates in which chimpanzees were still sitting, poked their fingers in, felt the faces of the occupants, and kissed them through the grating. Probably they had become pretty well acquainted, box to box, on the voyage. A half-grown youngster and the smallest of all, actually still a baby, came to us at once, held us by the leg, and the tot climbed up, kissed me and wanted to be hugged. Wherever we went, he would run after us, usually upright on two legs.

Zoo chimpanzees in training: Now all sit in a row. According to the latest report, zoo animals find such assignments great fun.

This would never do. The babies were supposed to stay with the adults. With chimpanzees, this is no problem in itself. All of them, even full-grown males, are kind to little ones. So we brought the little fellow to two females sitting at the edge of the woods. He did go to them; one embraced him and pressed him to her breast, but he soon came back to us. We didn't dare chase him away; he might have cried and become angry, and then the adults might come to his defense and bite us.

Carefully, from the water side, we opened the box of a large female known for ill temper. But the "ill-tempered" one was not at all interested in coming out. She only opened the door for a better view. As soon as the infant arrived, she pulled it into the box with her.

We were getting impatient, but the giant female didn't care. When she finally came out after a full quarter of an hour, she held the iron gate in one hand and looked over the surroundings very carefully, apparently concerned to keep the way open for a retreat. When she released the door for a moment, it slammed shut. Frightened, she went back and opened it again. Since we wanted her out of the way, so that we could put the empty boxes back on the boat, we sprinkled her with lake water; she went right back into the box. We don't see that very often at the zoo. Animals that have traveled for days or perhaps weeks in their transport quarters feel safe there and hesitate to venture out into strange surroundings.

Late in the afternoon, we returned with the motor boat. There were no chimpanzees in sight. The heaps of bananas, apples, and bread that we had left were untouched. Even if the chimpanzees do not find any fruit, they can live for weeks on leaves and shoots alone. That will be new and unaccustomed fare for them.

As we land, the big male comes out of the woods, greets us with a clamp of the hand, and is followed by the two small males. The females are nowhere in sight. Will the small males be able to find them?

Robert, the big black male, becomes more and more excited. His hair stands up, he becomes twice as wide, and begins to walk on two legs. We quickly retreat into the water and wade out to the boat.

Today is the 23rd of June; all our anthropoids are in their new home at last. They have spent more than five weeks, some of them even longer, in their boxes. We probably feel no less relieved than they do.

Next morning, we go on foot across the peninsula between our camp and the chimpanzee point of release. We have to climb a couple of mounds surfaced with broken rock.

The place where we released the chimpanzees is behind the next mound, I estimated. But as we enter the little wood, all at once there are two big female chimpanzees, including the Ill-Tempered One. It's too late. They stand up, they bristle – always a sign of excitement, sometimes aggression. They both charge us. There is no point in running away now. The giant female grabs me around the shoulder, opens her mouth, bares her big strong teeth, first takes hold of the calf of my leg with her jaws, then grabs my forearm. Finally she

Chimpanzee rehabilitation. (Left) The truck carrying the crates of chimpanzees has arrived on the shore of Lake Victoria. (Right) From the waterside, the boxes are opened one at a time, but it takes a while before the inmates leave their uncomfortable but now familiar quarters.

kisses me. All this has been merely cheerful greeting.

Knowing chimpanzees as we do, we know how easily excitement can turn into malevolence, and we hasten to the open clearing, and behold, the two females don't want to follow us into the sunlight; they stay in the wood. However, one of them grabs Dr. Walther's binoculars. He prudently offers no resistance but allows the strap to be taken off over his head. Robert, the big male, has joined us. The three nimbly open the leather case and take out the instrument. They study it closely, they look through it too, but from the wrong end. Dr. Walther tries to swap his handkerchief and a tote bag for the binoculars. He meets with no appreciation; to the contrary, they take possession of these items also.

We intend to withdraw unobtrusively into the woods in the other direction while the three chimpanzees are busy with the binoculars. But Robert notices and comes after us. He climbs up on Dr. Walther and takes the latter's chin between his jaws with their powerful canines. This looks desperate, but the intent is friendly. All the same, Robert's hair stands on end. Now he begins to stamp on the ground, his fur broadens and thickens, he races galloping between Dr. Walther's legs from behind, knocking him flat.

Our position is becoming a bit uncomfortable. Chimpanzees are adventurous, extremely active wanderers. Why should Robert not take hours walking us back to camp? What with their communication by voice and percussion, we might soon have the whole crowd upon us again. How are we to get rid of Robert? Go back to the females? But then they might attack us, or they might come with us too. We are all excited and speak together in half-whispers, which after all is hardly necessary. We return to our earlier salvation, the water.

As soon as we came back in the motor boat on the afternoon of the same day, all the apes were gone. The heaps of bananas were untouched. At the edge of the woods, well scrapped, peeled and disassembled, we found the case of the binoculars, and after some searching in the woods, the binoculars themselves, broken of course.

Among the eleven chimpanzees – three males and eight females – whose experience of release we have just shared, there was only one adult male. The two young males unfortunately did not attach themselves to the adults, but kept to themselves for the most part and would not hesitate to approach people during the first months, although after six months they would take flight at the sight of an unfamiliar boat. It took them a full year to become shy. It seemed rather unsafe to us to have only one adult male capable of reproduction in the group. So in 1966, two adult males, also from zoos, were flown singly from Europe to Rubondo, including the one-eyed Jimmy. In September 1967, 15 months after release, the male Robert showed up at the wardens' new second camp with two pregnant females. The animals tore open sacks of sugar and grain, strewed it all about, and became quite unruly. Later raids upon the camp were repulsed by the wardens with blank cartridges, and the chimpanzees did not reappear. In February 1968, two mothers with newborn infants were observed.

(Left) Some zoo-bred chimpanzees don't appreciate the value of freedom. At first, they are unwilling to part with their carrying case. (Right) This chimpanzee has opted for freedom. He leaves the proffered bananas behind and heads for the woods, his new, yet time-honored, habitat.

The Bonobo, or Pygmy Chimpanzee

by Alison and Noel Badrian

Pygmy chimpanzee mother and baby. The pygmy chimpanzees, or bonobos, as they may more properly be called, are not really much smaller than "ordinary" chimpanzees, but they are more slender in build; also, they have relatively longer limbs and less-protruding jaws.

(Opposite page) The "lord of creation," came forth from the phylogeny of the primates. All modern humans belong to the one species *Homo sapiens*.

In 1929, the German zoologist Ernst Schwarz, having examined some specimens at the Tervuren Museum in Belgium, described a new, smaller subspecies which he named *Pan troglodytes paniscus*. Strangely enough, the first good description of an anthropoid ape, by Nicolaas Tulp of Holland in the year 1641, is believed to pertain to just such a pygmy chimpanzee. In 1933, the Harvard zoologist Harold Coolidge, in studying the pygmy chimpanzee, arrived at the conclusion that the differences between this new form and the other subspecies were sufficient to rank the pygmy chimpanzee as a species in its own right. He gave it the name *Pan panicus*. Because of their physical traits, he considered the pygmy chimpanzees to be the most important of all chimpanzees to an understanding of hominoid phylogeny. He ventured to assert that "it (the species) is closer to the common ancestor of the chimpanzee and man than any living chimpanzee previously discovered and described." That Harold Coolidge's appraisal was quite correct has been confirmed by the unflagging attention to this species in research on human evolution.

The pygmy chimpanzee is not much smaller than other chimpanzees. Accordingly, many zoologists prefer the term "bonobo." The misleading name "pygmy chimpanzee" is based firstly on the fact that the bonobo is pedomorphic, that is, possesses juvenile traits in adulthood, for example a more rounded cranium. Secondly, the first individuals ever examined were among the smallest specimens described to date for this species of chimpanzee. The average body weight of a male bonobo is about 100 lb (45 kg), intermediate between the weights of two chimpanzee subspecies. However, the comparatively long limbs of the bonobo give the appearance of much greater slenderness. The bonobos are also far more mobile and demonstrate their physical coordination by acrobatic leaps and tow-legged "ropewalking" acrobatics 90 or 150 ft (30 or 50 m) above the forest floor.

Bonobos are confined to tropical forests. Their diet consists chiefly of fruit, but they also consume leaves, seeds, grasses, and other vegetable fare. Their diet is supplemented with caterpillars and other invertebrates, as well as the flesh of young duiker antelopes, rodents, or even snakes. In shallow streams and swamps, they also catch fishes and crabs. The live in communities of 50 to 120 individuals, occupying a range area of 7 to 22 mi^2 (20 to 60 km^2). The communities are composed of groups numbering 2 to 60 members. The size and composition of the groups vary continually as they merge and separate in the course of feeding and foraging.

By contrast with the other chimpanzees, ordinarily both sexes are represented in bonobo groups. Mutual grooming, which has not only a hygienic but also a social function, is widespread in adulthood between males and females as well as among females. Unlike the other chimpanzees, bonobo males will not groom each other. Female bonobos are sexually active longer than other female chimpanzees, and are homosexually active as well. Like mutual grooming, homosexual behavior not leading to procreation helps to strengthen bonds between members of the community.

Bonobos are rare in captivity; at the present time, they are to be found only in the Frankfurt, Antwerp, and San Diego zoos. How large the numbers of pygmy chimpanzees living in the wild may be is not known. Probably they are more severely threatened than is generally supposed.

HUMANS

Category
FAMILY

Classification: A family of the order Primates, suborder Catarrhini or Old World primates. Only one (now living) genus, *Homo*, with one species *Homo sapiens*. In the classification of the animal kingdom, hominids represent the highest level of evolution, and together with the gibbons (Hylobatidae) and the great apes (Pongidae) they constitute the superfamily Hominoidea.
The following synopsis deals primarily with the biological characterization of humans.

Head: Large brain case with correspondingly large brain; much convoluted cerebral cortex with high frontal lobes; foramen magnum displaced far towards center of base of skull; short face, having migrated under the skull; nose and chin protruding (nose as chamber for heating, moistening, and coarse-filtering of inhaled air, forehead as supplemental member for lower jaw); closed, parabolic dental arch with small canines.
Trunk: Vertebral column displaced towards center of body, which by position and multiply-curved form provides a spring axial rod to cushion the footfall; bowl-shaped pelvis to support the erect trunk; protruding buttocks due to pronounced development of muscles required to hold the trunk upright.
Limbs: Legs longer than arms; specialized foot for standing, with reinforced heel bone, arch, and powerful great toe, not averted; hands unspecialized compared to most primates, hence versatile in use.
Skin and Hair: Scant body hair; patches of hair on head, in arm pits, and in pubic area; beard in males; skin abundantly supplied with sweat glands and variable in color; female breast prominent due to fatty layer.
Reproduction and Stages of Life: Sexual activity not seasonal or limited to any time of year; duration of pregnancy 38 weeks; one young per birth as a rule; newborn slow to mature compared to other primates; prolongation of childhood and youth (female sexual maturity about 13 years, cessation of axial growth at 16–18 years); long life span after cessation of female reproduction; average life expectancy in modern highly-developed countries is 69 years for men, 76 for women (considerably lower in some countries).
Descent and Distribution: Earliest prehuman finds about 5 million years old; all essential traits of biped locomotion already present; steady enlargement of brain and increasing learning capacity as basis for cultural development; appearance of several prehuman groups; about 2 million years ago, transition to forms of early humans in Africa, then spreading to Asia and Europe; transition to anatomically modern humans about 100 000 years ago in Africa; thence distribution to nearly all habitats on Earth; production of present multiformity through adaptations to various living conditions, and creation of an artificial environment through culture and civilization.

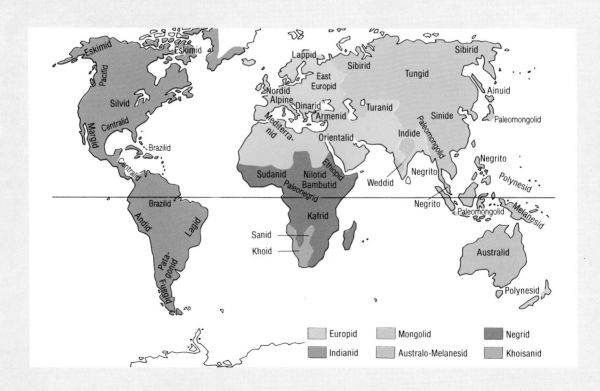

Hominidae
Hommes   French
Menschen   German

Habit and Habitat: By virtue of unique adaptability (omnivorous habits, diversified use of environment) and technology (tools, use of fire, clothing, shelter construction, etc.), progressive colonization of nearly all climatic regions and terrains, from the tropics to high northern and southern latitudes, except Antarctica and much of the Arctic; highly developed social behavior; family (association of several families in groups, extended or nuclear family) as basic unit of human societies.

Culture: Rapid brain development, biped locomotion (so that versatile hands are kept free), and refined intraspecific communication (verbal speech) as preconditions for rise of culture and cultural traditions; at first predators (hunters and gatherers) with use of fire, making of tools and weapons, cooperation and division of labor; then transition to agriculture and herding conjoined with settled communities and trade; more comprehensive social organizations and eventually founding of city states; first "high" civilizations in the last pre-Christian millennia.

The human cranium (yellow) gains in size at the expense of the facial skull compared to the anthropoid apes. Also, the foramen magnum (arrow) shifts downward as adaptation to erect posture. In the anthropoid apes, it opens obliquely to the rear.

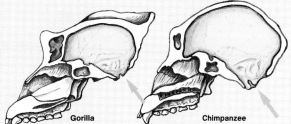
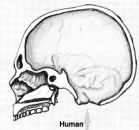

Gorilla   Chimpanzee   Human

Comparison of Hands. Making possible the precision work of the human hand, the thumb is set opposed to the other fingers so that its tactile tip will act against those of any other fingers. Unlike the chimpanzee's thumb (left), the human thumb can easily reach the last joint of the index finger.

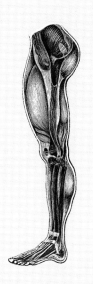

Pelvis and Leg. The anthropoid leg (right) exhibits some quadruped features, while that of humans has been remodeled for biped locomotion. The pelvis is broad and short, favoring the erect attitude of the trunk as well as attachment of the modified musculature. The knee joint allows full extension of the leg, necessary for standing upright. The leg musculature is no longer of about equal strength front and rear, but unequally distributed. The foot is long and narrow, and its skeleton forms an arch, important for allowing the sole of the foot to roll in walking or running. The extensors important for the biped gait are powerfully developed and follow an S-shaped course from the buttocks by way of the anterior thigh musculature to the posterior calf musculature.

Humans

Human Phylogeny
by Günter Bräuer

Accident and Controlled Research

"Light will also be cast upon the origin of man and upon his history." Just this one sentence in the work "On the Origin of Species by Natural Selection" by Charles Darwin, published in 1859, referred to the origin of humans. Although Darwin's theory of evolution suggested the conclusion that humans were descended from apelike ancestors, at that time fossil evidence of such an evolution was almost entirely lacking, or – like the Neandertal skeleton discovered near Düsseldorf in 1856 – not recognized as such. On the basis of findings in comparative anatomy, the zoologist Ernst Haeckel of Jena, in his work "Natürliche Schöpfungsgeschichte" [Natural history of creation] (1868), posited a connecting link between an extinct form of anthropoid apes and humans, naming it *Pithecanthropus.* Then the search began, punctuated by many almost incredible accidents, for the "missing link."

Thus the Netherlands physician Eugène Dubois, inspired by the ideas of Darwin and Haeckel, carried out excavations at Trinil during his military service on the Island of Java, and in 1891 he actually discovered the roof of a cranium of unusual shape and a tooth, at first ascribing them to an extinct form of chimpanzee. But when the following year he found a human-like thigh bone in the same deposits, he thought he had found the missing link, the form that walked upright. He therefore named the find *Pithecanthropus erectus.* At the scientific conferences of the time, Dubois' interpretation was widely rejected.

The South African anatomist Raymond Dart, when he announced the discovery of a missing link in 1925, did not fare much better. There were parts of a juvenile skull found at Taung in South Africa. Dart named it *Australopithecus africanus* ("southern ape of Africa"). His find, too, was taken for a mere chimpanzee by many scientists of the day. It was not to emerge until later that the Dubois and Dart finds were important evidence of human phylogeny.

Discoveries were so much a matter of chance, and there were so few fossils in any event, that it was only too understandable that a new find would often shatter the existing picture of early humanity. So it was when Louis Leakey, one of the pioneers of paleoanthropology, first thought in 1959 that he had discovered an early representative of the hominid line in the Olduvai Canyon of Tanzania. Although the skull was of uncommonly sturdy build and did not look very human-like, Leakey inferred from the tools found close by that it must represent a human ancestor. So he created the new genus *Zinjanthropus,* meaning "East Africa man" (Zinj is an old name for the East African coastal country) and added the specific name Boisei in honor of the London merchant Charles Boise, who had generously supported Leakey's researches. Hardly a year later, however, it turned out that

Excavations at Koobi Fora on northeastern Lake Turkana in Kenya.

HUMAN PHYLOGENY

Leakey's interpretation, already well known, was mistaken. For in the same Olduvai deposits, remains of a far more likely maker of the primitive stone tools had been found. He looked entirely different from *Zinjanthropus boisei*, was far more human-like in dentition, and had a bigger brain. This was the first evidence of the new species *Homo habilis*, the "man of skill."

These few examples, which could easily be multiplied, may suffice to show how far the study of early human was dominated for many decades by the search for so-called missing links, by new interpretations, controversies, and personalities of the scientists. This situation, and new finds often heralded as sensations rendering all past knowledge obsolete, has resulted in an impression, still persisting among the public, that it needed only a few bone fragments, more or less well preserved, and the scientists would hypothesize a family tree of human development from ape-like forefathers on the basis of surmises and suppositions.

That impression, however, has long ceased to correspond to the facts. Since the late 1960s, research on fossil human history has undergone an altogether dramatic expansion through many hundreds of new finds. This is the result in particular of a large number of systematic excavations, among them a series of international research expeditions that have been seeking our extinct ancestors systematically in suitable deposits over many years. These studies, carried out in many parts of the world, have led not only to an abundance of new fossil evidence from nearly all periods of human evolution, but also to a reliable picture of the general course of human origin, no longer in any way comparable with the guesswork that prevailed in the first half of this century.

Humans and Anthropoid Apes

The notion that humans evolved from "primitive, hairy apes" exploded like a bombshell upon publication of Darwin's work on the origin of species. It was an intellectuel and moral shock that is almost unimaginable today, and the commencement of decades of controversy, especially with ecclesiastical doctrines. Whereas Darwin himself shunned direct confrontation, he had an eloquent ally in the anatomist Thomas Henry Huxley, sometimes known as "Darwin's bulldog." As early as 1863, Huxley published his book, "Zoological Evidence of the Position of Man in Nature." In it he espoused the view that of all living creatures, the gorilla and the chimpanzee are most closely related to humans.

Today, more than 120 years later, not only further research in comparative anatomy and the fossil record, but also the discoveries of modern molecular biology and biochemistry, as well as genetics, impressively confirm Huxley's view. The available results of molecular biology agree in showing that humans are very similar to the chimpanzees and gorillas, and that the genetic distance from the orangutan is about twice that of humans from the African anthropoids.

Kamoya Kimeu of the National Museum in Nairobi, one of the most successful discoverers of fossil hominids.

Origins and Phylogenetic Subdivision of the Hominoidea

The more highly evolved apes are included in the superfamily Hominoidea. It is generally taken to include three families Hylobatidae (gibbons and siamangs), Pongidae (chimpanzee, gorilla and orangutan), and Hominidae (humans). While fossil primates first show up in the Upper Cretaceous of about 70 million years ago, the roots of the Hominoidea go back to the Early Oligocene, about 34 million years. The earliest primates with hominoid traits – in particular the peculiar five-cusped pattern of the lower molars – are the genera *Propliopithecus* and *Aegyptopithecus*, hitherto known only from the Fayum deposits in Upper Egypt. A first lower jaw of *Propliopithecus* had been found at the beginning of the twentieth century. The research by Elwyn Simons has added greatly to the number of Oligocene hominoid remains from Fayum since the beginning of the 1960s to date. Most of the material pertains to the form *Aegyptopithecus zeuxi*, closely related to *Propliopithecus*. Besides relics of teeth and skulls and other skeletal bones, a nearly complete skull was found in 1966 and three largely preserved facial skulls in 1981–82. *Aegyptopithecus* did not look anything like present Hominoidea, but resembled the baboon. It was only about the size of a fox or a cat, and traveled in the trees on all fours. Given our present knowledge of Oligocene primates, it is probable that the separation of the hominoid line took place rather more than 30 million years ago.

In the Early Miocene of East Africa – 22 to 17 million years ago – we have hominoid remains of many hundreds of individuals. Of the various forest-dwelling hominoid species of that era, one is especially well known, *Proconsul africanus*. By this time, we have not only many skull and jaw fragments but even much of the supporting and locomotive apparatus of this form. In past decades, opinions differed considerably as to the phylogenetic significance of *Proconsul africanus*. Some scientists considered it to be a quite primitive hominoid; others placed it well up in the chimpanzee branch. Today it is pretty well agreed that *Proconsul africanus* was a quite unspecialized hominoid. It was about the size of a savanna baboon, with distinct differences between the sexes, and traveled both in the trees and on the ground, probably for the most part on all fours but with a wide latitude of movement.

Besides these and other *Proconsul* species, some of them more highly specialized, there are other known hominoid genera of the Early Miocene in East Africa. These include the smaller forms *Micropithecus* and *Limnopithecus*. They exhibit a blend of hominoid, pithecine, and special features, and may perhaps represent the early gibbon branch. But despite certain resemblances between *Micropithecus* and the gibbon, the question of gibbon origins is still quite unclear at the present time.

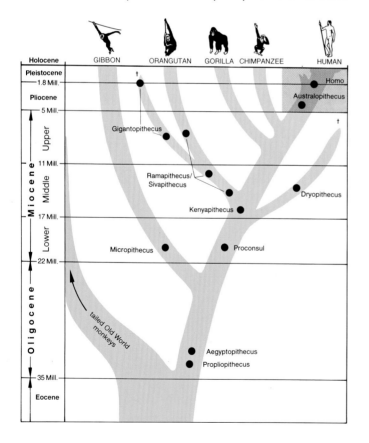

Family tree of the Hominoidea – humans and their nearest relatives.

Whereas hominoid evolution took place only on the African mainland towards the end of the Early Miocene, about 17 million years ago Africa and Arabia were joined to the Eurasian mainland (Europe and Asia) by continental drift, carrying these fragments of the earth's crust towards each other. Now African hominoids were able to spread into other parts of the Old World. Thus in Europe, about 15 or 14 million years ago, the genus *Dryopithecus* appears, strongly resembling *Proconsul* and very likely descended from the latter. Several species of *Dryopithecus* are known in Europe. Most of them seem to have died out during the Late Miocene, probably because of changes in climate and terrain.

Besides the dryopithecines, during the Middle and part of the Late Miocene there was another form, *Ramapithecus*, seen until recently as the beginning of the hominid line, the line leading to humans. The eponymous find – an upper jaw fragment about 8 million years old – was discovered in the Siwalik Mountains of northern India in 1932. The shorter face and the narrower "simian gap" (between incisors and canines) seemed to point in a hominid direction. Then in 1961, Louis Leakey found an upper and a lower jaw in 14-million-year-old deposits near Fort Ternan in Kenya, which he named *Kenyapithecus*, and which were usually assigned to ramapithecines on grounds of close similarity. The canine teeth did project above the plane of the bite but distinctly less so than in *Proconsul* or *Dryopithecus*. The simian gap was correspondingly smaller and the jaw as a whole less protruding. During the 1960s and 1970s, other findings in Europe and Asia were added to *Ramapithecus*, aged up to 15 million years. They were almost exclusively teeth and jaw fragments. What the skull or the rest of the skeleton must have been like was virtually unknown.

In addition to the smaller ramapithecines, a larger Miocene hominoid of Asia, *Sivapithecus*, has long been known. Like *Ramapithecus*, this form has thick enamel caps on the back teeth, but is distinguished by larger jaws and longer canines.

Beginning in the mid-1970s, the riddle of Miocene hominoid evolution was to be further solved. The anthropologist David Pilbeam and his associates found a large number of additional relics of *Ramapithecus* and *Sivapithecus* in Middle and Late Miocene deposits of the Potwar Plateau in Pakistan. Numerous remains of both forms were also found at Lufeng in China – actually in the same strata (horizons). These new finds, between 12 and 8 million years old, and for the first time including nearly complete facial skeletons, initiated a fundamentally new attitude towards *Ramapithecus* and *Sivapithecus* in the early 1980s.

The two forms in fact turned out to be very similar, and the differences in size may have reflected sex differences primarily. Hence it is probable that these are either two species of the same genus, or else two very closely related genera. More significant still was the realization, based on the more completely preserved finds, that these hominoids in many details, such as the form of the eye sockets or of the anterior segment of the upper jaw, exhibit striking resemblances to the orangutan. Meanwhile, therefore, it has come to be a widely accepted view that *Sivapithecus* and *Ramapithecus*

Proconsul africanus, a common ancestor of the great anthropoid apes and humans, is known today through a multitude of skeletal remains.

are in the line of descent of this sole living Asian great ape, and that therefore *Ramapithecus* is neither an early hominid nor a close relative.

This markedly novel view of course also affects the interpretation of the still somewhat older *Kenyapithecus* finds. Here again, the record has been profoundly altered by the latest studies in West Kenya. Whereas by 1980 only nine fragments of this form had been collected, the work of the British paleontologist Martin Pickford and the Japanese anthropologist Hidemi Ishida added more than 200 skulls and other skeletal remains, up to 16 million years old, within the ensuing four years. The new finds clearly show that in *Kenyapithecus* also the canine teeth of the male sex were definitely longer than had been supposed from the few specimens previously known. *Kenyapithecus* is certainly not a hominid, but a hominoid, still somewhat closer to *Proconsul* than the already more specialized *Ramapithecus* or *Sivapithecus*. *Kenyapithecus* may thus probably have been a starting point from which both the African hominoids (gorilla, chimpanzee, human) and *Sivapithecus/Ramapithecus* plus the orangutan originated.

Proceeding from the quite well-established proposition that this division into an Asian and an African line occurred about 15 to 16 million years ago, it is to be supposed from the major molecular biological resemblances between humans and African great apes that a period of about 5 to 8 million years was required for the separation into hominids and African great apes. On the basis of the still greater similarities between humans and chimpanzees, it must be supposed further that the gorilla line branched off in time somewhat before the separation into hominids and chimpanzees. This is suggested also by the distinct resemblances between the early australopithecines, now to be discussed, and the chimpanzees.

The Earliest Evidence of the Human Line

For the period between 8 and 4 million years ago, so important to the beginnings of independent hominid development, there are as yet only a very few finds, giving us no precise idea when or how the separation may have taken place. Two important developmental processes, however, may have characterized the hominid line during that interval; firstly, the very manifold anatomical changes in connection with the adoption of erect biped locomotion and, secondly, the increasing reduction of the canines and enlargement of the bite area of premolars and molars. As causes for both developments, adaptations to an altered habitat may be assumed, namely to a drier mosaic landscape of wood and savanna, as well as a diet containing more seeds, roots, shelled fruits, and tubers. Paleoecological studies – research on prehistoric environmental conditions – clearly show that the closing Miocene was characterized by violent climatic changes. The drastic temperature drop shrank the densely forested areas in broad parts of tropical Africa.

The new fossil remains older than 4 million years and showing signs of these hominid adaptations all derive from Kenya. The oldest, the Lukeino find, dating to about 6.5 million years ago, consists of only a single lower molar. An *Australopithecus*-like lower jaw fragment from Lothagan, in which the first molar and several roots are preserved, is 5 to 6 million years old.

Recently another hominid lower jaw remnant, between 4.2 and 5 million years old, was found at Tabarin near Lake Baringo. The two anterior molars present and the preserved jaw region show distinct resemblances to some finds of the species *Australopithecus afarensis*, to be dealt with in the next section. Lastly, there are two more fragments about 4 million years old; the fragment of a temporal bone from the Chemeron deposits, and the lower part of a humerus from Kanapoi, morphologically far closer to the human than to the chimpanzee form.

These few finds between 4 and 6.5 million years of age may be regarded either as earliest signs of the genus *Australopithecus*, or to be placed in its immediate vicinity. From about 4 million years on, the number and frequency of finds increase rapidly. From then on, all species of *Australopithecus* and *Homo*, as well as many groups distinguished as subspecies, are documented by numerous significant fossil finds.

Australopithecus afarensis

We have already seen that the Taung find, which Dart explains as a "missing link" between apes and humans, and to which he had given the name *Australopithecus africanus*, was followed by other finds of that genus, in South Africa and in the East African countries of Tanzania and Kenya. The *Zinjanthropus* found by Louis Leakey in the Olduvai Canyon also proved a near relative, if not a congener, of *Australopithecus*. Within the family Hominidae, many scientists subsume the tribe of *Australopithecus* in a separate subfamily of prehumans, the australopithecines. The word will be used here in that sense.

Remains of *Australopithecus afarensis*, the oldest well-known australopithecine species, were not discovered until 1973. The International Afar Research Expedition under the leadership of the French researchers, Yves Coppens and Maurice Taieb, and the American anthropologist Donald Johanson, at that time began systematic exploration of the Pliocene deposits near Hadar in the Afar region of Ethiopia. By 1977, some hundreds of hominid remains of more than 35 individuals had been found there, around 3 million years old. Especially well known is a skeleton, about 40 percent preserved, named "Lucy" after a Beatles song. At a single site, 197 fossil fragments were discovered, pertaining to at least 13 adults and children, possibly a living community of early hominids. About half a million years older were the jaw remains found by the well-known paleontologist, Mary Leakey, near Laetoli in North Tanzania in 1974. In the further survey of this area in 1977, she came upon footprints of these prehumans; they were dated at about 3.7 million years old, and testify that biped locomotion had developed by that time.

Comparative studies of prehuman remains at Hadar and Laetoli, 1100 mi (1800 km) apart, revealed definite resemblances between the finds in the two localities, and led Donald Johanson and the anthropologist Tim White to the conclusion that this was a new original australopithecine form. In 1978, they therefore instituted the new species *Australopithecus afarensis*, to which they assigned all the finds of both places. Some more remains found in 1981 south of Hadar near Belohdelie and Maka, likewise assigned to this species, are about 4 million years old.

What did these earliest prehumans, whose remains represent nearly all parts of the skull and skeleton, look like? How did they differ from the anthropoid apes, the Pongidae, on the one hand, and from the next-later species, now well-known for decades, *Australopithecus africanus*, on the other?

Apart from the Laetoli footprints, the numerous

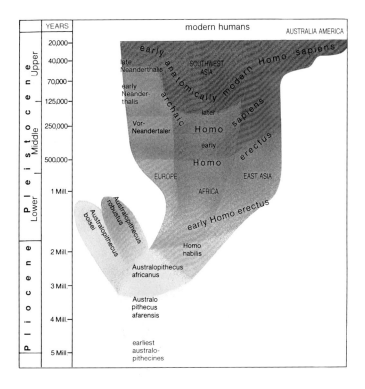

The development from the earliest australopithecines to modern humans.

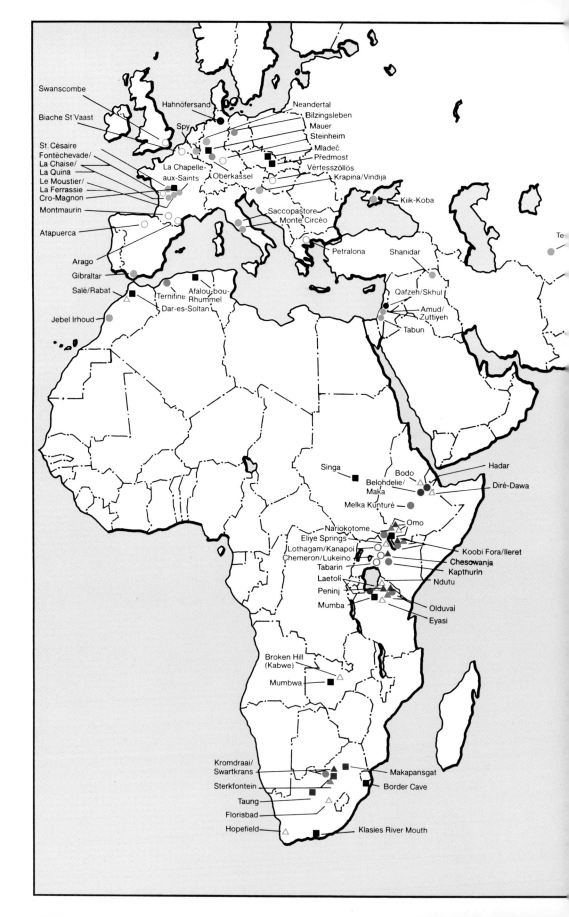

Africa is the cradle of human development. Only here has the earliest hominid genus *Australopithecus*, with its various species, been found. The development to *Homo habilis* and early *Homo erectus* was also confined to this continent. *Homo erectus* later spread from Africa to Southeast Asia and Europe. Only modern humans, whose origin was probably also in Africa, came at last to Australia and America.

skeletal remains of the Hadar finds definitely show that these prehumans already possessed the essential adaptations of the supporting and locomotive apparatus to erect locomotion. Unlike the anthropoid apes, these prehumans – as "Lucy" shows – possessed a human-like pelvic basin, in which the trunk rests during biped progress. The hip bone is therefore a good deal shorter, broader, and curved back. Whereas in the anthropoid apes the femurs are approximately parallel, in the australopithecines, as in humans, they converge towards the knee, so that the lower limbs draw close together, and the weight of the body is supported almost precisely on the standing leg. Despite these and other skeletal features showing an adaptation to two-leggedness, the locomotion of these australopithecines was not quite like that of modern human beings. Not only were the lower limbs comparatively shorter, but also the rather long, curved digits testify to more grasping ability, say for climbing trees. The rib cage, pelvis, and other parts also show features suggesting that these early hominids still spent some of their time, such as for sleeping, eating or defense, in trees, while at other times traveling on the ground in biped fashion. The skeletal and skull remains of *Australopithecus afarensis* show moreover that there were considerable differences in size between the sexes, comparable to those in modern gorillas or chimpanzees. While "Lucy" was about 44 in. (110 cm) tall and weighed 154 lb (70 kg), the largest male individuals attained 60 in. (150 cm) and 154 lb (70 kg).

If the cranial structure of *Australopithecus afarensis* is compared with that of the younger australopithecine species and the anthropoid apes, Afarensis manifestly has the most primitive cranium among the australopithecines, with the strongest resemblances to the chimpanzees. Not only does the small volume of the cranium, the "cranial capacity" (24–32 in.^{3} or 375–500 cm^3) approximately match that of anthropoid apes of like size, but also the state of the jaw proclaims close relationships to the anthropoid apes. Thus, by contrast with the situation in later australopithecines *(A. africanus, A. robustus)*, the anterior segment of the jaw is larger and more protruding. Even though the canine teeth of both sexes are considerably reduced, they often project somewhat above the plane of the bite. The upper canines then form a cutting system, much as in the anthropoid apes, together with the usually single-cusped first lower premolars. These and other features of the anterior dentition indicate that the increasing reduction of this part of the jaw as observed in the development of the australopithecines in favor of a further enlargement of the bite area of the premolars and molars, and hence also of the perpendicular chewing pressure, is as yet less far advanced than in the successor *Australopithecus africanus*. The different

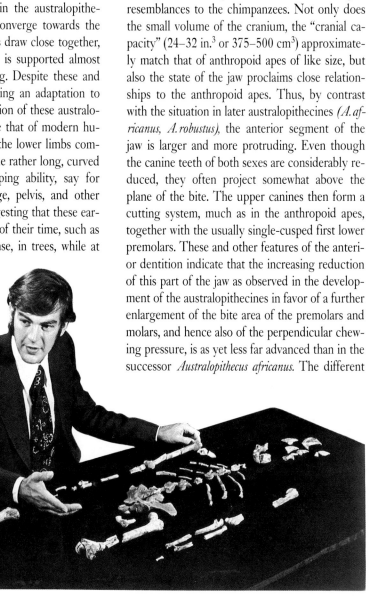

The celebrated nearly half-complete skeleton "Lucy" and its discoverer Professor Donald Johanson. "Lucy" lived about 3 million years ago in the Afar Triangle of Ethiopia.

use of the jaw in *Australopithecus afarensis*, probably more like that of the chimpanzee, also shows in cranial features, associated especially with the more horizontal direction of the chewing stress. Thus the lines of attachment of the temporal muscles are often drawn so far back that there is a direct connection with the lines of attachment of the other muscles of the nape of the neck. Likewise in the shape of the skull in back view, broadening sharply downward, and the very flat frontal bone profile, as in any number of other features, the primitive traits of these early prehumans are revealed.

Despite the original specimen pattern, there is as yet no agreement among specialists as to whether *Australopithecus afarensis* is rightly to be regarded as a separate species or whether it is rather a subspecies of *Australopithecus africanus*, as Philip Tobias believes, for example. Particularly as the number of finds becomes quite large, it becames more and more difficult to differentiate species clearly in the course of phylogeny, even though one is guided by variations in species living today. But there are numerous indications that *Australopithecus afarensis* is a more primitive form than *Australopithecus africanus*. Johanson, White, and others believe, furthermore, that *Australopithecus afarensis* was the original form of both *Australopithecus africanus* and *Homo habilis*.

Australopithecus africanus

Whereas *Australopithecus afarensis* can be traced up to almost 3 million years ago, *Australopithecus africanus* makes its appearance in South Africa at about that time; this was the species introduced by Dart in 1925 with the Taung child skull. Eleven years later, the physician Robert Broom confirmed its existence by the discovery of an adult skull near Sterkfontein. In the ensuing decade down to the present time, more skull and skeleton finds pertaining to a total of more than 80 individuals followed in the Late Pliocene cave deposits of Sterkfontein and Makapansgat. The strata of these finds are about 3 to 2 million years old; the Taung child is apparently more recent.

Whether *Australopithecus africanus* lived in East Africa also is uncertain. Although some partly well preserved skulls – notably in Kenya – have been identified as *A. africanus* by various investigators, others regard them as early representatives of the human genus *Homo*.

Australopithecus africanus averaged only slightly larger than *A. afarensis*. Again, there were clear differences between the sexes.

The numerous skeletal remains, including well-preserved hip bones and long bones, show the basic anatomical adaptations to biped locomotion, although these australopithecines certainly spent time in trees occasionally. The cranial capacity of *A. africanus* averages 28 in.^{3} (450 cm^3), slightly greater than that of *A. afarensis*.

By contrast with the latter and with the anthropoid apes, the frontal bone is more vaulted, and the cranium is enlarged in the vertex region as well. The temporal portions, however, bulge less markedly to the side. The jaws and cooperating skull parts showed that the chewing pressure was shifted somewhat more to the back teeth, and hence directed more towards the vertical than in *A. afarensis*. Thus the front teeth are further reduced and the premolars and molars relatively enlarged. The crowns of the canine teeth no longer project beyond the plane of the bite, the simian gaps are absent, and the lower first premolars are bicuspid and wider, better adapted to grinding food. The attachments of the temporal muscles generally come close together in the middle on the top of the cranium, then diverging evenly to the rear, whereas in *A. afarensis* they continue to run parallel farther back. The cheek bones arise farther forward in *A. africanus*, and the cheek bones themselves are likewise advanced. This provides more favorable chewing pressure conditions for the jaw muscles, which pull from the lower margin of the cheek bones towards the angles of the lower jaw and play a substantial part in grinding movements.

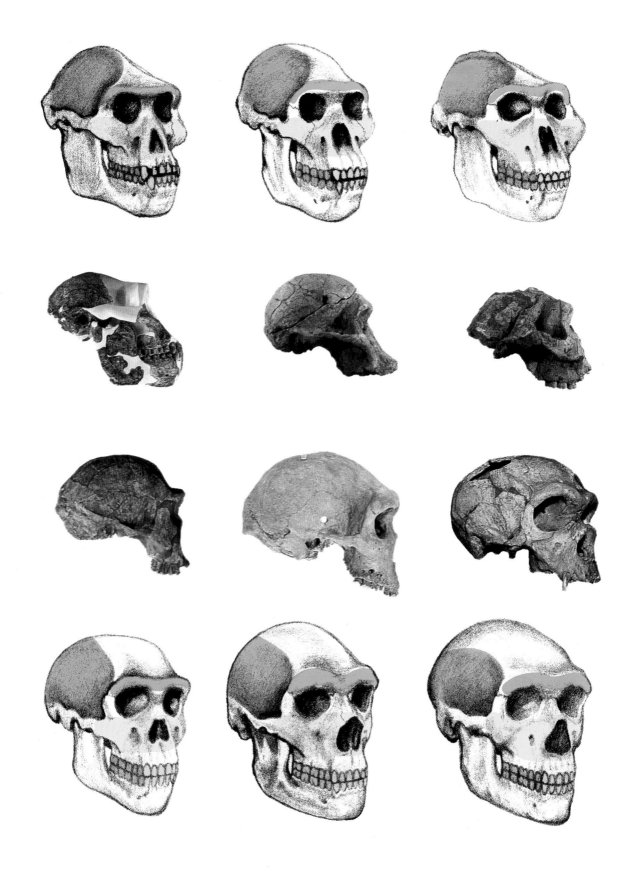

use of the jaw in *Australopithecus afarensis*, probably more like that of the chimpanzee, also shows in cranial features, associated especially with the more horizontal direction of the chewing stress. Thus the lines of attachment of the temporal muscles are often drawn so far back that there is a direct connection with the lines of attachment of the other muscles of the nape of the neck. Likewise in the shape of the skull in back view, broadening sharply downward, and the very flat frontal bone profile, as in any number of other features, the primitive traits of these early prehumans are revealed.

Despite the original specimen pattern, there is as yet no agreement among specialists as to whether *Australopithecus afarensis* is rightly to be regarded as a separate species or whether it is rather a subspecies of *Australopithecus africanus*, as Philip Tobias believes, for example. Particularly as the number of finds becomes quite large, it becomes more and more difficult to differentiate species clearly in the course of phylogeny, even though one is guided by variations in species living today. But there are numerous indications that *Australopithecus afarensis* is a more primitive form than *Australopithecus africanus*. Johanson, White, and others believe, furthermore, that *Australopithecus afarensis* was the original form of both *Australopithecus africanus* and *Homo habilis*.

Australopithecus africanus

Whereas *Australopithecus afarensis* can be traced up to almost 3 million years ago, *Australopithecus africanus* makes its appearance in South Africa at about that time; this was the species introduced by Dart in 1925 with the Taung child skull. Eleven years later, the physician Robert Broom confirmed its existence by the discovery of an adult skull near Sterkfontein. In the ensuing decade down to the present time, more skull and skeleton finds pertaining to a total of more than 80 individuals followed in the Late Pliocene cave deposits of Sterkfontein and Makapansgat. The strata of these finds are about 3 to 2 million years old; the Taung child is apparently more recent.

Whether *Australopithecus africanus* lived in East Africa also is uncertain. Although some partly well preserved skulls – notably in Kenya – have been identified as *A. africanus* by various investigators, others regard them as early representatives of the human genus *Homo*.

Australopithecus africanus averaged only slightly larger than *A. afarensis*. Again, there were clear differences between the sexes.

The numerous skeletal remains, including well-preserved hip bones and long bones, show the basic anatomical adaptations to biped locomotion, although these australopithecines certainly spent time in trees occasionally. The cranial capacity of *A. africanus* averages 28 in.^{3} (450 cm^3), slightly greater than that of *A. afarensis*.

By contrast with the latter and with the anthropoid apes, the frontal bone is more vaulted, and the cranium is enlarged in the vertex region as well. The temporal portions, however, bulge less markedly to the side. The jaws and cooperating skull parts showed that the chewing pressure was shifted somewhat more to the back teeth, and hence directed more towards the vertical than in *A. afarensis*. Thus the front teeth are further reduced and the premolars and molars relatively enlarged. The crowns of the canine teeth no longer project beyond the plane of the bite, the simian gaps are absent, and the lower first premolars are bicuspid and wider, better adapted to grinding food. The attachments of the temporal muscles generally come close together in the middle on the top of the cranium, then diverging evenly to the rear, whereas in *A. afarensis* they continue to run parallel farther back. The cheek bones arise farther forward in *A. africanus*, and the cheek bones themselves are likewise advanced. This provides more favorable chewing pressure conditions for the jaw muscles, which pull from the lower margin of the cheek bones towards the angles of the lower jaw and play a substantial part in grinding movements.

HUMANS

If the *Australopithecus* skull is compared with that of the chimpanzee, considered to be our nearest relatives among the great apes, then besides the enlargement of the cranium and the reduction of the jaw, some other important differences are noted. Some have to do with the erect gait, in which the head is more balanced over the trunk; the articulation of the head and the foramen magnum are farther forward, and the nape musculature is smaller and more nearly horizontal. It is also noticeable that even though the jaw still protrudes considerably, the bony supraorbital ridge is a good deal less developed and projects less than in the chimpanzee.

In lieu of the specific name *Australopithecus africanus*, the more general term "gracile australopithecines" is also much used. To be sure, this expression is justified only in comparison with the so-called "robust" australopithecines, such as *Australopithecus robustus*.

Australopithecus robustus and Australopithecus boisei

Only two years after Broom found the first adult *Australopithecus africanus* skull, he found remains of a new, larger and more "robust" hominid form quite nearby, at Kroomdraai. Broom considered this 1938 find to be more human-like than *Australopithecus africanus*, and accordingly named it *Paranthropos robustus* (the robust "near man"). In the following decades, this form became better

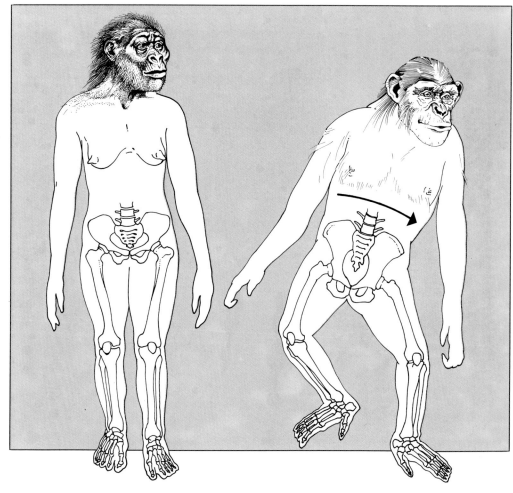

Preserved parts of the pelvis and leg bones of "Lucy" (left) show that in these early australopithecines, the essential adaptations to walking on two legs were already present. Owing to the placement of the knees close together, the body weight is placed almost directly above the particular supporting leg in walking. In the upright walk of a chimpanzee (right), its body will keep tilting from side to side because of bow-leggedness.

known through many additional finds – especially in Swartkrans – and close connections with the australopithecines were revealed. Most specialists today therefore join *A. robustus* to the genus *Australopithecus*.

To date, remains of more than 130 individuals of this "robust" South African australopithecine have been dug up. They are between one and two million years old, later than the "gracile" form. While *A. robustus* also possesses the fundamental biped adaptations, this form is distinctly larger and heavier – about 88–132 lbs (40–60 kg) – as well as more powerfully built than the "gracile." There are numerous features to indicate that these prehumans were highly specialized. Especially typical is the massive chewing apparatus; premolars and molars are considerably further reduced compared to those of the "gracile" australopithecines. The incisors and especially the canines have been further reduced and stand in a nearly straight row. These developments led to rather large, powerful lower jaws and correspondingly large chewing muscles. Hence the size of the cranium no longer sufficed for attachment of the temporal muscles, and usually an additional crest-like enlargement of the skull surface was required in the vertex region. With increasing growth of the grinding area of the back teeth in the course of australopithecine evolution, chewing pressure came to be directed more nearly vertically. Thus, by contrast with the early australopithecines, it was the anterior fibers of the temporal muscles that became more heavily taxed, accounting for the crest. Likewise, in combination with the special chewing condition, these hominids have a characteristic massive and very flat face, with flat cheek bones pointing forward, the prominent supraorbital ridge, and the very deep arches of the lower jaw. The skull and the rest of the skeleton plainly show that an extreme line of development was upon embarked here, characterized especially by an increase in stature and adaptation to coarse and hard vegetable fare, requiring strenuous grinding. The taller stature probably also accounts for the somewhat greater cranial capacity of the robust type, averaging 32 in.^{3} (510 cm^3).

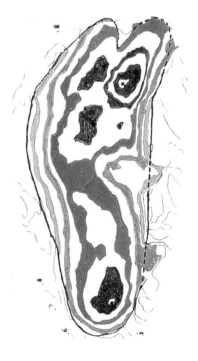

One of the footprints, about 3.7 million years old, in the volcanic ash near Laetoli, northern Tanzania (left). The same print (right) with photogrammetrically delineated contour lines (in different colors). It shows the weight distribution in walking to have been basically the same as in humans.

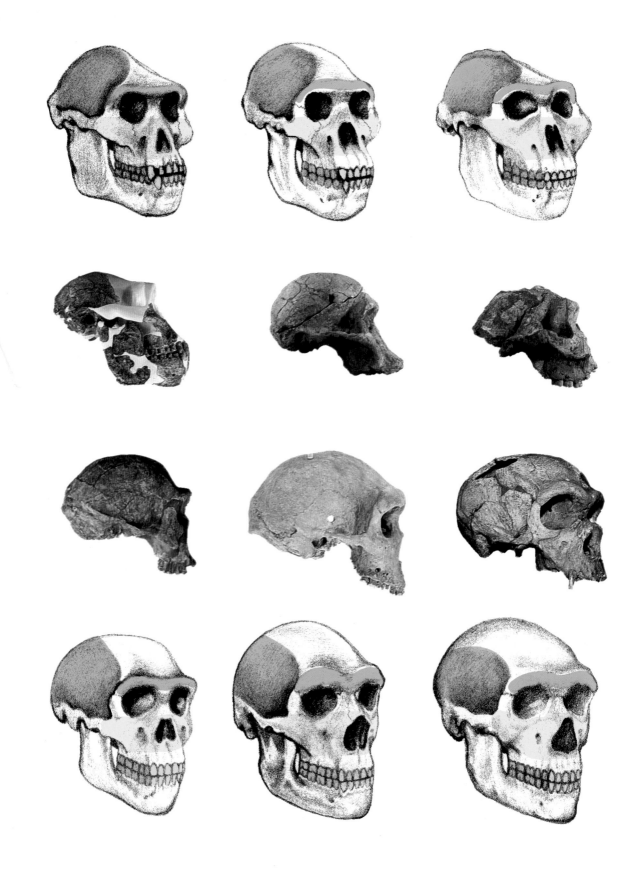

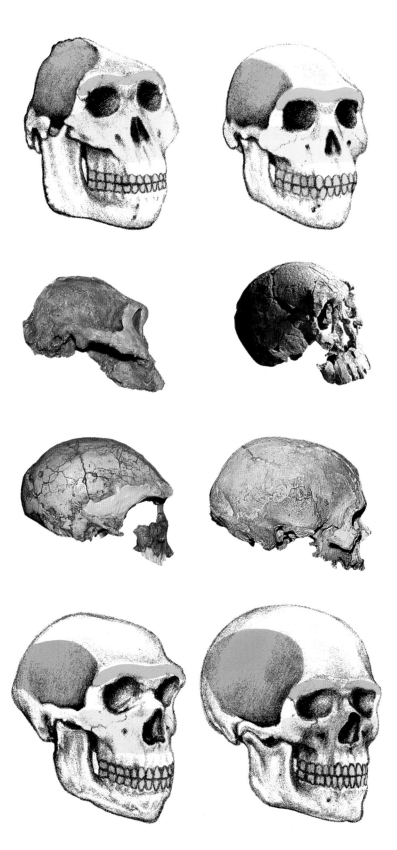

The principal fossil hominid forms (Top two rows, left to right) *Australopithecus afarensis, A. africanus, A. robustus, A. boisei, Homo habilis*. (Bottom two rows, left to right) *Homo erectus*, early archaic *Homo sapiens* of Africa, modern Cro-Magnon. Important sidelines of evolution, not in the direct ancestral line of modern humans, are the robust australopithecine species *A. robustus* and *A. boisei*, and the Neandertals. The different colors in the drawings indicate important skull regions. Besides the great enlargement of the cranium, note especially the changes in the chewing apparatus.
In the robust australopithecines *A. robustus* and *A. boisei*, the large chewing muscles (orange) are so greatly developed that a bony crest is formed.
The diverse forms of the middle face are decisively controlled by the dentition. From *A. afarensis* to *A. boisei*, the front teeth (violet) are increasingly reduced, while the premolars and molars (blue) grow larger. The supraorbital ridge (green) is another feature of fossil hominids. Only in modern humans is it replaced by two separated brows.

The principal fossil hominid forms. (Top two rows, left to right) *Australopithecus afarensis, A. africanus, A. robustus, A. boisei, Homo habilis.* (Bottom two rows, left to right) *Homo erectus*, early archaic *Homo sapiens* of Africa, Neandertals, late archaic *Homo sapiens* of Africa, modern Cro-Magnon. Important sidelines of evolution, not in the direct ancestral line of modern humans, are the robust australopithecine species *A. robustus* and *A. boisei*, and the Neandertals.

The different colors in the drawings indicate important skull regions. Besides the great enlargement of the cranium, note especially the changes in the chewing apparatus.

In the robust australopithecines *A. robustus* and *A. boisei*, the large chewing muscles (orange) are so greatly developed that a bony crest is formed.

The diverse forms of the middle face are decisively controlled by the dentition. From *A. afarensis* to *A. boisei*, the front teeth (violet) are increasingly reduced, while the premolars and molars (blue) grow larger. The supraorbital ridge (green) is another feature of fossil hominids. Only in modern humans is it replaced by two separated brows.

Robust australopithecines lived not only in southern but also in eastern Africa. The history of their discovery began there in the year 1959 with the celebrated *Zinjanthropos boisei* of the Olduvai Canyon in Tanzania, about 1.8 million years old. Other remains of this form followed, both in Olduvai and in other parts of Tanzania, as well as in Kenya and Ethiopia, scattered over a period of between two and one million years. Mary and Louis Leakey's "Zinj" first documented the East African Robustus, and their son Richard, the director of the National Museum in Nairobi, discovered another, complete skull of this form in 1968, eastward of Lake Turkana (formerly Lake Rudolf) in northern Kenya, a region that was to become a regular "gold mine" of fossil hominids in the coming years. There Richard Leakey and his associates found, among other discoveries, half of the skull of a female representative of the robust australopithecines. Evidently there were marked differences between the sexes of these hominids. The female skull, though quite similar to the male skull in general form, is definitely smaller, especially in the face and also at the muscle attachments on the cranium.

Comparing the East African and the South African Robustus, we find but few differences, and these attributable to the same tendency, namely the marked enlargement of the back teeth. In general, the East Africa form is the more extreme, sturdier, and larger. It has been referred to as the "hyper-robust" form of the robust australopithecines, and distinguished as a separate species *A. boisei* from the South Africa form *A. robustus*.

The robust australopithecines quite probably branched off from the "graciles," which themselves show a distinct enlargement of the premolars and molars. However, new questions about the origin of the robust line are raised by a skull about 2.5 million years old that was discovered only in 1985, west of Lake Turkana. It represents a blend of primitive Afarensis-like traits (for example,

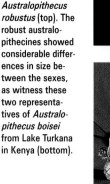

The dig at Swartkrans in South Africa, one of the important sites for *Australopithecus robustus* (top). The robust australopithecines showed considerable differences in size between the sexes, as witness these two representatives of *Australopithecus boisei* from Lake Turkana in Kenya (bottom).

highly prognathic jaw) and those of Robustus/Boisei (skull crest and large premolars). The roots of the robust line would thus seem to extend farther back than was presumed. *A. robustus* as well as *A. boisei*, judging by the fossil finds, seem to be branches of the hominid line that died out about a million years ago, at a time when the development of the genus *Homo* was already far advanced.

Homo habilis

The first hint that yet another hominid form lived in East Africa simultaneously with *Australopithecus boisei* came out of the Olduvai Canyon early in the 1960s. The thorough excavations of Louis and Mary Leakey, in deposits about 1.8 million years old, had brought to light not only thousands of rough stone implements and a large number of dwelling sites, but also skulls and other skeletal remains of an apparently more highly evolved hominid form. Louis Leakey was convinced that this new hominid, whose cranial capacity was about half again as great as that of the "gracile" australopithecine, was the maker of the primitive tools of the so-called Olduvai culture. Hence in 1964 this form was named *Homo habilis*, or "man of skill," thus placing it at the beginning of the line of development that eventually led to modern humans.

Although the pioneering research of the simiologist, Jane van Lawick Goodall, showed that chimpanzees too will modify objects into tools – for example, in order to angle for termites – only human beings use tools for making other tools. *Homo habilis* may well have been the first hominid species to collect stones systematically – often from several miles distance – and shape them into different types of tools (axe, knife, chisel, scraper etc.) by means of other stones. These simple stone implements, fashioned with a few well-directed blows, probably were put to various uses, such as jointing and carving game, and also to work pieces of wood and other objects. It is certainly to be supposed that the australopithecines also used stone and bone tools, but so far there has been no indication that they fashioned stones into tools as *Homo habilis* did. The first appearance of differently shaped stone implements of *H. habilis* is of such historic importance because it would seem to argue a more powerful brain. Studies on internal skull casts by Ralph L. Holloway show that the gross structure of the *H. habilis* brain more closely resembles that of modern humans than did that of the australopithecines. For the first time, too, there is a convolution possibly corresponding to the Broca speech center (so named after the French physician and anthropologist Paul Broca, who discovered the seat of the faculty of speech, in the previous century). Thus perhaps the neurological base for some capability of speech may have existed, but other anatomical details of the skull would indicate only a very primitive speech potential.

The *Homo habilis* finds at Olduvai were either much disrupted or fragmentary, but in the early seventies Richard Leakey offered impressive confirmation of the existence of that species. He and his associates, at Koobi Fora on the East side of Lake Turkana, found not only an abundance of stone implements but also numerous skull and

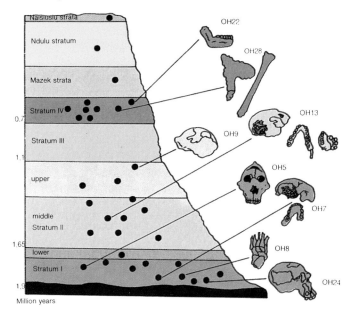

Geological profile of the Olduvai Canyon in North Tanzania, one of the most important sites of our remote ancestors. The deposits bracket a period of nearly two million years. Here many remains of the hominid species *Australopithecus boisei* (such as OH4; "OH" stands for Olduvai Hominid), *Homo habilis* (such as OH24), *Homo erectus* (such as OH9), and *Homo sapiens* were discovered.

other skeletal remains of these early people, including the largest *Homo habilis* skull so far, with a capacity of just about 51 in.^{3} (800 cm^3). The age of this find is about 1.8 million years. In the South of Ethiopia, at Omo, and at Sterkfontein in South Africa, remains of *Homo habilis* have also been found. All together, finds of this early human form span about the period from 2 million to 1.5 million years ago, while primitive stone implements go back as far as 2.5 million years.

There are many examples to show that evolution typically proceeds in the manner of a mosaic; that is, some parts of the organism change faster than others. In the cae of *Homo habilis*, developments affect especially the brain, and with it the brain case, the cranium. Besides the continued increase in size – compared to the "gracile" australopithecines – the greater frontal development, the less prominent supraorbital ridge and the rounded occiput are especially characteristic. In the facial region also, there are some essential differences from *Australopithecus africanus*. The premolars and molars are narrower and the front teeth larger relative to the back teeth. In other words, the dentition does not reveal the specialization to be found in *Australopithecus africanus* and especially *A. robustus*, but points rather to a mixed diet consisting of flesh and plants. Also, the jaws of *Homo habilis* are not so protruding, but retreat somewhat beneath the comparatively larger cranium. The form of the pelvis, legs, and feet suggests that *Homo habilis* had essentially the same biped locomotion as modern humans.

The fossil remains of *Homo habilis* show quite convincingly, overall, that a transition from the australopithecines to early humans took place during this period. Despite substantial agreement among *Homo habilis* finds, certain appreciable differences should not be overlooked. This is instanced by the two skulls ER 1470 and ER 1813 from Lake Turkana. The former has a large cranium, 26,950 ft^3 (770 cm^3), and a quite massive upper jaw as well, while the latter has a cranial capacity of only 49 in.^{3} (510 cm^3) and a definitely smaller jaw. It is not clear whether this variety chiefly reflects considerable sex differences, as many authors believe, or whether representatives of different hominid forms are involved. Thus the anthropologist Alan Walker believes that some of the *Homo habilis* finds represent a late form of *A. africanus*. Despite the higher development of *H. habilis*, there is a greater resemblance in skull form to the "gracile" australopithecines than to the successor species, *Homo erectus*. Some *H. habilis* specimens (for example ER 1813) also have traits – as in the supraorbital and occipital regions – that one might well imagine at the inception of the evolution of *Homo erectus*. Upon the whole, *Homo habilis* may well occupy an intermediate position, both in time and in skull structure, between

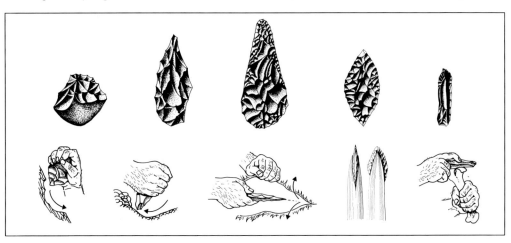

Roughly paralleling the enlargement of cranial capacity, there is a steadily improving technique of toolmaking. (left to right) Simple chopping tool of *Homo habilis*, earlier and later hand axe of *Homo erectus*, Neandertal spearhead, and knife of modern *Homo sapiens*.

the australopithecines and *Homo erectus*. The *Homo* line appears to have branched off from the Afarensis/Africanus line between 3 and somewhat more than 2 million years ago. Whether it emanated from the more primitive Afarensis form or from the somewhat more specialized Africanus is still in dispute. Clarification must await further discoveries.

Homo erectus

For many decades, attention regarding the early development of humans was directed upon East Asia. Here Dubois as early as 1891–1892 had discovered a very flat, primitive skull fragment and a thigh bone of human appearance on Java. He supposed that this form, designated *Pithecanthropus erectus* ("upright ape-man"), was to be placed between the anthropoid apes and humans, in particular because of a low cranial capacity of only 54 in.^{3} (850 cm^3). But what sort of fossil form this might really be did not become clear until the 1930s, when there were similar finds both in Indonesia and in China. Thus the paleontologist G.H. Ralph von Koenigswald, later with the Senckenberg Research Institute in Frankfurt, discovered a number of important fossils near Sangiran, Java, among them some fairly complete cranial and jaw parts of *Pithecanthropus* from the Lower Pleistocene.

About the same time, extensive excavations in caves near Zhoukoudian (formerly Choukoutien) near Beijing (Peking) led to the discovery of a large number of skulls, some well preserved, which the anatomist Franz Weidenreich described in detail. These Chinese finds, roughly 230,000 to 500,000 years old, were then ascribed to the species *Sinanthropus pekinensis*, described a few years before on the basis of three teeth, and known to the public as the Peking man. Comparisons made by Weidenreich and Koenigswald jointly in 1938 between *Pithecanthropus* and *Simanthropus* sufficed to show that this must be one and the same species, later to be named *Homo erectus*.

From the end of the 1950s to the present time, the number of East Asian finds of *Homo erectus* has continued to grow. Most of the Indonesian finds are from Sangiran. One especially well-preserved skull, about 600,000 years old, was found there in 1969. Skull remains from the Lantian region of China are only slightly older. Whereas until recently, on the basis of classical notions of the age of geological strata, it was supposed that the oldest *Pithecanthropus* finds dated back about 2 million years, more recent investigations by G. Pope show that the oldest East Asian hominids are probably not over 1 to 1.3 million years old. Since the greater age of about 2 million years was long accepted for early *Homo erectus*, who thus seemed to have lived at about the same time as the African *Homo habilis*, the difficult question arose as to the origins of the East Asian line of humanity. Were there as yet unknown Australopithecus-like hominids on that continent as well, who might have been ancestors of these early human beings? Even though it is still not possible to answer this question categorically in the negative, numerous recent discoveries and finds have rendered the proposition extremely improbable.

As a matter of fact, the oldest *Homo erectus* find yet known was in Kenya, and has a radiometric age of about 1.6 million years. It is an almost perfectly preserved skeleton, including skull, discovered only in 1984 during an expedition under the leadership of Richard Leakey and Alan Walker, near Nariokotome on the West side of Lake Turkana. The skeleton is that of a twelve-year-old child, no less than 66 in. (165 cm) in height, whose skull has the typical *Homo erectus* features to be described below.

Prior to this latest find of a *Homo erectus*, quite a number of skulls, some well preserved, and also other skeletal remains of this human form, were discovered in East Africa during recent decades. Richard Leakey and his associates, in the mid-1970s, found a nearly complete skull about 1.5 million years old and another somewhat younger cranium of *Homo erectus* on the East side of Lake Turkana. Some of the more familiar finds of this species include a very solid skull, about 1.1 million years old, discovered in the Olduvai Canyon in 1960, and several jaw bones from Ternifine, Algeria, dating back to about 600,000 years ago.

These and other finds in Africa convincingly document the very early presence of *Homo erectus* on that continent, as well as its occurrence there well into the Middle Pleistocene (Ice Age). In view of recent results on the dating of the finds and the fundamental agreement in skull structure between the African and the Asian fossils, it may be accepted today as highly probable that the origin of *Homo erectus* was in Africa and that this species spread as far as Java, which during the Pleistocene was joined to the mainland. Specialists are more puzzled by the question from which section of the known and quite diverse series of forms of *Homo erectus* the early *Homo habilis* may have developed, especially since the latter may actually be older than some *Homo habilis* finds.

Various temporally and spatially distinguishable groups or subspecies of *Homo erectus* lived in Africa between at least 1.6 million and roughly 400,000 years ago, and in East Asia between about 1.3 million and 200,000 years ago.

In Europe, the fossil record of human prehistory does not begin until about 500 000 years ago, with the jaw bone of Mauer near Heidelberg, while cultural relics of human occupancy, in the opinion of various paleontologists, might be a few hundred thousand years older. The Mauer jaw, discovered in 1907, is identified by most specialists as *Homo erectus*. This applies also to the skull remnants at Bilzingsleben in East Germany, the age of which is not quite clear but is probably over 300,000 years.

All told, the number of *Homo erectus* finds in Europe is very small, for as early as about 400,000 years ago, forms occur here that are regarded as archaic *Homo sapiens*. Between late *Homo erectus* and early archaic *Homo sapiens* there is no sharp boundary, but – as is shown by finds in Europe and Africa especially – a prolonged, mosaic-like transition. Hence the attribution of many finds, for example by Vértesszöllös in Hungary or Bodo in Ethiopia, to one or the other species is controversial.

Despite the great total length of time spanned by the species *Homo erectus*, the skull has a quite

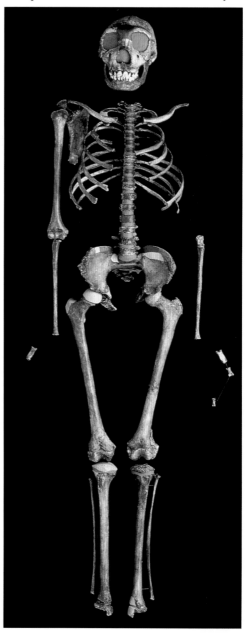

In 1984, this almost perfectly preserved skeleton of a *Homo erectus* was found in deposits aged 1.6 million years on the West side of Lake Turkana in North Kenya. It is that of a twelve-year-old boy, already about 66 inches (165 centimeters) tall. This is not only the most complete but also the oldest find to have been positively identified as *Homo erectus*.

clearly recognizable form and numerous typical features. The cranium is long, low and usually thick-walled. The occiput is angular in profile, with a more or less prominent inion. The side walls of the skull converge obliquely upward, giving it a tent shape as seen from behind. The supraorbital ridge protrudes distinctly forward; behind the ridge, the skull is sharply constricted. The face is low and sturdily formed.

The evolution of *Homo erectus* is characterized primarily by a considerably cerebralization. Whereas early representatives have a cranial capacity of barely 54 in.^{3} (900 cm^3; for example ER 3733 of Koobi Fora), the later forms of Zhoukoudian attained figures up to 74 in.^{3} (1225 cm^3). Differences between the Asian and African *Homo erectus* finds are limited to a few details of the skull. The recently discovered Nariokotome skeleton as well as other remains of limbs show that the stature must have been about the same as that of modern humans, but the bones generally were heavier.

There can be no doubt that *Homo erectus* played a decisive role in the development of humans. It was this species that spread first over Africa and then out to Asia and Europe, using cultural means of adaptation. *Homo erectus*, was a more skillful tool maker than its predecessors. They were typified especially by the hand axe, which in the course of time exhibits an increasingly perfected conformation. *Homo erectus* was also the first to master the art of kindling fire. This enabled them not only to penetrate more inclement regions but also to cook animal and vegetable food. However, their fireplaces may have served many other purposes: as a source of light, for protection against predators, and certainly also as social centers.

Many sites testify that *Homo erectus* was a successful hunter, often slaying larger animals. This required a high degree of planning and cooperation within the community. Cultural and intellectual changes would have been in a close feedback relationship to each other during these periods of development.

With a more highly developed culture, the individual's chances of survival became increasingly dependent on mental agility. The resulting pressure of selection was responsible for the significant cerebral development over the millennia, and also for other physical changes. A further enlargement and refinement of the associative centers of the brain in particular, whose function is to link ideas and impressions as well as memories, and changes in cranial form, finally gave rise to *Homo sapiens*, the rational animal.

Homo sapiens

About 30,000 to 40,000 years ago, the anatomically modern *Homo sapiens* appears in nearly all parts of the world – in certain places even earlier. The skull structure is not essentially different from that of people living today. The cranium is thin-walled and rounded all over. Supraorbital ridges are absent, the jaw has become smaller, and there is a well-developed chin.

Between late *Homo erectus* with its typical skull formation and anatomically modern *Homo sapiens*, there is a period of a few hundred thousand years, well-documented by a multitude of finds in Europe, Africa, and Asia. These fossils show a further gain in cranial capacity, more and more closely matching that of modern humans. In other traits, however, there are distinct transitions between the two species. The latter form is therefore referred to as "archaic" *Homo sapiens*. This form appears about 400,000 years ago in Europe and Africa and probably not until 200,000 years ago in East Asia. It is not clear whether the transitions from late *Homo erectus* to archaic *Homo sapiens* occurred more or less independently of each other in the three continents, or whether migrations and minglings of populations between continents were a major factor. The Erectus-Sapiens boundary, however, was probably no longer an interspecific boundary in the biological sense, that is, no reproductive barrier, but only a morphological one, manifested in skull structure. Archaic *Homo sapiens* developed differently in the several continents. In Europe development led to the Neandertals, whereas in Africa it led to anatomically modern humans. As to East Asia, there are still some enigmas, the number of fossils being scanty. The European line of archaic *Homo sapiens* exhibits three temporally successive stages of development: pre-Neandertal, early Neandertal and late Neandertal.

Our picture of the Neandertal predecessors has been clarified by many new finds since the mid-1960s. Formerly, perceptions were based primarily on two skulls about 250,000 to 300,000 years old, discovered in the early thirties at Steinheim, in Baden-Württemberg and at Swanscombe in England. Both seemed to have, besides archaic traits, other features closer to modern humans than to the Neandertals. Hence came the idea that since that early time there have been two parallel lines of development in Europe, one leading to the Neandertals, the other to anatomically modern humans. But the newer finds show that the Steinheim skull and the Swanscombe skull both fit into the spectrum of variation of the so-called pre-Neandertals, which taken together led only to the Neandertals.

Two especially well-preserved representatives of this group are the Arago skull (about 400,000 years old) of southern France and the Petralona skull (about 250,000 years old) of northern Greece. Like the Steinheim skull, they have a cranial capacity of about 72 in.^{3} (1200 cm^3). The supraorbital ridges are less pronounced than in *Homo erectus*, and the more nearly vertical side walls of the cranium (like Steinheim and Swanscombe) suggest a further enlargement of certain parts of the brain. In the facial region, the pre-Neandertals exhibit some variety of form. Thus Steinheim has a slight depression in the cheek that is absent in Arago and Petralona as well as in the late Neandertal. Somewhat less ancient late Middle Pleistocene finds in France, as at Biache St. Vaast or La Chaise, Suard, resemble the Neandertal more closely, so that some researchers would refer them to the latter and not to the pre-Neandertal.

But at least since the beginning of the last interglacial epoch – 125,000 years ago – the Neandertals settled Europe and adjacent territories until their mysterious end, about 30,000 years ago. The interglacial Neandertals are commonly regarded as early Neandertal or even pre-Neandertal. Important finds of this form have occurred in Gibraltar, Italy, Czechoslovakia, Yugoslavia and elsewhere (see map on page 496–497). Compared to the pre-Neandertals, the early Neanderthals show a further gain in cranial capacity, reaching an average of about 78 in.^{3} (1300 cm^3). The skull structure resembles that of the late Neandertal in essentials, but the cranium and facial skeleton are as yet less massive.

One of the first late Neandertal finds from the last glaciation was made in 1856 by a high-school professor, Johann Carl Fuhlrott, at Neandertal near Düsseldorf. Later, many other finds followed, almost throughout Europe and adjacent territories, representing remains of more than 300 individu-

New fossil finds are subjected to detailed scientific study, as here at the National Museum in Nairobi. (left) "True-to-life" casts are prepared in a special laboratory. (right) Often a skull must be reassembled by painstaking labor out of many bits and pieces.

als. The traditional and widespread notion that the Neandertals were aimlessly groping, dull-witted brutes has long been untenable. On the contrary, many studies of the physique and the intellectual sphere of the Neandertal people (burial practices, worship, etc.) have shown that this human being, with certain archaic features of the skull particularly, is very close to anatomically modern humans. Thus the Neandertal has a large cranial capacity averaging about 90 in.[3] (1500 cm^3). The skull is large, long, and low, and has a transverse oval shape as seen from behind. Over both eye sockets there is a continuous supraorbital ridge, though less prominent than in the pre-Neandertal. Especially typical is the shape of the facial skeleton. The mid-portion looks as though it were drawn forward. This peculiarity may be related to a heavier setting of the front teeth, which have increased in size. The facial skeleton is also characterized by a broad nasal orifice and by the absence of cheek depressions, giving the central face a particularly massive appearance. These and other traits indicate the Neandertal to be a much specialized form of archaic *Homo sapiens*, probably adapted specifically to glacial conditions in Europe. The same applies to the proportions of the limbs, similar to those of the Lapps and Eskimos, who live in cold climates. The West European Neandertals, because of their consistently rougher appearance, are sometimes referred to as "classic" Neandertal, in contradistinction to "eastern" Neandertal, apparently somewhat less specialized.

Despite the many finds, and 120 years of Neandertal research, the seemingly sudden end of this human form continues to pose many riddles. The Neandertal was followed in Europe, about 30,000 years ago, by anatomically modern *Homo sapiens*, with a skull structure radically different from the Neandertal type. As regards the causes of the Neandertal disappearance, we must depend for the most part upon surmise, but in the last few years there have been new discoveries relevant to the timing of that succession. Thus, in 1979 the latest Neandertal to date was found at St. Césaire in the Dordogne. It dates from about 32,000 years ago, and is associated with a Late Paleolithic blade manufacture rather than with the more typical mousterian culture a Paleolithic stage so named after finds at Le Moustier in the French Dordogne. The frontal bone find at Hahnöfersand near Hamburg may be somewhat older; besides distinctly modern features in the supraorbital region, it bears certain resemblances to the Neandertal. These two finds and other considerations indicate that Neandertal and modern human beings lived side by side for some thousands of years, and probably also interbred, before the Neandertal was finally displaced. The modern human being, who in all likelihood would not have evolved from the highly specialized European Neandertal, must in that case have entered Europe from elsewhere.

The search for the roots of anatomically modern *Homo sapiens* has recently led back to Africa again and again. Since the late 1960s, finds here have steadily increased in number, so that today the course of development of *Homo sapiens* is fairly clear. An early archaic form is followed by a late archaic form, from which anatomically modern humans proceed.

Early archaic *Homo sapiens* lived about 400,000 to 150,000 years ago. A long series of finds in various parts of Africa (Bodo, Eyasi, Broken Hill, Hopefield, etc.) shows an as yet quite primitive skull structure with a heavy supraorbital ridge, more or less comparable with that of certain early pre-Neandertals of Europe. The cranial capacity is generally around or above 75 in.[3] (1250 cm^3) and also the side walls of the cranium are more steeply set than in *Homo erectus*.

About 200,000 to 150,000 years ago, early archaic *Homo sapiens* gave way to the more modern late archaic *Homo sapiens*, who probably flourished until about 100,000 years ago and perhaps longer in certain areas. Finds in South Africa (Florisbad) and East Africa (Omo 2, Laetoli 18, Eliye Springs) may be ascribed to this form. The cranial

Three stages in development of human consciousness:

▷ *Homo habilis*, 1.8 million years ago, already capable of manufacturing multipurpose stone implements.

▷▷ *Homo erectus*, about 500,000 years ago. The possession of fire enabled settlment in cooler climatic regions, outside of Africa.

▷▷▷ Neandertals, about 60,000 years ago. Ritualized funeral observances bear witness to a high development of the imagination.

capacity in the known finds is over 81 in.^{3} (1350 cm^3), and there are numerous features of the skull that hint at the successor state, anatomically modern humans.

Somewhat more than 100,000 years ago, late archaic *Homo sapiens* is followed by anatomically modern humans. Important evidence includes the skull Omo 1 from southern Ethiopia as well as finds in South Africa (Klasies River Mouth, Border Cave). The clearing picture of *Homo sapiens* development suggests the conclusion that anatomically modern humans spread from Africa, where they lived far earlier than in Europe or in other parts of the world, in a northward direction, namely through southwestern Asia to Europe and western Asia, and eventually displaced the Neandertals living there ("Afro-European Sapiens hypothesis"). This more recent conception of development in the two continents seems to be supported by finds in the Near East as well.

In the region of southwest Asia, numerous Neandertal finds are known, as well as some suggesting mixtures between Neandertals and anatomically modern humans. Whereas a skull find at Mugharet-el-Zuttiyeh in Israel pertains to the last interglacial epoch, the other Neandertals of the Near East derive from the last glaciation, with an age between 70,000 and 50,000 years. A considerable number of individuals were excavated at Shanidar in Syria. In this case, fossil pollen analyses showed that the Neandertals had strewn flowering plants in bunches over a corpse. Other remains of Neandertals derive from Tabun and Amud in Israel. In contradistinction to the European Neandertals, the characteristic pattern of traits in the Near Eastern, though present, is less uniform and distinctive.

Successive to the Neandertal in time are finds that have long received much attention. Quite extensive skeletal material from the two sites of Skhūl and Qafzeh in Israel is probably between 40,000 and 60,000 years old. The skulls, despite a predominantly modern appearance, reveal much diversity. Whereas some – for example Qafzeh 9 – are anatomically quite modern, others (Skhūl 5 and Qafzeh 6) show archaic or Neandertal-like traits as well, such as a supraorbital ridge. The supposition, formerly common, that modern humans developed in southwest Asia directly from the Neandertals there, now seems quite unlikely, since here, as in Europe, completely modern humans followed Neandertals virtually without transition. The most probable explanation of these morphological diversities in the Near East might be the hypothesis of interbreeding between the local Neandertal and early moderns coming from East Africa, to the South. This period of interbreeding and displacement in southwest Asia, moreover, is prior to the corresponding phase in Europe. Anatomically modern humans in West Asia as in Europe, to judge by present evidence, seem most likely to have been of African origin.

Less clear is the course of development of *Homo sapiens* in southeast Asia and its relationship to developments in the western parts of the Old World. In the first place, there are comparatively few finds in the Asian region, and in the second place, most of them can be only very roughly dated.

Finds of archaic *Homo sapiens* are reported in China and recently also in India and Indonesia. Of the Chinese finds, the most significant are the nearly complete Dali skull discovered in 1978 and the skull pan found at Maba in 1958. The Dali skull would seem to be of an age probably between 200,000 and 125,000 years. Although the cranial capacity is barely 72 in.^{3} (1200 cm^3), the conformation and detailed features of the cranium and facial skeleton show distinct relations to other finds of archaic *Homo sapiens.* The Maba skull is probably somewhat younger, perhaps around 100,000 years old. Also, like Dali, it has a well-developed supraorbital ridge. Both finds show a slight carination, a trait often found in *Homo erectus* of China as well.

Also of great significance is a quite well-preserved skull found as recently as 1982 in the Narmada Valley of northern India, dating from the Middle Pleistocene or the very earliest Upper Pleistocene.

Studies thus far have shown that the find appeared most closely to resemble such European pre-Neandertals as Steinheim and Arago. Here again, however, the cranium is slightly carinate.

Archaic *Homo sapiens* material from Indonesia is more extensive. In the Ngandong deposits on the Solo River, a considerable number of more or less complete craniums were found early in the 1930s. Age estimates range from about 250,000 to 100,000 years. Although the Ngandong skulls show strong resemblances to *Homo erectus* of East Asia in general form, the cranial capacity is in the upper range of *Homo erectus*, and sometimes greater. Furthermore, several morphological features of the skull reveal a definite further development in the direction of *Homo sapiens*. Whereas probably most investigators see a form of archaic *Homo sapiens* in these finds, some are thought to represent a late form of *Homo erectus*. Here, as in Europe and Africa, there is the difficulty of positive attribution at the Erectus-Sapiens borderline. Although evidence of archaic *Homo sapiens* is scant in southeast Asia, one must nevertheless recognize a certain continuity of development in that region, and perhaps there are also affinities – as possibly in the case of the Narmada skull – with western forms.

While archaic *Homo sapiens* is documented up to about 100,000 years ago in China, and perhaps only until about 200,000 years ago in Indonesia, the ensuing period until the first appearance of anatomically modern humans is virtually barren, that is, lacking in any instructive remains.

The earliest anatomically modern finds in China (Liujiang, Zhoukoudian Upper Cave, Ziyang, etc.) are also for the most part imprecisely dated, but may pertain to the close of the Upper Pleistocene, which would make them about 20,000 years or less in age. Compared to modern Chinese skulls, these have a definitely sturdier appearance. There are also resemblances to early modern finds in Europe, as well as traits, such as carination, suggesting a certain regional development. The modern Indonesian finds at Wadjak probably also pertain to the close of the Upper Pleistocene and bear resemblances to the early moderns of China. The oldest anatomically modern find of southeast Asia may be a skull from Niah on Borneo, but its dating to 39,000 years ago is much disputed.

On the basis of the far-reaching similarities among early anatomically modern finds in China, Europe, and Africa, it is perhaps most probable at this writing that early modern humans spread from Africa through western Asia to East Asia and, as in Europe, interbred with and eventually displaced the local populations of archaic *Homo sapiens*, about whose appearance, however, we know next to nothing.

The races of human beings living today are so similar to each other, not only in skull structure but in many other features, that most authors posit a "monocentric" origin of anatomically modern humans. It therefore seems unlikely – and in any event not documented – that any independent local line of *Homo erectus* might have developed directly into modern humans, paralleling, that is, the Afro-European development. Hence the differences among races living today must have made their appearance within the past few tens of thousands of years.

In Australia and America, fossil human history begins with anatomically modern *Homo sapiens*. The oldest human finds in Australia, on Lake Mungo, are about 32,000 years old. Besides these and other not very "robust" skulls, to be sure, there are some among less ancient Australian finds, between 10,000 and 6000 years old (Kow Swamp, Cossack, Cohuna, etc.) that, despite a basic anatomically modern skull structure, do exhibit some seemingly archaic traits, especially in the supraorbital and frontal region. These manifest differences in finds from the close of the Upper Pleistocene and Early Holocene in Australia are perhaps accounted for by the manner of interbreeding between modern humans spreading southward and the existing archaic population. Certain archaic features might thus have been preserved in parts of the early Australian population. It cannot be ruled out, however, that besides more "gracile" modern man, perhaps 40,000 years ago, a more "robust" form still more closely resembling archaic *Homo sapiens* may have reached Australia from Indonesia.

America was probably first settled about 30,000 to 15,000 years ago, from Siberia. Lowering of sea level during the Ice Age created a land bridge to Alaska where the Bering Strait now lies. Although a large number of finds from North, Central, and South America are known, nearly all are hedged about with uncertainties as to their exact age. Some skulls might derive from the late Upper Pleistocene (for example, Los Angles, La Jolla). In the opinion of Milford Wolpoff, nearly all of these finds bear very strong resemblances to the American Indian skull conformation.

Although the phylogeny of humans must be traced chiefly on the basis of fossil evidence, of course there is one important aspect of humanity that should not be overlooked, namely the development of human consciousness. Information on this point is offered by studies of cranial casts, tool-making technology, signs of worship, earliest funeral observances, works of art, circumstances of life, and forms of society. All these data indicate that human consciousness has been developed and enlarged gradually. Hence it is difficult to say when humans first became able to see themselves in a world of differing creatures. The mode of life of early hunters and gatherers, however, insofar as we can describe it today (division of labor between the sexes, paternal role) suggests that those hominids already had a self-awareness beyond that of present-day human-like apes. What the extent of this was will no doubt always remain a mystery.

Modern Humans: Physical Form

by Rainer Knussmann

Characteristics of Modern Humans

Modern humans comprise one species, *Homo sapiens,* and was recognized as such in the classification of living creatures set up as the foundation of modern biology in 1758 by the Swedish naturalist Carl Linnaeus. As against the archaic forms of *Homo sapiens* that lived 100,000, 200,000 and more years ago, the anatomically modern human who appeared in Europe about 40,000 years ago (see the preceding article by G. Bräuer) may indeed be regarded as only a single subspecies *Homo sapiens sapiens.*

For Linnaeus, as at present, humans held their place among the primates, which include the prosimians, monkeys, and human-like apes. Because they proceeded from this group, their points of agreement with the animal as well as their peculiarities are best seen by comparison with the so-called lower primates. Their biological definition is based primarily on phenomena resulting from two decisive processes of human phylogeny: erect posture and brain development.

Erect Posture

The biped locomotion of humans contrasts with that of other animals, such as flightless birds (for example, the ostrich), saurians, and kangaroos in that the trunk is held vertically and the knee joint more or less completely extended. When the anthropoid ape stands on its hind legs, it also holds the trunk approximately vertical, but is unable to extend the knee and hip joints. It must stand knees bent, because the center of gravity is located far in front of the pelvis.

Design changes due to upright posture. With upright posture, changes came in nearly all parts of the supporting and locomotive apparatus. Whereas in quadrupeds the trunk is suspended from the

MODERN HUMANS: PHYSICAL FORM

In the tenth month of life, the human infant stands up on its legs, so that it is already taller than the beasts.

vertebral column as if from a horizontal bar, in humans the vertebral column shows a tendency to displacement towards the center, to serve as an axial column; although the spine is still located comparatively far back, the rib cage begins to curl around the advanced vertebral column. This development began with the anthropoid apes, if not with the American spider monkeys, for they are all hand-over-hand swingers, and hold their trunk more or less vertical, as humans do, during locomotion. Hence they likewise exhibit the same adaptations, though less pronounced.

In the case of humans, there is the added point that the vertebral column deviates from the quadruped pattern not only in position but in form: it describes a double S-curve, thus becoming a spring that cushions the impact of the foot, which would otherwise be transmitted directly from the heel to the brain because of uprightness. That the vertebral column in humans does not terminate in a tail, however, has nothing to do with erect posture. Lack of a tail is a kinship trait of humans and the human-like apes.

The deviations of humans from their nearest relatives as regards the rib cage and the spine are rather of a quantitative kind, but humans have a fairly unique form of pelvis. In the lower primates, the ischial blades are like plates at the posterior – or, in the quadrupeds, superior – body wall; in humans, transformed into a broad bowl, they have migrated beneath the erect trunk. At the same time, the large muscle of the buttocks, which spreads the legs in the lower primates, has been shifted and gained mass to serve as an extensor of the hip joint; the convex rump, by which humans are distinguished from all the other primates, is thus necessary to hold the trunk erect over the legs. The human rump is further reinforced by deposition of fat, as in some breeds of domestic animals. The interpretation of this fatty layer as a seat cushion seems rather dubious, inasmuch as the baboons and macaques have developed ischial callosities far better suited for sitting on a hard or uneven surface.

Another adaptation to biped locomotion is the great length of the human leg, as the length of the anthropoid arm is an adaptation to swinging hand-over-hand. Also, the femurs in humans, attached laterally to the pelvis, converge downward, so that the knees are not separated, as in the apes, but brought together towards the center. This so-called physiological knock-kneed condition has the mechanical advantage that, in erect locomotion, the center of gravity of the body is located almost exactly over whichever foot is planted at the time.

The structure of the foot, along with pelvic shape, is a principal peculiarity of humans due to erect posture. The lower primates' flat foot became an arched foot in humans, its posterior abutment consisting of the reinforced heel bone and the anterior abutment of the powerful great toe. In order to perform its function, the great toe, spread aside in the lower primates, had to be moved towards the axis of the foot. The other toes were shortened, allowing the foot to roll forward. Thus the grasping foot of the simians became the specialized standing and running foot of humans.

Advantages and disadvantages of erect posture. Erect posture gave early humans the advantage of a broad view over the tall savanna grasses. Perhaps more important, however, it freed the hands from functions of locomotion. Thus, even in transit, the hand could be used to "handle" hunting weapons, for example. Incidentally, the human hand has remained far less specialized than that of many primates, so that it is not committed by adaptation to some particular task, but is available for versatile use. Compensating for the posterior shift of the shoulder blade, the hand has been brought into a favorable position in front of the body by rotation of the humerus.

But neither the widened field of vision nor the freedom of the hand would have been the incentive for erect posture. Rather, it is to be supposed that with climatic change, the primeval forest receded, and the tree-dwelling human ancestor, who like the anthropoid apes had already begun to evolve towards swinging hand-over-hand, was urged in the direction of a terrestrial habit. In that situation, the hand-over-hand set of adaptations to erect posture tended not to prompt a retrograde evolution towards the quadruped but a progressive evolution towards complete uprightness on two legs. This may have been reinforced by the benefits of better reconnaissance and the tool-holding hand to impose the biped habit.

Alternatively, however, it is conceivable that the forest did not so much retreat, as that some human ancestors at the edge of the forest simply adopted terrestrial life in order to take over new habitat. In any event, the development began with a tree dweller having some adaptation to swinging hand-over-hand.

Such adaptations may still be noted today in the skeleton of the human arm, and during the transitional phase, as well as at birth, humans have longer arms than legs, like the arboreal apes. The rapid axial growth of the legs begins after birth. It is a general rule of nature that traits phylogenetically acquired late will appear last in embryonic development, in ontogeny, following a recapitulation of the preceding stages of phylogeny. Thus, in the human fetus the great toe is averted, and by way of so-called atavism, that is, an isolated throwback to phylogenetically earlier times, an averted great toe may even persist into adulthood. A tail, too, is present in the embryo.

Theoretical considerations, as well as a comparison of the human and the simian arm, render inheritance of erect locomotion from a terrestrial quadruped extremely unlikely. Such a route would have had to pass by way of "semi-erect" locomotion, mechanically quite disadvantageous because of the leaning trunk, and hence unlikely to have displaced quadruped locomotion.

Skull and Brain

Human erect posture not only called for adaptations in the trunk and limbs but also influenced the structure of the skull. In typical quadrupeds, for example the tree shrew, the facial skeleton is anterior to the cranium, and with increasingly erect posture of the trunk it migrates beneath the cranium. For if a quadruped were simply to be set up on its hind legs, its face would be directed upward. Hence by an indentation of the base of the skull, the facial skeleton was rotated under the cranium, a tendency noticeable in the arboreal apes but carried to completion only in humans.

At the same time, the foramen magnum, far posterior in the quadrupeds, has been moved far towards the center of the base of the skull, so that the vertebral column also moves towards the center under the skull, and the latter can be held in equilibrium without effort. Even in humans, the vertebral column has not come all the way beneath the center of gravity of the head, as witness the fact that upon falling asleep sitting down, the head will drop forward.

Besides erect posture, the skull configuration is determined primarily by its most important functions: to accommodate the brain and sense organs, and to ingest and masticate nourishment.

Cranial capacity. The indentation of the base of the

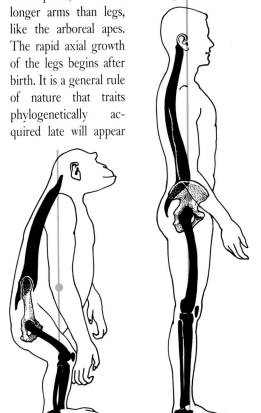

Adaptation of the human body to an erect gait: In the chimpanzee, the center of gravity is unfavorably located, so that in standing the legs cannot be extended. In humans, the carriage of the head and the path of the vertebral column may permit more favorable mechanical conditions.

skull as a result of erect posture did not only place the facial skeleton beneath the cranium but at the same time elongated the top of the skull and added volume to the cranial cavity. Since in human phylogeny, according to the fossil record, erect posture preceded the development of the brain, it is reasonable to suppose that it gave impetus to the unique development of the brain (cerebralization) in humans. Then, to be sure, a sharp increase in the height of the cranium was necessary in addition. But this further development also may have been decisively influenced by erect posture; for owing to the liberation of the hands, each

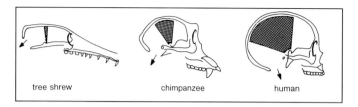

Median section of skull of tree shrew, chimpanzee, and human. By increasing indentation of the base of the skull, the facial skeleton, while becoming smaller, is shifted beneath the cranium. But the curvature of the base of the skull is accompanied by an elongation at the margin of the cranium and hence an enlargement of its volume; this is indicated by the shaded area. The arrows represent the direction of attachment of the spinal column.

Averted great toes – a trait reminiscent of the lower primates.

individual, enabled by higher intelligence to make better use of the hand as a sovereign tool, gained an enormous selective advantage.

Despite its outstanding capability, the human brain is not unique as to size. True, the range of brain volume in humans – aside from malformations – does not overlap that of any other primate: humans average about 78–84 in.^{3} (1300–1400 cm^3), or interior of cranium including cerebral membranes and fluid 84–90 in.^{3} (1400–1500 cm^3) with rare extremes of 51 in.^{3}

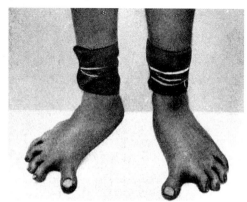

(850 cm^3) below and about 132 in.^{3} (2200 cm^3) above; orangutan and chimpanzee about 24 in.^{3} (400 cm^3), gorilla around 30 in.^{3} (500 cm^3) with maximum of 45 in.^{3} (750 cm^3); but when the comparison is extended to other vertebrates, humans are surpassed by very large species such as elephants and whales. Since there is obviously a relationship to body size, brain weight has been taken as a fraction of body weight; but even on those terms, humans are not in the lead (humans 1:45 to 1:50; capuchin monkey 1:30; tarsier 1:40; house mouse 1:40; dolphin 1:38; sparrow 1:29; chimpanzee 1:110; elephant 1:560; sperm whale 1:10,000; and crocodile 1:5000). The total volume of the brain is not in itself as decisive as the structure of the brain, that is, which parts of the brain are especially developed. Thus the extent of the cerebral cortex is increased by folding in the ascending series of primates, reaching its peak in humans. At the same time, the frontal lobe accounts for 29 percent of the surface of the cerebrum in humans, a larger proportion than in any other primate (chimpanzee 17 percent, prosimians about 8 percent).

Long-distance senses. The olfactory center of the primate brain lags in development, and, unlike the other mammals, the primates are not olfactory animals. Hearing has also become less important; the primate external ear has become a nearly functionless, instead of being an erectile and directional acoustic receiver in typical mammalian fashion. The sense of sight, on the other hand, has been refined and perfected.

All these changes had been largely completed in the simians, which is to say that the developments in question took place on the prosimian level. Here the eyes moved from the side of the face – where they also appear in the human embryo – to the front and center, permitting binocular vision. At the same time the orbits were increasingly protected by bony formations, until in the apes they occupy sockets that are open towards the front only. In humans, this opening tends to have a rectangular outline, whereas it is rounded in the an-

MODERN HUMANS: PHYSICAL FORM

thropoid apes and lower primates. However, human beings do not differ much from their nearest relatives in acuity of vision. This applies also to color vision, which has reached a high level of development in the simians, by contrast with other mammals.

Dentition. The facial skeleton could not have been moved back so far beneath the cranium by mere indentation of the base of the skull. There was also a marked shortening of the "snout," so that the proportion of the cranium to the facial skeleton was shifted in favor of the former by a reduction of the latter, as well as by enlargement of the former.

The slenderer conformation (gracilization) of the jaws was achieved by a reduction of the dentition. This reduction did not come about through a lessening of the number of teeth – only the last molar ("wisdom tooth") is occasionally absent – but through finer teeth. Humans were able to afford this loss of biting capacity because they were able to prepare their food and because they created more effective weapons of defense. Thus, in particular, a reduction of the anterior dentition, especially the canine teeth, was possible, and the rectangular arrangement of the teeth with a gap to accommodate the large canine tooth of the other jaw ("simian gap") was replaced by the closed parabolic dental arch.

With the reduction of the snout, the nose and chin remained in place; that is, these features were less sharply reduced than other parts. Alongside the nose too, the jaw receded sharply, leaving an actual depression. The prominent nose by no means implies any very keen sense of smell, but with the shortened muzzle, it preserved the space required for preheating, moistening and coarse filtration of the air inhaled. This human "nosiness" is not unique either, but far surpassed by the bright red nose of the proboscis monkey.

The prominent human chin was needed as a stabilizing element at the point of greatest liability to fracture, with increasing gracilization of the lower jaw. On the other hand, the prominent supraorbital ridges that provide an attachment for chewing pressure in the African anthropoid apes were expendable. In humans, the chewing pressure is transmitted directly into the vault of the cranium, which they bear above the jaws.

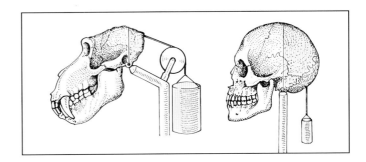

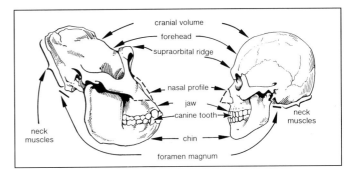

Body Covering

Hair. Next to erect posture and brain development, the most conspicuous human trait is the reduced coat of fur. Thus, the English zoologist Desmond Morris called humans the "naked apes." Not that they are altogether naked; fine short hairs are to be found on most parts of the body, and in some males they form a considerable mat on the chest. The human tendency to hairlessness is also not unusual among mammals; some burrowing forms, bats, and marine mammals are more or less completely devoid of hair.

The great anthropoid apes have a comparatively sparse coat, anticipating the direction of human evolution. They exhibit some manifestations that

(Top) Skulls of baboon and human in normal attitude of the head. To hold the skull in this attitude requires a large force in the baboon, a small one in the human. (Bottom) Skull of an African anthropoid ape (gorilla) and human - both male - with principal points of contrast.

form a link between the "hairlessness" of humans and the furred lower primates. Prenatally, humans past through a hairy stage; premature infants may show remnants of this coat of hair on the shoulders, upper arms and the upper area of the back. As an atavism, isolated skin areas may be haired in the manner of fur even in adulthood, often in conjunction with a dark birthmark. Finally, the phenomenon of "goose pimples" signifies a bristling of the coat – although there are no bristles.

Human "nakedness," compensated by trappings of civilization (housing, clothing, heating), is mitigated also by some patches of hair. On the scalp, the most thickly furred area in most of the lower primates, hair is plentiful, except for the balding of many males. Other patches of hair appear with sexual maturity, the axillary and pubic hair, and in males also the beard. While the axillary and pubic hair is perhaps associated with scent glands, the beard may be a signal identifying the male.

Signal and outer soft parts. In possessing flat nails instead of claws and in the configuration of skin ridges as in the familiar fingerprint, humans are consistant with the other Old World primates, but there are differences with respect to the cutaneous glands. Humans have considerably more sweat glands than the simians; only the chimpanzees approache them in this respect. The abundant supply of sweat glands enables the naked unprotected skin to be kept soft and supple by moistening. Simians also have more sweat glands in hairless body areas. In contradistinction to the sweat glands, the scent glands are reduced in humans; they are to be found essentially only in the pubic and ischial regions, and especially in the armpit. The appearance of hair in this particular location at sexual maturity may be conducive to the release of scent.

The mammary glands in humans, as in all the simians, are a pectoral pair. In some prosimians, there is an inguinal pair of milk glands as well, as in many other mammals (for example, dogs, swine). The primordial mammary glands in prenatal development are also located here in humans. Occasionally women, especially in pregnancy, produce supernumerary nipples in corresponding locations. The human mammary glands are more cushioned with fat than those of other primates. The rounded, prominent female breast consequently represents a distinguishing human trait, the most conspicuous female secondary sex characteristic. Breasts comparable in size can be formed in anthropoid apes only during lactation, owing to special development of glandular tissue.

Another pronounced human trait is the markedly everted lips with their covering of mucous membrane. Compared to the great anthropoid apes, human females also present a marked development of the labia majora. The hymen, which has been thought peculiar to the human female, is to be found in other primates also, for example, the gorilla. As to the male sex organs, humans do not differ markedly from the anthropoid and other apes except in internal features; humans have no penile bone (although the penis is not movable by musculature in the apes either), and the human prostate gland completely encircles the urethra (often causing trouble in older males).

Genetics and Protein Structures
The nucleus of each cell contains the genetic strings or loops (chromosomes) consisting of rows of genes, the vehicles of genetic information. The number of chromosomes and their size and ap-

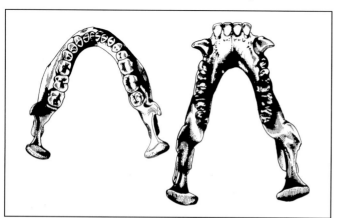

Lower jaws of anthropoid ape and human. In the human (left), the dental arch is rounded; in the chimpanzee (right) it is angular and furnished with a gap accommodating the canine tooth of the upper jaw when the mouth is closed.

pearance are characteristic of each organism. Humans have 23 pairs of chromosomes (numbered from 1 to 22, plus the sex pair, XX in the female, XY in the male). The great anthropoid apes have 24 pairs, and the gibbons from 22 to 26 pairs of chromosomes, depending on species. Humans thus fit in well among the higher apes.

By means of modern staining techniques, darker and lighter segments can be rendered visible in the chromosomes. This banding permits a detailed comparison of human chromosomes with those of the great apes. It turns out that half of the human chromosomes, except for small missing terminal portions, coincide outwardly with those of the African hominoids (gorilla and chimpanzee). In the other chromosomes, there are segments here and there that are reversed in order, or missing in the apes, thus shifting the parallelism of subsequent (transverse) bands. The banding comparison also shows that the human Chromosome 2 evidently arose through the fusion of two anthropoid chromosomes; this accounts for the difference in number of chromosomes between humans and the great apes.

The genetic substance of the chromosomes (that is, the chain of genes) consists of deoxyribonucleic acid (DNA). In total quantity of DNA in the cell nucleus, humans rank close to the apes (humans 7.3, gorillas 7.5, chimpanzees 8.0, orangutans 8.5, and gibbons about 6 picograms). But the individual genes also show much resemblance between humans and the African great apes, as human genetics has shown by various methods in recent years. One such technique proceeds from the proteins, each of which is determined in structure by a corresponding gene. Each protein is made up of a chain of amino acids, and each amino acid is coded in three successive elements of

Variety of humans (I). (Upper left) Europid Lapps of northern Norway, with compact build and round, comparatively low face in adaptation to the cold climate. (Upper right) Maoris of New Zealand, members of the Polynesid race, the most distant outriders of the Europid. (Lower left) Europids of Yugoslavia; the man and woman on the right represent the Dinarid race, while the woman on the left shows definite Mongolid traits. (Lower right) Europid of India, a representative of the gracile Indid race, among whom pedomorphic (child-like) features are of frequent occurrence.

the DNA string. Thus the sequence of amino acids in a protein, which can be ascertained by biochemical analysis, reflects the sequence of DNA building blocks in the corresponding gene. An example is the vital substance hemoglobin (the red pigment of the blood), which takes care of oxygen transport in the blood. Its protein part consists of four interconnected amino acid chains, comprising a total of 574 amino acids. There is no difference between human and chimpanzee in the precise sequence of the amino acids, and humans and gorillas differ only at one point in each chain. Taking all proteins of so far known structure together, humans and the African great apes differ in less than 1% of the amino acid positions, and hence in less than 1% of genetic substance.

Reproduction, Development, and Life Span
Discoveries in reproductive biology. In common with the Old World monkeys and apes, humans have a female menstruation period of several days. In many Old World monkeys, especially baboons and mangabeys, and also in chimpanzees, swellings occur in the genital region at the time of ovulation (and consequently likelihood of conception) in the mid-part of the cycle, and female estrus coincides with that phase. Neither occurs in human beings, but ovulation is accompanied by a slight rise in morning temperature, and according to statistical surveys, sexual intercourse and orgasm may be more frequent than at other times. Genital swelling is absent also in the gibbon, which is monogamous as a rule, and usually also in the orangutan; in the gorilla it is vestigial.

In many mammals, sexual activity is limited to some particular season. The primates, however, show little seasonal change, not manifest in the human-like apes except for seasonal variations in the birth rate. Such differences in statistical frequency can also be detected in humans; the birth rate peaks at the beginning of the year, and conception is consequently in the Spring (May), although custom (festivals) may occasion additional local peaks or shifts or various kinds. That lower primates generally give birth at night is reflected in the anthropoid apes and in humans by a somewhat higher frequency of births by night than by day.

The depth of implantation of the ovum on the placenta and hence the supply of nutrients to the embryo increases within the primate series, and humans are close to their near relatives in this respect. Single births as a general rule, and the infrequent occurrence of identical or fraternal twins, is also a trait that humans and apes as well as most lower simians have in common. As to the size of the birth canal relative to the head size of the newborn, the human condition is less favorable than that of the apes, because of the large cranium.

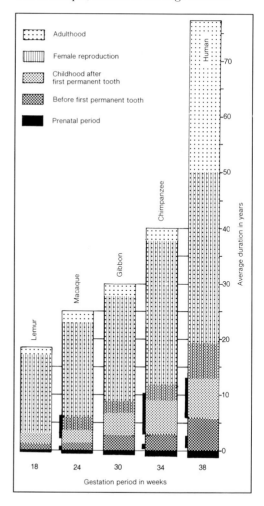

Duration of parts of the life cycle in humans, great apes, gibbons, simians and prosimians. For the macaques, chimpanzees, and humans, vertical bars at left edge of column indicate periods of emergence of deciduous and permanent dentition.

Rates of development and life spans. The prenatal stage of development (gestation period) in mammals is loosely correlated with body size, the primates having comparatively long gestation periods for their size. Among the primates in turn, the larger species have longer gestation periods; at 38 weeks humans fit unremarkably among the other primates and are very close to the large anthropoids (orangutan 39, chimpanzee 34, and gorilla 38 weeks). Nevertheless, the human newborn is well behind the lower primates in maturity; the baby gorilla is also less mature than other primates at birth. The lower primates, such as the ungulates and large carnivores, mature early, whereas humans and gorillas – like many smaller rodents and small predators – are still helpless at birth. Humans, in other words, are born too early, which is why the Swiss zoologist Adolf Portmann refers to them as "physiologically premature" – but this trait is shared with the gorilla, a fact not known at the time of Portmann's observation.

Human postnatal development is likewise slower than that of other primates, even much slower than that of the gorilla, which is not much different in this from the chimpanzee and orangutan. Thus, the age of female sexual maturity is 9 years and axial growth ceases at 11 years, as against 13 and 16–18 years, respectively, in humans. The difference in duration of female reproductive activity is not great (chimpanzee about 30 years, human 35 or more). However, the life span after the female menopause is much prolonged; lower primates survive that phase by only a few years, whereas in humans the time remaining before death is a quarter century or so.

The prolongation of childhood and adolescence in humans is interpreted as adding to the learning period, and hence as closely connected with mental humanization. But in the case of a "late bloomer," and moreover a creature dependent on learning, a prolonged childhood calls for a corresponding extension of the life span beyond the last childbirth, so that the last child can be reared. It is true that civilization and medicine may also have helped to prolong the human life span, but rather by improving the life expectancy of the newborn (especially by lowering infant and child mortality) than by extending the limit of old age; there were aged individuals in ancient and prehistoric times, but a far smaller proportion of individuals attained that state.

The retardation of development throughout the period of growth and maturation of the human individual has occasioned some scientists to see a principle of humanization in this phenomenon (the "retardation" hypothesis). Thus, the Netherlands anatomist Louis Bolk, more than half a cen-

The vertical profile of the feminine face contrasts with the larger, rougher, and more protruding male profile. (Drawing below) The Konrad Lorenz "baby" configuration, which spontaneously triggers a nurturing urge. That configuration is represented by the rounded heads on the left in each instance, accompanied by contrasting types.

tury ago, spoke of human "fetalization," meaning the retention of formerly prenatal conditions into adulthood. As a matter of fact, modern humans in many respects resemble the infantile pattern (large cranium, flat face) of prehumans, and their adult state in turn resembles the juvenile stage of our animal ancestors. However, this is not a pervasive principle that could account for all the changes that have taken place in the course of humanization; for example, not the adoption of erect posture.

The Variety of Modern Humans

In breadth of intraspecific diversity, humans have been compared with the domestic animals. The circumstance that they have made themselves a domestic animal by creating an artificial environment (civilization) is doubtless not the only cause of human variety. For the animal primates also exhibit high variability compared to other mammalian groups. Thus, it would appear to be a general trait of the primates.

Differences within populations
Time of life. A fundamental factor of variety in any population, or horde, is variation in age. The difference between the child and adult images is great especially in "late blooming" species, for the childlike appearance that Konrad Lorenz described in 1943 as the "baby" configuration must prompt adults to provide care. Besides, in species that live in groups, the childlike traits will protect the juvenile from adult aggression. Accordingly, the human infant does match the baby configuration: head large in relation to the body, with a small, rounded, flat face surmounted by the vault of the large cranium. This picture is completed by large eyes, snub nose, pursed lips (narrow mouth opening with upper lip everted in the middle and lower lip retracted), fat cheeks, and short, softly rounded limbs. These typical infantile proportions proclaim the child as such even at a distance, without benefit of scale.

At primary school age, the "baby" configuration becomes less pronounced, and the adult proportions of the head and the adult features appear at puberty. With long, gangling frame and limbs, the adolescent in fact exhibits the opposite image from the infant. But in adulthood, fullness of figure increases again, though to an extent subject to extraordinary individual differences. The fact that each individual's approximate age is apparent in the adult stage is not so much due to corpulence as to loss of tissue tone and increase in number and prominence of expressive wrinkles.

Sex. A second factor of intraspecific differences in all mammals is sex. As in the other mammals, the basic human design is feminine, but in embryos having a Y-chromosome the gonads develop into testes instead of ovaries, and the male hormone (testosterone) produced by the testes modifies the individual in the masculine direction. Embryos with XY chromosomes in which testosterone fails to be produced because of some genetic disorder or fails to act for any reason always yield an outwardly female individual, even having a vagina; but instead of the ovaries, testicular tissue is present in the abdominal cavity.

Thus, the sexes are not absolutely distinct entities; the circumstance that the male represents a modification of the female determines a certain equivalence between sex organs. The ovaries

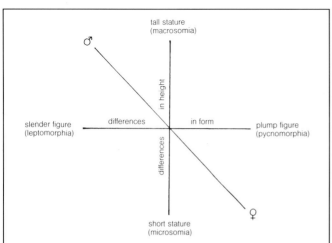

Coordinate axes of physical types, from slender (left) to plump and from tall (top) to short. The typical male is tall and on the slender side, the typical female shorter and plumper.

correspond to the testes, the labia majora give rise to the scrotum, and the penis is a much enlarged clitoris. The phylogenetic community of female and male will also produce hermaphrodites in rare instances. Especially in the secondary sex characteristics – all female-male differences beyond the actual sex organs – there is not simply a feminine and a masculine type, but an entire spectrum of variation between an extreme feminine and an extreme masculine pole, the particular individual approaching one or the other pole, more or less. In all secondary sex characteristics, there are only mean or frequency differences between the sexes, and their latitudes of variation overlap.

Variation in body structure according to sex differs in degree from species to species. The gorilla and orangutan show marked differences in secondary sex characteristics, but humans and the chimpanzees and gibbons less so. There nevertheless remain distinct physical differences between female and male. Many of these differences may be related to two principles:

The first principle is that of sex-related function. The female body must be adequate to the maintenance of offspring. Thus, the female mammary glands form a suckling apparatus of which the male has no trace except for non-functional nipples. Besides, the wide pelvis and higher abdomen in the woman serve to accommodate the fetus, and the wider pelvic opening facilitates childbearing. In consequence, the sacrum is more inclined, leading to a more pronounced curvature of the spine in the lumbar region. Besides, the wider hips, in relation to stature, increase the inclination of the thighs, because the knees must be brought together beneath the body. The male, on the other hand, whose functions were those of a hunter and protector of the horde in primitive society, has broader shoulders and a larger rib cage with correspondingly large organs.

The heavier skeleton and larger stature of the male might also be interpreted as an adaptation to such original functions. Conversely, however, the more delicate skeleton and smaller stature of the woman may be understood in the light of the second principle of sex differences. This manifests itself in some degree of incorporation of the above mentioned "baby" configuration in the female image. The woman retains childlike features to a greater extent; she has a smaller and rounder face in relation to the cerebrum, with wider eye slits, and smaller nose, as well as a more vertical forehead, shorter legs relative to the trunk, and soft, rounded form due to a thicker subcutaneous layer of fat. The "baby" configuration may have been echoed in the female for protection from the greater strength of the male.

Some other secondary sex differences are not to

Variation of humans (II). (Left) The Australian aborigines represent the last remnant of an ancient basic stratum of *Homo sapiens*. (Right) Mongols of Mongolia - Tungids, the purest representatives of the Mongolids.

be related to the two principles mentioned. In the first place, there is the slenderer neck of the female, as well as the distribution of fatty tissue, tending to emphasize the buttocks, hips, thighs, and breasts, and in males the belly. Perhaps the female configuration may be accounted for by mating preference. Still, there are sex differences that can hardly be ascribed to partner preference, especially in the case of inconspicuous traits such as the degree of attachment of the ear lobes, which is greater in women on the average. Here there is not much to suggest in the way of explanation; at most a surmised link with other more important sex traits.

The difference in quantitative proportions of various types of tissue in the male and female body leads to differences in body functions, such as exertion. In power output of the motor apparatus, woman represents the weaker sex, both in short-term maximum power and in endurance. Whereas instantaneous exertion depends on mass of muscle, duration is a question of circulation and breathing. Relative to body volume, the female has a smaller volume of blood, and her blood has a lower content of hemoglobin. Lung capacity and air utilization are also less. So the female lung requires about 26 qt (24.5 l) of air to extract 1.06 qt (1 liter) of oxygen, and the male lung only about 22 qt (20.6 l).

The female is, however, by no means inferior in all physical functions, nor in motor apparatus in general. She has no less dexterity than the male, and if anything has more motor precision of the hands. She surpasses the male in mobility of the joints. Besides, her thicker subcutaneous layer of fat provides a more favorable balance of heat; fat is a good insulator, the female requires less energy to maintain body temperature, and her skin feels cooler than the male's because it gives off less heat.

Above all, the female has the greater vital force, manifesting itself in a longer life span. More male than female fetuses perish in the womb; infant mortality of boys is higher than that of girls. Also, the life expectancy of the female newborn is up to ten years more – depending on the country – than that of the male (West Germany 75 and 68; East Germany 74 and 69; Austria 75 and 67; Switzerland 76 and 70; United States 76 and 68; Japan 76 and 71; and the Soviet Union 74 and 64 years, respectively). The difference in average life span may be due in part to greater stress exposure on the part of the male; but even in countries where females work as slaves of the men, they live longer, if only marginally. There are a very few countries, with especially adverse hygienic conditions, where the female life expectancy – doubtless chiefly because of childbed mortality – is a few months less than that of the male.

Physical types. Even at the same time of life and

Variety of humans (III). (Left) Paleo-mongolid woman and child of Thailand. The woman has a typical mongoloid eye formation and otherwise a pedomorphic face. (Right) Indians of Paraguay with faintly Mongolid but many Europid characteristics.

within the same sex, all populations show great variety of physical build. In past decades, attempts were made to catalogue this variety in physical typologies. In modern physical anthropology, where data are processed mathematically in their interrelationships, physical types are understood to be the poles of series of variation as they emerge from growth tendencies in various aspects (basically, sexuality is one such series). These series of variation yield type systems in which the individual can be assigned a position.

The simplest type system (see illustration, page 530) involves a variation series from slenderness (leptomorphism) to plumpness (pycnomorphism) and another from tallness (macrosomia) to shortness (microsomia). The former series reflects differences in form, namely in proportions, and the latter differences in size, namely in dimensions.

The leptomorph retains the physique of puberty throughout life. The pycnomorph retains more infantile proportions through puberty and beyond: softly rounded form, comparatively short legs, round face with vertical profile. In adult middle age, considerable flesh is taken off. The corpulent physique is a variant such as is encountered among domestic animals – a typical phenomenon of domestication. In the wild, the pycnomorphic individual would succumb to natural selection because of "inactive" mass (fatty tissue) and consequently more heavily burdened circulation compared to the leptomorphic competitor.

Social groups. Members of a given social group often also present more physical resemblance to each other than to members of other social groups. Thus there are occupational types, such as, for example, the nimble tailor or shoemaker and the athletic longshoreman – always with numerous exceptions. Such occupational types are not alike in all populations, and they may also change with time.

The emergence of occupational types – as well as sports types (long-distance runner, basketball player, weight lifter) – may be due to formative effects of environment or to so-called screening. Environmental influences associated with the activity in question will leave their mark, for example, weather in agriculture, trade-related eating habits of confectioners, or effects of training upon ballerinas. Screening, on the other hand, means that individuals of a certain given physique may choose a certain occupation because they feel that this physique equips them for that job. Thus, the dock worker requires strength, which demands an athletic build, which the tailor does not require; sensitive fingers will serve him better.

Formative environmental influences and screening are also causes of average differences between social strata. The most obvious physical feature to

Variety of humans (IV). (Left) Brazilid Indian of the Amazon, where in some tribes the male sex also exhibits some comparatively feminine features. (Right) Paleonegrid woman of Namibia, with typical Negrid features (for example, thick lips, broad nose).

be mentioned in this connection is stature. In Central Europe, the spread between upper stratum (academics, professionals) and lower stratum (unskilled labor) is about 2 in. (5 cm) for males and 1.6 in. (4 cm) for females. Clearly, screening is an important factor in bringing about this difference, on the basis of ancient principles of vertebrate behavior: bigger is synonymous with higher ranking, and the downward glance is, therefore, associated with a sense of superiority, making for self-assurance; the upward glance connotes a sense of inferiority. The expressive and dramatic value of stature renders it easier for tall people to rise in society; stature lends them assurance and makes an impression of authority. Statistical data indicate that such residues of instinct are still operative in humans. The upwardly mobile are taller than those who have remained in the stratum below, and in professional groups with levels of rank, the tall tend to be promoted, so that – matching their salary level – on the average, police captains are taller than lieutenants, these in turn are taller than sergeants, and sergeants are taller than patrolmen. The same holds true in hospitals, where head nurses are taller than subordinate nurses.

Geographical Classification (Races)

Multiplicity of traits and its causes. Humans have settled almost the entire Earth. That they have been able to do this is due in part to civilization, that is, an artificial environment that they can create (clothing, shelter, implements for procuring food, fire). But humans have not only adapted an environment to themselves, they have also adapted their bodies to the environment. Over the millennia, selection has brought forth suitable variants in different habitats and climatic regions. Accidental factors have also shaped development differently in different places. Interbreeding has enriched multiplicity of form. Thus, populations today differ in the frequency of particular genes; in some instances, new genes may even occur. The variety of form of populations by virtue of such differences is referred to as racial variability.

The geographical distribution of traits is not always easy to explain in detail. This may be because the environmental influences, as factors of selection, are complex, and in the case of some traits (such as outward physical characteristics, but not blood characteristics), the environment can act on the human individual directly.

As to geographically conditioned variations in stature, there is some correlation with mean annual temperature. Like other warm-blooded animals, human inhabitants of colder zones are generally taller than those of warmer ones (Bergmann's rule). In Europe, the average stature decreases from North to South. The reason for this is that with increasing bulk, the ratio of the surface emitting heat to the volume generating heat shifts in

Variety of humans (V). (Left) Women dancing in North Cameroon, representatives of the Nilotid race, with somewhat masculine proportions. This slender, long-legged physique is an adaptation to a hot savanna climate. (Right) Pygmies of Central Africa, a dwarf Bambutid race whose adult members retain childlike proportions.

favor of the latter. The relationship of mean annual temperature to body weight is even closer. This relationship emerges more clearly if the relative humidity is factored in; the greater humidity in the tropics, for example, leads to a far heavier heat load than in the subtropics despite lower temperature, and hence to a greater advantage for the smaller body with its relatively large surface area.

Some of the many exceptions to Bergmann's rule are accounted for by another climatic rule: The lower the temperature, the more compact and short-legged the figure (Allen's rule). This also tends to reduce the surface area emitting heat (a spherical shape would be the optimum). This mode of adaptation to cold, coupled with a thick subcutaneous layer of fat, has been adopted especially by populations in unusually cold areas, like the Lapps in northern Scandinavia, the inhabitants of the Siberian tundra, and the Eskimos. The short-leggedness of Mongolids is probably attributable to their origin in the cold region of Central Asia. On the other hand, distinctly long-legged peoples live in the hot savanna areas in the Sahel zone in Africa and in Australia.

The high geographical variability of breadth of nose is closely linked to atmospheric humidity (Thomson-Buxton rule). Specifically, within each racial group, it is found that cool, dry air leads to a narrower nose, for it has the advantage of better moistening and warming of the air inhaled in contact with the nasal mucous membranes.

Within a racial group, there is also a definite relationship between skin color and intensity of ultraviolet radiation (Gloger's rule). The adaptation to ultraviolet dosage by means of darker skin provides a screen in regions of strong sunlight, whereas in regions of weaker light the rays absorbed by a paler skin help ensure formation of an adequate quantity of vitamin D. In terms of biological kinship, however, differences in color are of minor significance. Great diversity of skin and fur color is to be found within the same species among apes also (for example, chimpanzees).

A second area in which geographical distribution of traits may be recognized as the result of processes of adaptation controlled by selection is the manner in which characteristics of the blood are expressed. Such variants are of course associated with differences in resistance to diseases. This accounts for the pattern of distribution of the AB0 blood group system in terms of the interaction of several epidemic infectious diseases.

Members of blood group A are susceptible to smallpox; therefore, that blood group is rare in regions where the disease was widespread for centuries (India, Pakistan, much of Africa). The more benign course of syphilis in members of blood group 0 led to a high frequency of this group among the American Indians because that venereal disease appeared in North America in pre-Columbian days. On the other hand, individuals with blood group 0 seem to be more susceptible to plague; in any event, this group is infrequent in India, in Turkey and in Lower Egypt, which were ancient centers of the plague. Since blood groups A and 0 are rare in India, consistently with the above account, blood group B necessarily predominates.

Racial groups. Geographically conditioned variations are more or less continuous. Racial distinctions are therefore largely artificial. They represent an attempt to summarize radial diversity by classification. Basically, only a few points emerge clearly, more specifically three:

1. The white, Europid, or Caucasian racial group, which also occupies North Africa and extends into the Middle East, with outposts perhaps all the way to Polynesia (mingled with Australo-Melanesid and Mongolid elements). They are typified by a medium-long or short skull, strong facial features, narrow nose, thin lips, hemispherical breasts, medium-long legs, smooth to wavy head hair, abundant beard and body hair, light skin, tendency to lighter hair and eye color, and medium development of sweat glands. Their most typical embodiment is found in northern Europe.

2. The yellow or Mongolid racial group, inhabit-

ing Asia with the exception of the Near and Middle East. They are characterized predominantly by a short skull, flat face, narrow eye openings with the epicanthine fold, medium width of nose, medium-thick lips, shell-shaped breasts, short legs, straight head hair, sparse beard and body hair, yellowish to brownish skin, dark hair and eye color, and moderate sweat gland development. Their center of concentration is Central Asia.

3. The black or Negrid racial group, indigenous to Africa south of the Sahara. The skull is medium-long to long, the nose broad, lips thick, breasts tapered, comparatively pronounced lumbar depression, long legs, curly head hair, moderate beard and body hair, dark color of skin, hair and eyes, and pronounced development of sweat glands. The most typical representatives are found in the western Sudan.

Between any two of these three concentrations of racial diversity, there are all degrees of mixing. The following are blends of Europid and Mongolid elements, best listed as additional racial groups because of long independent development and consequent typical constellations of features.

4. The Indianids of America, with medium-long or short skull, sometimes strong, sometimes softer facial features, medium breadth of nose, medium thickness of lips, hemispherical or tapered breast, medium-long legs, wavy to straight head hair, sparse beard and body hair, reddish brown skin, dark hair and eye color, and moderate development of sweat glands. As the latest wave of immigration, the Eskimos present the most pronounced Mongolid cast with a definite tendency towards narrow eye openings and epicanthine fold.

Another group is difficult to interpret as a blend of racial groups mentioned above. There are many resemblances to the Negrid group, but there is much to suggest that this is a remnant of an original *Homo sapiens sapiens* strain. Thus the following is to be listed as a fifth racial group.

5. The Australo-Melanesids of Australia, New Guinea, and Melanesia. Their characteristics are a long skull, rugged facial features, inclined forehead, broad nose, thick lips, tapered or bud-shaped breast, long legs, wavy to curly hair, dark brown skin, and dark hair and eye color. Possibly the Ainu of northern Japan, the Negritos of Indochina and the Philippines, and the Weddids of the Indian highlands and Sri Lanka (Ceylon) may represent splinters of this ancient stock – the Ainuids with admixture of Mongolid, the Weddids of Europid elements.

Lastly, in southern Africa there are the Hottentots and Bushmen, who appear to represent a mixture of Negroid and Mongolid traits. But since their resemblances to the Mongolids may not be due to kinship with them but rather to parallel development, and there are specific traits as well, this special division will be regarded as a racial group also:

6. The Khoisanids, with medium-long to long

Variety of humans (VI). (Left) Young Hottentot woman of Namibia with typical features of the Khoisanid race. (Right) Hottentot, also of Namibia, with bronze skin, frizzy hair and flat face, typical of the Khoisanid group.

skull, flat face, narrow eye openings and Hottentot fold, broad nose, thick lips, tapered breasts with tendency to axillary position, deep lumbar depression (in females, sometimes massive fat deposition on buttocks), medium-long legs, curly head hair in separate patches, sparse beard and body hair, yellow-brown to bronze skin, dark hair and eye color.

The division of racial groups into individual races and the demarcation of the groups from each other are largely subjective. Therefore, no detailed classification will be presented. Instead, the manifold racial typology will be examined from some general biological aspects.

The question arises as to what extent the several races have passed through different intensities of phylogenetic evolution, in the sense that some races have attained a greater adaptation, usually associated with greater specialization ("progressiveness"), while others have remained more primitive. Possibly the Australo-Melanesids represent a racial group in which comparatively many traits of the first representatives of *Homo sapiens sapiens* have been retained (a remark not to be misinterpreted in terms of racial ideologies). The other racial groups are hardly distinguishable in degree of progressiveness. Specifically human expressions of characteristics, remote from the animal primate stage, are not concentrated in any particular racial group, but occur as individual traits, now in this racial group, now in that. Thus, the narrow, prominent nose of the Europids is a progressive trait, but so are the flat face of the Mongolids and the thick lips of the Negrids.

Within the great racial groups, there are some individual races that exhibit a more childlike image, and others in which individual development leads farther away from juvenile proportions. Thus the Paleomongolids of Indonesia and the Bambutids of Central Africa, with their rounded figures and short legs, exhibit comparatively childlike traits, while the long-legged Nilotids of the Upper Nile and the hook-nosed Dinarids of the eastern Alps and the Balkans represent the opposite type.

Sexual typology is also differently pronounced in the several races. Some races as a whole vary more towards the feminine (for example, Paleomongolid, Brazilid, Alpine), others more towards the masculine (for example, Dinarid, Nordid, Silvid). In many respects the difference between Nordid and Mediterranid echoes the difference between masculine and feminine; some Mediterranid male skulls are difficult to distinguish from female Nordid ones.

Variety of humans (VII). (Left) Albino Negrid. Thick lips, broad nose and curly hair identify the Negrid type. Skin color alone is not a reliable racial criterion. (Upper right) Bushmen of Namibia, adapted to civilization ("aculturated"): right, a Sanid man with delicate features; left, a man with coarse features suggesting Negrid admixture. (Lower right) Mixed family of Namibia: man ³/₄ Negrid, ¹/₄ Europid (German); woman ¹/₂ Negrid, ¹/₂ Europid (English; child manifests Europid inheritance more strongly than his parents.

Although leptomorphism and pycnomorphism occur in all races, some tend more to one or the other pole of the basic typological variations in physique. Markedly leptomorphic races include the Dinarids and the Nilotids; the Alpine and the East Europid races are stockier and thus more pycnomorphic. Races also differ markedly in stature. There are no real giant races, but there are dwarf races (mean height of males under 5 ft or 1.5 m). Such pygmies include the Bambutids of Central Africa, which have childlike proportions, and in normal proportions, the Negritids, which live as widely scattered stocks in Indochina and the Philippines, as well as isolated Melanesid strains on New Guinea (Tapirids). The Bushmen of South Africa (Sanids) are only slightly taller than the true pygmies.

Biological Roots of Human Behavior
by Irenäus Eibl-Eibesfeldt

The Standpoint of Human Ethology

The science that deals with the biology of human behavior is human ethology. Its pioneers are two Nobel laureates, Konrad Lorenz and Nikolaas Tinbergen, who laid the foundations, beginning in the 1930s by clarifying the key terms "innate" and "instinct."

Before that, and to some extent into the 1960s, it had been more or less settled that nothing was innate to humans except some elementary reflexes. It was only by processes of learning that they would acquire all the knowledge and skills necessary for survival. Humans, it was supposed, could be educated in any direction with equal ease.

The ethological picture, that is, the picture gained by comparative research in ethology, does not by any means oppose that view in principle. We have consistently emphasized that we regard a human being as "by nature" a creature of culture. To be sure, the biologist brings phylogeny, or evolution, into the discussion as an additional dimension, posing the question of the significance of biological inheritance in human behavior, and investigating the role that may still be played today by the innate in human behavior. We want to know to what extent human behavior may also have been pre-programmed by phylogenetic adaptations. Sometimes this is interpreted to the effect that biologists are primarily interested in the "animal" heritage in human behavior. It should be stated, therefore, that our inborn behavior patterns are by no means always an inheritance from the beasts. Many are extremely ancient, but much is specifically human, that is, peculiar to our species.

By introducing the phylogenetic dimension, biology has an important contribution to make to human self-understanding and, therefore, also self-control. Biological inheritance is potentiality, not destiny. Of course, inborn inclinations, perceptual compulsions, and intellectual compulsions may constitute dangerous restrictions of freedom. Ignorance is bondage, knowledge is liberty. In this sense, biological research has much to contribute to the cause of emancipation.

Far from standing at a fixed terminus of development, we know that we are at best an intermediate stage on the way to a higher humanity. The real "missing link" on the way to humanity is ourselves, Konrad Lorenz once said. We are thus offered prospects for the future, but denied the security of a rigid view of the world. We must no longer think of ourselves as the pinnacle of creation, but as in a state of becoming, with every possibility of disaster as well as triumph.

Survival and Adaptation

Any structures or modes of behavior that further the fitness of a creature, fitness being measured by survival in its own or closely related posterity, are called adaptations. To make no adaptations is to perish.

Since adaptations reflect environmental conditions, the adaptive system must at some time have faced the environment in question and received "information." We know today that animals have two reservoirs of information, inheritance (genome) and individual memory. To these are added "artificial organs" – books, electronic data banks etc. – in the case of humans.

Adaptations brought about through the genetic mechanism of modification of species are called phylogenetic adaptations. The information gathered in the course of phylogeny is then stored in the genome.

The adaptation of the individual by leraning also plays an important role, in the higher vertebrates especially, although even learning is guided into certain paths by phylogenetic adaptations, so that animals mostly learn what will contribute to their fitness. In this case, information is stored in the central nervous system of the individual. One learns in the first instance from one's own experience. In higher mammals and in birds, social learning by example is important also. Social learning allows knowledge to be transmitted from one animal to another, and that is tradition. With the development of verbal language, humans have even been able to transmit tradition in abstraction from live demonstration. Whereas the Japanese red-faced macaques must show the inexperienced how to wash sweet potatoes, human beings can transmit culture independently of the object, for example by explaining that potatoes should be washed before eating and how it is done. This makes possible the transmission of much knowledge and the production of much culture. With the development of writing and other techniques for storing information, this process has been further accelerated.

In humans, cultural development has thus taken on enormous importance as a new process of adaptation. Adaptations can be made more quickly, and the extermination of those with maladapted traits is not required. It suffices to abandon the

People, people everywhere! A mark of *Homo sapiens* distinguishing them from all higher animals is their great ecological adaptability - their ability to maintain themselves successfully under all sorts of environmental conditions. Beginning in Africa, humans have spread little by little over all continents, and settled in the most inhospitable regions. People have pushed into red hot deserts, like these camel riders in the Sudan (upper left), and inhabit the ice-cold arctic, like the igloo-building Eskimos (lower left). They build their villages and monasteries on bleak Alpine peaks in Tibet and Nepal (upper right), and though essentially terrestrial, they have conquered the open ocean as a source of nourishment (lower right).

maladapted behavioral recipe – say a mistaken opinion, a device that doesn't work, or an outdated practice. This can well be done on the basis of insight. Still, many cultural rules and practices are not developed, retained or abandoned on the basis of insight. Over long intervals, cultural evolution too is shaped by natural selection, and people cling to the familiar, often without knowing why.

The Term "Innate"

Modes of behavior are called "innate" or "inborn" if the neural networks underlying them are brought to functional maturity in a process of self-development well insulated from disturbing outside influence, in accordance with programmed instructions laid down in the genome. The term "phylogenetically adapted" is sometimes used as a synonym for "innate," alluding to the source of the adaptation.

It has variously been stated that the distinction between inborn and acquired is impossible in practice, so that neither contribution can be exactly determined. Experiments involving privation of experience are not accepted as proving anything, since after all a creature can never be deprived of all experience. Even in the egg or in the womb, it is in an environment that acts upon it. To this it must be answered that modes of behavior as adaptation presuppose that the adapted system should at some time have confronted those sectors of its environment to which it proves to be adapted. It must have acquired information essential to the adaptation. That such acquisition of information may occur in the course of phylogeny or of individual development has been stated. Now if that information is withheld from the maturing creature, and it nevertheless exhibits the adapted behavior in question, innateness or phylogenetic adaptation has been proved.

So if we wish to know whether a bird must learn its song or whether it is inborn, it need only be raised in a soundproof environment, beginning with the egg. If it nevertheless sings its specific notes, the experiment proves that adaptation is phylogenetic.

Phylogenetic Adaptations in Human Behavior

Expressive movements

Human beings have a large number of innate behavior programs. In the newborn baby, 21 distinct reflexes can be identified. In addition, there are a number of other capabilities. Thus a newborn baby has a richly graduated repertory of vocal utterances with concealed signal value. Besides various forms of crying, there are sounds of mood, contentment, interaction, discomfort, and sleep.

The newborn baby will direct its eyes towards the mother's face when she speaks, as though it meant to look at her. To be sure, this is essentially an "unconscious" operation of fixation, largely independent of visual feedback. Even the congenitally blind will fix upon the source of sound. Mothers experience this as a sign of attachment. The infant's response promotes the mother's emotional link to her child.

There is a vast repertory of facial movements, which decisively influence our social relations by serving as signals. Children born deaf and blind will laugh, smile, and weep with the same facial expressions and vocalizations as sighted infants that can hear, even though there is no example for them to follow. Furthermore, cultural comparison shows that facial expressions exhibit an astonishing uniformity across cultural boundaries. Many facial gestures are quite ancient, such as the "playful" face, which has its counterparts in the Old World simians.

Gestures and bodily attitudes also include universals, expressive movements that are extremely widespread. At the crawling age, children will point with the outstretched index finger, and they do it in all cultures that I know of. Human beings have even evolved a special muscle for doing this.

When frightened, human beings will cover the head protectively with one or both hands. Child-

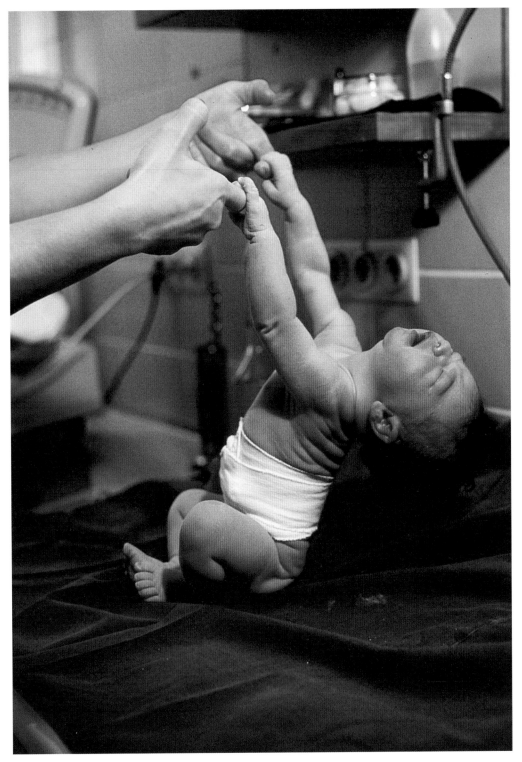

Human baby with "simian" reflex. The handclasp of the newborn is not learned, but an inheritance from our animal ancestors; for baby monkeys, the ability to cling to mother's fur or to the branches is important to survival.

ren will do this as they retreat, crying, from a fracas. By this reaction human beings protect the fragile head from injury. Another universal is the head-shoulder reaction to fright. The shoulders are pulled up and the chin is drawn in. This protects the vulnerable sides of the neck. Let these examples suffice.

Nevertheless, humans avail themselves of their innate behavior patterns voluntarily. They can "feign" expressions, as every actor shows us. They can also suppress and so conceal spontaneous expression, and there are cultural differences in readiness to show emotions. Besides, there are also culturally evolved behavior patterns in facial expression and especially in gesture.

The movements attendant upon negation are interesting, because they illustrate a form of cooperation between cultural convention and innate programming. In most cultures, one shakes the head in negation. We find this behavior pattern among many Papuan peoples, the Polynesians, the Bushmen of the Kalahari, the Yanomami of South America, and many others. This behavior may be interpreted as a repetitive movement of lateral avoidance by way of rejection, perhaps also as a shaking off. Then there are other forms of negation. Greeks and other Mediterraneans will raise the head high, thrown back, closing the eyes and then sometimes turning the head to one side. At times this "No" sign is limited to a quick backward jerk of the head while closing the eyelids. This form of negation is culture-bound. But the movement from which it derives is universal. Haughty rejection is everywhere expressed by lifting the head, closing the eyes, and exhaling, in symbolic rejection of sensory stimuli emanating from the scorned source.

The Ayoreo of Paraguay express denial in still another way. They wrinkle the nose, close the eyes, and purse the lips as if pouting. This "No," certainly culture-bound, is also based on a universal gesture of rejection; people everywhere rebuff unwanted sensory impressions by wrinkling the nose and closing the eyes.

In all three cases, then, a preexisting gesture of rejection becomes an accompaniment to the spoken negative.

Perception

Studies in the psychology of perception tell us that certain perceptions take place quite unavoidably, often contrary to better information. If on a windy night of full moon we look up at the sky, we seem to see the moon flying in front of the clouds. Our knowledge that in fact it is the clouds that are passing by does not save us from the illusion. It is evidently built into our perception as phylogenetic experience that objects normally move in a stationary matrix, in other words that generally the surroundings are at rest and objects move. To apprehend this immediately has survival value. In our perception, therefore, there are built-in assumptions, enabling us for example to estimate distances or the size of objects and to recognize the same object at different distances. The sensory impressions are processed according to a preassigned program, and the occasional deceptions tell us what the programming principles are. Gestalt psychology has explored a number of principles of this kind.

Further, our perception is often so intertwined with motor operations that certain stimuli will

Physical contact, eye contact, and "expressive sounds" – from the point of view of the ethologist or student of comparative behavior, these are the most important expressive movements, underlying the intimate bond between mother and child. This relationship is most highly and richly developed among human beings, but the stage immediately preceding it is the manifold mother-child relationship among apes, especially the human-like apes. The human baby has a repertory of no less than 21 distinct innate reflexes.

trigger specific modes of behavior. Infants fourteen days old will react to a dark shadow expanding uniformly on a screen as though an object were approaching threateningly. They blink, lift the hand protectively, and turn away. In other words, from certain visual expressions, they expect certain tactile impressions, even when it can be stablished that they have never experienced any such disagreeable consequence.

When newborn infants are presented with certain facial expressions (sticking out the tongue, opening the mouth wide, pouting protrusion of the lips, smiling extension of the lips), they will respond by converting what they see into movements of their own, for example sticking out the tongue in imitation when they see the example. In other words, prior to experience of their own, they know how to respond to certain visual impressions with certain behavior patterns. Apparently, in the course of phylogeny, "detectors" have been tuned to certain stimuli and so "wired" to the motor system that when the stimuli occur, they elicit certain responses, for example parrying gestures when objects come at you. We call these innate trigger mechanisms.

They play a particularly important part in our social behavior. Thus, a friendly, smiling face is easy to imitate engagingly by an upward curve of the line of the mouth. If the same line is turned 180 degrees, a mournful face results. The very fact that we respond to such simple tricks of facial expression with emotion is a sign that not only the facial movements but also the interpretation is inborn. Consistent with this, people of other cultures interpret facial expressions as we do. Konrad Lorenz pointed out in 1943 that we respond to certain infant traits with strong emotions. Rainer Knussmann has dealt with this in detail in his article.

The ideal pattern of sexual partnership may also have been preassigned in outline by innate trigger mechanisms. It may also be noted that so far as we can see, there are some universal esthetic preferences or "prejudices" as to physical beauty. These include traits by which *Homo sapiens* contrasts with non-human relatives, such as a slender neck and delicate features. Among the Eipo in the western hill country of New Guinea, a narrow bridge of the nose is thought handsome, even though that ideal is seldom realized there. Such traits are emphasized in the art of both primitive and advanced cultures, apart from contemporary directions in art than often deliberately distort the human image. Innate trigger mechanisms in the form of perceptual preferences thus influence our designs in the realms of art and fashion.

It may be noted that humans tend to emphasize the shoulders fashionably by means of clothing and adornment. Upon examination of the direction of the hair on the back, it is found to go upward, so that when there is considerable body hair, erect tufts appear on the shoulders. This may have enlarged the silhouette of our still hairier ancestors, especially when agitated, when they were already walking erect. In an aggressive mood, the erector muscles of our hairs also contract – we raise hackles that we no longer possess, and experience the reaction as a shudder. Now it would seem that the perceptual preference has outlived the loss of the coat of hair, and we indulge the preference by artificially emphasizing the shoulders.

In interhuman communication, the eyes play a preeminant part. We perceive them with some degree of ambivalence or "ambiguity." Doubtless on the basis of innate trigger mechanisms, we experience two eyes as threatening, especially when fixed upon us for any length of time. On the other hand, eye contact is indispensable to friendly intercourse.

In the service of communication, the human eye has developed some peculiarities, such as the white of the eyeball, which contrasts with the iris and enables us to follow every movement of the eye. Experiments have shown that eye spots, in a pair and presented on the horizontal, will attract special attention, and also that we perceive very fine degrees of pupillary movement, which in turn signify assent or positively accented perception and interest by enlargement of the pupils.

The ambivalent effect of eye contact has a phylogenetic background. Eye contact among vertebrates is menacing. Predators glare at their prey, and we observe that mimic forms of vertebrate eyes have often been developed as deterrent signals (for example, eye mimicry of butterflies). The fixed gaze, then, would be primarily a threat. We human beings too are wont to depict eyes and eye motifs on amulets and phylacteries. In human behavior, we observe the occurrence of the aggressive stare. Quarrels are often acted out in the form of "staring contests." The impulse to expression of friendly communication by eye contact, as noted by Dian Fossey in gorillas, might have been the personal mother-to-child bond in higher primates.

Besides visual signals, acoustic and olfactory signals also play a role, and we may also have been prepared for perception of some of these signals by inborn trigger mechanisms. Thus, people speak to small children, when they want to encourage friendly interaction, in a tone about an octave above the normal speaking voice. "Baby talk" like this is found among such diverse people as the Eipo, Yanomami, Kalahari Bushmen, and Europeans. Small animals or the aged and infirm may be addressed similarly, and lovers often use baby talk between themselves.

The significance of olfactory signals has not been much investigated. However, we respond to scents that serve as pheromones in the mating habits of mammals. From puberty until menopause, women have a lower olfactory threshold for musky substances than men. Moreover, this olfactory threshold fluctuates with the cycle. At the time of ovulation, women have an especially keen sense of these odors.

The pheromone androstenol is produced in the testicle of the boar, and enters the fatty tissue and the saliva by way of the bloodstream. This substance tranquilizes the sow in estrus, so that she will allow the boar to mount her. The human male produces similar substances, secreted in axillary perspiration and elsewhere. In experiments, these substances positively affect response. That is, under the influence of the pheromone, people are perceived as warmer and more sympathetic. Chairs sprayed with androstenol repel men but attract women.

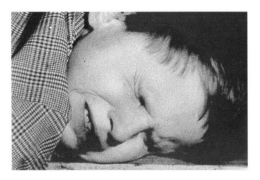
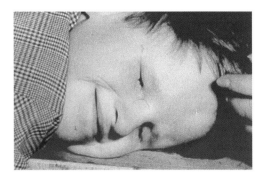

Facial expressions show astonishing agreement in all human beings; some of them can be derived from our phylogenetic past. This little girl, born blind and deaf, behaves no differently from a child that can see and hear; she responds with a smile to friendly contact (right) and cries when she gets a bump (left).

Models

Models or patterns are central nervous structures in which information (knowledge, experience) is stored for comparison with incoming messages. They can have been acquired in the course of phylogeny, or like trigger mechanisms, acquired by learning processes. Thus there are some birds that must first hear their songs from a social model, in order to learn from the remembered pattern. Interestingly enough, they come to this learning with some foreknowledge. If songs of different species of birds are played for them on tape, they will choose their specific song as a model. In other words, they know by phylogenetic experience what to imitate.

Such models likewise determine knowledge of correct behavior, the right example, including moral norms in human beings. Some of these norms may rest on a phylogenetically given program. Wolfgang Wickler reported some remarkable research on the subject in 1971.

Deviation from the norm induces distress, whether the norms in question are acquired or inborn. Endorphine (endogenous morphine) may be a factor in this.

Emotions

Emotions are experienced subjectively. Observers can at first perceive only the expressive movements, that is, the forms of expression of certain states of excitation. However, they know what is going on within themselves when they behave similarly in comparable situations, and we human beings, by means of language, can also discuss our feelings. There is evidence and testimony that these emotions are among the universal. They are presumably innate in us, for one can perhaps "indoctrinate" someone with the object of love or hate, but not with the emotion itself.

Urges

In observing animals, one finds that they exhibit fluctuations in readiness to act, to which there are no corresponding fluctuations in external environmental conditions. They are now disposed to hunt, now hungry, thirsty, sexually excited, or aggressive. These are often deviations from the norm of a psychological equilibrium, as in the case of hunger and thirst of vertebrates, causing the animal to seek food or water. Here, internal sensory stimuli report the deviation from the norm. Often, of course, other causes will set an animal in motion. For example, there are "tonic" stimuli that cause physiological changes in the course of time which in turn affect readiness to act.

Also remarkable in this connection are hormonal reactions to activity. When a man wins a game of tennis, within 24 hours the testosterone level will rise appreciably. When he loses, it declines. Studies of male medical students have shown similar reactions to passing and failing examinations. The testosterone level in the male is also raised following heterosexual intercourse, but not after masturbation.

Finally, there are promptings independent of external and internal sensory stimuli, based on processes in the neurons (nerve cells). Konrad Lorenz has repeatedly described how animals long kept from engaging in certain behavior will respond with increasing randomness to internal stimuli and finally even go through certain pattern of behavior idly. This phenomenon was mysterious until Erich von Holst's discoveries of the endogenous automatic basis of vertebrate locomotion. Von Holst showed first in the eel and then in other fishes that there are spontaneously active automatic cell groups in the central nervous system of these animals, coordinating their activity so that even without any influence from the outside, an orderly pattern of movement comes about. Thus, a dynamic conception displaced the preceding doctrine of reflexes that saw all behavior as a response to incident stimuli.

From all that has been said, we see that impulses may be actuated in very different ways. There is no unitary principle of causation and no monistic doctrine of incentive. "Urge" merely recognizes that behavior need not be a purely reactive response to triggering stimuli, but that there is readiness to act that is generated within the organism itself. Such readiness may exert reciprocal promoting or inhibiting effects. In the latter case we speak of conflicts. They may cause modes of behavior of different tendencies to be superimposed according to the intensity of one or the other, so that the individual vacillates between two endeavors, or so that neither of the mutually inhibiting urges manifests itself in appropriate action, but quite different modes of behavior are released in the form of acts of "preterition."

The "playful" face may be derived from playful biting and playful ("grooming") nibbling, as this girl of West New Guinea illustrates in playful interaction with a baby. The girl responds to the mouth-to-cheek contact with an (exaggerated) offer to bite, and then nibbles the infant tenderly, waits with a playful expression for its response, and then gently bites again.

Learning ability

Learning is so channeled by phylogenetic adaptation of all kinds – models, drives, etc. – that ordinarily what is learned promotes fitness. This means that there are species-specific learning abilities, for what will promote fitness varies from species to species. Thus phylogenetic adaptations regulate what is to be associated with what, and what is to be placed in what context. Nausea, for example, is associated with what one has ingested some time before, but not with the events perceived at the time of being nauseated. Physical pain, on the other hand, say an electric shock, is associated with the sensory stimuli occurring simultaneously (light, sound, etc.). Stimuli of punishment operate differently in different functional contexts. An injured creature will avoid the situation in which the injury was incurred. But sometimes behavior will be reinforced by stimuli of punishment. If a young rhesus monkey is punished with electric stimuli whenever it follows its mother or takes refuge with her, the reaction is intensified. The youngster only tries all the harder to join the mother. It is the same with human beings, as we shall have occasion to note.

It should be mentioned that there is also a special incentive to learning that we call curiosity. It prompts us to court novel situations so as to learn from them, and also to explore our environment in play. Besides, higher mammals and some birds have developed the ability to detach actions from their motives. A dog playing can combine modes of behavior that would preclude each other if serious, for example hunting behavior and combat behavior. By the ability to disconnect actions from the imperatives of a real-life situation, higher mammals can create a field of low tension, a sort of free space for experimentation with their own capabilities and for dialogue with the surroundings.

In that dialogue, people also make exploratory use of aggression, as a form of social inquiry. Children often challenge not only their peers but also adults by shoving, hitting, taking something away from them, or otherwise teasing and annoying in an aggressive manner. From the responses of those challenged, they learn what limits are imposed by the culture and how various persons should be dealt with. If no line is drawn, the inquiry becomes more pressing. This has sometimes not been understood; ist has been thought that a liberal education should set no limits. It was not realized that the child's behavior is often a question, and that a child becomes insecure when the rules of social intercourse are not communicated. This does not mean that all these attempts to chart the latitude of social action should be blocked immediately, or that the setting of limits must be achieved by punishment. Social inquiry is positive behavior, an effort for the sake of information. But one must realize that it demands an answer, and if the answer is not given soon enough, it may eventually have to be given in a restrictive way.

Exploratory aggression is not limited to childhood. There are phases of development in which it regularly recurs, for example as rebellion in puberty, and more or less whenever a new level of rank is attained, and one is unsure what liberties may be taken.

Exploratory aggression as inquiry also plays a part in symmetrical relationships. Partners challenge each other, partly in order to map the limits in their own relationship. At the same time, they thereby test the firmness of a bond, as if to ask what will be put up with, what is the threshold of tolerance, where do I stand. The bond between partners is generally confirmed by certain modes of behavior of mutual consideration and recognition, and this confirmation is constantly required. If it is omitted – for example because one partner is overworked – inquiry results, and may likewise be amplified into aggression if no response in terms of reassurance is forthcoming.

Strategies of Interhuman Behavior

A person's behavior may also be described as a sequence of steps leading to a certain goal. One and the same goal may be reached by different routes, that is, by way of different sequences of steps, with points of decision at which a choice between several alternatives is open. The various possibilities can be seen as a network of paths leading to a goal, and the preferred sequences of action can be called strategies. It is interesting that when we compare cultures we discover a number of fundamental interaction strategies. How one represents one's self in order to gain esteem, how one proceeds so as to obtain something from someone, how one reinforces a bond, establishes friendly contact, or parries aggression – all this is done according to a uniform basic pattern. There would seem to be a limited number of possibilities.

Since we human beings are dependent on social bonds, we value connections with our fellow humans as an important asset, and as we all know. Even as small children we defend such bonds against siblings, and later, with the emotion of jealousy, against any rivals. Hence when someone indicates readiness to sever the bond, the threat is a highly effective one. Now we can state that children in all cultures we know of are equipped with this weapon. The series of photographs on pages 550–551 shows children of the Yanomami culture, not at all close to us, but it is easy for us to understand, and similar pictures might just as well have been taken in a kindergarten in the United States or in Europe. The strategy these children are following everywhere in non-verbal behavior can nevertheless be verbalized – and that is a human specialty. An offended person can say something like, "I'm not speaking to you"; and such a phrase too is to be found in other cultures. The verbally expressed threat to break off relations has the same effect. The offenders will usually quit their offensive behavior – often they will exert themselves to bring about a reconciliation, either directly or through third parties. What happens in disputes between individuals may also be observed in disputes between groups, even between governments. On the diplomatic level, one sulks by threatening to break off relations, showing that our repertoire of available behavior patterns with which to shape our social contacts is somewhat limited. Effectiveness of the threat to break off of course depends on the actual existence of a bond between the parties. In the anonymous society of large cities, such bonds between fellows are often

A gesture of denial: Greeks and some other Mediterranean people will emphasize a verbal "No" by a dismissing gesture of the hands, palms towards the person addressed, throwing back the head, averting the glance, closing the eyes, and finally turning the head aside.

lacking, and this may be one reason – among many others – for the rise in crime. Aggression-parrying signals in the absence of any personal bond often misfire and fail to take effect.

On the ontogeny of the strategy discussed above, there are recent observations by L. Murray. He would instruct mothers to hold perfectly still for one minute, without moving out of the infant's field of vision, and then to resume paying attention to it. During the period of impassiveness, the infants would try to get the mother's attention, and act distressed by her lack of interest. Then when the mother restored contact, they would turn away pouting and refuse to respond for a time, acting manifestly insulted. It is hardly to be supposed that these babies were acting consciously.

I therefore assume that there is something like a universal grammar of social behavior, governing both verbal and non-verbal social interactions. This hypothesis bridges the gap between verbal and non-verbal behavior in the social sphere, and opens the way to investigating the rules of social conduct that hold for both modes and for human beings in general.

Thus we have found in our comparative culture surveys – as well as our observations of children in kindergardens, – that the gesture of intention to take away is always interpreted as aggression, never as entreaty. If you want the other to give you something voluntarily, you must ask in a way that clearly recognizes that person as the possessor; the rule of possession must be observed. The possessor must not be compelled to surrender, but must be left free to give or withhold. Non-verbally, this is done by holding out the hand, a gesture familiar to chimpanzees as one of supplication.

Among human beings, this strategy can also be verbalized, and just as the direct seizure is perceived as an act of aggression and prompts resistance, so people will restist if asked in a tone of command. Of course, if there is a wide discrepan-

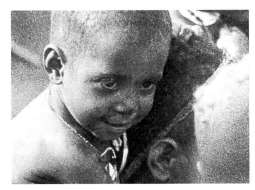
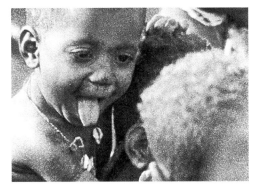
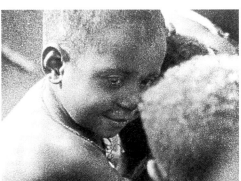
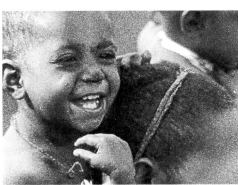

Social inquiry: A little Eipo girl, not quite three, testing an infant's reaction to a protruding tongue.

cy of rank, surrender may follow anyway, but then the demand amounts to dispossession. If one wants to be obliged voluntarily, decency requires a request which the other may deny. If one Eipo wants another's feather, they will mention casually in conversation that they like that feather. The other may then bestow it or else explain why they are unwilling to part with that feather, perhaps because it was a gift from a good friend. A general rule of adoption of various strategies, it seems to us, is expressed in these examples. If there are several possible strategies at hand – in this case, demand, request, or covert inquiry – one first tries the one that leaves the most possibilities open, in this case the veiled inquiry. True, the direct request may sometimes succeed more often, but if it is clearly rebuffed, then usually the case is closed, and there is not much further prospect of coming by the object peacefully. Instead, the request is experienced as aggressive, weakening the bond. Along this line, W. A. Corsaro states that children who want to join a game less often ask the question, "May I play too?" even though it will succeed in most cases. More often, they will first participate as a marginal player, and so gradually work themselves into the game. The advantage of this strategy is that the goal is reached more surely, if more slowly. A positive denial that would bar further access has been avoided.

Awaiting developments takes time, but in this case it is the strategy that will leave the most alternatives open. Within the rules of social conduct that are given as phylogenetic adaptations, innate sequences (expressive movements), cultural behavior patterns (rituals), and finally words may replace each other. With this multiplicity of forms of expression, the common fundamental rule is easily forgotten.

In friendly approach, as upon greeting, a combination of acts of self-display is usually observed with acts of camaraderie and conciliation. When, for example, a Yanomami male enters the host's

A conflict and its solution between two Yanomami children (upper left to lower right): The girl is about to climb a pole. A boy wants to take it over and use it himself. The girl is aware of his intention from experience, and tries to placate him by smiling pointedly. That doesn't stop him; he hauls off. The girl responds by averting her face and pouting, with the result that the boy goes away. Thus refusal of contact is an effective means of parrying aggression.

village as an invited guest, he performs a round dance with extremely warlike movements. His demeanor is haughty and overbearing. He dances in full battle array, and repeatedly draws his bow. While he thus behaves as though almost unapproachable, children dance at his side waving green fronds in their hands. The message is clear. Translated into words, it says, I come as a friend, but not subserviently. The encounter is tinged with a certain ambivalence. A relationship of superiority and inferiority might all too easily be established. To prevent this beforehand, people are very much aware of themselves on such friendly occasions, keeping countenance. At the "Schützenfest" in the Tyrol, when the rifle companies parade in the square of the host village, we observe a rather similar self-presentation. But we also note that the color guard is attended by children or young girls. If we consider the events around a modern visit of state, we find much the same; the hosts in greeting the distinguished guest strike a military pose, firing salutes, marshalling a guard of honor, and performing rituals, but at the same time the guest is offered flowers, usually by a young girl. And when two good friends in our culture meet, they will greet each other sometimes with a firm clap on the shoulder – an equivalent for a firm handclasp, but with some element of a test of strength. This self assertion is accompanied by friendly words, friendly smiles, nods and other tokens of affection. These various performances are quite different in outward appearance, but they follow the same general rule.

We are only beginning to investigate these phenomena, but I believe that in the hypothesis of a grammar of social conduct congenital with human beings we hold a key to understanding certain cross-cultural manifestations. It seems to me especially important that this hypothesis will allow fields of verbal and non-verbal behavior, heretofore viewed separately, to be viewed from a single, common perspective. I do not at all mean to say

that when human beings speak, they are merely translating instinctive behavior. That would be a drastic oversimplification. Our social behavior may of course be codetermined to a considerable extent by phylogenetic adaptation. That does not mean that we are like rail vehicles, with narrowly limited options. Culturally, humans can transcend innate predispositions; in fact they require the restraint of culture to control their motivational life.

It is equally important to realize that the human being is not a product of upbringing alone; certain capabilities, inclinations, urges and patterns of behavior are standard as phylogenetic adaptations. We may view them positively as the common inheritance that links people across cultural barriers. In the course of further research, it might be found that some of these phylogenetic adaptations no longer perform their functions at the present time. In that case, the only path to a new adaptation is by way of cultural precepts. They will then be repressive, in a certain sense. But we must not forget that it was only by dint of culture that human beings have been able to fit into the large number or "ecological niches" that they now occupy. Their universality consists precisely in that they are able to readapt themselves culturally to requirements from time to time. It would of course be desirable for this readaptation to be guided by knowledge about our nature.

Human ethology proceeds on the premise that, as stated, humans have a limited number of interaction strategies at their disposal that may be distributed worldwide. The differences that strike a comparative observer of culture arise from the fact that we human beings are able to interchange different actions as equally expedient, including non-verbal with verbal behavior. We can to a certain extent translate instinctive behavior into linguistic behavior. Words are linked to key stimuli, and actions are performed by speech. The confusing multiplicity of verbal and non-verbal possibilities for interaction that result are based, in our view, on a limited number of elementary interaction strategies. Their discovery would open a way to work out a universal grammar of human social behavior, applying to verbal and non-verbal action alike.

Verbal Behavior

Only humans possess a language of words. They are phylogenetically prepared for its acquisition by a number of adaptations. Thus, children grasp the basic rules of grammar at the very beginning of learning to speak, as is shown by their errors of analogy ("gooder" for the comparative of "good"). Learning to speak is to some extent a necessary part of the plan of normal development, and the brain structures that enable us to do it are quite well known. Language is an acquired code of signals that can be made into statements (sentences) by means of a likewise acquired system of rules of linguistic grammar. By means of this code, we can make statements to people about present, past, and future, refer to things in their absence, pass on

"Holding out the hand," an unmistakable begging gesture without words, is well known to our nearest relatives, the chimpanzees. It means that one would like a share but respects the other as the owner of the thing coveted. One is placed under "moral" suasion, but is meant to feel that one compliance is voluntary.

Among human beings also, asking works better than grabbing; good manners pay off. An Eipo boy eats a slice of taro. His little sister tries to take it away from him, but he draws back. The mother intervenes. She breaks the piece into two halves and gives both of them back to the boy. Now he willingly gives away one of them.

directions or formal instruction, and transmit them independently of the object.

There have been many attempts to explain the evolution of language. If we begin with function, the social function takes precedence. In traditional cultures people talk about who gave what to whom. People tell each other about origins and destinations, people chat about other people, quarrel, prattle affectionately with children, or converse during mutual grooming. Verbal instruction comes to play a greater part in advanced cultures. During development demonstrative words are at first linked to gestures, but social information also appears very early, and quite soon social behavior is translated into speech.

Certainly the ritualization of social interactions by means of speech has been a major factor. One who could wage and resolve a conflict in words was certainly at an advantage over one who had to settle everything on a level of non-verbal behavior, by threats and combat. Verbalization of social confrontation has certainly helped to harmonize group life.

Universal modes of perception and thought are also crystallized in terminology. In studies made by Volker Heeschen and this writer among the Eipo in the western mountain country of New Guinea, some remarkable behavioral constants made their appearance. We found astonishing correspondences to our own tongue – correspondences that presumably reflect general human modes of perception and attitudes.

Freedom and Self-Control

In play, as we have seen, higher mammals can detach their actions from their motives, thus creating a stress-free field for unfettered experimentation. That is perhaps the root of our own far more developed ability to stand back, inwardly, and make "clear" decisions that are less loaded with emotion. This detachment from emotion, whereby behavior has become more accessible to us, is probably due to the development of the cerebral cortex. However, this has placed us in a position not only to detach the motivational but also to suppress it. All this makes self-government and self-control possible. They count as virtues in all cultures, incidentally, and are practiced in a variety of ways, even to extremes of asceticism. This faculty gives us the power to act against our nature, a prerequisite for cultivated behavior. In many spheres, it is true, human behavior is bound and prescribed by phylogenetic adaptations, but these bonds by no means suffice to control life in society. Here there is a danger and an opportunity. A danger because humans are often off balance in the realm of social behavior, and are especially inclined to escalation in the area of aggression – an inclination they share with other primates. Then again, this absence of restraints lends itself to cultural response as well as to individual adaptation through learning and hence various forms of ecological niching

Self-assertive display and conciliatory gestures both belong to the ritual of a friendly encounter among the Yanomami – and not among them only. This warrior performs a boastful dance in full war regalia on entering his host's village. He thus indicates that he is not to be held subservient. But at the same time, he announces his peaceful intentions by bringing with him an innocent child dancing by his side, waving leafy springs of greenery.

as well. Cultural evolution can thus provide an opportunity for rapid readaptation.

However, the development of the cerebral cortex was only a first important step towards mastery. A second decisive step was the division of labor between the two cerebral hemispheres, creating an emotional, artistic right brain and an analytical, mathematical, language-oriented left brain. The two halves are linked by the trabecula at the base, and so they function as a unit. In individuals with severed trabecula, it is found that the two brains perceive and operate separately and in slightly different ways. This division of function enables people to give more weight to the one or the other of our "personalities" by differential activation of the

hemispheres. When we switch to objectivity by activating the left brain, we can, as it were, see our other self distantly reflected in this mirror. The power of conscious thought presumably came into the world with the evolution of speech, which translates action among other things, thus requiring a certain ability to observe one's self and gain distance from the self.

Social Behavior

Humans are without any doubt social creatures and for this, too, they carry with them a number of phylogenetic adaptations, among them a rich repertory of expressive movements.

Friendly, bonding modes of behavior evolved in mammals – and independently in birds – with the elaboration of care for the young. In this sense it is a landmark in the behavioral evolution of the vertebrates. The mothers develop nurturing modes of behavior, and the offspring develop signals to trigger them. These mother-child signals brought the concept of friendliness into the world. Beyond this, a number of birds and mammals have developed a personal relationship between mother and offspring. The ability to recognize each other individually served to bond the partners together in all cases where mothers care for their young over a long period. Basically, personal bonding is what we call love.

The derivation of friendly modes of behavior from nurturing behavior is well documented by both culture-to-culture and animal-to-human comparisons. Thus, both chimpanzees and human beings engage in mouth-to-mouth feeding as child care behavior. In both species, however, this "kiss" feeding is practiced also as an expression of affection, and ritualized into kissing. Chimpanzee friends sometimes greet each other by embracing and kissing.

Ambivalence of Interhuman Relationships

Among many vertebrates, the conspecific is a bearer of signals that release behavioral tendencies of acceptance and approach as well as of rejection and distance. When these features are perceived simultaneously, there is conflict between the two behavioral tendencies.

The same is true of human beings. We know that infants begin to be "shy" at 6 to 8 months. Whereas until that time they beamed friendliness at everybody who paid any attention to them, they now distinguish clearly between the familiar and the unfamiliar. Known individuals trigger friendly acceptance. Strangers, however, are presented with a blend of acceptance and apprehensive rejection. Usually the strange individual is smiled at. After a few moments, the child will turn to its mother and bury its head in her lap, only to re-

Kissing originated from mouth-to-mouth feeding, which occurs not only among chimpanzees but also among people. "Kiss" feeding became a ritualized gesture of affection.

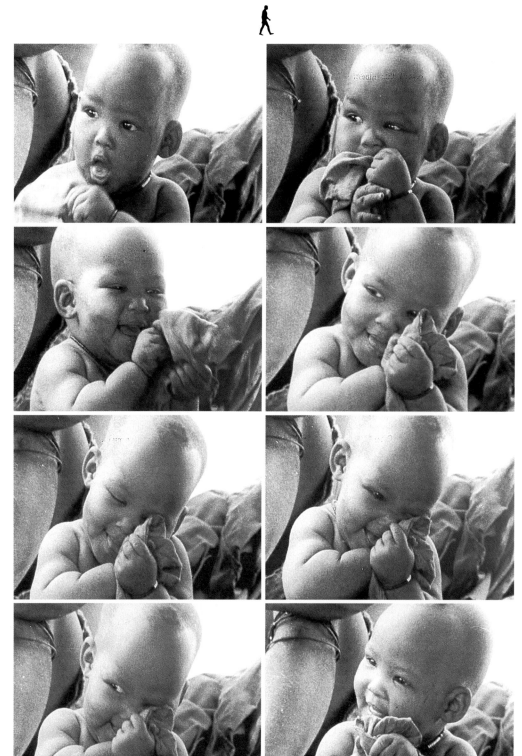

Vacillation between acceptance and rejection is expressed in the behavior of this female Bushman infant. Upon eye contact, she brings her hand to her mouth as if to conceal the lower part of the face, smiles and turns away; then she resumes eye contact, looks away again, thus hovering between acceptance and rejection. Note the superposition of withdrawal (posture) and approach (eyes) in the last picture but one (lower left).

sume friendly eye contact soon after. Reactions of acceptance and rejection alternate, or are superimposed simultaneously, for example when the child turns away, but at the same time smiles and maintains contact out of the corner of the eye. If the stranger keeps aloof, usually friends are made quickly; but if that person approaches, not giving the child enough time for familiarization, its behavior will often shift to fear. Then it will cling crying to its mother, and if the stranger tries to pick the child up, it will protest with every sign of panic. All healthy children exhibit this pattern of response, they fear the conspecific stranger even if they have never experienced anything but good from a stranger.

Comparison of cultures tells us that this ambivalent response to strangers is a universal pattern. Everywhere, that is, from a certain age onward, fellow humans are perceived as bearers of signals that provoke responses of both acceptance and rejection. The ability to respond to conspecific signals triggering anxiety may be attributable to processes of maturation in perception. There is an instructive parallel to this in the case of rhesus monkeys. G.P. Sackett raised rhesus monkeys individually in cages in which they could see neither their own reflections nor other monkeys. They were daily shown a selection of slides that they could project themselves after presentation by pressing a key. The slide would then appear for 15 seconds, and within five minutes this self-presentation of a particular slide could be repeated. The slides showed landscapes, fruits, and monkeys of the same species. It was found that the pictures of monkeys aroused great interest. The number of self-presentations of such slides rose more sharply than those of slides not representing conspecifics. The monkeys would also utter sounds of social contact, and they would approach the images with clear invitations to play. As it were, they regarded them as conspecifics even though they had had no experience with any. Among the slides there was one representing a monkey engaged in threatening behavior. This picture too was at first among those that triggered positive reactions. At the age of about two-and-a-half months, there was a striking change. Thenceforth the picture of the threatening monkey triggered sounds of fear, self-embracing and withdrawal, and the number of self-presentations of this slide decreased abruptly, while the other monkey pictures continued to be accepted. Since these monkeys had had no social experiences with conspecifics, it must be supposed that the recognition of threatening gestures is attributable to processes of maturation in the perceptual apparatus.

We are familiar with particular signals that trigger acceptance as well as rejective behavior in human beings. An especially significant one is eye contact. On the one hand, it is certainly perceived very positively. It is interpreted as acceptance, an announcement that the channels of communication are open. In interhuman communication, of course, the other must not be looked at so long that the eye contact is experienced as a stare. This too, as intercultural comparison shows, holds everywhere. We know from experiments that eye contact is exciting, for example causing the pulse rate to increase; a brief intermission is tranquilizing. Thus, we are able to control our level of excitement by means of eye contact alone. Besides visual signals, there are others that we perceive ambivalently, for example olfactory signals.

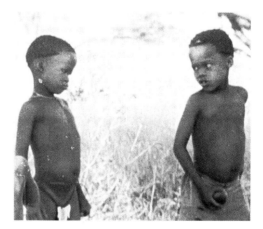

Staring as threat: The Bushman boy (right) intimidates the girl by a "hard stare" to such an extent that she drops her head, pouting in token of subordination, to avoid the other's gaze.

However, the situation has not been by any means completely researched. At present we can only say that people respond to certain signals from some people with friendly acceptance, to others with rejection or even aggression. An infant's parents are certainly likewise bearers of traits that in themselves trigger negative conspecific reactions, and we can demonstrate this if we include the inconspicuous movement of an infant in the dialogue with its mother. However, we human beings are so constituted that personal acquaintance greatly attenuates the effect of the anxiety-triggering signal, so that friendly responses of acceptance predominate. Behavior, one might say, is shifted in the direction of confidence in dealing with acquaintances. But in encounters with strangers, the anxiety-triggering conspecific signals take full effect, and accordingly behavior shifts in the direction of diffidence.

This simple social reaction pattern has far-reaching consequences for our social life together. It leads to a tendency for people to gather in small, closed groups and to confront strangers with some degree of rejection. For ages, humans lived in small groups in which everybody knew everybody else and in which personal relations of confidence therefore prevailed. This did not change until the rise of mass societies. Today we spend much of our time with people whom we do not know. As a result, in the metropolis, another hu-

The shy beauty: The dichotomy between rejection and acceptance is reflected in the face of this Himba girl. The series of pictures (from upper left to lower right) shows the girl's reaction to a compliment by this writer. First eye contact is followed by a suppressed smile and closing the eyes, turning away, corner-of-the-eye contact, again turning away, then a smile, and last, renewed eye contact.

man being becomes a "stressor," because distancing signals, not attenuated in their effect by acquaintanceship, are acting upon us constantly. The resulting social irritation may be read off from a variety of signs. Thus, city people hurry past each other and try to attract as little attention as possible, to minimize surfaces of friction. In the male, this may be expressed by unassuming everyday attire. The extent to which the male in a large society conceals his individuality in public becomes clear only upon comparison with primitive peoples. In a closed group, the male makes an impressive display with adornments and weapons, quite unembarrassed. He can afford this, for his self-assertion will not provoke his friends. Finally, in mass society people wear a mask-like expression. Outwardly, they appear anywhere from indifferent to self-assured. They avoid showing any weakness by look or gesture, perhaps because this might lead someone else to take advantage. This masking of expression can become so habitual that people finally cannot lay the mask aside even within the family.

Now we could not and would not break out of the large society, which after all offers many advantages. The question we should reasonably ask is whether we could not fashion life in such a society so as to be more human. We could start out from the very positive readiness of people to open friendly relations with other people. This is an old

biological heritage. It began, as previously mentioned, with the development of care for the young. With it, a family-oriented ethos developed, which today we transfer to the large group and through which we can identify even with people whom we do not know. However, a prerequisite for such a bonding, and therefore, aggression-mitigating identification, is growing up in an intact family, imbedded in a stable small community. Only thus do we acquire the interhuman trust that permits us to recognize possible friends even in strangers.

But just those advantages are increasingly destroyed today by mobility, collapse of the extended family, and also by various measures of urban administration and planning. We live, contradictory as this may sound, quite solitary amidst the togetherness of many. That is, we do meet many people, but those whom we number among our familiy and friends live so widely scattered that we communicate at best by telephone or letter, and otherwise remain limited to the nuclear family. And our cities do not invite us to build up new contacts. The problem is slowly being recognized. Beginnings have been made towards more human mass housing, with provision for human social requirements, for both privacy and interaction.

Even in mass housing, the provision of certain facilities has set social processes in motion towards formation of a community. In Alt Erlaa near Vienna this was done by installing pools on the roofs of the large apartment blocks. Most tenants would come regularly and social encounters were thus encouraged. Architecture conducive to community formation of this kind is worthwhile. From such individualized communities, life in the big city could begin to be humanized in a way essential to the survival of a liberal democratic social order. If people fear their neighbors, even subliminally, they tend to seek security in ideologies or charismatic personalities. This too is an ancient inheritance, evolved in the mother-child relationship. As we know, young animals, for example young rhesus monkeys, will take refuge with the mother even if they are punished for it. They seek her protection all the more. Later, rhesus monkeys will transfer this behavior to high-ranking group members, who are also points of refuge, even if they are a source of anxiety. Much the same is true of human beings. We know that abused chil-

"The Lonely Crowd" - a familiar title that has become a slogan of our time. The isolation of modern big-city dwellers is impressively illustrated.

dren are strongly attached to their parents so that it is often difficult for social agencies to separate them from their abusive parents. But such social agencies have existed only recently, and for most of human history a child had no choice but to adhere closely to its parents if it was to survive. We exhibit the same childlike need for dependency in emergencies by our tendency to entrust ourselves to high-ranking leadership personalities. Then our critical faculties often fail us.

The Family
The discussion of ambivalence in interhuman relationships shows that people are programmed at first as small groups, even if culturally they eventually manage to live in big, anonymous societies. There has been much dispute about the original form of small human groups. Among our nearest relatives, the chimpanzees, we find small groups of several males and females with no fixed mating relationships. It has been thought that such small groups with unregulated sex relations perhaps also represented the original natural state of humans. Very likely, however, they were typical of prehuman stages only. The human community is composed of family groups. Husband and wife as a rule form a lasting partnership, with a division of labor, over many years. Parents and children remain in touch for generations, so that ordinarily three generations (grandparents, parents, children) are found together in any large family group. This group in turn is integrated in a larger network of cousins and in-laws.

The initial family was surely maternal, and mother and child are still attuned to each other in a special way by a number of adaptations, among human beings as well as animals. Six hours after birth and following only a single presentation, mothers can distinguish their infant from other infants by odor, which fathers experimentally have not been able to do. The infant also learns quickly to recognize its mother. At only six days of age, it will more often accept a cloth with its own mother's body odor than a cloth with a stranger's odor.

Mothers delivered without anesthesia show strong affection for the baby immediately after birth. They will stroke it and also resort to words in seeking eye contact. The infant in turn is in a wide-awake condition, turns its eyes towards the source of sound when the mother speaks, and tries to get at the breast.

If mothers are permitted thoroughgoing contact with their babies immediately after birth, they

The small closed group, in which the human individual feels sheltered (left), is the primeval counterpart of mass society, estranged from nature. But even in highly civilized countries, appropriate architecture can permit personal forms of living, a humanization of community life, and even some proximity to nature. (Top) Façade of a terraced apartment house in the Vienna working-class district of Favoriten. (Bottom) Façade in the residential park of Alt Erlaa, Vienna.

show a stronger affection for the little ones in the ensuing weeks. Mothers who have been denied this contact will develop the same bonds, but with some delay. It is only after some months that there are no longer any appreciable differences between these two groups of mothers. We may therefore suppose that early mother-child contact is important, presumably securing the bond against interference.

Mammals may be divided into two large groups according to their manner of nursing the young. Some nurse at rather long time intervals, others at short intervals, more or less continuously, while the young are awake. Mammals that maintain long intermissions in nursing have a milk of higher protein and fat content than those which nurse continually.

Mother's milk is of the "dilute" type, so that in terms of the physiology of nutrition, humans may be identified as nursing continually. Mothers in hunter-gatherer societies are observed to nurse their children on demand, several times an hour. This has noteworthy consequences. In nursing, the hormone prolactin is secreted in response to stimulation of the nipple. Upon stimulation of the female nipple for 5 to 15 minutes, the prolactin content of the blood plasma increases to between 2 and 20 times the baseline. This suppresses ovarian function, so that Bushman women do not become pregnant during their three-year nursing period. Among the Australian aborigenes, on the other hand, who nurse at longer intervals and often leave children behind with aunts or other "baby sitters" during their community excursions, babies follow at much shorter intervals. Infanticide is common among these groups.

The bond attaching the child to its mother develops gradually. The baby doesn't care who first provides acceptance and warmth. Quite soon it will prefer the person who cares for it. After a few months, the personal bond to one who is now definitely preferred has been consolidated. The child will also establish personal bonds with other individuals also, for example the father and older siblings, but these bonds are qualitatively different. In pain or fear, children tend to seek out the mother, both among us and among primitive peoples that have been studied. If a mother leaves her infant, it often protests against the separation and shows definite signs of distress. But infants will react in the same way if the father goes away in the morning and leaves them with the mother. I have observed this in primitive societies, for example with the Yanomami.

We occasionally read that among primitive peoples, a child is socially integrated into the community and shows no special personal preference. According to Margaret Mead, for example, this is so among the Samoans. The information is inac-

curate, based on hasty observation. In such traditional societies, the children are imbedded in a small community whose representatives are in touch with them many times a day. The bonding with these different persons, however, varies qualitatively, and there are preferred individuals after all.

Dependable reference persons, as is shown by the research of René Spitz and John Bowlby, are crucial to a child's normal development. They are the only means by which children develop a "basic trustfulness." If they are denied this opportunity, deficits of "socialization" will result.

Among so-called primitive peoples, mothers customarily nurse their children for a long time, regularly and at short intervals, as is more in keeping with the biology of humans as mammals than the practice of civilized countries. In terms of the physiology of nutrition, the human being is a "long-suckling" animal.

It has sometimes been asserted that bonding to one or a few reference persons should be prevented if a person is to become fully accepting of the community. A firm relationship of the child to one person would be an obstacle to full identification with others. Against this, I have advocated the view that it is only through personal relations within the family that one really becomes capable of strong emotional identification with others. For within the family, a human individual develops that family ethos which then extends to the group, using – significantly – kinship terms to prefer even to group members who are not related. For example, we quite often refer to nonsiblings as brothers or sisters.

In child development, the father plays an important part, not merely as a result of modern social history. Among primitive peoples also, fathers play with their children and caress them. Strong attachment is reinforced by a similar attachment on the part of the child. This is true specifically of extremely warlike peoples such as the Yanomami, the Himba, or the Eipo. Fathers have the same repertory of affectionate modes of behavior as mothers, and they are about equally well able to recognize the expressive movements of their children. There are differences, however, in patience and endurance, as well as in the kind of intercourse with the child, that suggest a supplementary role of the father.

The question as to what extent motherliness is based on biological inheritance is at present under dispute. Extreme advocates of the feminist movement maintain that the differences in role between the sexes are only socially conditioned. Elisabeth Badinter in fact regards motherliness as an invention of late civilization. Among primitive peoples, she believes, the child is treated as a small adult, and of small account, because infant mortality is high. That could only be written by one who never lived in a primitive community and never saw how much the death of an infant is mourned.

The claim often heard today that the sexes have been assigned their roles exclusively by culture is false. That doesn't mean that a woman must necessarily see her ultimate fulfillment in motherhood, or that there are no other possible careers open to her; it only means that as a mother, a woman is performing a function specially suited to her feminine nature and not to be socially devalued.

In discussing these matters, one should first free oneself from value judgments. There is no fundamental superiority of one sex over the other. If differences are found, the question becomes how they are to be interpreted as adaptation and how they arose. There are without doubt differences, for example, in the physiological realm.

If one inquires what functions men and women perform in different cultures, some clear tendencies can be identified. Men generally exert themselves in defense of the group. There is no culture in which women wage war. That "advance" has been reserved for the technological age. Hunting big game has always been a male pursuit. In collective activity, gardening or agriculture, women and men work side by side in more or less the same way. On the whole, men undertake the tasks that call for heavy physical effort. The women's share is usually such that they will not go very far from home, and seldom for long periods. Food

In traditional community life, the child's mother is a fixed, dependable reference person to whom it is constantly attached - not just emotionally, but physically. Such attachment is a prerequisite for normal child development and the establishment of "basic trust."

preparation is predominantly, yet not exclusively, the woman's job, as are many domestic "projects" such as making clothing, ornaments, and the like also. Finally, as a rule women assume the chief burden of child care. In interactions with other groups, primarily men will represent the group to the outside. Correspondingly, in the religious sphere men play the part of intermediaries to higher powers. Such is not the case everywhere; we are only points of emphasis in discussing the division of labor. In modern mass society, many traditional norms of sexual role behavior, whether culturally or biologically based, may have lost their adaptedness and consequently stand in need of revision. That is one of the justified concerns of the women's movement today.

Some feminists are quite radical in their rejection of the traditional female role. Simone de Beauvoir maintained as early as 1951 that woman is bound to motherhood and the body, like an animal, and consequently is dependent upon the male, like a parasite. Only by vocational activity could woman gain her independence; that was the key to emancipation. The care and rearing of children are community functions, and marriage should rest on free consent, subject to cancellation at any time.

What is ominous in these views is primarily the generosity with which the children's rights are disposed of. Here the denial of the fact that mother and child are phylogenetically attuned to each other, and that children need a fixed reference person for their psychological welfare and success, has led to quite inhuman proposals.

The results of ambitious experiments undertaken in abolishing the traditional roles of the sexes in order to effect the emancipation of women are instructive. Thus, in the kibbutzim, the rural community settlements in Israel, woman was to be freed, by full integration into vocational life, from the economic overlordship of the male, and by community child-rearing, from the slavery of motherhood. Children were entrusted to special attendants, and would come home to their parents for playtime in the evening. Marriage was tolerated, but it was hoped that "bourgeois" marriage would be outgrown some day. In 1954, Melford Spiro, an American sociologist, studied a kibbutz organized on these principles and found that its inhabitants had realized the ideal and were living

Loving fathers: For the psychological development of the child, paternal affection and acceptance is of vital importance. In forms of society that we tend to call "primitive" or "backward," it is a matter of course.

by it. The family, he concluded, was not a product of nature, but expendable. However, when he visited the kibbutz again a generation later, he found to his astonishment that the women who had grown up in the kibbutz were no longer submitting to having their children cared for primarily by other people, and they were again placing a higher value on marriage. Whereas in the founding period women had assimilated themselves to men outwardly in clothing, they were now returning to feminine attire. They were withdrawing from the masculine occupations, and a good many of the women reaching adulthood showed more interest in family affairs than in politics. Spiro, who at the time of his first study had been quite confident of success for the experiment, had also recorded play behavior at this time. A review of the data of that survey showed that the children, though brought up under the same conditions, with the same toys, played differently according to sex. The boys had taken men for their models, the girls took women. In 1979, this led Spiro to the conclusion that psychological sex roles are also decisively codetermined by precultural, biological, factors.

Biologically based sex differences exist in the realm of sexual behavior, both as regards perception and in concrete behavior.

Thus, men and women differ in their excitability by visual impressions. The key stimuli to which males respond have been better investigated. It is occasionally held that women are less readily aroused by visual stimuli than men, but there are also statements to the contrary. There are certainly differences in responsiveness to the touching of erogenous zones. The area of the nipples is mildly sensitive in the woman, but by no means generally distinguished as erogenous in the male. We have already noted differences in olfactory perception.

In human sexual behavior, then, sex-linked adaptations on a hereditary basis are not an insignificant factor. But it is equally certain that experiences of early childhood have a strong influence, and in extreme cases will even submerge and largely eliminate innate dispositions.

Studies of child behavior have brought to light an abundance of sex-typical patterns. Newborn girls smile spontaneously and with eyes closed more than do boys. Later they exhibit more voluntary communicative smiling. Newborn boys are more restless than girls and more inclined to make faces. As early as 2 to 2½ years of age, boys are more aggressive than girls in word and deed, and will stray farther from their mother. European children and Bushman children behave very much alike in these respects.

Among the Bushmen also, boys exhibit more physical aggressiveness and will go farther away from camp. Children have a tendency to associate with children of the same sex at a very early age, and are somewhat shy of the opposite sex. Thus there is a preference for forming play groups of the same sex; this may be observed also among children of other cultures, for example the Kalahari Bushmen. The boys like to engage their peers in games of fighting and hunting, while girls play dancing games together. Interest in objects among the Bushman children is likewise directed at different items according to sex.

Comparative study of cultures tells us that marital partnership with division of labor occurs everywhere. Of 849 societies, 708 (83.5 percent) allowed polygamy; 137 societies (16 percent) are by law monogamous, and 4 are polyandrous. This would suggest that monogamy is somewhat exceptional. However, the picture is misleading, because even in polygamous societies, men are mostly married to only one woman. The primeval practice of promiscuous sexual intercourse does not exist anywhere.

Humans are adapted to matrimony in a remarkable way. Whereas among the other primates sexual behavior is linked to brief estral cycles, human sexual appetites have been released from that limitation. The woman is prepared to accept the man at other times than during her days of fertility and thereby afford him satisfaction, strengthening the bond. She moreover experiences an orgasm that binds her to the man – to my knowledge likewise a peculiarity among the primates.

A close cennection between nurturing behavior and sexual behavior is also seen in certain reactions of the female breast. Like nursing, sexual excitement will cause erection of the nipples and a flow of milk, and in some individuals stimulation of the nipples will by itself induce orgasm. Probably the strong sex urge of the male is also to be regarded as an adaptation in the interests of bonding. Certainly human sexuality has an important function in the pairing bond over and above its reproductive function.

The Small Group

For the greater part of its history in time, humanity has lived in small groups of 30 to 50 individuals as hunters and gatherers. We still have a model of this situation today in the few remaining hunter-gatherer communities, the Bushmen, and some Indians of the South American forests. These small groups are associations whose members are usually interconnected by kinship. There is as yet no occupational division of labor at this stage, but there are individuals who put themselves forward as soldiers, hunters, or healers and enjoy special honors accordingly. There is a striving for rank and respect, probably an early primate heritage. However, the striving is dealt with in various ways. The Kalahari Bushmen will not tolerate ambition. If a male boasts of his hunting success, he is re-buffed. In other traditional cultures, however, ranking orders tend to be built upon such behavior, as among the cattle herding peoples, who employ a hierarchy to prevail militarily over those who steal their cattle. For human beings as for others, those of high rank occupy the center of attention; they are "regarded" most by all other members of the group; all are guided by them.

They are chosen as group leaders on the basis of a number of positive social qualities such as the ability to settle disputes, organize games, and assist the weak. Aggression is by no means paramount as a quality of leadership. In this respect, ranking based on qualities of leadership differs from relations of mere dominance, in which the stronger oppress others against their will, and which is represented in the animal kingdom by the pecking order in a flock of chickens.

A prerequisite for the establishment of an order of rank, besides the need for esteem, is a readiness to follow and submit – a readiness that has its own

Distribution of roles among primitive peoples: In primitive societies, men represent the group to the outside, as hunters (left), as medicine men in combat with demons of disease (upper right), and as warriors (lower right).

perils. Too little or too much obedience can disrupt the harmony of community life. Too much obedience has led to inhuman action in past and contemporary history.

In this connection, reference may be made to a study by Stanley Milgram (1966) on the average person's obedience to authority. His subjects were to administer increasingly painful electric shocks (15–450 volts) for wrong answers in a learning experiment. The whole experiment was a sham, but the subjects did not of course know this. Authority was represented by the director of the study, who would order the subjects to persevere when they wanted to stop. When the "victims" feigned only moderate discomfort, 26 out of 40 subjects administered the strongest shocks. Readiness to obey increased with proximity to the "learners," but it was only when the subjects could themselves select the intensity of the shocks that the average (supposed) voltage declined to 75 volts.

The members of a small group, even in the hunter-gatherer stage, regard each other as allied in a special way. They distinguish between themselves as an "in-group" and others, and often set themselves apart in an accentuated manner, say by peculiarities of dialect, custom, and style of adornment or attire. Within the group, a strong pressure to conform prevails. Those who deviate become a target for mockery and other forms of aggression, enforcing adherence to the norm. If adherence is not forthcoming, expulsory responses emerge. Such intolerance may have been adaptive in the small group, where accurate prediction of the behavior of each member was valuable. But in modern pluralistic society, outsiders are often especially valuable citizens because of their special capability, and tolerance is a desideratum. It must be encouraged by educational work.

Despite their distinctness, the small groups by no means exist for themselves alone. They are allied with others in manifold ways. This may come about by covenants between groups or by matrimonial connections. When a !Ko Bushman marries a woman from another village, he thereby

Correspondingly, women have jurisdiction over the domestic sphere. They prepare food (upper left), care for children, and at the same time weave or spin (lower left), gather wild fruits or dig edible tubers out of the ground with a stick like the one held in this Bushman woman's hand (right).

gains access to his wife's family's hunting and gathering grounds, an important asset in times of need. Among the !Ko, forthermore, several groups are confederated into a "nexus." In emergencies, members of a nexus stand by each other, they assemble for certain rituals, and marriages within this community are preferred. Among the !Kung Bushmen, there is a highly developed system of exchange on a personal basis. Each adult is in communication with a certain number of exchange partners, and the result is a mutual assistance contract, a sort of Stone Age social security.

Territorial Behavior and War

Among vertebrates, including mammals, territoriality is the rule. Individually or in groups, at least at certain seasons, animals hold a territory and defend it against conspecific strangers.

Although territorial disputes run like a red thread through human history, it has been believed, by a kind of wishful thinking, that this discordant trait originated with the tilling of the soil. At the stage of the hunter and gatherer, humans had lived in free, open societies, having no possessions worth defending. Today we know that this conception does not correspond to reality. Even hunter-gatherer people are territorial. They thereby secure their means of livelihood.

Together with territorial boundaries, humans have also developed special forms of group defense, whence the form of intergroup aggression that we call war ultimately emerged. At this stage, it is a product of cultural evolution. It makes use of some of the emotional urges of combativeness but at the same time suppresses others, for example compassion, through which aggression is kept from degenerating into destructiveness. One persuades oneself and the other members of the group that the enemy is not human, or is human but not of the same species. The confrontation is, as it were, shifted to an interspecific plane, where the rules of fairness and humanity do not hold. That is what the Yanomami Indians are doing when they speak of their enemies as though they were prey. Also, killing enemies and thus serving the group is made into a virtue. A filter of biological rules that prohibits killing is superimposed with a filter of cultural rules that mandates killing. This war ethos is a cultural innovation, linked to family-oriented defense but carrying it radically further.

The development of war is promoted by the development of long-range and eventually long-distance weapons. A rocket can be launched by people who could never bring themselves to kill an individual face-to-face. People began using weapons against each other in this way very early. Representations of warriors shooting arrows at each

Practicing role behavior in primitive communities: Children early join into groups of their own sex. Whereas boys prefer fighting and hunting games, girls prefer dancing together (right).

other are to be found in cave paintings of the Middle Stone Age.

Since warlike aggression is a product of cultural evolution, and at variance with our conscience and the strong bonding tendencies of human beings, it would seem quite possible that war could be conquered by cultural means, provided international agreements could solve the problems, without bloodshed, that have been settled by warfare in the past.

Future Prospects for Humanity

About ten thousand years ago, when the development of agriculture and animal husbandry began, more people could maintain themselves in a given space than before. Growth of population, division of labor in society, trade, and the founding of cities followed. Greater power resources could be deployed for armed hostilities. Enforced or voluntary associations were favored, and led to the formation of settled communities. However, the tendencies of life in small groups continued to govern community behavior.

Modern societies are essentially anonymous communities, presenting us with some problems on that account but also with new opportunities. Among other things, it is only the large society that makes possible the cultural achievements of the technologically civilized community. That community, to be sure, is in crisis. Increasing scarcity of raw materials, famine, overpopulation, environmental destruction, and war have become everyday topics. We worry about our future. What is at stake is survival. But what does survival mean? What seems to endanger it? And what concretely could be done against perceived dangers?

Survival means, firstly, survival in posterity, in other words genetic survival. Of course ideas may also survive if we commit them to paper, as may other cultural goods. But people are certainly right te seek not only cultural but also biological survival. This need not be achieved as heretofore in adversary ways. We are becoming increasingly aware that the various nations today are imbedded in a global community and dependent on cooperation – and that therefore, besides group interests, there is an interest of humanity.

Survival in one's own descendants is only one desirable goal. The biologist knows that species often flourish for hundreds of thousands, even millions, of years, finally to reach new evolutionary levels in the transformation of species or to die out. If a species gains differentiation in the course of modification, we speak of evolution. If it loses differentiation, we speak of involution. It is often attended by one-sided specialization, cutting off any further "evolutive potential." Developments

Feeding as bonding behavior occurs not only among the brute creation but also in human society, for example here on Bali, where partners at the "tooth filing" ceremony put dainty morsels into each other's mouths.

that further the stream of life emanate as a rule from universalized rather than specialized species. What endangers these various forms of survival, and how can the dangers be met?

The dangers that threaten us directly, arising from overpopulation and the associated danger of destruction of the environment and exhaustion of non-renewable resources, have been noted often enough. Yet we confront them with a certain helplessness, and environmental destruction, including the poisoning of waters and a suspect alteration of the atmosphere, proceed apace. Furthermore, the present world population is living and multiplying at the expense of irreplaceable fossil fuels. Since no substitute has yet been found, the possibility of a collapse of population is discernible that might overshadow all precedent. Effective measures against that eventuality are impeded, if only by the fact that states compete economically and by power politics. This competition leads to technological innovations that are subject to natural selection. Here cultural evolution in a sense mimics biological evolution. To be sure, it does so also by striving for growth in every area. That is the lemming strategy; if food is abundant, the lemmings multiply until their supplies are exhausted and population collapse is inevitable. Mass death and mass migration are the consequences. The ecosystem recovers, and so does the lemming population, beginning with those "fittest" which survived the catastrophe. That the many perish in times of crisis is of no concern to nature. For nature, all that counts is that population pressure forces distribution into marginal areas, and thus also forces new adaptations and drastic selection for fitness.

We human beings of course experience the value of the individual, since in us creation has, as it were, become conscious of itself. We endeavor to avert catastrophes by planning, and of course often it doesn't work. But even though we are overtaxed if we try to plan cultural evolution as a whole, we can nevertheless introduce rules of the game that may avert excessive extremes, and adopt programs essential to our continued biological and cultural evolution.

A rational approach to nature and a rational population policy, however, as has been said, are impeded by interstate competition. States or groups

Humans as "territorial" creatures: Building sand castles on the beach is nothing but a playfully modified modern form of marking and defending territory.

of states, in their effort to secure their own existence, strive for the maximum of power. Behind it all is the fear of being outrun by the adversaries striving for superiority. Hence the arms race of our time. Yet on all sides there is the unmistakable desire to build up friendly relations in the international sphere and to abolish the rule of force. What stands in the way of this universal aspiration?

In the first place, we are called upon to condemn war, even though we are all descendants of successful warriors. All our ancestors were conquerors, for of the conquered not much remains. War can indeed be an effective means of enforcing self interest and group interests, but conscience impels us to seek other solutions. The longing for peace is much older than the atomic bomb, even though peace based on deterrence has provided a new powerful incentive.

An important prerequisite for solving these problems is that we begin to think on a different time scale. Politicians think from one election to the next; fathers and mothers of families have a wider horizon in the sense that they are also concerned for the welfare of their children and grandchildren. For times farther into the future, imagination quite often fails us. Yet our survival depends crucially on our developing an ethic of survival spanning future generations, that we commit ourselves emotionally and intellectually on behalf of the as yet unborn. It should be possible to achieve this by education. We have succeeded in developing, from the family ethos, a group ethos enabling us to identify with people we do not know. There are signs that we are at the point of developing a future-oriented morality.

The objective of survival alone, however, is not sufficient in my opinion. The objective should include preservation of the potential of evolution. Many species perish through specialization. They were perfectly adapted to certain environmental conditions, and consequently were unable to adapt to further environmental change. New developments, as has been mentioned, have always emanated from species that were less specialized and more generalized. The chances for human beings are therefore unique. We are extreme universalists, physically and mentally. There are certainly mammals that swim or run faster, or are better climbers, than we are. But no other animal can hope to defeat a human being if the requirement is to sprint 300 ft (100 m), dive into a pool, bring up three specified objects from a depth of 12 ft (4 m), swim 300 ft (100 m), climb a rope on the far bank, and then hike 6 mi (10 km) – a hexathlon invented by Konrad Lorenz to illustrate human versatility. Add to this excellent sensory equipment, a highly developed intellect, the grasping hand with which we create the artificial organs of technological civilization, and language. Each of us is such a universalist, and that is the foundation of our

Group aggression of hunter-gatherer peoples: This cave painting shows Bushmen who have stolen livestock from the Bantus (at the right in the picture) and are defending their prize against the plundered.

prospects for the future. Among factors that threaten our universality, which is rooted in personal freedom, are primarily institutions created by human beings themselves. Institutions tend to develop a life of their own, to escape from our control, and to turn from servants into masters.

A special danger threatens the dynamics of organization in the field of administration, because it directly threatens the freedom of the individual, seeking to integrate and subordinate the human being to a greater whole. Here there is a general biological trend; the incorporation and subordination of parts developed into higher forms of organization, organisms, or insect societies. Configurations of this type are found to be all but perfectly

HUMANS

Plume of pollution, the symbol of advancing destruction of the environment in the Industrial Age.

adapted. At the same time, their capacity for further evolution is restricted. If it were possible culturally to bring up a human being completely integrated in the state, undoubtedly biological development would follow – and together with the loss of freedom and universality, the chances of further evolution and intellectual independence would also be curtailed.

So we can indeed recognize some dangers that threaten our survival. But will that help us to guide our destiny by insight?

The economist and Nobel laureate F. A. von Hayek believes that we should leave everything to the self-regulating forces that have determined past cultural evolution; that one cannot plan culture. Basically that may be accepted. But we can set ourselves goals, and one such goal would be the survival and, at least culturally, the further evolution of an independent, socially and creatively intelligent, responsible, cooperative human type. In my opinion, the necessary condition would be social human beings, though endowed with versatility as individuals, and that condition is biologically given. Its preservation should be one of our concerns.

To reach that objective, other objectives are called for. The "open society" demanded by Karl Popper, in which many ideas flourish side by side in a climate of readiness to understand, would be one such objective. To this I would add the preservation of cultural multiplicity as another intermediate objective. Each people, each culture, is experimenting with a survival strategy proper to itself, and in their totality, cultures augment the latitude of human adaptation.

It is important to bear in mind that all these propositions are based on assumptions that might prove false. In principle, therefore, there must be a readiness to discard assumptions or hypotheses that have failed, and to set new goals. Here we address a difficulty, an obstacle to rational planning, namely our reluctance to abandon hypotheses when they prove untenable. Prehistoric people already used hypotheses as frameworks for orientation on the way to certainty. Within the scope of their explanation of the world, they were then able to act, no longer slaves of destiny. For example, the Bushmen explain disease as invisible arrows planted in the patient's body by enemies or demons. It follows from that explanation that these same arrows could be removed by enchantment. Anxiety is assuaged. Hypotheses often evolve into systems of belief, which then serve for group identification. The sinister tendency to promote economic and sociological hypotheses to the rank of ideologies is familiar enough. Our intelligence governs when we deal with problems of the world outside our species. When social problems are to be solved, we emotionally defend preconceived opinions as though our identity were threatened.

Dogmatic attitudes are also promoted by the human tendency to perceive and to think in pairs of opposites. This is indispensable to our orientation in the world, and is implicit in our preassigned programs of perception and thinking. We distinguish between good and evil, and are inclined to polarize the virtues themselves. There are such virtues as courage, zeal, self-sacrifice, loyalty, often practiced one-sidedly at the expense of the humane virtues. In reaction against impending catastrophe, the virtues of altruism are cultivated at the expense of civic virtues, indeed elevated into a program, to the point where these very virtues are drained of their value and meaning, and converted into vices. The period of European self-aggrandizement has been followed by a period of self-abasement. Continual self-abasement results in an erosion of identity; it leads to self-destruction.

It would certainly not be of much help to the world if specifically the occidental civilization of self-resolution were to collapse, for with all its darker aspects, which are undeniable, it is after all, to paraphrase Popper, the world's most self-critical and self-reformative civilization. Only in this civilization has the moral demand for personal freedom been largely recognized and realized.

(Opposite page) With their curiously cunning faces, the sloths are among the most popular representatives of the order Xenarthra, inhabitants of the New World exclusively. The picture shows a three-toed sloth.

NEW WORLD EDENTATES

Category
Order

Classification: An order of mammals, comprising three suborders and four families; formerly – and sometimes still today – referred to as the order Edentata (toothless, or having few teeth).

Suborder Vermilingua
Family Myrmeciphagidae (hairy anteaters)
Suborder Pilosa (sloths)
Family Bradypodidae (three-toed sloths)
Family Choloepidae (two-toed sloths)
Suborder Cingulata
Family Dasypodidae (armadillos)

Hairy Anteaters
(3 genera with 4 species)
Body length: 6.4 in.–52 in.; 16–130 cm
Tail length: 6.6 in.–36 in.; 16.5–90 cm
Weight: 11 oz to 77 lb; 300 g to 35 kg
Distinguishing features: Usually thick, rough coat of hair; elongate to tubular snout with small mouth slit; toothless; long wormlike tongue; sticky secretions from large salivary glands; sturdy limbs with highly specialized digits and claws; tail of great anteater with long fringe of hair, hair otherwise sparse and short; small eyes; short, round ears; vision and hearing limited, sense of smell highly developed.
Reproduction: Gestation period 180–190 days; one young per birth.
Life cycle: Hardly known; sexual maturity of tamanduas at about 12 months; life span in nature not known.
Food: Almost exclusively ants and termites.
Habit and habitat: Chiefly diurnal (giant anteater) or nocturnal; arboreal or terrestrial; solitary or in twos (usually mother and offspring); in forests, sometimes also in more open terrain.

Sloths
(2 genera with 5 species)
Body length: 1.5–2.8 ft; 44.5–85 cm
Tail length: 0.6–3.6 in.; 1.4–9 cm
Weight: 4.9–17.6 oz; 2.25–8 kg
Distinguishing features: Compact build; rough coat of hair usually inconspicuous in color, often with greenish sheen due to growth of algae; small round head with small ears; long slender limbs; two or three fingers and three toes, fused and bearing sickle-shaped claws; five teeth in upper jaw, four in lower; chambered stomach; simple womb; two pectoral nipples; testes in abdominal cavity; senses of smell and taste very highly developed.
Reproduction: Gestation about 4–9 months or longer, sometimes with delayed implantation of embryo; one young per birth; birth weight where known, 11–14 oz (300–400 g).
Life cycle: Weaning presumably at about 4 weeks; sexual maturity not earlier than one year; life span up to 40 years.
Food: Almost exclusively vegetable, predominantly leaves.
Habit and habitat: Diurnal or nocturnal; solitary; arboreal with slow-motion-like climbing movements; good swimmer; in tropical forests (coastal to mountain).

Xenarthra
Xénarthres FRENCH
Nebengelenktiere GERMAN

Armadillos
(8 genera with 20 species)
Body length: 4.8–40 in.; 12–100 cm
Tail length: 1–20 in.; 2.5–50 cm
Weight: 3 oz to 110 lb; 90 g to 50 kg
Distinguishing features: Compact build; skin ossification in the form of variously shaped plates and girdles on dorsal side and head; sparse to dense bristles or hairs, especially on ventral side; narrow, wedge-shaped to elongated head; 6–25 similarly shaped teeth in each half of jaw; very sturdy skeleton; short limbs, usually with five digging claws or powerful talons; 9–12 thoracic vertebrae, fused in region of cervical and lumbar vertebrae; simple digestive organs and womb; testes in inguinal groove; copulating member markedly long; generally two pectoral nipples, four in one genus; limited ability to regulate body temperature (lower than in most mammals, 90–95° F or 32–35 °C); good hearing and sense of smell.
Reproduction: Gestation 2–4.5 months; in some species, delayed implantation of embryo; usually two young per birth, four in one genus (always identical quadruplets).
Life cycle: Weaning time not known; sexual maturity at about 6 months or later; life span presumably 12–16 years.
Food: Insects and their larvae, small vertebrates, occasionally plant parts.
Habit and habitat: Active primarily in twilight and at night; terrestrial, good digger; usually in excavated burrows; chiefly solitary; in woods, steppes, and savannas, also elevations up to 9600 ft (3000 m); nine-banded armadillo also in cultivated land.

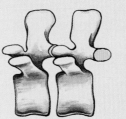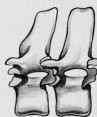

Additional vertebral joints (above). The name Xenarthra refers to a vertebral peculiarity not found in any other mammals. Normally the mammalian vertebrae are interconnected by only a single pair of joints (lavender). The last thoracic vertebrae and the lumbar vertebrae of the xenarthrans have two additional pairs of joints (blue). The picture shows the connection of two lumbar vertebrae in humans (left) and in the giant anteater (right).

Toothlessness (below). The earlier designation of the New World xenarthrans as Edentata (toothless) and their inclusion in a common order with the Old World scaly anteaters and aardvarks has proved unwarranted. Simplification of the dentition to the point of total loss of teeth is not a sign of close relationship. The reduction of the teeth is due to similar adaptations to the diet. Thus, toothlessness was repeatedly attained by unrelated anteating forms independently of each other (spiny anteater, pangolins, hairy anteaters). The figure shows skulls of a sparingly toothed sloth *(Bradypus)* and above it, a toothless anteater *(Tamandua)*.

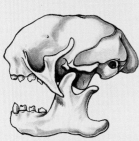

Modification of the pelvis (right). There are several features of the skeleton that mark the special position of the xenarthrans in relation to other mammals, as for example, the occurrence of septomaxillary bones in the skull or the complete ossification of the thoracic ribs and their articulation to the sternum. Among the most conspicuous is the modification of the pelvis into a synsacrum, much as in birds. The units of the pelvis are firmly fused and merge with the sacrum and the anterior caudal vertebrae to form a massive bony complex. The picture shows the synsacrum of a larger hairy armadillo.

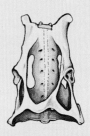

New World Edentates

Introduction
by Walburga Moeller

Abundant fossil finds show us the high degree to which the Xenarthra typified South America in their heyday, and the few surviving relatives of this group, once so rich in species, are still considered to be characteristic of South American fauna. The tubular, elongated head of the anteater, the bony skin armor of the armadillos, and the slow-motion, hand-over-hand progress of the sloths seem aboriginal, and cast a peculiar spell upon the observer.

When the French zoologist Georges Cuvier worked out a new classification of the mammals late in the eighteenth century, he included spiny anteaters, the duck-bill platypus, scaly anteaters, and aardvarks in a group to which he gave the name Edentata (lacking in teeth); later he added the hairy anteaters, armadillos, and sloths. Almost 90 years passed before studies in comparative anatomy revealed certain common features of the New World "edentates": their pelvis is joined to the sacral vertebrae by way of not only the ilium but also the ischium. Furthermore, additional vertebral joints, one pair each on the last thoracic vertebrae and on the lumbar vertebrae, point to close relationships among hairy anteaters, armadillos, and sloths, and separate them from all other mammals living today.

The name Xenarthra refers to this peculiarity of the spine, and modern taxonomy places these three groups of animals together, different though they are outwardly. Their internal organs show predominantly archaic features, but also special adaptations. In principle, we find poorly developed dentition; but only the giant anteater is completely toothless. The central nervous system is primitive in structure, with anterior lobes of the brain not highly convoluted. The neencephalon of the armadillos is comparatively small, while the olfactory lobe and cerebellum are relatively large. In terms of the external structure of the cerebrum, the armadillos are the most original family, followed by the sloths; the anteaters represent the highest level of development in the series.

The present range of xenarthrans extends from the extreme South of Patagonia throughout South and Central America and into the United States as far as Kansas and Missouri, embracing such various terrains as the Alpine, tropical rainforest, open woodlands, savannas, pampas, and croplands.

Different as these animals appear, zoologists have placed them together in the order Xenarthra. What connects them is not visible from the outside: common pecularities of skeletal structure and blood composition. Shown left to right as representatives of the three suborders: a hairy anteater, a sloth, and an armadillo.

Phylogeny
by Erich Thenius

The phylogeny of Xenarthra seems to have been essentially clarified, but many problems remain unsolved. Originally, several mammalian stocks were grouped together as "Edentata," their most conspicuous common trait being toothlessness or at least drastic reduction of dentition. This trait is certainly connected with diet. Species specialized in eating ants or termites (for example, hairy, scaly, and spiny anteaters) are mostly toothless. Studies of anatomical morphology have shown that there are three different groups among the modern "edentates", which are not closely related: Xenarthra of the New World, the scaly Pholidota of the Old World, and the aardvarks (Tubulidentata).

While the aardvarks are aberrant descendants of archaic ungulates (Condylarthra), the Xenarthra and probably also the Pholidota are regarded as descendants of early placental mammals, to be grouped with the other placentals (Edentata; *Paratheria thomas*). This explains various primitive common features of xenarthrans and pholidotes. The xenarthrans may have been distributed worldwide in the Cretaceous, which would explain the occurrence of xenarthrans in the Early Tertiary of Europe *(Eurotamandua)* and Asia *(? Ernanodon)*.

The xenarthrans are a natural group, as serology and morphological anatomy confirm (for example, additional vertebral joints, placental type, structure of primitive skull). Hence their division into two separate orders, Cingulata and Pilosa, by M.C.McKenna in 1975 seems unnecessary. The phylogenetic origin is indeed undetermined, since the paleanodonts of the North American early Tertiary, once regarded as the parent group, is related to the Pholidota. Derivation of the group from the extinct Taeniodonta (=Ganodonta) – known as the "Ganodonta theory" – has long since been untenable.

The North American paleanodonts (for example, *Metacheiromys, Palaeanodon, Epoicotherium*) show, by a simplified dentition (reduction in number of teeth and in enamel, simple cylindrical molars) and distinct skeletal adaptations to digging, a number of resemblances to the xenarthrans. Apart from geological age, the absence of skin ossifications and additional vertebral joints, and the presence of a twofold bony connection of the pelvis to the vertebral column, make it impossible to trace the Xenarthra back to the Palaeanodonta. On the other hand, there are similarities with the Old World pholidotes that, according to R.J.Emry, do not only represent common primitive features. The similarities between the palaeanodonts and the xenarthrans are due firstly to common primitive features and secondly to independently

NEW WORLD EDENTATES

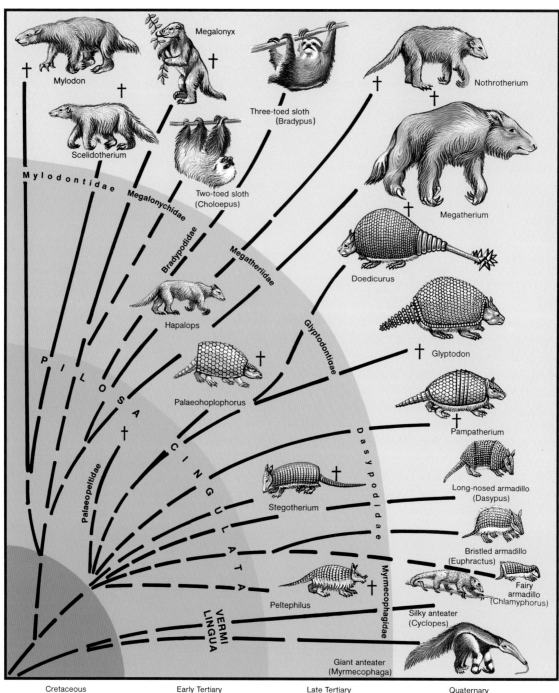

Family tree of the Xenarthra, a very ancient group of animals. The three suborders Vermilingua (anteaters), Cingulata (armadillos), and Pilosa (sloths) separated early – one reason for their difference in appearance.

acquired characteristics (insect eating, digging adaptations).

The classification of the Xenarthra is disputed. Usually, the armored Cingulata (= "Loricata", armadillos in the broad sense) and the "unarmored" Pilosa (sloths and anteaters) are separated. This dichotomy does not do justice to the special position of the hairy anteaters, whose early branching has recently been confirmed by fossil finds. Hence the need to follow R. Hoffstetter in separating them as the Vermilingua. If anything, a division into Cingulata plus Pilosa and Vermilingua would be justified.

The oldest remains of xenarthrans *(Utaetus)* are found in the Late Paleocene *(Riochiquense)* of South America, although somewhat more complete finds are known from the early Eocene. They are primitive members (Dasypodidae) of the Cingulata, already possessing skin ossifications. The simple pin-shaped molars have enamel caps, which are lost completely in later xenarthrans – when they are not completely toothless. Many fossil forms are known in the Tertiary and Pleistocene, indicating that the xenarthrans of the New World were once far more diverse and numerous than at present. In South America, which was isolated almost throughout the Tertiary, a wide variety of forms have emerged through adaptive evolution ("adaptive radiations") without appreciable competitive pressure, among which the most remarkable are the so-called giant armadillos (Glyptodontidae) and the giant sloths (Gravigrada).

The Glyptodontidae, which in the Quaternary produced true gigantic forms with rigid body armor and a bony tube for the tail, ranged into Central America and southern North America in the Pleistocene. These craetures (for example, *Glyptodon, Panochthus, Doedicurus*) were vegetarian, with a dentition of long, rootless, tripartite molars. Later glyptodonts were probably herbivores in open country, as is suggested by the highly articulated jaw, the form of the lower jaw, and the chewing musculature. As evidenced by the enormous olfactory lobes, these were highly macrosomatic forms, with a well-developed sense of smell. Several stocks can be distinguished (Palaeohoplophorinae, Hoplophorinae, "Sclerocalyptinae", Doedicurinae, and Glyptodontinae), whose sparse remains have been found in the early Eocene *(Casamayorense)*, documenting the great geological age of the glyptodontids *("Glyptatelus")*. The Oligocene species of *Glyptatelus* and the Propalaeohoplophorinae (including *Propalaeohoplophorus*) were small to medium sized. Among the numerous hoplophorine species, characterized by the cylindroconical tail tube, a form arose in the middle Miocene *(Friasense)* that eventually led to two giant Ice Age forms, *Hoplophorus* and *"Brachyostracon"*. Much the same is true for the somewhat later Doedicurinae, which also produced giant forms *(Doedicurus)* in the Pleistocene. The skull is high, the snout is very short, and the joint is set unusually high in the lower jaw. These giant armadillos, among which large forms of glyptodontines developed (for example, *Glyptodon*) in the very late Tertiary and in the Pleistocene, represented the true ungulates. These forms – even-toed and odd-toed – reached South America during the Ice Age by way of the Isthmus of Panama which formed in the very late Tertiary. The giant armadillos died out in the Quaternary; the last species became extinct in the early Holocene.

Other Cingulata that should be mentioned are the extinct Palaeopeltidae (for example, *Palaeopeltis*), the Steotheria (including *Stegotherium*), the Pampatheriinae (including *Pampatherium*), and the Peltephilinae (including *Peltephilus*) of the Tertiary and Quaternary in South America. The peltephilines are also referred to as horned armadillos, because of horn-shaped skin ossifications on the skull. In the Stegotheria, the dentition, reduced to a few tiny, peg-like molars, is limited to the most posterior section of the prolonged facial skeleton. The Pampatheria produced very large forms in *Holmesina* and *Pampatherium (= Chlamydotherium)*, of which the former ranged into the central United States.

Several lines are distinguished *(Dasypus, Euphractus, Cabassous, Tolypeutes, Priodones* and *Chlamyphorus)* among the recent armadillos (Dasypodidae). *Chlamyphorus* and *Calyptophractus* (="Burmeisteria") are the highly specialized subterranean fairy armadillos, whose phylogenetic origin is still doubtful. *Dasypus* did not spread to North America until the Late Pleistocene.

The Pilosa are here understood to include the extinct Gravigrada ("giant sloths") and the tree sloths (Tardigrada). Numerous species of Gravigrada were distributed in Tertiary and Quaternary South America. In the Late Tertiary, they spread as "island hoppers" to the Antilles (for example, *Megalocnus, Acratocnus,* and *Paulocnus* in the Quaternary of Cuba and Haiti), while some forms (for example, *Megalonyx*) went as far as Alaska in the Ice Age. Two main branches, the Mylodontoidea (Mylodontidae) and Megatherioidea (Megatheriidae and Megalonychidae) can be distinguished, unless the geologically oldest Gravigrada *(Orophodon, Octodontherium)* of the Early Oligocene *(Deseadense)* are separated under the name of Orophodontoidea (syn. ? Paragravigrada). Other remains of (early) Oligocene Pilosa have been described as *Hapaloides* and *Holomegalonyx*.

In the Early Miocene, at least ten genera flourished among which *Hapalops, Schismotherium, Planops,* and *Nematherium* are some of the most well-known. They are members of the Megalonychidae, Megatheriidae, and Mylodontidae, documenting the division of the edentates into branches. The Mylodontidae include the Scelidotheriinae in the Plio-Pleistocene *(Scelidotherium)* and the Mylodontinae (the Pleistocene *Mylodon),* very large forms distributed from Patagonia to-

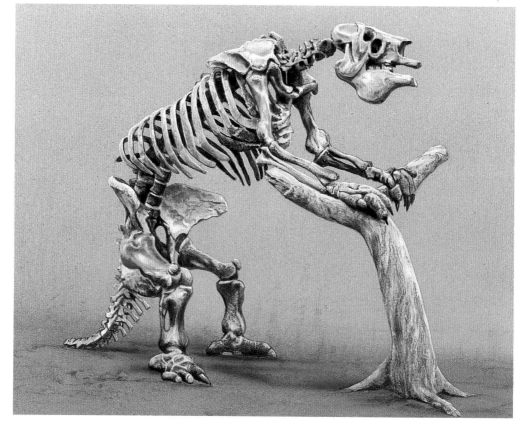

An extinct giant: Skeleton of a late Ice Age xenarthran, the giant sloth *Megatherium americanum* of Argentina. These animals stood up to 10 feet (3 meters) in height.

wards North America. The same pattern was followed by the giant Ice Age forms among the Megatheriidae (for example, *Nothrotherium, Eremotherium, Megatherium*) and the Megalonychidae *(Megalonyx)*. Some megatherians did not die out until the early Holocene (up to 8500 years ago), as is documented by cave finds in North America. They may have been exterminated by contemporary human beings. These caves yielded not only skeletal remains of giant sloths but also remnants of fur, and droppings, which serve as clues to their diet (the Gravigrada were markedly vegetarian).

In contrast to the terrestrial Gravigrada, modern representatives of the Pilosa, namely the tree sloths, are arboreal. Adaptations for their habit extend from reverse direction of hair growth and algae living in the fur, through the structure of the limbs, to leaf eating. The exclusively arboreal habit of the tree sloths is an adaptation to inundated forest terrain. Considerable differences in form between the two recent genera *Bradypus* and *Choloepus* suggest that the tree sloths, usually combined as Bradypodidae, are not a natural group but derive from two different ancestral groups. In that case, following G.J.Scillato Yane, *Bradypus* would be derived from (Proto-)Megatheriidae and *Choloepus* from (Proto-)Megalonychidae. Hence, *Choloepus* has recently been regarded as a member of the Megalonychidae, which are otherwise known only from fossil remains.

As termite eaters and anteaters (Myrmecophagidae), the Vermilingua are highly specialized representatives of the xenarthrans, whose phylogenetic position is controversial. It was a great surprise to specialists when G.Storch, only a few years ago, identified an anteater *(Eurotamandua)* in the Middle Eocene of Europe. They were surprised not only because of its occurrence outside the neotropics (South and Central America) but also because of the high degree of specialization of that form. Besides certain primitive features (for example, closed zygomatic arch, complete clavicle, supratemporal bone of primitive structure) and characteristics of the Xenarthra (additional vertebral articulation, sacro-ischium, shoulder blade with two plates), *Eurotamandua joresi* possesses a number of highly specialized features (toothlessness, arboreal diggers) that indicate a long phylogenetic separation. This would mean that the recent tree-living anteaters of South America *(Tamandua* and *Cyclopes)* are not – as has been supposed – comparatively recent descendants of terrestrial ancestors, but that they themselves have a long geologic history. *Palaeomyrmedon* of the late Miocene *(Huayqueriense)* of South America which close to *Cyclopes* according to S.E.Hirschfeld, documents the separation of *Myrmecophaga* and *Tamandua* lines as early as the Oligocene.

Modern Xenarthrans
by Walburga Moeller

Hairy Anteaters (Family Myrmecophagidae)

A black-brown sweep of hair moves slowly, stopping from time to time, through the dry, hard grass of the savanna in the hill country of central Brazil. The GIANT ANTEATER *(Myrmecophaga tridactyla)* is on the hunt for food. It holds its long, tubular head obliquely downward. Let us follow its trail as quietly as we can; we don't want its keen sense of hearing to detect our presence. Within 30 or 35 ft (10 or 12 m), sometimes even less, we can approach unnoticed, and recognize the tasseled tail and black and white marking that optically masks the body outline even in glaring sunlight, providing effective camouflage.

How will this shy solitary creature find a meal here? There are termite hills, the dried grassy plain extends to the horizon, and only small isolated groups of trees or shrubs are seen. Where the plants have been trodden down to leave a narrow wake behind the anteater, we see small circular holes in the turf, close together, as if made with an auger. Sniffing intently, it locates ant nests, unsheaths the great sickle-shaped claw of its forepaw, and carefully makes one or two incisions in the soil with the tip, as with a scalpel. The opening is enlarged somewhat by circular movements of the slender snout, and finally the tongue darts in several times, picking up a goodly number of ants and ant larvae. The tongue which can protrude as much as 22 in. (55 cm), is not sticky, as has sometimes been reported, but is wet with saliva. The tongue must transport the insect prey into the small mouth opening, assisted by minute horny papillae directed towards the rear, and which are also found at the cheeks. The retractor muscle of the tongue is highly efficient (motion pictures taken at a termite hill have shown giant anteaters executing 160 tongue strokes a minute). The ingested insects are masticated when they reach the extremely muscular stomach. The daily ant or termite requirement amounts to about 35,000 insects. Of course, not all anthills are attacked indiscriminately. Close study has shown that the more chemical deterrents an ant species can deploy, the less the anteaters like it. Their keen sense of small enables them to be very selective.

We remain nearby for an hour, and within that time we count 49 feeding sites in an area of 16 by 16 ft (5 by 5 m). The anteater proceeds very cautiously when seeking food. It sniffs the anthills, but does not destroy them, and takes only a small portion of the population. This instinctive technique is biologically sound because the animal is conserving its source of food. Termite hills are not destroyed by giant anteaters either, as was once believed. They expose just enough of the termite nest so that the prey can be reached with the tongue.

Some years ago, Wolf Bartmann, director of the Dortmund Zoo, studied giant anteaters in the wild at regular intervals. He selected a tract of 26 mi^2 (72 km^2) in Brazil in which the anteater population density was 1.2 individuals per square kilometer (meaning that 6 anteaters occupy 5 km^2; 1 km^2 = 0.36 mi^2). According to his observations, each individual has its own territory, with preferred rotation and sleeping place. The territories overlap considerably, but the animals are always found in the same places at certain times of day, often only a few yards from their position the day before. To make sure of this territoriality, Bartmann tagged some individuals and after three years, these anteaters were still observed in their old territories. Some individuals travel only a few hundred yards a day, others traverse up to about 1 mi (2 km). The giant anteater sleeps in a flat trench that it likes to prepare in the shade of trees or in the bottom of a depression. Covered with its broad tail for protection from wet and cold, it hardly stands out from the surroundings. About 10 a.m., it leaves its sleeping place at a slow trot, and spends most of the day on its feet. The extensive range does not exclude thornbush steppes, swamps, park forests or tropical rainforest. Wide watercourses are no obstacle, for giant anteaters are good swimmers. If a puma or jaguar pursues them, they go into a clumsy gallop, and only when cornered will they climb trees, but their enemies are faster and better armed. The anteater's weapons – used only for defense – are its extremely

The giant anteater, unlike its smaller relatives, is strictly terrestrial, seeking food chiefly by day (left). Within its range, it frequently encounters bodies of water, which it crosses easily by swimming. Even wide rivers are no obstacle (right).

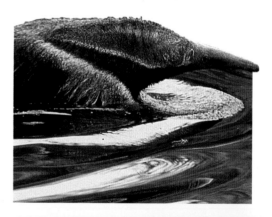

muscular forelimbs, with long sharp claws 4 to 6 in. (10–15 cm) in length on the second and third digits. For defense, the anteater stands up, supported on its hind legs and tail, and throws its arms around the adversary.

Carl Hagenbeck experienced the great physical strength of a giant anteater in England in 1864. Riding with the beast in a carriage on his way to set sail from Southampton for Hamburg, he was not alarmed. "There I sat with my four-footed neighbor, who suddenly tried to grab me with his clawed forepaws. First he had aimed at my legs, which he gripped so tightly that I had trouble getting him loose. During the entire ride, I had to contend with new attacks, and this was no easy matter, for the animal had the strength of a giant. I was completely exhausted when we finally arrived in Southampton."

David Attenborough reports, in the book *Life on Earth*: "... the giant anteater is far less innocent than he looks. The story is told that once the dead bodies of a jaguar and a giant anteater were found intertwined on the savanna. The anteater had been terribly mauled by the teeth of the jaguar, but its claws were sunk deep in the jaguar's back, and even in death their grip had not loosened."

The maned dog, which shares the anteater's habitat in large areas of South America, is of course a threat to the young, especially when, at the age of six weeks, they take their first steps. The anteaters have no litter nest. Immediately after birth, the young, often with embryonic membranes still attached, will climb up to ride on the mother's back. Multiple births have not yet been noted.

After a century of experience in zoological gardens, giant anteaters are no longer difficult to keep. However, this does not mean that they will necessarily reproduce. Worldwide, the successful breeding of giant anteaters is still a rare event. As a glance at the "International Zoo Yearbook" will show, out of an average of three to four young born, at most two a year have made it over a period of many years.

In the mid-1970's friends of animals and mammalogists took interest when, after three adult giant anteaters had been kept at the Dortmund Zoo only quite briefly, a baby was born and raised. After a 44-year interval, this was the first successful breeding at a zoo in German-speaking territory, and it was not to remain an isolated occurrence. Within ten years, Dortmund had eight births to report, involving three different pairs. The first breeding female raised a second infant, and the firstborn of another anteater was raised in the Director's household when it turned out that maternal instinct was falling short.

Giant anteaters get along quite well in an enclosure, and lend themselves to being kept in groups. A spacious outdoor run with heated Winter shelter is a natural meadow hedged by trees and shrubs, with bathing pool, sandpile, and wallows where lowland tapirs, capybaras, and jabiru storks are found. The giant anteaters, known to be solitary animals, may often be observed "tonguing" their neighbors in the enclosure, and both the

Giant anteater (*Myrmecophaga tridactyla*)
Silky anteater (*Cyclopes didactylus*)

▷ A giant anteater on the prowl in the dry grassy plains of the Brazilian highlands. With tubular head pointed obliquely downward, and sniffing incessantly, it searches the ground for hidden anthills. As soon as it finds one, it is ripped open with the sharp sickle claws of its forepaws. Then the long, mobile tongue, wet with saliva, is inserted and used to extract the delicious ants.

capybaras and the tapirs find this "grooming" agreeable. When the lactating nipples of a lowland tapir were discovered by the giant anteater, routine tonguing was not enough; the nipple fitted like a button into the small oral opening, and the source of extra nourishment was tapped.

Since there is no pronounced social behavior among the anteaters in the strict sense of the world, contact between the sexes occurs only at the time of estrus. When a female is in heat, the male gets wind of it and follows her trail. There is no limitation, even in the wild, to any particular time of year. The male first shows his interest by careful touching and laying his head on her back. Playfully, by small approaches, the two get acquainted, and their behavior is coordinated in the form of harmless blows and pinches until mating finally ensues. At the male's first attempts to mount from behind, the female will pace forward, but finally drops to her knees. However, that's not the right position. A brief tussle follows, and the male urges his partner on until she is lying on one side. Mating may take 5 to 10 minutes and may be repeated on the same or the following day. But after that, the partners separate, and each goes their own way.

When a female is in heat in the enclosure, there is dissension among the males, so that they must be separated. The female not infrequently takes the more active role in courtship. The gestation period is 180 to 190 days. Even shortly before giving birth, a female's pregnancy is not apparent, and in the zoo there may be surprises.

Wolf Bartmann was in luck again when on a routine round near the South America exhibit he heard shrill whistling sounds, the unmistakable abandonment call of a very young anteater. The newborn was being rejected by its mother, and would not have sur-

The elongated tubular snout (right) and the extensible tongue (below) are radical adaptations of the giant anteater to its feeding habit. The tongue is a highly efficient "tool," for it can lap up some 35 000 ants and termites in the course of a day.

vived the day. As it was, this little bundle, weighing 2.5 lb (1.16 kg), was rescued in time and brought into the warm house. This healthy male baby immediately exhibited the grasping reflex; the strong little beast tried to hang on by its pointed claws anywhere on the human body and to bury its slender head as if in the mother's coat. Taking hold is the ultimate necessity for survival in the wild, otherwise the newborn loses touch with the mother, and this release is triggered instinctively with respect to any foster mother. As soon as contact was interrupted even very briefly, the animal would wail at high pitch until he had "cloth contact" again. He was happiest when being carried around, and it took much patience to accustom him to his box with toweling and wool blanket. Floor heating and a red light maintained 93°F (34 °C) in his vicinity, approximately the body temperature of the mother, which when nursing would also cover the baby with her bushy

tail. A special nipple was found for the small mouth slit, with which he could take his formula only with 4 to 6 in. (10–15 cm) of tongue hanging out to one side. Supported by two practiced keepers, he had his formula 5 times a day, corresponding in fat and protein content (13% and 11%) to his mother's milk. In four weeks, he doubled his weight. This monthly gain continued for half a year. Eating food from a dish was unattractive for a long time, although the young that had grown up with their mother in the park would join her in eating by about the 12th week of life. So additional nutrients were added to the formula.

At 6 to 8 weeks, he began to explore his surroundings. Around the house, he stuck his wet tongue into every accessible nook and cranny, and the kitchen especially – his temporary abode – was a never-failing adventure for his delicate nose. The fragrance of vanilla was especially attractive – why not try putting some vanilla sauce in the dish? That was a breakthrough, Wolf Bartmann writes. The half-year-old anteater, who had gained almost 22 lb (10 kg) on the bottle, smelled the vanilla sauce, emptied the theretofore despised receptacle, and thenceforward subsisted on an approved mixture of milk, whole-grain porridge, dog kibble, egg yolk, and honey. Now it was time to think of putting the little one outside; he was moved to the Winter shelter, and during the day he spent hours getting acquainted with three half-grown conspecifics.

Frequent diarrhea of giant anteaters was a concern to keepers until examination of feces in the wild revealed the importance of roughage. With its concentrated insect fare, the anteater occasionally picks up small fruits and above all an appreciable quantity of earth, sand, tiny twigs, and small stones. Thus, an admixture of sandy earth (about 25% of the quantity food) finally turned things around for healthy digestion in the zoo.

Another flashback to the story of our baby anteater. When he was hardly a week old, he was inclined to play, especially after feeding. In cloth contact with his keeper, he would take hold everywhere with his sharp claws; touching his head always made him duck and raise his arms over it protectively, as has been observed in playful combat between mother and baby also. Little pouncing and prancing leaps were favorites at two weeks. The little anteater liked to groom his fur, still short and marked on much the same pattern as the adult, with hind foot and forepaw. Before curling up in his box to sleep, he would scrape a couple of times on the deck and then flap his tail over his body. In strange surroundings, it always took a long time before he would move. But as soon as anyone approached him, he would try to climb up on their legs or arms, and it was difficult to disengage oneself. Self-assurance had increased by 8 weeks. It now sufficed to have a keeper's legs alongside when exploring the garden, which was scanned with claws, snout, and tongue. With forearm raised threateningly and erect mane, he would

The giant anteater babies are carried on the mother's back. Even half-grown youngsters, almost as big as the mother, continue to be tolerated.

▷ The little ones hold on very tight lest they be lost in the wide-open spaces.

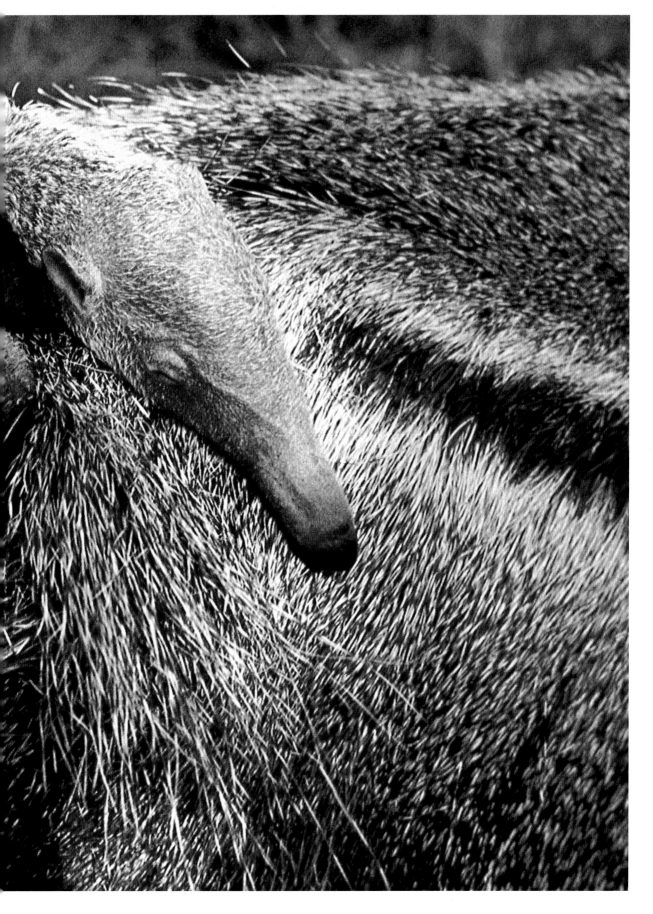

keep the house dogs at a distance. Astonishingly enough, at the first sign of ants, the whole program of behavior was launched. "With circling lunges the nose drills into the turf. With audible snuffling the nest is discovered, a few digging movements, alternately with right and left clawed forepaw, expose it, and already the tongue licks forth from the opening of the mouth in rhythmic sequence," Bartmann writes.

In natural development, detachment from the mother begins with short, careful excursions. At any slight disturbance, the youngster will return to the mother in panic, and cannot climb onto her back fast enough, sitting astride at the base of the tail. In that position, the markings of mother and young fit each other seamlessly.

Adolescent animals often try their climbing skills on the treetrunks of the enclosure with their protective wire mesh. Their play urges are focussed entirely upon the mother; first the youngster dances around her and challenges her with mock attacks. The mother tolerantly accepts all reactions; careful and protective, she often lays her head on the youngster's back. At nine months old, it still enjoys being carried about, but she tries to avoid his clutches, for a load of 65 lb (30 kg) on her back brings her to her knees. So long an attachment to the mother as in the zoo would hardly occur in the wild. At about one year, a giant anteater is fully grown, but still needs six months or so to attain full weight of from 110 lb (50 kg) to at most 132 lb (60 kg).

Giant anteaters are very popular in zoos. Out of 89 individuals worldwide at the present time, 19 live in the Federal Republic of Germany, and 9 in the Dortmund Zoo alone. In its natural range, hunting of the giant anteater has taken on dangerous proportions, affecting Bolivia in particular. As yet there are strict legal protective measures only in Brazil and French Guiana. What has become a threat to the survival of this species is the clearing of land by humans. In Guatemala, agriculture, highway construction, and grazing have contributed to the destruction. Permanent habitat for the giant anteaters in their natural surroundings is to be found only in national parks and preserves.

Somewhat overshadowed by their larger relatives are the silky anteaters or TAMANDUAS *(Tamandua)*. They are less noticeable when they seek the nests of ants and termites in the branches of the trees or sleep in a forked branch, even by day, using their muscular winding tail for anchorage. Thanks to adaptability regarding habitats, the two tamandua species inhabit not only the moist highland and lowland forests but also thornbush savannas, park terrains, dry deciduous forests, and the llanos, the almost treeless plains of South America. They are almost everywhere more abundant than the giant anteater, whose range approximately coincides with their own. Food supply and predators probably determine when and how long the tamandua will be at large on the ground. When nature wardens in Venezuela took a census, they found 64% of the animals counted in trees (three-quarters of these in evergreen hardleaf), 20% in savanna or

Unlike the giant anteaters, the tamanduas live chiefly in trees, where they climb about at ease in search of ant or termite colonies. They will also eat small fruits now and then.

croplands, and 5% in deciduous forest. As the highway network keeps growing, more and more tamanduas perish on the roads as they wander about the country. The population density in the llanos and forests of Venezuela is estimated at three individuals per square kilometer ($1 \text{ km}^2 = 0.36 \text{ mi}^2$).

The tamanduas move deliberately, on solid ground as well as when climbing. The curling tail as "fifth hand" delays their progress when climbing – some time is always spent in testing for stability. Only half the sole of the hind foot makes contact; and in front, only the outer edge of the hand, with claws retracted, makes contact. The claw of the predominant third finger is about 2 in. (5 cm) long, and the second and fourth fingers also have powerful claws; only the first finger is much reduced. On the hind foot, the first of the five digits is somewhat smaller. With a similarly specialized forepaw, the giant anteater can gallop pretty well, while the tamandua seems rather helpless on the ground. Hans Krieg describes it in his South America travelogues as "a cute little elf, with gleaming yellowish-white coat, and dark stripes on the back like suspenders." Such is the appearance of an individual that is upright in defense, arms outspread, uttering a soft, breathy hiss. The attacker need not fear an embrace, but the extremely sharp-pointed claws can inflict severe damage.

Tamanduas are wantonly killed by human beings, who make no use of the flesh or fur, far more frequently than by predators. The Indians very rarely eat tamandua. "Caguaré," they call him, which in the Guarani language means "forst stinker," and refers to the disagreeable odor exuded when the posterior glands produce a musky secretion that adheres to the skin. According to Brunhild Encke of the Krefeld Zoo, this penetrating odor of the tamandua will disappear only after a hot bath.

Within the genus, we distinguish the NORTHERN TAMANDUA *(Tamandua mexicana)* and the SOUTHERN TAMANDUA *(T. tetradactyla)* as two separate species. The common boundary of their ranges extends along the chain of the Andes in Ecuador and Venezuela. The northern tamandua occurs from southern Mexico throughout Central America as far as the Pacific slopes of the Andes in Colombia and Ecuador and in northwestern Peru; it always has black markings on the trunk. On a pale white, yellowish-brown or golden background, it wears a black "vest" with wedge-shaped neckline in back. The range of the southern tamandua begins southeast of the crest of the Andes. The fur coloring of the southern tamandua, however, is not uniform throughout its range. Completely black or dark brown individuals are found in Peru, on the eastern foothills of the Andes in Ecuador, and in the delta of the Amazon, while a vest-like marking occurs only in the southeastern part of the range. Transitional forms exist in other areas.

The southern tamandua generally has longer ears and a broader skull than the northern species. For clear identification of the two species, Ralph M. Wetzel, after years of study throughout the range of the genus *Tamandua*, worked out criteria whereby each subspecies of the southern tamandua can be distinguished from the northern varieties.

The coat of both species is dense and short, the under side and the tip of the tail being hairless. There are as yet no long-term observations concerning reproductive behavior. After a gestation period of about six months, a single young is born. It is carried around on the mother's back for about three months. Close physical contact between mother and baby is interrupted about the 10th or 12th

The southern tamanduas have an extensive range, in which several color variants have developed. Besides the predominant light-colored form, all-black specimens also occur in certain regions.

week, when the youngster begins to find food for itself. Occasionally mother and baby have been found curled up together in a burrow; apparently, nursing mothers continue to alternate between arboreal and terrestrial habit. A Dutch observer who had a tamandua briefly in his keeping reported the following interesting piece of drinking behavior: Once the animal was back in its accustomed environment, it immediately climbed a tree and took a long drink of water from the leaf cups of a bromeliad.

In zoos, tamanduas are a rarity. In the 1930s, an individual in the London Zoo survived for four years on a diet of milk, eggs, and chopped meat. For decades, that success remained the exception. When the Krefeld Zoo started keeping tamanduas in 1968, it had very scanty experience to draw upon. With favorable conditions and sympathetic care in his own house and garden, Dr. W. Encke solved the problem. After a brief phase of home life, individuals would live for over seven years at the Krefeld Zoo. The restlessness of the tamanduas, due in the wild to their specialization in very small prey, requires a field of action in order for them to thrive in captivity. Brunhild Encke tells how a tamandua was busy for hours with pulpwood, cardboard boxes, and paper, shredding it all with his sharp claws, and then laying down to rest in the litter. In the absence of opportunity for exercise the animals will curl up and go to sleep after meals, and quickly put on fat. A supplemented diet of whey, honey, fruit, and a little meat was helpful. The food must be suited to the large salivary glands, which become inflamed if secretion is insufficient. Such inflammation is dangerous and among the causes of premature death. Hence mixed feed in the form of small bits to which earth, gravel, and wood have been added as ballast is recommended.

There was extensive playful roughhousing between older and younger individuals at the Krefeld Zoo. The return of a briefly isolated male precipitated regular greeting ceremonies in the group. With visible excitement, he was sniffed and licked over, and in turn attempted to mount his mate. She had spontaneously come into heat; he embraced her and pressed her against his abdomen. At the Krefeld Zoo, permanent matings were observed. "The bond between mates is such that no other male will try to mount a female in heat," Brunhild Encke writes.

Female tamanduas come into heat every four weeks. The signs are sluggishness, poor appetite, and more need to sleep. Whether the periodic skin changes observable in the giant anteater as well are related to the reproductive cycle has not been established, but is likely enough. At regular

The size of a squirrel, but by no means as agile, is the silky anteater. In climbing, it takes safe hold with its curling tail and pads on the hands and feet (above). As a specialist feeding primarily on ants and termites, the silky anteater has a very long tongue (below).

intervals, the skin of the legs and below the neck takes on a brownish-red color and forms "goose pimples." The tamanduas showed good orientation and an excellent sense of smell; they responded to their names, and seemed generally sensitive to soft sounds. Vision is inferior to the other senses.

After sixteen years of effort for good tamandua living conditions, breeding at the Krefeld Zoo succeeded in 1984 with the first live birth in Europe. The three-day-old tamandua weighted 1 lb (450 g), with a body length of about 9 in. (22 cm), and tail length of about 7 in. (18 cm). Surprisingly, immediately after birth the baby would not maintain physical contact with the mother only, but climbed about on the other two inhabitants of the enclosure as well. It was therefore indicated to leave the mother and baby alone in their accustomed quarters, as Paul Vogt writes. During the mother's rest periods, the baby would find the pectoral nipples and nurse on its back between her front legs. Gaining weight steadily, it thrived very well. A uniform temperature of 79–82°F (26–28 °C) was found favorable. With an average body temperature of 95.7–96.3°F (35.4–35.7 °C), distinctly lower than that of many mammals, tamanduas are very sensitive to major fluctuations in the ambient temperature.

A skillful climber in the tropical forests of Central and South America, the Pygmy anteater *(Cyclopes didactylus)* is wholly adapted to arboreal life. Its hands and feet are the most specialized within the family. It will not voluntarily come down to the ground; it sleeps all day curled up in a ball in a fork or hollow of a tree, and limits its periods of activity entirely to twilight and darkness. This squirrel-sized animal, with curling tail of about the length of its body, shows a special preference for the kapok tree *(Ceiba pentandra)*. It affords excellent protection against enemies, for the animal's silky soft, golden-yellow to silver-gray fur is hardly distinguishable from the kapok. Thus the animal is protected from attack by harpies, owls, and other predatory birds. If attacked by serpents or cats, the silky anteater sits up and strikes with its sharp claws, while anchoring itself securely with its muscular, curling tail, the underside of which is hairless towards the tip. On the forepaw, only the second and third digits are developed, and fixedly fused; the powerful curved claws – that of the third digit considerably larger – are retracted against the palm. The latter, as a prominent, callused pad, serves for opposition in grasping

(Left) Tamandua mothers also carry the baby around on their back. (Right) In feeding, the tamandua is not exclusively limited to its tongue. Often it will bring food to the mouth with its front claws, having moistened its hands with saliva.

branches. The four digits of the hind foot are firmly joined, except for their almost equally large claws, by a callused sole. The sole surface broadens on the inside to make an elastic rounded cushion, again serving for securing a grasp, in the case of the hand. The scientific name *Cyclopes* (Greek *kyklos* "circle," Latin *pes* "foot") is evidently an allusion to this feature.

The pygmy anteater's mouth opening is comparatively larger and the snout shorter than in the tamandua. It has the habit of conveying its prey – bees or wasps as well as ants and termites – to the mouth with the claws of its forepaw; the worm-like tongue is also employed in eating. First it wets the paw abundantly with saliva. Like all species of the family, the pygmy anteaters are toothless. As in the case of the tamandua, distinct tooth sockets or their vestiges indicate that the teeth have been lost in the course of phylogenetic development. The only known utterance of sound is a soft, piping hiss. The silky anteater has very sharp hearing and can perceive frequencies around 10 kHz (with a range from 0.6 to 62 kHz at 46 db). The rearing of the young is not only the female's responsibility. Both parents regurgitate crushed insects to feed the baby. The silky anteater is the only species of the family having two pectoral and abdominal nipples. There are as yet no reliable data on the birth of more than one young.

The pygmy anteater is supposedly not rare in its range. Its flesh is scarcely eaten, but even these harmless creatures are quite often killed by humans. A pygmy anteater lived for eighteen months in the Berlin Zoological Garden.

Sloths (Suborder Pilosa)

Sloths have round, flat faces that seem rather human and attractive to many people, but the deliberateness of their movements is not esteemed by humans who seem to admire speed. Thus, we marvel at the gibbons, and are astonished at the lightness and elegance of their swinging locomotion through the trees; it makes us aware of our own limitations.

The sloths through their millennia of development, have if anything retarded their motion in time and space. Their deliberateness has allowed them to adapt to their surroundings, and has afforded them a large measure of safety. Their tenacity of life was noted by Alexander von Humboldt in numerous experiments early in the nineteenth century; they resist extinction more effectively by remaining in the tree tops than the agile and alert capuchin monkeys in the same habitat.

The sloths hang in the treetops by the curved claws of their four limbs. They have excellent camouflage and for the most part escape the great birds of prey. In the struggle for survival they are undoubtedly successful; in large parts of South and Central America today, they account for up to 25% of the mammalian biomass. Humans have become their worst enemy, by destroying their habitat with indiscriminate clearcutting in the forests or by running them over with cars as they slowly cross the highway.

In books of natural history, the sloth has long been pictured as a freak of nature. Many imaginative drawings and descriptions, distributed before Alexander von Humboldt's research voyages to South America, probably go back to the earliest reports of these strange creatures. The observations noted down by the Spaniard Gonzalo Fernández de Oviedo y Valdés are rather amusing today. In 1526 he wrote (freely translated in part): "The sloth takes the whole day for fifty paces, is about as long as it is broad, and has four thin legs with long nails, which cannot support the body. It prefers climbing trees. Eight, ten, even twenty days it will remain on the topmost branch, and what it eats nobody knows. Having myself had it by me, I am convinced that it lives upon air alone. I have never seen anything uglier or more useless than the sloth."

The author of a French travelogue of 1555 evidently took quite a different view. He represents it with a bear-like body and a friendly, human face, and opines that to investigate so marvelous a form would be going too far, for only the Creator himself knew why He had made the sloth thus and not otherwise.

The human facial features of the sloth probably also moved Carl Linnaeus in the eighteenth century to place the sloths as relatives of monkeys, apes, and humans among the primates. That high rank was to be granted them only briefly, however, for so much laziness so close to the "elect" was considered unbecoming. Eventually the sloths were endowed with an order of their own, the Bruta ("the unready ones").

The sloths were deemed an incompleted work of nature by the English author Oliver Goldsmith in 1825, and the French zoologist Georges Cuvier added in 1837 that nature had been pleased to produce something imperfect and grotesque in jest. William Beebe dealt at length with the three-toed sloth in his papers of 1925 on the tropical forest, and came to the conclusion that it was well adapted to life on Mars, where the year had 600 days. In its total self-absorption, it stood even below the reptiles, and had no right to live on Earth. Evidently humans do not find these creatures very accessible, and so are brought to absurd judgments.

A group of American scientists began to shed light on the habits of some species of sloths on Barro Colorado Island in the late 1930s. Colonies of 50 to 100 individuals were observed for long periods, and the descriptions of nineteenth century naturalists were thereby corrected and extended.

Slowly, thoughtfully, almost in slow motion, the sloths sway through the branches of trees in the tropical rainforest. Belly up and back down, they move like sleepwalkers from tree to tree. Yet this form of locomotion, unique in the animal kingdom, is not due to "sloth," but to an exceptional environmental adaptation. Nature has equipped them with a unique coat of fur, in which two species of blue-green algae of the genera *Trichophilus* and *Cyanoderma* flourish. Living conditions in the fur, favored by the hot, humid climate, allow the algae to multiply very rapidly, lending a greenish sheen to the sloth's coarse gray-brown pelt, thus providing excellent camouflage among the branches. Between sloth and algae there is a true symbiosis. Also, the sloth's fur is worn "upside down"; parting does not run along the spine, but along the ventral midline. Rain, therefore, does not collect in the fur of the belly, but drains off to either side.

We find sloths almost everywhere in the forested regions of Central and South America (except in the southernmost portions). Whereas many mammals of comparable size are long since extinct in these areas, the sloths have been able to survive. What they lack in alertness and speed is compensated by advantages of physical structure and outward appearance.

The modification of the limbs – the arms are longer than the legs in some species – and the narrow hands and feet that terminate in long sickle-shaped claws, render the sloths extremely capable among the branches. Fingers and toes are fused and reduced in number to two or three. On-

Slow-motion progress. Hanging from a limb by the curved claws of its extremities, the three-toed sloth climbs about in the trees with great deliberateness. This "upside down" posture is perfectly normal for sloths as evidenced by the apparently wrong parting of the hair. The hair falls from the chest and belly to the back which has the advantage of allowing the rain to run off unimpeded to either side.

ly when the food supply fails will sloths move to a neighboring tree by way of the branches. If that option is not available, they will hesitate for a long time before abandoning their tree. Their sureness of grasp on the limbs is contrasted with their extreme helplessness on the ground. Lying prone, they will laboriously drag themselves forward yard by yard. When they have arrived at the next tree, a rest period will be followed by a heavy meal. They have no incisors and simply crop the vegetable fare with their hard, horny lips.

The chewing surfaces of their small enamelled teeth are worn from grinding plants, but the open pulp cavities permit the teeth to grow in continuously. The three-toed sloths have the most unvaried diet. They have a stomach of highly complex structure – the right and left halves are each divided into several compartments. In the left half, thick horny membranes form three chambers that are incompletely separated from each other. The largest, into which the esophagus opens, comprises about a third of the stomach area and contains pregastric glands which produce secretions that decompose leaves. In the right half of the stomach, which is shaped like a horseshoe, there are two chambers that communicate only through a narrow opening. One chamber contains chiefly fundic glands, and is therefore referred to as the pepsin stomach; the strong muscle wall of the second chamber with its ridges and papillae covered by horny skin is reminiscent of the stomach of the giant anteater. The two halves of the stomach can be closed off from each other. There are as yet no adequate studies of the exact route of food through the several divisions of the stomach. The material

(Top) Male three-toed sloth with dorsal markings. (Middle) Hand of a three-toed sloth, with long, slightly curved claws. (Bottom) Three-toed sloth.

▷ The tree-toed sloth is an acrobatic climber.

▷▷ Mother sloth with baby.

Leaves and buds of many species of trees are the staple in the three-toed sloth's diet. The clawed hands are used quite dexterously to hold food.

probably remains in the stomach for a very long time, for the intestine is remarkably short. There is only a vestige of a cecum.

Some of the important internal organs in sloths have undergone a unique transposition. The liver has rotated about 135 degrees to the right towards the back, and is completely covered by the similarly displaced stomach, so that it does not touch the abdominal wall. The spleen and pancreas follow this rotation, and are not located on the left as in other mammals, but on the right, near the exit from the stomach. The spleen serves primarily as a blood reservoir. It is difficult to interpret this finding since a blood reserve, to be tapped in case of very great exertion, is not required. Respiration and circulation follow a very easy rhythm, as do all of the sloth's life processes. Sometimes they do not evacuate for a week.

Sloths have a variable temperature, fluctuating between 75 and 91°F (24 and 33 °C), depending on the outside temperature. On the average, sloths spend 15 hours a day sleeping; they drop their head on their breast and hang up their arms and legs quite close together, thus protecting themselves against needless loss of body heat. Sloths do not drink; their need for moisture is supplied by their juicy leaf diet and occasionally licked dewdrops. Hans Krieg found that the bladder is uncommonly large. Probably it too serves as a moisture reservoir. Unfortunately, sloths are enthusiastically hunted, especially in Brazil, for their flesh has little fat and is similar to mutton in flavor; also, the strawlike fur is desired for cool saddlecloths and the claws are worked by the Indians for necklaces.

Today five species of sloths live in Central and South America. They were long regarded as members of one family, chiefly on the basis of many similarities in outward appearance. More recent studies, however, especially of extinct terrestrial species, have revealed two geologically ancient lines of development indicating that the living sloths are not a natural group. Two families have been established, the three-toed sloths, with three species, and the two-toed sloths, with two species. In number of cervical and thoracic vertebrae, in dentition, and in coat, and down to the microstructure of the individual hairs, there are substantial differences between the two.

THREE-TOED SLOTHS *(Family Bradypodidae)*. The family of three-toed sloths, distributed from Central America to northern Argentina, contains only the one genus, *Bradypus*. The best-known and commonest species is the TRUE THREE-TOED SLOTHS, or ai *(Bradypus tridactylus)*. The name "ai" is derived from their two-syllable cry, heard especially in March and April, the mating season. Besides this long-drawn-out "A-iii," a distinct sound is heard when the creatures are unhappy, and – especially from the young when clinging to the mother's belly – a gutteral purring sound.

The facial expression of the three-toed sloth is more rakish than phlegmatic.

The habitat of the three-toed sloth is the tropical forest. Their diet is more limited than that of their cousins, and therefore they will remain longer on a given tree that affords them the right choice of

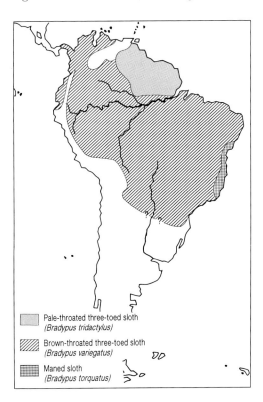

leaves. They live a solitary life and are both diurnal and nocturnal.

It is astonishing that the ai are good swimmers and seem to like the water. If they are attacked on the ground, they try to grasp the adversary with their claws and to inflict serious injury. If they are threatened while in tree branches, they are able to move with surprising speed. They develop a speed of flight one would not have credited to them. Their chief enemies – apart from humans – are large snakes and birds of prey, especially the harpy *(Harpia harpyja)*, beneath whose roost sloth bones are often found.

Three-toed sloths have always been only transient guests in zoological gardens. In the Brazilian forest, however, it is not difficult to tame the three-toed sloth. For several years, sloths were house

The baby sloth's cradle is its mother's belly. It is carried about in this unusual position for several months.

and garden companions of Hermann Tirler, who had the opportunity to observe their family life as well, for his pair blessed him with offspring. It turned out that the parents remained solitary. The male took no part whatever in raising the young. Even the mother seemed often to forget her baby when climbing; the youngster who clings tightly to her breast often gets caught with no clearance. Left entirely to its own devices, it is ever on the alert, and at the instant when a branch threatens to brush it off from its "ambitious mother," it lets go, scrambles "around the obstacle by itself, and goes aloft in a flash, there to head off the conveyance and hop aboard again."

The other two representatives of the genus hardly differ from the ai in habits, but they do in distribution. The BROWN-THROATED THREE-TOED SLOTH *(Bradypus variegatus)* lives over an enormous range extending from Guatemala and Honduras to northern Argentina. The MANED SLOTH *(Bradypus torquatus)* is confined to eastern Brazil. While the first-mentioned species still appears to be quite abundant, the maned sloth is seriously threatened. Even in the Pocodas-Antas Preserve, only 15 individuals were counted in an area of 18 mi^2 (50 km^2). Outside the preserve, the diminished populations are severely endangered by hunting, and especially by the irresponsible clearing of the Brazilian coastal forest.

TWO-TOED SLOTHS *(Family Choloepidae)*. This family also comprises a single genus, *Choloepus*, with only two species. The two-toed sloths differ from their relatives not only by the number of toes but by some other external peculiarities as well: The nostrils are larger and farther apart; the eyes are larger and more prominent; the small ears are hardly visible under the long coat; the distinctly shorter neck is less mobile; the tiny tail is hidden in fur; and the long slender limbs are only slightly longer in front than behind (the three-toed sloth's forelegs are longer by about 35 percent). The two-toed sloths are also chiefly nocturnal and less specialized. They move more frequently from one tree to another, and eat not only leaves and buds but also slender twigs, blossoms, fruits, and even root tubers and small prey. On the ground, they are exposed to predators, for example jaguars, ocelots, and other felines.

The commoner of the two species, the TRUE TWO-TOED SLOTH, or UNAU *(Choloepus didactylus)* inhabits the forests of northern South America in as yet comfortably large numbers, down to the Amazon basin. The consistently somewhat smaller, HOFFMAN'S TWO-TOED SLOTH *(Choloepus hoffmanni)* is less well known, although its range is more extensive, from Nicaragua to Peru and central Brazil.

For decades, the two-toed sloths have been kept in many zoos. They quickly become accustomed to a mixed vegetable diet, and will readily accept even boiled eggs. An unau pair that lived in the Prague Zoo for more than 15 years was very tame: at feeding time they would climb to meet their keeper and would take lettuce, sprouted wheat, and fruit from his hand. Males and females are difficult to distinguish; the scrotum is not externally visible. There is probably no fixed mating season; births have been recorded in all months of the year except April, September, and November. In mating, the animals simply hang from a branch by their arms and turn to face each other. The gestation period is stated by Zdenek Veselovsky to be five months and twenty days. Birth also occurs in the climbing tree. The female giving birth will often hang at full length, anchored only by the arms. The fully developed young is born head first without an amnion. Breathing hard, it immediately begins to help itself until it is able to cling to the mother with its claws. Only after the umbilical cord has been bitten through will the female retract her legs.

The newborn measures about 10 in. (25 cm) and weights 11 to 14 oz (300 to 400 g), its woolly fur is 0.4 in. (1 cm) in length on its dark-colored back; and the eyes are open. The dentition is complete, the teeth already having their brown coloration and attaining their full size within the first year. In the first four weeks, the baby sloth remains hidden in the fur of the mother, who hardly moves during that time. Then it begins to take an interest in its surroundings; it releases the claws of its forepaws from the mother's coat to snatch at nearby branches and to sniff anything within reach. The youngster voids feces and urine into the mother's fur at intervals of up to eight days, and she removes the excretions very thoroughly. At ten weeks, the baby takes part int he mother's meals for the first time, but in so doing it continues to cling to her by its legs. At nine months, it will try standing on its own feet, or rather hanging by its own hooks. Each attempt to return to the

Sometimes sloths have no choice but to come down to earth when they want to move from one tree to another. They do so most reluctantly, because they are quite helpless on the ground. Sliding on their bellies, they make progress with great difficulty.

▷ When sloths take a nap, they do not sit or lie down like most animals, but hang themselves on a limb by their sickle-clawed hands and feet and pull their heads in.

▷▷ The arboreal sloths are by no means shy of water. Actually, they prove to be quite able swimmers.

mother is rebuffed by her, very decidedly. If the youngster persists, she will even drive it away by biting at it.

Adult size and weight are not attained until the age of two and a half years. In captivity, two-toed sloths have a long life expectancy. An unau at the San Diego Zoo lived to at least 27 years and 9 months, and a Hoffmann's two-toed sloth there was no less than 31 years and 8 months. A life span of 30 to 40 years is thought probable for both species.

Armadillos (Family Dasypodidae)

Among the mammals living today, ossification of the dermis to form a bony skin armor occurs only among the armadillos. The belts or bands referred to by the name of the suborder, Cingulata, which differ in number form species to species, are open ventrally and only remotely suggest belts or girdles. The Spanish name *armadillo* "armored one,"

Two-toed sloth (Choloepus didactylus)
Hoffman's two-toed sloth (Choloepus hoffmanni)

adopted in the English-speaking world as well, is more apt. Unlike turtles and tortoises, the coat of mail is interrupted by several folds of skin at the "midriff", providing more mobility.

Immediately after birth, the armadillo's body is completely covered with epidermal scales. However, the newborn remain outwardly soft and pink for only a short time. In the first phase of growth, ossifications are formed beneath the scaly coat, which takes on a gray, yellow or brown color. The ossifications are dermal formations, which gradually enlarge on the dorsal parts of the body to hard, polygonal bony plates. Differing greatly in size and shape, the plates join to form a fixed armor for the head, shoulders, and pelvis, and several bands around the body. The ossifications also harden on the anterior surfaces of the limbs and on the tail, while on the ventral surfaces they are considerably reduced. The ventral skin of the adults is hairy to varying degrees. In some species, hard bristles penetrate the bony plates and extend between the horny plates that form on them. All species of armadillos have assemblages of glands in small posterior skin pouches. Besides, the bristled armadillos have so-called dorsal glands with a yellowish secretion having a disagreeable odor. These glands are clearly distinguishable at two to four small orifices in the pelvic armor. The ossified exoskeleton of the armadillos affords protection from predators and is an effective adaptation to burrowing for many species.

Another peculiarity of the armadillos, otherwise found only among toothed cetaceans, is the great variation in number of teeth between species and within the same species. The jaw consists of similarly shaped cylindrical teeth (homodontism) with open pulp cavities. The teeth, therefore, always continue to grow. Without enamel, surrounded merely by a jacket of cement, the teeth are quickly worn down by use. The largest and sturdiest are the teeth of the bristled armadillos, which often eat carrion and subterranean plant parts as well as insects, snails, and worms. Solid food constituents are extremely important to the health of bristled armadillos. Malformations commonly observed in captive individuals were traceable to an excessively soft diet, for the teeth were not ground down sufficiently and, by continuing to grow, injured the jaw and the gums.

Insects, mushrooms, and fruit are preferred in the diet of the softer armadillos. At birth, the teeth are covered with enamel. In the embryo, an enamel organ and four to six primordial incisors have been found, which are later absorbed. Of the eight buccal teeth, seven double-rooted milk teeth are replaced in the nearly adult animal. This change is limited to the soft armadillos.

The largest species living today, the giant armadillo, has up to 100 teeth, but the dentition is too weak for masticating food. Ants and termites, upon which the animal chiefly feeds, are picked up with the strap-like tongue which is closely set with papules to which the small prey adhere. Much as in the anteaters, the salivary glands in the lower jaw are very large. The viscous saliva collects in so-called "reservoirs" for instant availability when needed. The armadillo tongue has fewer taste buds than that of most mammals. However, the sense of smell – indispensable for finding food – is especially well developed, as the development of the forebrain clearly shows. The sense of sight plays a subordinate role for this predominantly crepuscular and nocturnal animal.

Especially in the United States, and now also in the leading universities of South America, armadillos are used extensively for physiological experiments. In the common long-nosed armadillo, for example, oxygen consumption is considerably lower than in domestic cats or rabbits of comparable size. It can hold its breath for up to six minutes, even while digging rapidly, which keeps dirt out of the respiratory passages. Wide bronchi and air passages serve as an air reservoir.

The nine-banded armadillo is a good swimmer; it paddles in the water like a dog. It is able to lower its specific gravity and gains buoyancy in the water by swallowing air and inflating the stomach and intestine.

As in the anteaters and sloths, the circulatory system of the armadillos features very finely branched arteries, observable especially in the limbs, like a close-knit fabric, that provide a better supply of oxygen during extraordinary exertion.

On the other hand, the regulation of body temperature in all xenarthrans is reminiscent of the

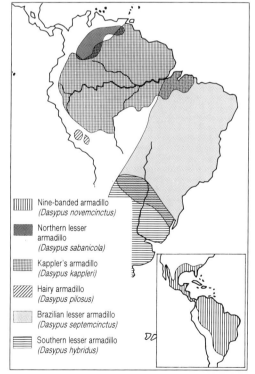

Armadillos look like armored reptiles at first glance, but the sparse, bristly hairs betray a mammal. As in the case of the larger hairy armadillo shown, the shields, each composed of polygonal bony plates, are detached from each other and for the most part are independently movable. This bony exoskeleton protects the armadillo from enemies and facilitates the burrowing activity vital to many species.

reptiles: to a certain extent, body temperature depends on outside temperature. In the nine-banded and southern three-banded armadillos, the body temperature remains at 90°F (32 °C) so long as the outside temperature stays between 61 and 64°F (16 and 18 °C). But if it goes down to 52°F (11 °C), the body temperature may fall by 5.5°F (3 °C) within four hours. The animals do not survive prolonged frost.

SOFT ARMADILLOS *(Supergenus Dasypodini)*. The NINE-BANDED ARMADILLO or TATU *(Dasypus novemcinctus)*, the best-known and most widely distributed representative of this group, exhibits a reproductive peculiarity that was noticed years ago by the Brazilians. Alfred Brehm dismissed the observation as legend, but today there is no longer any doubt of the special embryology of soft armadillos. Three species regularly produce identical multiple births. Thus, the common long-nosed armadillo always has quadruplets, all of the same sex, while two smaller species of the genus *Dasypus* even give birth to identical octuplets or duodecuplets. Upon the discovery of this trait (polyembryology), these armadillos drew the attention of geneticists. Here they had an opportunity to study variations in the fourfold, eightfold, or twelvefold expression of characters attributable to the same genetic information.

Ever since the nine-banded armadillo was identified in the United States in the middle of the nineteenth century, it has very rapidly gained ground and respect. It is less specialized in its dietary requirements, has a greater tolerance for temperature fluctuations, and is generally less limited in

habitat than its relatives, which have not extended their range as far North. Today, the northern boundary of the range runs from New Mexico and Colorado in the West to South Carolina in the East. The decline of natural enemies such as the coyote, red wolf, puma, ocelot, and lynx, to name a few, has favored its northward expansion. Low temperatures, that is, prolonged periods of frost,

(Top) The ventral surface of the armadillos is not armored and therefore vulnerable. **(Bottom)** These small armadillos already bear great similarity to their adult relatives. Soon their soft, pink, leathery skin will be transformed into a firm, brownish coat of mail. The bands referred to in names such as "nine-banded armadillo" are plainly visible.

Giant armadillo
(Priodontes giganteus)

Greater fairy armadillo
(Chlamyphorus retusus)

Lesser fairy armadillo
(Chlamyphorus truncatus)

are perhaps the only insurmountable barrier to the further spread of this armadillo. Once it had been released by humans in what was a new territory, it multiplied at an astounding rate. Unless the use of insecticides overtaxes the armadillo's chief source of food, expansion to the North and East will continue. Armadillos do well in slightly swampy, wooded terrain, and do not prefer any particular soil type.

In Central and South America as well, this is perhaps the commonest armadillo species. It makes a burrow on the edges of creeks and rivers, near trees and shrubs. Usually there is a passage 6–8 in. (15–20 cm) in diameter, dug up to 18 ft (6 m) long and 6 ft (2 m) deep into the ground. These passages end in a chamber that is lined and padded with dry plant parts. Often several passages lead to this central nest, though one passage is definitely preferred; the others serve merely as retreats. Occasionally armadillos use their pelvic shield to push their bedding and nest materials for their young into the burrows, and then they are seen to disappear into the earth backwards. After heavy rains, the house is cleared out and reupholstered. Rotting foliage and grass are dumped near the entrances.

In the hot Summer months, soft armadillos leave their burrows only in the evenings and at night, but if it is cool, they will take advantage of the noonday Sun to warm themselves while foraging. With lowered head, in the manner of swine, armadillos sniff the soil for small prey. A characteristic is the continually changing direction of their moderately rapid gait, with which they traverse about a half a mile (one kilometer) in the course of an hour. Worms and snakes are detected by the armadillo's acute sense of smell at depths up to 8 in. (20 cm) in the ground and scraped up with the sharp claws of the forelegs. When they sense a large predator, armadillos go into a fast gallop, and human beings cannot catch them within a short distance. At the last moment, the fleeing animal will often save itself in an abandoned burrow if it has strayed too far from home.

Their armor protects them from dense, thorny growth; their pursuers are not as fortunate, and ear injuries will testify to a hasty flight into the underbrush. The worst enemy of nearly all armadillos today is increasing highway traffic.

The tender white flesh of the tatu is today esteemed by many, even outside of Brazil. Concerning the hunting of this prized delicacy, Hensel writes, "Even two strong men cannot pull the tatu out of his hole provided it is narrow enough so that the animal can brace itself with its feet and back. Then it must be remembered that its tail tapers towards the tip and is difficult to grasp. But if one holds the tatu by the tail as tight as possible while the other clears some earth away with his knife so that he can grasp a hind leg, the tatu gives way."

GIANT and NAKED-TAILED ARMADILLOS *(Supergenus Priodontini).* With a body length from 2.5 to 3.3 ft (80 to 100 cm) and about 20 in. (50 cm) of long, strong, armored tail, the GIANT ARMADILLO *(Prio-*

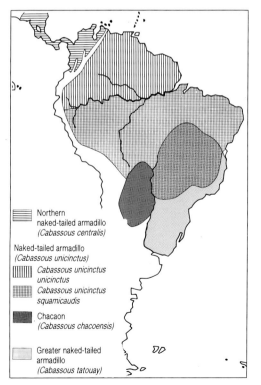

▷ The southern three-banded armadillo is rightly called the "ball armadillo". When alarmed, it instantly rolls itself up in a hard ball that even fairly large predators cannot pry open.

dontes giganteus) is the largest member of the entire armadillo family. It is widespread almost everywhere in South America East of the Andes, from Venezuela to northeastern Argentina, but despite protective measures, stocks are constantly declining because of overhunting and land clearing and development.

The giant armadillo feeds chiefly on termites. Supported on its hind legs and muscular tail, it attacks the stony fortresses of these insects and picks up the inhabitants with its sticky tongue. Its tracers are unmistakable. "These are big holes," Hans Krieg writes, "that one could crawl into. Sometimes the earth is merely ploughed up. In some places I found entire woods dotted with the marks of this work of destruction." Meeting the giant armadillo itself is a rare occurrence, for it is active chiefly at night and avoids settled areas or cattle country. If it occasionally strays near human settlements, it is pursued zealously because its search for insects, worms, and spiders will damage the fields. Besides, its flesh is prized.

Giant armadillos have rarely been kept in captivity. The first to be exhibited in a European zoological garden lived at the Berlin Zoo in 1937. Years ago, Agatha Gijzen told me of an interesting mistake. Through the animal trade, the Antwerp Zoo acquired a giant armadillo with young. The animals soon became acclimated and there was no difficulty, in keeping them, only there was a mystery. Although they had all been fed regularly for some months, the young were not getting any bigger. Finally the "young giant armadillos" were unmasked as adult naked-tailed armadillos. The two genera are not only very similar in appearance; their gait too is much the same.

At a run, only the tips of the claws of the forelimbs touch the ground, while the hind limbs are set down on the full sole.

This in itself suggests that the giant armadillo, the sole species representing the genus *Priodontes*, and the four species of NAKED-TAILED ARMADILLOS *(Cabassous)* are closely related. The most important external distinguishing characteristic – apart from smaller size – is the larger, protruding external ears and, of course, the "naked tail." The lack of tail armor, incidentally, is unique in the family. Not that the tail is completely naked, for on closer inspection, it is found to be studded with inconspicuous, thin, irregularly arranged platelets.

In habits and behavior as well, there are strong similarities between the giant armadillo and the naked-tailed armadillos. All of them appear to be

Most armadillos are efficient diggers, and spend their rest periods in underground burrows which they excavate by themselves.

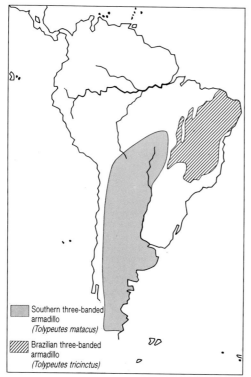

nocturnal, efficient burrowers and diggers; they are all insect eaters with a preference for ants and termites.

THREE-BANDED ARMADILLOS *(Supergenus Teolypeutini)*. In both species, the SOUTHERN THREE-BANDED ARMADILLO *(Tolypeutes matacus)* and the BRAZILIAN THREE-BANDED ARMADILLO *(T. tricinctus)*, the pectoral and pelvic shields have nearly spherical curvature. It is startling to see these beasts, tripping along on the tips of their front claws, suddenly transform themselves into a ball. The Spanish-speaking na-

tives allude to this ability by the names *bolita* "little ball" or *tatu naranja* "orange armadillo". Like the European hedgehog, the three-banded armadillos contract the skin musculature of the trunk when alarmed, bringing their head and tail shields together. "Rooting and digging," Hans Krieg writes, "the bolitas run through the open grassland even by day, even dig little burrows, first try to run away if alarmed, and then, with a hissing breath, roll up into a ball. At the Krefeld Zoo, I had an opportunity to pick up the "bolita" that had been living there for many years. It instantly pulled its legs under its pectoral and pelvic shields and clapped together into a solid sphere that could be rolled like a ball. After 50 to 60 seconds, the seams slowly opened, the little fellow got its front claws and half the sole of its hind feet back on the ground, and scuttled off in a straight line like a wind-up toy."

According to Hans Krieg's observations, domestic dogs are not able to open the hard ball. Probably this is generally true of the coyote, and the maned wolf in the wild as well.

HAIRY ARMADILLOS *(Supergenus Euphractini)*. In his book on the *World Travels of a Naturalist*, Charles Darwin devoted the following lines to the LITTLE ARMADILLO *(Zaedyus pichiy)*, the smallest of the *peludos* ("hairy ones"): "The pichi prefers a very dry soil, and the sand dunes of the coast, where all water is lacking for months, are his favorite abode. As soon as one was spied, it was necessary to get down from horseback quickly to catch him, for where the ground was soft, he would dig himself

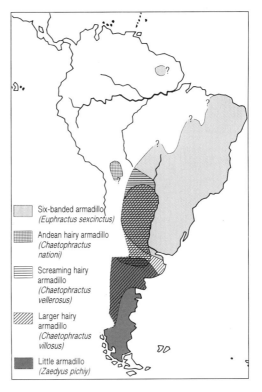

A nine-banded armadillo searching for food. With lowered head, it sniffs the earth for small prey and vegetable food, in the manner of a truffle pig. With its fine sense of smell, it can detect insects, worms and snails as much as 8 inches (20 centimeters) below the surface. The prey is then scratched up with the sharp claws of the forefeet and devoured.

New World edentates (Xenarthra)

Nomenclature English common name Scientific name French German	Approximate Size Body length Tail length Weight	Distinguishing Features	Reproduction Gestation period Young per birth Weight at birth
Giant anteater *Myrmecophaga tridactyla* Grand fourmilier; tamanoir Großer Ameisenbär	3.3–4.3 ft; 100–130 cm 2.2–3 ft; 65–90 cm 66–77 lb; 30–35 kg	Dense, long, hard, gray-brown body hair with conspicuous dark and light markings; imposing, very long-haired tail brush; long cylindrical head; claws 4–6 in. (10–15 cm) in length on 2nd and 3rd digit of forepaws, retracted when running	180–190 days 1 2.6 lb; 1160 g
Southern tamandua; yellow tamandua *Tamandua tetradactyla* Tamandua à quatre doigts Südlicher Tamandua	1.7–2.3 ft; 52–67.5 cm 1.3–2.3 ft; 40–67.5 cm 7.7–13.2 lb; 3.5–6 kg	From solid beige-golden brown in Surinam and northern Venezuela to black, vest-like marking in the southeast of the range of distribution, with intermediate patterns; greatest variability in Amazon basin; upper limit of auditory range, 55,000 Hz	About 6 months 1 Not known
Northern tamandua *Tamandua mexicana* Tamandua mexicaine Nördlicher Tamandua	Much like southern tamandua	Consistently black trunk markings; generally, shorter ears and narrower skull than southern tamandua	Much like southern tamandua
Silky anteater *Cyclopes didactylus* Myrmidon Zwergameisenbär	6.4–9.2 in; 16–23 cm 6.6–11.8 in; 16.5–29.5 cm 10.1–17 oz; 300–500 g	Silky, soft fur; golden-yellow to silver-gray; long curling tail; comparatively short snout; two thoracic and abdominal nipples	Not known
Pale-throated three-toed sloth *Bradypus tridactylus* Bradype; paresseux tridactyle; ai Dreifinger-Faultier; Ai	1.5–1.9 ft; 44.5–55.5 cm 1.2–3 in.; 3.1–7.5 cm 6.6–11 lb; 3–5 kg	Male with escutcheon-like marking on back between shoulder blades; throat and forehead yellow or white	120–180 days 1 per year Not known
Brown-throated three-toed sloth *Bradypus variegatus* — Braunkehl-Dreifinger-Faultier	1.6–2.3 ft; 50–70 cm 1.5–3.6 in.; 3.8–9 cm 4.9–12.1 lb; 2.25–5.5 kg	Male with escutcheon-like marking on back; throat brown; face and forehead white with dark stripes over eyes; top of head nearly black	Much like pale-throated three-toed sloth
Maned sloth *Bradypus torquatus* Bradype à collier Kragenfaultier	1.6–1.8 ft; 50–54 cm 1.9–2 in.; 4.8–5 cm 7.9–9.2 lb; 3.6–4.2 kg	Long black or dark brown fur at neck, falling over nape and shoulders like a mane	Much like pale-throated species
Linné's two-toed sloth; unau *Choloepus didactylus* Paresseux didactyle; unau commun Zweifinger-Faultier; Unau	2–2.8 ft; 60–85 cm 0.6–1.3 in.; 1.4–3.3 cm 8.8–17.6 lb; 4–8 kg	Hair uniformly dark brown at throat and neck; 6–8, usually 7, cervical vertebrae; 24–25 thoracic vertebrae; ear length 0.8–1.4 in. (20–35 mm)	About 8–9 months 1 10.7–14.3 oz; 300–400 g
Hoffmann's two-toed sloth *Choloepus hoffmanni* Unau d'Hoffmann Hoffmann-Zweifinger-Faultier	1.8–2.4 ft; 54–72 cm 0.6–1.2 in.; 1.4–3 cm 9.9–14.7 lb; 4.5–6.7 kg	Forehead, cheeks, throat, and neck light brown or golden-yellow; 5 teeth in upper jaw, 4 in lower jaw; the first pair of teeth are canines; 6 cervical vertebrae; ear length 0.8–1.5 in. (20–37 mm)	Probably much like unau
Common long-nosed armadillo *Dasypus novemcinctus*, 3 subspecies [?] Tatou à neuf bandes Neunbinden-Gürteltier	1.1–1.9 ft; 35.5–57 cm 10–18 in.; 25–45 cm 5.8–13.8 lb; 2.65–6.25 kg	Scales black in middle of back, yellow-brown; laterally and on tail; abdomen sparsely haired; usually 8 fully mobile dorsal bands, ninth band mobile at the side only; 7, 9 and 10 bands occur; crown-shaped marking on head shield; usually 8 teeth to each half of jaw, of which 7 are replaced as milk teeth	About 140 days (up to 14 weeks ovulation cycle) 4 (quadruplets of one sex) Not known
Brazilian lesser long-nosed armadillo *Dasypus septemcinctus* Tatou à sept bandes Siebenbinden-Gürteltier	8–12.2 in.; 20–30.5 cm 6.8–8 in.; 17.5–20 cm 3.2–3.9 lb; 1.45–1.8 kg	Usually 6–7, rarely 8, mobile dorsal bands; eighth band mobile at sides only, otherwise attached to pelvic shield; armor darker at sides than in 9-banded species; 7–8, rarely 6, teeth in each half of jaw	About 140 days 4, 8, or 12 Not known
Nothern lesser long-nosed armadillo *Dasypus sabanicola* — Nördliches Siebenbinden-Gürteltier	10.1–12.6 in.; 25.3–31.4 cm 7–8.4 in.; 17.5–21 cm 2.2–4.4 lb; 1–2 kg	Similar to Brazilian species	Not known
Southern lesser long-nosed armadillo *Dasypus hybridus* — Südliches Siebenbinden-Gürteltier	11.2–12.4 in.; 28–31 cm 6–7.4 in.; 15–18.5 cm 4.4 lb; 2 kg	Similar to Brazilian species	Not known
Kappler's armadillo; greater long-nosed armadillo *Dasypus kappleri* Tatou de Kappler Kappler-Weichgürteltier	1.7–2.3 ft; 51–67.5 cm 1.1–1.6 ft; 32.5–48 cm 18.7–23.1 lb; 8.5–10.5 kg	Dorsal armor sparsely haired; 7–8 mobile dorsal bands; 5 digits on forepaws, 5th digit very small or absent; front of hind shanks bear 2–3 rows of claw-like horney shields, up to 0.8 in. (2 cm) in length; no crown-shaped marking on head shield 7–9 teeth in each half jaw; ear length 1.6–2.2 in. (40–55 mm)	Not known
Hairy long-nosed armadillo *Dasypus pilosus* Tatou poilu Pelzgürteltier	1.5 ft; 44 cm 12.4 in.; 31 cm Not known	Entire dorsal armor densely haired, abdomen less dense; 10 mobile dorsal bands, 11th band fused to pelvic shield in middle of back; long, narrow, hairless head shield with crown-shaped marking; ear length 2 in. (5 cm)	Not known

COMPARISON OF SPECIES

Life Cycle Weaning Sexual maturity Life span	Food	Enemies	Habit and Habitat	Occurrence
Age not known Age not known In captivity, nearly 26 years	Ants and termites; beetle larvae	Puma, jaguar	Predominantly diurnal in ground dweller; good digger and swimmer; in savannas, swamps, moist and parkland forests, and thornbush steppes	Endangered by hunting and agriculture
Age not known Age about 12 months In captivity, over 9 years	Ants, termites, small fruits	Predator cats, large birds of prey	Predominantly nocturnal tree and ground dweller in rain forests, mountain country, llano, deciduous forests, and forest margins	Still fairly abundant; in Venezuela, 3 individuals per km^2 (1 km^2 = 0.38 mi^2)
Much like southern tamandua	Much like southern tamandua	Same as southern tamandua	Much like southern tamandua	Still fairly abundant
Not known	Ants and termites; bees and wasps	Birds of prey, predator cats, snakes	Nocturnal tree dweller in tropical forests; skillful climber	Apparently still fairly abundant
Age not known Age not known Probably 30–40 years	Leaves and buds of very diverse tree species	Large birds of prey, snakes, predator cats	Tree dweller in tropical forests; active nocturnally and by day; good swimmer	Fairly abundant
Much like pale-throated species	Leaves and buds of diverse tree species	Same as pale-throated species	Much like pale-throated three-toed sloth	Still quite abundant and widely distributed
Age not known Age not known Up to 12 years	Leaves	Same as pale-throated species	Diurnal and nocturnal dweller in treetops in tropical rain forest of coastal regions; solitary; good swimmer	Seriously threatened; protected by law in Brazil
Age not known About 1 year or later Probably 30–40 years	Leaves, fruits, tubers	Predator cats	Predominantly nocturnal tree dweller in tropical forests; good swimmer	Still fairly abundant
Probably much like unau	Much like unau	Same as unau	Much like unau	Widely distributed and apparently still quite abundant
Age not known Age 6–12 months 12–15 years	Insects, larvae, worms, snails, small vertebrates; mushrooms, fruits, plant parts	Puma, lynx, maned dog, coyote, black bear	Predominantly nocturnal in swamps, rain forests, pampas, prairie, and cultivated terrain; adaptable; 0.36 individuals per km^2 in parts of Brazil, 50 per km^2 in coastal region of Texas (1 km^2 = 0.38 mi^2)	Widely distributed and quite abundant
Age not known Age 6–12 months Not known	Insects, larvae, worms, mushrooms, fruits	Maned dog, coyote	Predominantly nocturnal; prefers open grassland (pampas)	Still quite abundant
Not known	Probably much like Brazilian species	Not known	Predominantly nocturnal	Still fairly abundant
Not known	Probably much like Brazilian species	Not known	Predominantly nocturnal; in savannas and open grasslands at elevations between 190 and 320 ft (60 and 100 m)	Still fairly abundant
Not known	Probably much like Brazilian species	Not known	Predominantly nocturnal	Widely distributed
Not known	Insects	Not known	Predominantly nocturnal; in highlands (moist eastern slopes of the Andes) at elevations up to 10,200 ft (3200 m)	Not known

NEW WORLD EDENTATES

Nomenclature English common name Scientific name French German	Approximate Size Body length Tail length Weight	Distinguishing Features	Reproduction Gestation period Young per birth Weight at birth
Giant armadillo *Priodontes maximus* Tatou géant Riesengürteltier	2.6–3.3 ft; 80–100 cm 1.6 ft; 50 cm 110 lb; 50 kg	Largest species of armadillo; long, powerful, armored tail; dark brown back armor with 11–13 mobile bands; forefeet with 5 claws, 3rd claw sickle-shaped, up to 8 in. (20 cm) in length; hind legs, hind feet with 5 fused digits and 5 short claws; long worm-like tongue	Not known 1, rarely 2 Not known
Naked-tailed armadillo *Cabassous unicinctus*, 2 subspecies Tatou à onze bandes Nacktschwanz-Gürteltier	11.6–17.8 in.; 29–44.5 cm 3.6–8 in.; 9–20 cm 3.5–7.9 lb; 1.6–3.6 kg	Armored tack with 10–13 mobile bands; tail not armored; usually 9 teeth on each half of jaw; ear length 1–1.2 in. (25–30 mm)	Not known
Northern naked-tailed armadillo *Cabassous centralis* — Nördliches Nacktschwanz-Gürteltier	12–15.2 in.; 30–38 cm 5.2–7.2 in.; 13–18 cm 4.4–7.7 lb; 2–3.5 kg	Armored back with 10–13, usually 12, mobile bands; slender tail, not armored; no scales on cheeks and posterior of ears; ear length 1.2–1.5 in. (31–37 mm); walks on tips of forefeet and soles of hind feet; quite fast over short distances	Not known
Greater naked-tailed armadillo *Cabassous tatouay* — Großes Nacktschwanz-Gürteltier	1.4–1.6 ft; 41–49 cm 6–8 in.; 15–20 cm 7.5–14.1 lb; 3.4–6.4 kg	Largest *Cabassous* species; tail not armored; scales on posterior sides of tubular external ears; ear length 1.6–1.8 in. (40–44 mm); 7–10 teeth each half of upper jaw, 8–9 each half of lower jaw	Not known
Chacoan naked-tailed armadillo *Cabassous chacoensis* — Chaco-Nacktschwanz-Gürteltier	12–12.2 in.; 30–30.5 cm 3.6–3.8 in.; 9–9.5 cm Not known	Smallest *Cabassous* species; 12 mobile dorsal bands; short ears 0.5–0.6 in. (14–15 mm) do not overlap first row of plates in shoulder shield as in all other species of the genus; anterior margins of external ears have fleshy thickening	Not known
Southern three-banded armadillo *Tolypeutes matacus* Tatou à trois bandes Kugelgürteltier	1.2–1.5 ft; 35–45 cm 3.6 in.; 9 cm Not known	Forelegs usually with only 4 digits, fifth often absent, occasionally also the first; 9 teeth in each half of jaw; able to roll up into a ball; runs on "tiptoe", claws of forefeet, soles of hind feet	Not known 1 per year Not known
Brazilian three-banded armadillo *Tolypeutes tricinctus* Apar de Buffon Dreibinden-Kugelgürteltier	1.2–1.5 ft; 35–45 cm 3.6 in.; 9 cm Not known	Thoracic and pelvic shields highly arched; hairless armor, only whitish bristles at edge; very small ears; forefeet with 5 digits and claws, third claw powerful and larger; 8 teeth on each half of jaw; otherwise like southern species	Not known 1 per year Not known
Yellow- armadillo; six-banded armadillo *Euphractus sexcinctus*, 4 subspecies Tatou à six bandes Sechsbinden-Gürteltier; Weißborsten-Gürteltier	1.3–1.6 ft; 40–49 cm 4.4–9.6 in.; 11–24 cm 6.6–17.6 lb; 3–8 kg	Largest species of bristled armadillos; 6–8 mobile dorsal bands; sparse beige-yellow bristles; 9 teeth on each half of upper jaw, 10 on each half of lower jaw; ear length 1.3–1.9 in. (32–47 mm)	60–64 days 1–3 3.4–4.1 oz; 95–115 g
Larger hairy armadillo *Chaetophractus villosus* Tatou velu Braunhaar-Gürteltier	1.1–1.5 ft; 32–44 cm 4.4–4.6 in.; 11–11.5 cm Not known	Largest *Chaetophractus* species; hair of dorsal armor not very dense, dark brown to black-brown; thick hair on abdomen and limbs; ear length 0.8–0.9 in. (20–24 mm)	About 2 months Usually 2 Not known
Andean hairy armadillo *Chaetophractus nationi* [?] — Anden-Borstengürteltier	Not known	Dorsal armor light yellow-brown and covered with long light hairs; head shield about 2.5 in. (6 cm) long and wide; 7 mobile dorsal bands	Not known
Screaming hairy armadillo *Chaetophractus vellerosus* — Weißhaar-Gürteltier	9.6–10 in.; 24–25 cm 4–4.2 in.; 10–10.5 cm Not known	Smallest *Chaetophractus* species; horny plates of dorsal armor with light and dark brown pattern; long, fairly dense white fur on limbs and abdomen; head shield distinctly triangular; ear length 1.2–1.4 in. (29–34 mm)	Not known
Pichi *Zaedyus pichiy* Tatou nain Zwerggürteltier	To 10 in.; 25 cm 3.8–4.8 in.; 9.6–12 cm Not known	Smallest species of bristled armadillos; head shield and dorsal armor dark brown with yellow or white edges; under parts set with yellow-white bristles; ear length about 0.5 in. (13 mm)	Not known
Lesser fairy armadillo; pink fairy armadillo; lesser pichiciego *Chlamyphorus truncatus* Chlamyphore tronqué Kleiner Gürtelmull; Schildwurf	4.4–6 in.; 11.4–15 cm 1–1.4 in.; 2.5–3.5 cm 3.2 oz; 90 g	Soft dorsal armor, joins head shield; no shoulder shield; about 24 dorsal "bands" of very thin bony plates and horny scales; hard pelvic shield perpendicular to axis of trunk; dense hair on ventral parts and limbs; coat of hair under dorsal armor	Not known
Greater fairy armadillo *Chlamyphorus retusus* Chlamyphore de Burmeister Burmeister-Gürtelmull	6.6–7.6 in.; 16.5–18.9 cm 1.4–1.5 in.; 3.5–3.8 cm 3.6 oz; 100 g	Soft dorsal armor completely fused to body; no shoulder shield; 24 mobile dorsal bands; pelvic shield perpendicular to axis of trunk; sparse hair on armor; dense hair on ventral parts and limbs	Not known

COMPARISON OF SPECIES

Life Cycle Weaning Sexual maturity Life span	Food	Enemies	Habit and Habitat	Occurrence
Age not known Age not known In captivity, over 4 years	Termites, ants, and other insects, worms, spiders, snakes, carrion	Not known	Predominantly nocturnal in forests and savannas; solitary; digs caves for shelter	Reduced by hunting, development, and agriculture; protected in many countries; still abundant in Bolivia
Not known	Ants, termites, and other insects	Not known	Nocturnal; industrious digger	Widely distributed
Not known	Ants, termites, and other insects	Not known	Nocturnal; industrious digger; digs caves for shelter	Widely distributed
Not known	Ants, termites, and other insects	Not known	Found in open, dry or moist bush country; good digger; fairly good runner	Widely distributed
Not known	Ants, fermites, and other insects	Not known	Probably much like the other species	Widely distributed
Not known	Insects	Not known	Active nocturnally and by day in open grasslands and bush forests; does not burrow	Still fairly abundant
Not known	Inscects	Not known	Much like southern species	Not known; considered rare
Age 1 month Age 9 months Not known	Insects, small vertebrates, carrion, roots, fruits	Not known	Active nocturnally and by day; preference for savannas and forest margins; good digger; inverted U-shaped burrow entrances (8.5 × 7.5 in.; 21 × 19 cm)	Especially abundant in Brazil
Age not known Age not known In captivity, nearly 19 years	Presumably, much like yellow armadillo	Not known	Active nocturnally and by day in open country; good digger	Still quite abundant
Not known	Presumably, much like yellow armadillo	Not known	Active nocturnally and by day in mountain grasslands at elevations of 11,200–12,800 ft (3500–4000 m); good digger	Not known
Not known	Presumably much like yellow armadillo; much vegetable fare in winter	Not known	Active nocturnally and by day in open country; good digger	Still fairly abundant
Not known	Insects, worms, and other small prey; carrion	Not known	Active nocturnally and by day; good digger	Still quite abundant and widely distributed
Not known	Insects, predominantly ants, and larvae; worms, small seeds	Not known	Predominantly subterranean in loose, usually sandy soils in dry terrain with high temperatures; also in low, thorny bush forests; good digger; reportedly active by night or at dusk	Rare and endangered; as yet no protective measures
Not known	Insects, their larvae, worms, small seeds	Not known	Inhabits dry terrain; habits barely studied	Rare and endangered; as yet no protective measures

The soft armor, looking as if cut short behind, and the woolly ventral surface are the most striking characteristics of the two species of fairy armadillos. The drawing shows the greater fairy armadillo.

in so rapidly that the hind legs would almost have disappeared before one could dismount."

The hairy armadillos are still frequently encountered everywhere in Brazil, Paraguay, and down to the southern tip of Argentina. Very similar in outward appearance, the five species form a series of sizes, like organ pipes. They are the best diggers among the armadillos, and regularly leave their burrows even during the day to hunt. Usually they do not return to their underground sleeping places, but dig themselves a new refuge. The entrances, matching their armored shape, are flatter and often wider than those of the soft armadillos. Extending 3 or 6 ft (1 or 2 m) into the ground, they are enlarged into a chamber at the end. The peludos dig into the earth using their powerful claws and muscular forelimbs, then throw it out to the rear with their hind limbs. They are solitary; only females are observed with their young for a few weeks.

(Above) Three-banded armadillo in a ball. (Below) Southern seven-banded armadillo, still fairly common in its range, but not much is known about its habits in detail.

The SIX-BANDED ARMADILLO *(Euphractus sexcinctus)* gives birth to one to three young after a gestation period of 60 to 64 days. In the first days after birth, when the body is still soft, the mother reacts very aggressively to any disturbance, and will sometimes carry her offspring into a different burrow. Between the 22nd and 25th days of life, the eyes of the young open, and after a month the peludos will have quadrupled their birth rate of about 3–4 oz (95–115 g). In captivity, young six-banded armadillos will take solid food about this time, and at nine months they reach sexual maturity.

Mating of hairy armadillos does not seem to be linked to any particular season, for pregnant females were found in January as well as in September and October. Possibly there may be several births in a year. According to observations by Rengger, males and females meet as though by chance on their nocturnal excursions, nuzzle each other a while, mate several times in succession, and then go their separate ways. The carrion smell of a cadaver may indeed assemble some numbers of the otherwise unsociable peludos.

When Hans Krieg traveled the Argentine province of Santa Fe, bounties were offered on the LARGER HAIRY ARMADILLO *(Chaetophractus villosus)* because its excavations were devastating freshly ploughed fields. Krieg availed himself of the bounty for the benefit of science. I chanced upon the population of these armadillos, from the vicinity of Galvez in the Brazilian state of Minas Gerãis, when seeking materials for phylogenetic studies of armadillos at the Stuttgart Museum of Natural History in the mid-1960s.

Today, the numbers of this once abundant species, as well as of the smaller pichi, have declined, but the hairy armadillos are not yet listed as endangered.

The establishment of exhibits of nocturnal animals in zoos has brought greater popularity to the armadillos. Today the visitor does not see them huddled together in a warm corner, but can see soft and hairy armadillos out foraging on the ground. In the six-banded armadillos, I noticed less play of the external ears than in other armadillos. Peludos tend to put on weight in captivity. They may sometimes be seen sleeping on their backs in the enclosure. They will stretch out their legs and tremble like dogs sleeping.

FAIRY ARMADILLOS *(Subgenus Chlamyphorini)*. If the phylogenetic development of the armadillo species living today is pictured as a branching diagram, the fairy armadillos, so far as we now know, are doubtless the latest branch. What would lead a taxonomist to that conclusion? First of all, the higher specialization of these least of all the armadillos. Two species are known, the lesser fairy armadillo *(Chlamyphorus truncatus)*, discovered in 1824 by Richard Harlan near Mendoza in central Argentina, and Burmeister's greater fairy armadillo *(Chlamyphorus retusus)*. Hermann Burmeister came upon the greater fairy armadillo in Bolivia in 1859 when an Indian showed him this peculiar animal as a mummy. After the founded the Museum of Natural History in Buenos Aires, he described this species in 1863, and it first entered the German scientific literature als *Burmeisteria retusa*.

The transformation of the pelvic region outlined in other armadillos reached its greatest development in this genus. The pelvis armor is not separate from the skeleton as in the other species, but is firmly attached to the spine and the pelvic bones. In underground passages, the posterior armor provides effective protection against intruders. The pelvic shield is almost perpendicular to the line of the back, so that the animal looks truncated. The old Spanish name *Juan calado* "lacy Jack" alludes to the conspicuous white fringe of hair bordering the circumference of the posterior shield.

The range of the lesser fairy armadillo extends from the center of Mendoza into the provinces of San Juan, San Luis, and La Pampa. Preferring dry, sandy terrain with thornbushes or cactus, and dependent on high temperatures, it is limited in range by the Andes to the West, by increasing humidity (borderline about 16 in. or 40 cm annual rainfall) to the East, by the Rio Colorado to the South, and by excessively hard soil to the North. The lesser fairy armadillo leaves its underground passages infrequently and briefly. The legs, with their heavy digging claws, are hardly lifted, but pushed forward, leaving a characteristic trail. Af-

The larger hairy armadillo owes its name to the sparse but distinctly visible dark brown hairs of its dorsal armor.

(Opposite page) The pangolins are an Old World counterpart of the armadillos of the New World. Yet despite certain resemblances between the two groups, they are not very closely related and belong to different orders of mammals.

ter only a few yards, the small body will turn in a circle several times, the horny snout will plunge repeatedly into the sand, and a burrow is dug rapidly with the forelegs. Two little heaps of sand or earth mark the spot. Friedrich Kühlhorn, who examined the stomach contents of some individuals, found chiefly ants and spores of club mosses; insect larvae are reportedly eaten as well.

Advancing civilization and cultivation of the pampas are the most formidable enemies of the lesser fairy armadillo today. Much the same is true of the greater species, which is equally endangered and perhaps still rarer. We know hardly anything about its habits, and not even its range is accurately known. In any case, it lives farther North than its smaller cousin, inhabiting essentially the Gran Chaco region of western and central Bolivia, Paraguay, and northern Argentina.

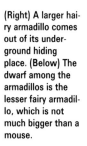

(Right) A larger hairy armadillo comes out of its underground hiding place. (Below) The dwarf among the armadillos is the lesser fairy armadillo, which is not much bigger than a mouse.

PANGOLINS

Category
ORDER

Classification: An order of mammals, with only one family living today, containing one genus of four African and three Asian species. Formerly the pangolins were united with the xenarthrans in the order Edentata, which is no longer used in modern classifications.

Family Manidae
Genus Manis (pangolins or scaly anteaters)
Body length: 12–32 in.; 30–80 cm
Tail length: 14–28 in.; 35–70 cm
Weight: 4 lb (1.8 kg; African tree pangolin) to 66 lb (30 kg; giant or ground pangolin)

Distinguishing features: Body and tail covered with large scales, overlapping like roof shingles (a body covering unique among mammals); under parts and insides of legs hairy; forefeet with digging claws, hind feet clawed; all extremities have five digits; all species are plantigrade; body capable of rolling up more or less into a ball (sleeping position and for protection from enemies); long, powerful, highly mobile tail (aid in climbing and support when standing upright on the ground); small, tapered head; small external ears and eyes (good hearing, vision subordinate in importance); skull lacks zygomatic arch; lower jaw is reduced to two strips of bone; mouth is toothless; long, wormlike tongue; large salivary glands (supplying the tongue with a sticky coating); clavicles absent; sternum has peculiar sword-shaped process; 11–16 thoracic vertebrae, 5–6 lumbar vertebrae, 2–4 sacral vertebrae, and 21–47 caudal vertebrae; lumbar vertebrae do not have articular processes; muscular stomach with pavement epithelium (horny skin layered like pavement); large olfactory lobes (sense of smell highly developed); posterior glands secrete scents; primitive cerebrum; bicornuate uterus; two pectoral nipples; inguinal testes.

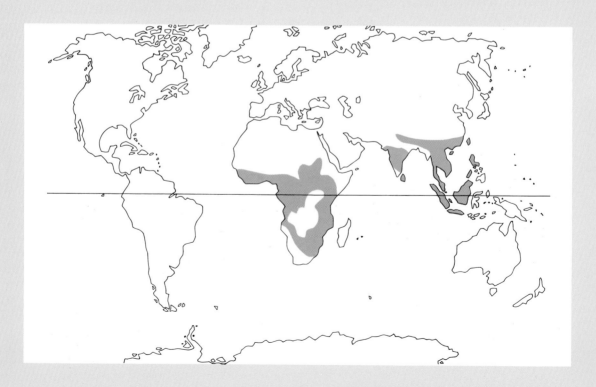

Pholidota
Pangolins FRENCH
Schuppentiere GERMAN

Reproduction: Gestation, where known, 140–150 days; one young per birth as a rule; birth weight 3–18 oz (90–500 g).
Life cycle: No reliable information on weaning or sexual maturity; life span in nature not known, in captivity up to 4.5 years.
Food: Termites and ants; in larger species, possibly other insects also.

Habit and habitat: All species are nocturnal, except the predominantly diurnal long-tailed pangolin; arboreal or terrestrial; most are good climbers, some also swim; rest periods in burrows or in or under trees; solitary, unsociable, with individual territories; size of territory, where known (African tree pangolin), 38–63 acres (15–25 hectares) for males, 8–10 acres (3–4 hectares) for females; marking of home territory with posterior gland secretion and urine, also with droppings in the case of arboreal species; sense of smell indispensable for intraspecific communication and locating insect food; scant expressive behavior, hardly any vocal utterance; habitat diverse according to species (tropical rainforests, savannas, steppes).

Toothless skull. The head shape of all ant-eating mammals, regardless of relationships, exhibits the same adaptive features. The skull is elongated and the narrow jaws, which bear small primordial teeth in the embryo, become entireley toothless.

Digging claws. A specialized diet of ants or termites leads to several typical adaptations, including powerful forelegs with digging claws. Note right forefoot of Chinese pangolin, viewed from outside and inside (at left).

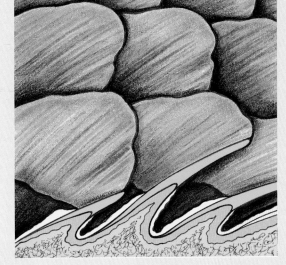

Massive horny scales, covering nearly the entire surface of the body of the Pholidota and giving this order of mammals its name, are produced by cornification of the epidermis, as are nails or hair. The epidermis rests on dermal papillae (blue) which consist of living cells, not yet cornified (red), which grow constantly, migrate to the surface, and there become cornified. The epidermis cornifies in the depressions between the dermal papillae to form a loose, supple stratum corneum (yellow), much as in the unspecialized skin of most mammals. On top of and at the tip of the papillae, the processes of epidermal cornification lead to the formation of hard scales (green).

Pangolins

Phylogeny
by Erich Thenius

The phylogenetic origins of the pangolins are as yet largely obscure. The combination of primitive and highly modified traits, some of them related to diet (for example, toothlessness, signs of primordial teeth in the embryo only, chewing musculature, reduction of zygomatic arch, elongated facial skeleton, worm-like tongue, stomach, forelimbs as digging tools, and prehensile tail) implies a long independent history. Resemblances to the xenarthrans pertain to common primitive traits or to convergence (parallel lines of development). Whether the palaeanodonts of the early Tertiary in North America are in fact relatives of the pangolins, as R.J. Emry supposes, seems uncertain. In any event, an early separation from primitive placental mammals is to be assumed.

Two of the four African species of pangolins. The African tree pangolin is the most frequently seen (here demonstrating the use of its prehensile tail) (left). Less common is the quite similar long-tailed pangolin, with larger scales and darker skin and hair (right).

Today, the pangolins, represented by only the one genus, *Manis*, are limited to the paleotropics (tropical regions of the Old World). In the early Tertiary, pangolins were also at home in Europe (for example, *Eomanis, Leptomanis, Necromanis*) and North America *(Patriomanis)*. *Eomanis* in the middle Eocene was already highly specialized (horny scales, toothlessness with a diet of ants and termites), arguing an early Cretaceous separation from the other mammals. Among recent species, the African forms are more specialized in almost every way than the Asian.

Modern Pangolins
by Urs Rahm

Within the class Mammalia, the pangolins constitute the order Pholidota. They were formerly associated with the xenarthrans (order Xenarthra: armadillos, hairy anteaters, and sloths). Their skin covering, from which the pholidotes take their name, is unique among the mammals; sometimes they are referred to as "living pine cones" or "pineapples." The French name *pangolin* is derived from the Malay *pëngulin* "roller"; in French the term *fourmilier* "ant bear" is often used in error, and occasionally in English "armadillo." The correct English designation is "scaly anteater."

In all pangolins, overlapping scales, like roof shingles, cover the dorsal surfaces of the head, neck and trunk, and the flanks, and tail. In newborn individuals, the scales are still soft and do not overlap. The number of scales remains constant throughout life. Parts of the head, throat, neck, inside of legs, and belly are scaleless and sparsely covered with hair. In form and origin, the horny scales are derived from the dermis, covered with epidermis, and cornified. The scales are bilaterally symmetrical, flattened from the top down, and directed towards the tail. They differ from reptile scales by their horny coating and in that they are not renewed by molting. In the case of the pangolins, loss of the horny scales by wear is replaced from the stratum germinativum in the dermis. In other words, the scales are not "cemented hairs" but are comparable in microstructure with nails, and in form with reptilian scales. In the climbing species, the under side of the tip of the tail has a naked area presumed to be a tactile organ.

Of all mammalian skulls, that of the pangolins has the least projection. There is no zygomatic arch nor any bony crest for attachment of muscles. The halves of the lower jaw are reduced to narrow slats. The snout is elongated and the mouth opening is constricted. The jaws are toothless (only embryos have some dental primordia). As we shall see, these are adaptations to the special feeding habit of the animals. The eyes are small with thick eyelids. The external ears are much reduced. The forelimbs and the hind limbs are short, with five digits on all four feet. The giant (ground) pangolin, and Temminck's (ground), or Cape pangolin, both strictly terrestrial, have broad sole cushions and blunt claws on the hind feet. The forefeet are clawed, with a large digging claw. The limbs of arboreal species are more slender with comparatively long claws, an important aid in climbing. Besides skin glands, the pangolins have glands in and around the posterior that play an important part in intraspecific relations throughout the life of the animal.

The pangolins show adaptations to their mode of life and to their diet of ants and termites in internal structure also, especially the skeleton. There is no clavicle. The long-tailed pangolin, with its 46 or 47 caudal vertebrae, has the largest number of all mammals.

The sword-like process of the sternum (the xiphisternum) has undergone a modification in all pangolins. In the Asian species, it is fairly long and terminates in a laterally broadened blade resembling a shovel. In the ground pangolins, the process forms a long rod of cartilage extending to behind the end of the rib cage. In the African tree pangolin and the long-tailed pangolin, the process consists of two extraordinarily long cartilagenous rods extending outside the diaphragm, first toward the rear and then, describing an arc, toward the head again.

This peculiar structure of the xiphisternum is related to the enormous elongation of the tongue, which plays an important part in feeding. The tongue is extraordinarily long, measuring 6 to 7 in. (16 to 18 cm) in the small species and 16 in. (40 cm) in the giant pangolin. According to the species, it is either round or flattened. In resting

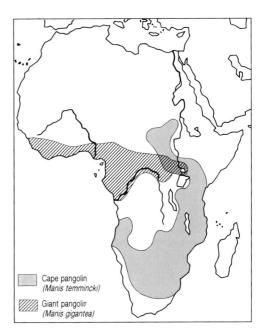

▷ Like this African tree pangolin, all species are able to curl up more or less completely in a ball when danger threatens or for sleeping.

position, the tongue is retracted into an invagination extending into the thoracic cavity. The muscles required for this are attached to the xiphisternum. Huge salivary glands, the size of a goose egg in the giant pangolin, supply the tongue with a sticky saliva to which ants and termites adhere. Since the mouth is toothless, the stomach must masticate the chitinous exoskeleton of the insects; the stomach does the chewing, as it were. Instead of an internal mucous membrane, the stomach has a cornified, stratified "pavement" epithelium a forming horny layer in the posterior part of the stomach. Opposed to this is a muscular organ that is also studded with horny teeth. Mucus glands and a large gastric gland also contribute to digestion. Thus the stomach is uniquely modified for processing insects. Swallowed sand and bits of stone merely assist the grinding acitivity of the stomach. The intestine exhibits no remarkable features; in the African tree pangolin it is 6 ft (2 m) long.

Examination of the African tree pangolin has shown that its body temperature fluctuates between 86 and 95 °F (30 and 35 °C), depending on the outside temperature. Pangolins are poikilothermic to a certain extent, with an average body temperature of 89.9 °F (32.2 °C), below that of highly developed mammals. The situation is quite similar in the two-toed sloth and in the tenrecs.

The pangolins are very strongly adapted to an especially limited diet. Certain peculiarities of physical structure, such as the form of the tongue, the toothless mouth, and the design of the stomach, betray a specialized feeding habit. All pangolins consume termites and ants almost exclusively. A few reports indicate that at least the large species occasionally take other insects. Observations of living pangolins as well as analyses of droppings and stomach contents suggest that some species prefer ants, or termites, but reports are conflicting. It is quite possible that the differences in feeding simply depend on the local supply from time to time, or on seasonal variations. Concerning the

Pangolins are solitary. All alone, a tree pangolin comes to drink in the evening, having slept all day in its burrow.

Malayan pangolin, for example, some zoologists write that it eats mostly termites, others that its sole food is ants and their pupae. The well-known myrmecologist A. Forel identified eleven different species of ants in the stomach of a giant anteater. This species, however, takes termites also.

The large claws of the forefeet enable the giant anteater to dig for termites in the soil. It scrapes funnel-shaped holes around the hill until it encounters termites. Around the large hills of the giant termines, *Macrotermes*, a system of passages and ditches results. The termite hill is rarely destroyed completely; the colony survives and rebuilds the hill. A pangolin can therefore visit the same termite constructions again after a suitable

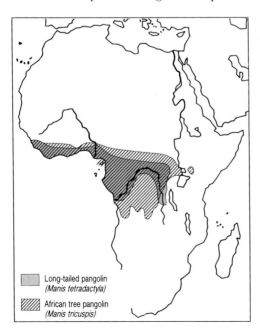

interval. Observations by R. Sweeney, studying the Cape pangolin in the Sudan, indicate that this species distinguishes between "good" and "bad" termites; termites of the genera *Trinervitermes, Macrotermes, Odontotermes, Microcerotermes, Amitermes, Ancistrotermes,* and *Microtermes* are eaten. The animals readily detected the preferred species before opening the hill.

Species of ants living concealed in the ground were also located. When the pangolins found old animal droppings inhabited by termites, they seized the pellets with their forelimbs and turned over on their backs. In that position, they would hold the pellet between the claws above the belly, raise the head and lick off the exiting termites. The African tree pangolin feeds on ground and tree termites, and does not refuse tree ants of the genus *Crematogaster*. Our observations and those

of Elisabeth Pagès have shown the long-tailed pangolins to have a pronounced preference for tree ants of the genera *Crematogaster, Camponotus,* and *Catalacus.*

When breaking up tree nests, the pangolins anchor themselves in the branches with hind legs and prehensile tail. The claws of the forefeet then dig precisely to the left and right of the mouth, which gradually advances into the cavity formed in the nest. The worm-like tongue is constantly in motion and stabs into the narrow passages of the termite nests. The quantity of insects consumed daily by the African tree pangolin is 5 to 7 oz (150 to 200 g). All pangolins drink water regularly, lapping it up with their tongue.

Natural food is hard to procure for pangolins in captivity, even in their home country. Substitutes are not always accepted, and the animals will perish when they have consumed their body reserves. The simplest to keep appear to be the Asiatic species and the giant pangolin. Several zoos have developed recipes for substitute provisions. Besides vitamins and minerals, the essential ingredients are raw and cooked finely chopped meat, chopped boiled eggs, boiled rice, cornstarch, ant pupae, and crushed meal worms. Keeping pangolins on such substitute food is very difficult, even now.

We have very little information on the life, habits, and behavior of the Asian pangolins in the wild. The studies of E. Pagès and other in West Africa, have provided a good deal of information about the three African forest-dwelling species.

The pangolins occur only in the tropics and the transitional zones (subtropics) bordering the tropics to the North and South. The presence of ants and termites in sufficient numbers is crucial. The Indian pangolin prefers hilly regions, whereas the Malayan pangolin occurs in sandy open country. The African tree pangolin, long-tailed pangolin, and giant pangolin inhabit the rainforests of Africa, while the Cape pangolin is at home on the steppes and savannas. All species are nocturnal except for the long-tailed pangolin. The three Asian forms are predominantly terrestrial, but they like to climb trees – and often do. They dig burrows in which they spend the day. The Chinese pangolin digs tunnels up to 9 ft (3 m) long which terminate in a den. The two terrestrial African species, the giant pangolin and the Cape pangolin, also dig burrows. The former often uses the passages it digs around large termite hills as sleeping places, or hides among platform or stilt roots of large trees. The long-tailed pangolin sleeps

(Left) Pangolins have a remarkably long, mobile tongue, which they use - like the armadillos - for collecting ants and termites, but also, as may be seen, for lapping up water. (Below) Before this Indian pangolin can use its tongue for feeding, it must break open the solid termite hill. For this it uses the claws of its forefeet.

in hollow trees, liana curtains, or epiphytes. It often suns itself stretched out on a limb.

Concerning activity and territory size, we have accurate data only on the African tree pangolin. E. Pagès has tagged individuals of this species with small transmitters and tracked them by radiotelemetry. The territory of a male covers 38 to 63 acres (15 to 25 hectares), that of a female 8 to 10 acres (3 to 4 hectares). Rest periods are spent in hollow trees, epiphytes, or excavated termite hills. The sleeping places are located 30–40 ft (10 to 15 m) above the ground. Sometimes individuals spend the day in holes in the ground. The African tree pangolin spends most of its active period in searching for termites on the ground. When an animal senses an enemy, it climbs a tree until it is out of reach. The females are active for 3 to 4 hours every night, traversing an average of 1300 ft (400 m). The far more enterprising males are about for 2 to 10 hours, traveling some 2200 ft (700 m) or more. During the rainy season, activity diminishes and the animals may be inactive for several days.

The females traverse their territories following a zig-zag or circular course, usually returning to their previous shelter. Thus they utilize only part of the territory. The result of this type of motion and markings with the secretion of the posterior glands is that females rarely meet. If a female leaves her home ground and encounters fresh marking of another female, it returns into its own territory. The males survey much of their territory each night, the path traversed being longer and more linear. This permits the males to encounter females, since the male territory overlaps several female territories. If the female is not ready to mate, the meeting with the male is of short duration.

The African pangolins are known ot be good swimmers and to cross rivers with great ease. The giant pangolin will hold only its head above water, the body being submerged. Swimming movements resemble those of the "doggy paddle," and the animals soon tire. The two tree dwellers swim rapidly with undulating movements. E. Pagès notes that the African tree pangolin fills its stomach and intestine with air (270 ml) when swimming, thus helping it to float.

Pangolins are plantigrade. The large digging claws of the forefeet are held inward by terrestrial species, which walk on the outer sole. The arboreal forms bend the claw downward when traveling on the ground. In walking on the ground, chiefly the hind limbs are used; often distances will be traversed on the hind feet only. At a fast pace, the tail is lifted from the ground. It is surprising how rapidly pangolins move over the ground (the Cape pangolin 160 ft or 50 m per minute, the African tree pangolin 180 ft or 60 m). From time to time, they halt, rise with the support of the tail, sniff the air, and reconnoiter.

The African tree pangolin and the long-tailed pangolin are extraordinarily good climbers. Tree trunks are climbed lineman fashion, front and hind legs taking hold at the same time. The long, muscular and highly mobile tail is very useful in climbing. The tip of the tail with its bare spot constantly explores the substrate, seeking a purchase. The two tree dwellers and the Malayan pangolin can hang by the tip of the tail. If they find no

Chinese pangolin (Manis pentadactyla)
Malayan pangolin (Manis javanica)
Indian pangolin (Manis crassicaudata)

branch to grasp with the forefeet, they will climb up their own tail. In captivity, pangolins are regular escape artists.

All pangolins can roll up more or less completely, the African tree pangolin and the long-tailed pangolin being able to form a ball. They adopt this posture when sleeping and for defense. The only enemies of the pangolin are humans, the leopard, and the python. Burrows provide some protection during the day, and at night the color of the scales is excellent camouflage. The scales and rolling up behavior are good defenses against small and medium-size predators. On the other hand, the scales are no protection against the bites of soldier ants and termites. Insects lodged under the scales are rendered harmless when the pangolin crawls between branches and the like, thus crushing them.

Although the pangolin's hearing is acute, it has little function in intraspecific relationships; no well-distinguished sounds are uttered; the only sound is an aggresive snort. The sense of sight is better developed than one might suppose from the small eyes, but it plays a minor part for the nocturnal species. Pangolins have virtually no facial gestures (only closing the eyes or putting out the tongue). The sense of smell is important for finding insects. It plays a substantial role in communication with conspecifics; most important information is transmitted by this means. Tree branches and other places are marked with urine. Terrestrial species mark with droppings also, together with secretions from the posterior glands. The glands in and around the posterior are very important. They serve for territorial marking, but they also provide important information concerning the psychophysiological state of a pangolin to its conspecifics. These scents, for example, prevent aggressive males from approaching each other too closely in the wild. They also facilitate the finding of a mate, and are important in the mother-offspring relationship. The functions of the skin glands, that is to say of the body odor, have been little studied. The urine is employed in defense; if a pangolin is unrolled by force, a stream of urine strikes the attacker. At the same time feces are voided, "perfumed" with glandular secretion.

The African pangolins give birth to one young at a time; it is said of the Asian species that they may have two offspring. In the newborn, the scales do not yet overlap, and their color is lighter than in the adult. For the Cape pangolin, a gestation period of 140 days is reported; that of the African tree

The Indian pangolin, like its two Asian cousins, is chiefly terrestrial, although it will climb trees readily and often. On the ground, all pangolins are plantigrade quadrupeds, but they rise on their hind legs for fast running.

PANGOLINS

Pangolins (Pholidota)

Nomenclature English common name Scientific name French German	Approximate Size Body length Tail length Weight	Distinguishing Features	Reproduction Gestation period Young per birth Weight at birth
Temminck's ground pangolin; Temminck's pangolin; cape pangolin; *Manis (Smutsia) temmincki* Pangolin de Temminck Steppenschuppentier	20 in.; 50 cm 14 in.; 35 cm 33–39.6 lb; 15–18 kg	Large, gray-brown to dark brown scales; skin white with fine dark hairs	140 days 1 10.7–14.2 oz; 300–400 g
Giant ground pangolin; giant pangolin *Manis (Smutsia) gigantea* Pangolin géant Riesenschuppentier	2.5–2.6 ft; 75–80 cm 1.6–2.2 ft; 50–65 cm 0.7–1.1 oz; 20–30 g	Large, gray-brown scales; white skin and sparse hairs	Not known 1 14.2–17.8 oz; 400–500 g
Long-tailed pangolin *Manis (Uromanis) tetradactyla (longicaudata)* Pangolin à longue queue Langschwanz-Schuppentier	12–16 in.; 30–40 cm 2–2.3 ft; 60–70 cm 4.4–5.5 lb; 2–2.5 kg	Scales larger than in three-pointed pangolin, dark brown with yellowish edge; black skin and hair; long prehensile tail	Not known 1 3.6–5.4 oz; 100–150 g
Three-pointed pangolin; African tree pangolin *Manis (Phataginus) tricuspis* Pangolin à écailles tricuspides Weißbauch-Schuppentier	14–18 in.; 35–45 cm 16–20 in.; 40–50 cm 3.9–5.3 lb; 1.8–2.4 kg	Comparatively small, brown-gray scales; dorsal scales have three pronounced points (hence "tricuspis"), often worn down in older individuals; white skin and hair; prehensile tail	150 days 1 3.2–5.4 oz; 90–150 g
Indian pangolin *Manis (Phatages) crassicaudata* Pangolin indien Vorderindisch-Ceylanisches Schuppentier	2–2.2 ft; 60–65 cm 1.5–1.8 ft; 45–55 17.6–19.8 lb; 8–9 kg	Large, pale yellow-brown scales; brown skin and hair	Not known 1–2 8.2–8.6 oz; 230–240 g
Malayan pangolin *Manis (Paramanis) javanica* Pangolin de Malaisie Javanisches Schuppentier; Malaiisches Schuppentier	1.6–2 ft; 50–60 cm 1.6–2.6 ft; 50–80 cm Not known	Amber-yellow to brown-black scales; white skin with fine light hairs	Not known 1–2 Not known
Chinese pangolin *Manis (Manis) pentadactyla* Pangolin à queue courte Chinesisches Schuppentier; Ohrenschuppentier	1.6–2 ft; 50–60 cm 12–16 in.; 30–40 cm 15.4–19.8 lb; 7–9 kg	Dark brown scales; gray-white skin and hair	Not known 1–2 Not known

The African giant pangolin (portrait, top), which may measure nearly 60 inches from tip to tail (below), is a distinctly terrestrial inhabitant of the rainforest.

pangolin is about 150 days. When a baby Chinese pangolin nurses, the mother lies on her back or side. While resting, she holds it pressed to her abdomen. When the mother is feeding, the offspring is left alone. The newborn African tree pangolin climbs up on its mother immediately after birth and finds the pectoral nipples. It is already able to hold on by the claws of the forefeet, either to the mother or to a limb, although it cannot yet walk. While sleeping and when alarmed, the mother rolls up around the young, which rolls up also. Beginning in the first few days, the little one makes small excursions with the mother, riding at the base of the tail and holding on to the scales (the same mode of transport is found in other species). The mother occasionally leaves the baby behind on a limb and picks it up after a while. The youngster is bound to accompany its mother in the search for food quite soon, as she cannot convey the insect food to the den. Weaning begins at

COMPARISON OF SPECIES

Life Cycle Weaning Sexual maturity Life span	Food	Enemies	Habit and Habitat	Occurrence
Not known	Termites, ants	Leopard, python	Nocturnal ground dweller on steppes and savannas; day spent in burrows	Severely threatened
Not known	Termites, ants	Leopard, python	Nocturnal ground dweller in rain forests; digs burrows for daytime shelter	Not abundant
Not known	Predominantly ants	Leopard, python	Predominantly diurnal in trees of rain forests; sleeps in hollow trees and epiphytes	More rare than three-pointed pangolin
Not known	Termites, ants	Leopard, python	Nocturnal on ground and in trees in rain forests; territory size 37.5–62.5 acres (15–25 hectares) for male, 7.5–10 acres (3–4 hectares) for female	Most common African species
Not known	Termites, ants, other insects	Leopard, python	Nocturnal on ground and in trees; prefers hilly terrain	Endangered
Not known	Termites, ants	Leopard, python	Nocturnal on ground and in trees on open, sandy terrain	Endangered
Not known	Termites, ants	Leopard, python	Nocturnal ground dweller, but good climber; digs tunnels	Endangered

the age of three months. The juvenile first eats insects to be found between the mother's scales, then it picks up insects while the mother is breaking open a termite nest, and later it digs for itself. The choice of food and the digging activity seem to be innate. At the age of five months, before the next birth, the young pangolin leaves its mother. Females readily adopt young of other mothers. Mechanical and olfactory contact with the young is important. In the case of the African tree pangolin, the long tails are intertwined during mating, and the pair lie ventrally opposed.

The only available observations on life span are those made in captivity. Since pangolins are very hard to keep, the results must be accepted with reservations. Indian pangolins have survived for four years and a few months in captivity in their home country. All other species have had briefer survival times. As dietary specialists, these animals are extremely difficult to keep outside of the tropics.

The long-tailed pangolin, whose tail accounts for about two-thirds of its over-all length, lives in trees. It climbs the tree trunks of the rainforest with the help of its prehensile tail in the style of an inchworm.

The African tree pangolin is at home on the ground and in the trees of the rainforest alike. The termites preferred as food are found at night chiefly on the ground, but the sleeping places to which it retires during the day are mostly hollows of trees, epiphytes, or excavated termite nests far aloft.

Pangolins are often infested with external and internal parasites. Ticks dig in at the base of the scales. On the three forest-dwelling African species, the true ticks (family Ixodidae) *Amblyomma cuneatum, Rhipicephalus senegalensis, Ixodes rasus,* and *Aponomma exornatum* were found. The Indian pangolin is attacked by *Amblyomma javanense.* Intestinal parasites of African pangolins include the tapeworms *Metadavainea aelleni, Raillietina rahmi,* and *R. anopolocephaloides.* The protozoan *Eimeria tenggilingi* has been found in the Malayan pangolin. The following nematodes were indentified in the Chinese pangolin: *Chenofilaria filaria, Leipernemia leiperi, Trichochenis meyersi,* and *Dipetalonema fausti.*

In Asia, the flesh of pangolins is a delicacy, and among the Chinese it is held to be an aphrodisiac. The African pangolins are also eaten by the natives. However, the giant pangolin is taboo for some tribes. The animals, defenseless against humans, are caught in snares, killed with clubs or, for example in Indonesia, hunted with dogs. In Africa and Asia, the claws and scales are still in use today as medicine, philters, and charms. The Cape pangolin is called *bwana mganga* "bwana doctor" in East Africa because every part of its body is said to possess healing virtues. That may be one reason why this species is much endangered and listed as threatened. Scales worn on a neck or wrist chain impart great strength. Finger rings made out of scales avert the evil eye. Pulverized scales are good for nosebleed. Formerly in Indonesia, pangolin hides were made into shields that would deflect arrows. In 1925, Indonesia exported skins and scales corresponding in number to between 7000 and 10,000 individuals. Since 1931, exportation has been prohibited, but the trade in pangolin scales still flourishes in Asia. Today the survival of the three forest-dwelling species of Africa is additionally jeopardized by extensive clearing of the rainforest.

All Asiatic pangolins are threatened, and so listed in Appendix II to the Washington Convention, which means that they may be imported to and traded in Europe only with a permit from the authorities in the country of origin. The Cape pangolin of Africa is especially threatened with extinction, and is consequently included in Appendix I. Traffic in these animals or their parts (for example, scales) is completely prohibited; exceptions may be made only for scientific zoological parks, museums, and research institutions.

Not a fallen pine cone, but a rolled-up "pine cone creature," as the pangolins have been called.

APPENDIX

References

Ankel, F.: Einführung in die Priamtenkunde. Grundbegriffe der modernen Biologie, Bd. 6. Stuttgart, 1970.
Berger, G. and E. Tylinek: Das grosse Affenbuch. Hannover, 1984.
Bischof, N.: Das Rätsel Ödipus. Die biologischen Wurzeln des Urkonfliktes von Intimität und Autonomie. Munich, 1985.
Bramblett, C. A.: Patterns of primate behaviour. Palo Alto, 1976.
Campbell, B. G.: Humankind emerging. Boston, Toronto, 1985.
Chalmers, N.: Social behaviour in primates. London, 1979.
Charles-Dominique, P. et al. (Eds.): Nocturnal Malagasy primates: Ecology, physiology and behavior. New York, 1980.
Clutton-Brock, T.H. (Ed.): Primate ecology. London, 1977.
Coimbra-Filho, A.F. and R.A.Mittermeier: Ecology and behavior of neotropical primates. Rio de Janeiro, 1981.
Deluce, J. and H.T. Wilder: Language in primates. Berlin, 1983.
Doyle, G. A. and R.D.Martin (Eds.): The study of prosimian behavior. New York, 1979.
Eibl-Eibesfeldt, I.: Die Biologie des menschlichen Verhaltens. Grundriss der Humanethologie. Munich, 1984.
Epple, G.: The behavior of marmoset monkeys (Callitrichidae), Primate Behavior, 4: 195-239, 1975.
Fossey, D.: Gorillas in the mist. New York, 1983.
Franzen, J.L.: Die Primaten als stammesgeschichtliche Basis des Menschen. In: Kindlers Enzyklopädie 'Der Mensch', Bd.I, 557–597, Zürich 1982.
Goodall, J.: The Chimpanzees of Gombe: Patterns of behavior. Cambridge (Mass.), London, 1986.
Grande Enciclopedia Illustrata degli Animali, Mammiferi. Milan, 1981.
Grzimek, B.: Vom Grizzlybär zur Brillenschlange. München 1979 Einsatz für Afrika. Munich, 1980.
Hershkovitz, P.: Living New World monkeys (Platyrrhini). Chicago, 1977.

Hill, W.C.O.: Evolutionary biology of the primates. London, New York, 1972.
Hrdy, S.B.: The langurs of Abu. Cambridge (Mass.), 1977.
Janschke, F.: Orang-Utans in Zoologischen Gärten. Munich, 1972.
Johanson, C.D., Edey, M.: Lucy. Die Anfänge der Menschheit. Munich, 1982.
Jolly, A.: Die Entwicklung des Primatenverhaltens. Stuttgart, 1975.
Kawamura, S.: Die Ausbreitung einer Subkultur bei Rotgesichtsmakaken (Macaca fuscata). In: W.Wickler, U.Seibt: Vergleichende Verhaltensforschung. Hamburg, 1973.
Kleiman, D.G. (Ed.): The biology and conservation of the callitrichidae. Washington, 1977.
Knussmann, R.: Vergleichende Biologie des Menschen. Lehrbuch der Anthropologie und Humangenetik. Stuttgart, New York, 1980.
 Anthropologie. Handbuch der vergleichenden Biologie des Menschen. Stuttgart, New York, 1987 ff.
Koehler, A.: Intelligenzleistungen und Werkzeuggebrauch bei Primaten. In: Kindlers Enzyklopädie 'Der Mensch', Bd.I, 598–643, Zürich, 1982.
Kummer, H.: Sozialverhalten der Primaten. Berlin, Heidelberg, New York, 1975.
Lawick-Goodall, J. van: Wilde Schimpansen. Reinbek, 1971.
Leakey, R.E.: Die Suche nach dem Menschen. Wie wir wurden, was wir sind. Frankfurt am Main, 1981.
Lethmate, J.: Problemlöseverhalten von Orang-Utans (Pongo pygmaeus). Berlin, Hamburg, 1977.
Lindburg, D.G. (Ed.): The macaques: Studies in ecology, behavior and evolution. New York, 1980.
Macdonald, D. (Ed.): The Encyclopaedia of Mammals, 2 Vol. London, 1984.
Mackinnon, J.: The ape within us. London, 1978.
 In search of the red ape. London, 1974.
Moss, C.: Portraits in the wild. New York, 1983.
Moynihan, M.: The New World primates. Princeton, 1976.

Napier, J.R. and P.H.Napier: The natural history of the primates. London, 1985.
Niemitz, G.: Biology of tarsiers. Stuttgart, 1984.
Nowak, R.M. and J.L.Paradiso (Eds.): Walker's Mammals of the World. 2 Vol. Baltimore, London, 1983.
Pages, E.: Etude éco-éthologique de Manis tricuspis par radiotracking. Mammalia, 93, 613–641, 1975.
Petter, J.J., R.Albinac and Y.Rumpler: Faune de Madagascar. Paris, 1976.
Premack, D. and A.Premack: The mind of an ape. New York, 1983.
Preuschoft, H., D.Chivers, W.Bockelman and N.Creel (Eds.): The lesser apes. Edinburgh, 1984.
Reader, J.: Die Jagd nach dem ersten Menchen. Eine Geschichte der Paläanthropologie von 1857 bis 1980. Basel, 1982.
Rijksen, H.: A field study on Sumatran orangutans. Wageningen, 1978.
Rijt-Plooij, H. v.d. and F.X.Plooij: Schimpansen. In: Die Psychologie des 20. Jahrhunderts, Bd.VI, 177–188. Zürich, 1978.
Sauer, E.G.: Zur Biologie der Zwerg- und Riesengalagos. Z.d. Kölner Zoo, 17: 67–84, 1974.
Schultz, A.H.: Die Primaten. Lausanne, 1972.
Sussman, R. W. (Ed.): Primate ecology: Problem oriented field studies. New York, 1979.
Struhsaker, T.T.: The red colobus monkey. London, 1975.
Terborgh, J.: Five New World primates. Princeton, 1983.
Thenius, E.: Die Evolution der Säugetiere. Eine Übersicht über Ergebnisse und Probleme. Uni-Tb. 865. Stuttgart, New York, 1979.
Tiger, L. and R.Fox: Das Herrentier. Steinzeitjäger im Spätkapitalismus. Munich, 1971.
Vogel, Ch.: Der Hanuman-Langur, ein Parade-Exempel für die theoretischen Konzepte der "Soziobiologie"?. Verh. Dtsch. Zool. Ges., 73–89, 1979.
Zimmermann, E., P.Zimmermann, and A.Zimmermann: Soziale Kommunikation bei Plumploris (Nycticebus councang). Z.d. Kölner Zoo, 22: 25–36, 1979.

Authors of this volume

Dr. Alison Badrian, born 1946 in Dublin, Ireland. Instructor at the State University of New York (Stony Brook). Principal fields of study: behavior and environment of the pygmy chimpanzee.

Dr. Noel Badrian, born 1948 in Johannesburg, South Africa. Instructor at the State University of New York (Stony Brook). Principal fields of study: behavior and environment of the pygmy chimpanzee.

Monica Borner, born 1945 in Zurich, Switzerland. Project Director, Frankfurt Zoological Society, Arusha, Tanzania. Principal fields of study: nature and animal preservation, animal surveys, vegetation surveys, population dynamics.

Prof. Dr. Günter Bräuer, born 1949 in Klagebach-Altena, Germany. Professor of anthropology at the University of Hamburg. Principal fields of study: human phylogeny, population biology, prehistoric anthropology.

Prof. Dr. Irenäus Eibl-Eibesfeldt, born 1928 in Vienna, Austria. Director, Human Ethology Research Station, Max Planck Institute for Behavioral Physiology, Seewiesen, Upper Bavaria; Professor at the University of Munich. Principal fields of study: human ethology, zoology.

Dr. Dian Fossey, born 1932 in San Francisco, California, United States. Until assassination in December 1985, Director of Karisoke Research Center, Ruhengeri, Ruanda. Principal field of study: behavior of the mountain gorilla.

Prof. Dr. Dr. h.c. Jane Goodall, born 1934 in London, England, Director, Gombe Stream Research Centre, Tanzania. Principal field of study: behavior of chimpanzees and baboons.

Prof. Dr. Dr. h.c. Bernhard Grzimek, born 1909 in Neisse, Silesia, Germany; died 1987 in Frankfurt-am-Main. Former Director of the Frankfurt Zoological Garden and Professor at the University of Giessen; Honorary Professor at the University of Moscow; President of the Frankfurt Zoological Society for Protection of Endangered Animals throughout the World and for the Conservation of Nature; Trustee of the National Parks of Tanzania and Uganda; editor-in-chief of the journal *Das Tier (The Animal)* and editor/publisher of this Encyclopedia.

Prof. Dr. Dietrich von Holst, born 1937 in Danzig, Germany. Professor of animal physiology at the University of Bayreuth. Principal fields of study: social behavior and social stress, chemical communication in mammals.

Prof. Dr. Jan A.R.A.M. van Hooff, born 1936 in Arnheim, Netherlands. Professor of ethology at the National University of Utrecht (Netherlands). Principal field of study: ethology of social behavior of primates.

Prof. Dr. Klaus Immelmann, born 1935 in Berlin, Germany; died 1987 in Bielefeld, Germany. Professor of behavioral physiology at the University of Bielefeld. Principal fields of study: behavioral research, particularly early development of behavior (imprinting).

Prof. Dr. Kosei Izawa, born 1939 in Tokyo, Japan. Professor at Miyagi University of Education, Sendai (Japan). Principal field of study: primatology, particularly New World monkeys and Japanese macaques.

Marvin L. Jones, born 1928 in Philadelphia, Pennsylvania, United States. Registrar of the Zoological Society of San Diego, California, and Director of the Studbook for Orangutans. Principal fields of study: life span of wild animals in captivity, organization and administration of zoos.

Prof. Dr. Rainer Knussmann, born 1936 in Minz, Germany. Professor of anthropology at the University of Hamburg. Principal fields of study: population biology, sexual anthropology, constitutional anthropology, paleoanthropology, anthropological primatology.

Dr. Kurt Kolar, born 1933 in Vienna, Austria. Professional writer on zoology, Chairman of the Austrian Zoological Society. Principal fields of study: behavior of prosimians and of psittacines, household animal care, keeping of animals in municipal parks.

Prof. Dr. Hans-Jürg Kuhn, born 1934 in Heidelberg, Germany. Professor of anatomy at the University of Göttingen and Director of the German Center for Primates, Göttingen. Principal fields of study: primatology and comparative anatomy.

Dr. Kathy MacKinnon, born 1948 in Newcastle-upon-Tyne, England. Consultant to the International Union for Conservation of Nature and Natural Resources (IUCN). Principal fields of study: ecology of the tropics, organization and administration of national parks.

Dr. Walburga Moeller, born 1937 in Aachen, Germany. Free-lance biologist. Principal field of study: translation of scientific and popular scientific literature.

Priv. Doz. Dr. Ewald Müller, born 1946 in Pforzheim-Hohenwart, Germany. Academic associate to the Department of Biology at the University of Tübingen. Principal fields of study: adaptive strategies in the energy and water economies of mammals, influence of environmental factors.

Dr. A. George Pook, born 1949 in Salisbury, England. Ethologist and instructor in biology at Cheltenham (England). Principal fields of study: behavior (especially communication) of primates, in particular New World monkeys.

Prof. Dr. Holger Preuschoft, born 1932 in Hanau, Germany. Professor of anatomy at the University of Bochum. Principal fields of study: functional morphology, primatology, anthropology, phylogeny.

Prof. Dr. Urs Rahm, born 1925 in Basle, Switzerland. Director of the Museum of Natural History in Basle. Principal fields of study: biology and ecology of mammals, in particular of Africa and Europe.

Dr. Cornelia Schäfer-Witt, born 1955 in Wiesbaden, Germany. Scientific associate of working group on primate ethology at the University of Kassel. Principal fields of study: comparative research on social behavior of primates.

Dr. George B. Schaller, born 1933 in Berlin, Germany. Director of Wildlife Conservation International (of the New York Zoological Society), Bronx, New York. Principal fields of study: behavior and protection of large mammals.

Prof. Dr. Erich Thenius, born 1924 in Abbazia, Italy (now Opatija, Yugoslavia). Professor emeritus of paleontology at the University of Vienna. Principal fields of study: vertebrate, in particular mammalian, paleontology, zoogeography.

Prof. Dr. Christian Vogel, born 1933 in Berlin, Germany. Professor of anthropology at the University of Göttingen. Principal fields of study: evolution of humans and non-human primates, ethology and sociobiology of primates.

Priv. Doz. Dr. Christian Welker, born 1948 in Berlin, Germany. Scientific associate, working group on zoology and comparative anatomy, primate ethology, at the University of Kassel. Principal fields of study: comparative research on social behavior of primates.

Dr. Paul Winkler, born 1951 in Lünen, Germany. Scientific officer of the Deutsche Forschungsgemeinschaft (East Germany), Behavioral Ontogeny Concentration Program, Göttingen. Principal fields of study: ecology and population dynamics of primates.

Jürgen Wolters, born 1951, Emmerich, Lower Rhineland, Germany. Director of the Primate Station of the School of Biology at the University of Bielefeld. Principal fields of study: primatology, applied ethology.

Illustration Credits

Photographs

T. Angermayer, Holzkirchen, Upper Bavaria: 39, 58 center, 109, 125, 131, 281, 624 bottom and 625; 626 top
Anthony Verlag, Starnberg, Upper Bavaria: 541
W. Bartmann, Dortmund: 586, 587, 589
Bavaria, Gauting, Upper Bavaria: 304, 305, 531 right (Eckebrecht), 533 left (Schmid-Tannawald), 534 left, 542 (Alexandre), 560, 563 (Pundsack)
Biofotos, Farnham (England): 13, 222 top, 242, 243, 332 left, 414 left
H. and U. Borner, Thalwil (Switzerland): 368, 372, 402, 404 top, 418, 419, 420, 421
C. K. Brain, Pretoria: 504 top
G. Bräuer, Hamburg: 501 left, 502, 503 photos top 4th and 5th from left, bottom 1st, 2nd and 4th from left, 504 bottom
D. L. Brill, Fairburn, United States: 508
D. Bryceson, Tanzania National Park: 27
B. Coleman, Uxbridge, Middlesex: 57 bottom right (Cubitt), 62 top, 66 (Myers), 78 (Barlett), 79 (Burton), 91 (Lyon), 96 (Sauer), 119 (Williams), 149 (Marigo), 150 left (Williams), 161 left (Williams), 209 bottom, 211, 214 right (Cubitt), 278 (Freeman), 285 (Reinhard), 287 (Kaufmann), 314 left (Compost), 342 top (Thoma), 363 top (Williams), 369 right, 383 right and left, 406-407 (Price), 426 (Cancalosi), 456-457 (Campbell), 463 (Davey), 478-479 (Albrecht)
U. H. Day, London; 501 right
I. Eibl-Eibesfeldt, Seewiesen, Upper Bavaria: 544 left and right, 546 all, 548 all, 549 all, 550 all, 551 all, 553 all, 554, 556 all, 557, 558 all, 559 all, 561 lower left, 562, 564 all, 566 all, 567 all, 569 left and right
B. Eichhorn, Wiesbaden: 348
Field Museum, Chicago: 582
D. Fossey, Ruhengeri, Ruanda: 449
H. Glück, Vienna: 561 top and bottom left
O. Gollnek, Hamburg: 529 top
J. Goodall, Kigoma (Tanzania): 376, 379, 392 right, 470-471, 474 bottom, 476 left, 480, 481, 552, 555
W. Götz, Ammerwil (Switzerland): 237, 241, 246-247 all
B. Gray: 466
H. N. Hoeck, Constance: 378 top and bottom, 436 top, 448, 453, 455, 460
D. von Holst, Bayreuth: 4 left, 5, 6 left and right, 8 left and right top and bottom, 9 left and right, 10, 11 left and right, 12
J. A. R. A. M. van Hooff, Utrecht: 257 right
Jacana, Paris: 20 (Varin, 127 right, 139, 240, 348 center, 365
D. C. Johanson, Berkeley, California: 502-503 photo upper left
Keystone, Hamburg: 498
H. P. Klinke, Munich: 520
R. Knussmann, Hamburg: 533 right, 536 left and right, 537 all
E. Koning: 375 top
H. Kummer, Zurich: 245 bottom, 252 top and bottom, 284
H. van Lawick, London: 375 bottom, 377, 392 left, 465 bottom, 475, 476 right
J. Lethmate, Ibbenbüren: 384-385 all, 386-387 all, 388-389 all, 393

APPENDIX

J. Mackinnon, Haddenham (England): 80, 81, 103, 209 top, 212, 213 bottom, 222, 233, 236, 297 bottom, 303 top, 325, 329, 333 bottom, 334–335, 342 bottom, 354, 408 bottom, 409 top
W. Mayr, Grossenrade: 183, 191 left and right, 200, 201 left and right
E. Merz, Sélestat (France): 205, 228 all
R. Mittermeier, Washington: 35 top right, 36–37, 42 left, 47 top right, 48, 49, 50 top, 53, 57 bottom left, 59, 69 top, 135, 136, 137 top left and bottom, 138, 140–141, 143, 153 left, 156, 160, 161 right, 162 top, 171 top and bottom, 176, 184 left, 189 right, 204, 223 left, 232, 296 top, 303 bottom, 336–337, 360 left, 361 right, 404 top, 486
Musée de l'Homme, Paris: 502–503 top 3rd and 5th photo from left
R. Noe/B. Sluijter, Utrecht (Netherlands): 22 right, 238 bottom, 244 top, 245 top, 283
Okapia, Frankfurt-am-Main: 1 (McHugh), 19 right, 25 (Russkinne), 38 center, 87 top, 220 right, 224 (McHugh), 271, 275 right, 275 left, 293 right and left (Kowei), 324, 348 top, 348 center, 366, 369 left, 373, 374, 394, 396, 400, 403, 412–413 (McHugh), 415 top (McHugh), 415 bottom (Tijima), 424 left (McHugh), 428 top, 431 right, 433, 442 left, 444 left, 447 top and bottom, 454, 461, 462 (Fossey), 477, 483, 484 right and left, 485 right and left, 594 (Root), 595 bottom (Root), 614 bottom (Foott)
Prenzel-IFA, Munich: 122, 527 top left, 539 top left (Aberham), 570 (Ostgathe), 527 top left, 527 top right, 539, 570
H. Preuschoft, Bochum: 294 top and bottom, 295, 331 left and right, 340
U. Rahm, Basle: 638 top and bottom, 639, 640
H. J. Roersma, Vlaardingen (Netherlands): 244 bottom
D. M. Rumbaugh, Atlanta, Georgia: 398
G. B. Schaller, New York: 441
D. Schilling, Munich: 223 right, 231, 333 top, 341, 343, 352–353 all
C. R. Schmidt, Zurich: 423, 428 bottom
Seaphot, London: 17 right, 182, 230 top, 249 top, 532 left (Stevenson), 624 top
P. K. Seth, Delhi: 299
Silvestris, Kastl, Upper Bavaria: 7 top (Lummer), 17 left (Denzau), 18 (Nature), 19 left (Varin), 21 (Dani/Jeske); 21 top right, 22 left (Varin), 23 left (Varin), 23 right (Visage), 24 (Angermayer), 28 (Wothe), 31 (Visage), 35 top left (Kerneis), 35 bottom (Nature), 38 top (Visage), 38 bottom (Visage), 38 bottom (Varin), 41 (Lummer), 42 bottom (Visage), 43 (Visage), 54–55 (Visage), 56 top right (Varin), 56 bottom left (Chaumeton), 56 bottom right (Visage), 57 top (Lane), 58 top left (Visage), 58 right (Visage), 60–61 (Visage), 63 (Visage), 64–65 (Visage), 67 left (Nature), 67 right (Visage), 68–69 (all Visage), 71 top (Visage), 75 (Visage), 76 (Visage), 77 top (Hosking), 82 (Balleau), 84 (Höfels), 85 top (Devez), 85 bottom (Lummer), 89 (Tercafs), 90 (Nature), 92 right (Varin), 100–101 (Fogden), 105 (Fogden), 117 (Dani/Jeske), 123 (Kerneis), 127 left (Nature/Gohier), 128, 129 (Fogden), 132 (Nature/Lanceau), 137 right (Lummer), 142 (Dubois), 144 left (Visage), 145 (Dani/Jeske), 146–147 (Dani/Jeske), 150 right, 152 (Wothe), 153 right (Wothe), 154–155 (Wothe), 157 (Varin), 158–159 (Ziesler), 161 center (Varin), 162 bottom (Dani/Jeske), 164–165 (Dani/Jeske), 166–167 (McHugh), 168 (Hladik), 169 (Dossenbach), 178 (Nature), 181 (Varin), 184 right (Silvestris), 185 (Visage), 186 (Ziesler), 188 (Dani/Jeske), 190 right (Wothe), 192 (Dani/Jeske), 194 right (Nature/Lanceau), 196 (Ziesler), 197 right (Visage), 208 (Frederick), 208 left (Gohier), 210 left (Nature/Gohier), 210 right, 214 bottom (Lummer), 215 (Lane), 216 left and right (Ziesler), 217 (Silvestris), 218 (Dani/Jeske), 219 (Wothe), 220 left (Bertrand), 222 center (Wothe), 225 (Prenzel), 226–227 (Bertrand), 229 bottom (Visage), 229 top (NHPA), 230 top (Gronefeld), 230 bottom (Gronefeld), 238 top (Dani/Jekse), 239 (Angermayer), 248 (Angermayer), 250–251 (Varin), 253 (Gronefeld), 254–255 (Visage), 256 (Angermayer), 257 left (Angermayer), 262 (Wothe), 263 left (Ziesler), 264 left (Wothe), 264 top and bottom right (Ziesler), 265 (Meyers), 266–267 (Wothe), 272–273 (Angermayer), 277 (Dani/Jeske), 282 bottom (Robert), 282 top (Angermayer), 286 (Bertrand), 288–289 (Bertrand), 290 (Gohier), 292 (Gohier), 296 (Ziesler), 297 left (Ziesler), 300–301 (Ziesler), 302 (Ziesler), 306 (Wothe), 307 top (Wothe), 307 bottom (Denzau), 308 left (Denzau), 311 bottom (Ziesler), 313 (Ziesler), 314 (Nature), 316 (Lane), 319 (Ziesler), 328 (Gronefeld), 330 (Labat), 332 bottom (Varin), 338–339, 344 (Lummer), 345 (Varin), 346–347 (Gronefeld), 348 center (Agence Nature), 356 (Nature/Lanceau), 357 (Dani/Jeske), 360 right (Dani/Jeske), 361 left (Dani/Jeske), 362 left (Bertrand), 362 bottom, 363 left (Dani/Jeske), 364 right and left (Dani/Jeske), 367 (Dani/Jeske), 370 (Hosking), 371 (Dani/Jeske), 380/381 (Dani/Jeske), 390–391 (Dani/Jeske), 401 (Dani/Jeske), 404 bottom (Wothe), 405 (Dani/Jeske), 408 top (Varin), 409 left (Dani/Jeske), 410 and 411 (Dani/Jeske), 416 left (Bertrand), 416 right (Lummer), 417 (Dani/Jeske), 424 right (Dani/Jeske), 425 (Dani/Jeske), 427 right and left (Lummer), 430 (Nature), 431 bottom (Dani/Jeske), 434–435 (Dani/Jeske), 436 bottom (Lummer), 439 (Dani/Jeske), 440 (Dani/Jeske), 442 right (Dani/Jeske), 443 (Dani/Jeske), 445 (Bertrand), 452–453 (Dani/Jeske), 458–459 (Bertrand), 464 (Dani/Jeske), 465 top (Bertrand), 467 left (Bertrand), 467 right (Varin), 468–469, 472–473 (Wothe), 474 top (Varin), 527 bottom left (Lindenburger), 527 bottom right (Wothe), 531 left (Dani/Jeske), 532 right (Lindenburger), 534 right (Gronefeld), 539 bottom left (Bonington), 539 top right (Daily Telegraph), 539 bottom right (Daily Telegraph), 568 right and left (Gronefeld), 572 (Moog), 575 (Dani/Jeske), 578 (Seaphot), 579 right and left, 584 right (Varin), 584 right (Dani/Jeske), 588, 588 left (Gronefeld), 590–591 (Dani/Jeske), 593 (Dani/Jeske), 595 top (Patzelt), 596 left (Patzelt), 596 right (Fogden), 598 (Wothe), 599 right top (Fogden), 599 left top, 599 bottom (Fogden), 600–601 (Dani/Jeske), 602–603 (Dani/Jeske), 604 and 605 (Dani/Jeske), 606 (Fogden), 607 (Dani/Jeske), 608–609 (Fogden), 610–611 (Fogden), 613 (Nature), 614 top (Gohier), 616–617 (Sundance), 618 (Lane), 619 (Patzelt), 627 left (Devez), 630 left (Janou), 630 right (Gens), 632–633 (Devez), 634, 635 left, 635 right (Ferrero), 637 (Wothe), 641 (Hosking)
H. Sprankel, Giessen: 98–99
T. Stack: 249 bottom
F. Strohecker, Quickborn: 490, 491, 510 left and right
Transvaal Museum, Pretoria: 502–503 top 2nd and 3rd photo from left
Chr. Welker, Kassel: 133, 144 right
A. Young, Washington: 134, 177, 202–203
ZEFA, Düsseldorf: 487
G. Ziesler: 110–111, 126, 130, 213 top

Designs for Drawings

G. Bräuer, Hamburg: 492, 495, 496–497, 500 (after J. Jicha), 502–503, 505 (after A. Zihlman)
M. Klima, Frankfurt-am-Main: 3 (left top after Eimerl, DeVore 1966; right after P. Napier), 15 (left after Encyclopaedia of Mammals 1984), 33 (left top after P. Napier, right top after Eimerl, DeVore 1966; bottom after Berger, Tylinek 1984), 107 (left top after P. Napier), 121 (after Hershkovitz 1977), 207 (top after P. Napier, bottom after Berger, Tylinek 1984), 327 (top after P. Napier, bottom after Eimerl, DeVore 1966), 359 (bottom after Berger, Tylinek 1984), 498 (bottom after Eimerl, DeVore 1966), 577, 629
E. Thenius, Vienna: 29, 30, 40, 113, 114, 580

Drawings

H. Bell, Offenbach: 3, 15, 33, 107, 121, 207, 327, 359, 489, 577, 629
E. Bierly: 623, 626
M.-D. Crapon de Caprona, Bielefeld: 29 (after Thenius), 30 (after Thenius), 82 right, 83 (after Müller), 85 top right (after Time-Life), 85 bottom right (after Berger, Tylinek), 87 bottom, 505 (after Bräuer), 524 top (after Knussmann), 528 (after Knussmann), 529 bottom (after Lorenz), 530 (after Knussmann)
H. Diller, Munich: 4 (left), 7 bottom, 46 left and right, 47 top left and bottom, 50 center and bottom, 62 bottom, 70, 77 bottom, 94, 96 right, 97
W. Eigner, Hamburg: 189 left, 190 left, 261, 263 right, 268, 269, 270, 274, 308 right, 311 top, 317, 322, 323
K. Grossmann, Frankfurt: 71 bottom, 112, 115, 116 top and bottom, 117 right and bottom, 118 top and bottom
F. Wendler, Weyarn, Upper Bavaria: 26 (after Napier & Napier), 40 (after Thenius), 86, 114 (after Thenius), 180 (after Pook), 197 left (after Crapon de Caprona), 351 (after Preuschoft), 492 (after Bräuer), 493 (after Pickford), 495 (after Bräuer), 496–497 (after Bräuer), 500 (after Bräuer), 506 (after Hess, Müller-Beck), 512–513 (after Bräuer), 514–515 (after Bräuer), 516–517 (after Bowen), 523 (after Knussmann), 525 top and bottom (after Knussmann), 526 (after Knussmann), 580 (after Thenius)

Maps

All range and distribution maps were prepared by G. Oberländer, Munich.

Index

Aegyptopithecus 492
African tree pangolin *(Manis [Phataginus] tricuspis)* 634, 636, 638
Agile gibbon *(Hylobates lar agilis)* 350-351, 354
Agile mangabey *(Cercocebus galeritus)* 258, 262, 263
Allen's bush baby *(Galago alleni)* 80, 94
Allen's (swamp)monkey *(Allenopithecus nigroviridis)* 280-281, 284
Allenopithecus nigroviridis (Allen's [swamp] monkey or swamp guenon) 280-281, 284
Allocebus trichotis (bushy-eared or hairy-eared dwarf lemur) 48, 72
Alouatta belzebul (black and red howler) 174
Alouatta caraya (black howler) 148, 174
Alouatta fusca (brown howler) 148, 174
Alouatta palliata (mantled howler) 144, 148, 174
Alouatta seniculus (red howler) 144, 148, 174
Alouatta villosa (Guatemalan howler) 148, 174
Anathana ellioti (Indian tupaya or tree shrew) 6, 10
Andean hairy armadillo *(Chaetophractus nationi)* 622
Angwantibo *(Arctocebus calabarensis)* 78, 94
Anteaters (Myrmecophagidae) 583-597
Anubis *(Papio anubis)* 237-245, 258
Aotus trivirgatus (night or owl monkey) 123-126, 172
Arctocebus calaberensis (angwantibo or golden potto) 78, 94
Armadillos (Dasypodidae) 612-626
Assamese macaque *(Macaca assamensis)* 221, 234
Ateles belzebuth (long-haired spider monkey) 160, 163, 176
Ateles fusciceps (brown-headed spider monkey) 163, 176
Ateles geoffroyi (black-handed monkey) 160, 163, 166, 168, 176
Ateles paniscus (black spider monkey) 161, 163, 176
Australopithecus afarensis 495-499
Australopithecus africanus 490, 495, 499-500
Australopithecus boisei 500-505
Australopithecus robustus 500-505
Avahi *(Avahi laniger)* 69, 74
Avahi laniger (woolly lemur or avahi) 69, 74
Aye-aye *(Daubentonia madagascariensis)* 71, 74

Banded leaf monkey *(Presbytis melalophus)* 312, 320
Barbary ape *(Macaca sylvanus)* 205, 222-224, 234
Bear macaque *(Macaca arctoides)* 230-231, 236
Bearded saki *(Chiropotes satanas)* 138, 172
Beeloh *(Hylobates klossi)* 344, 350, 354
Black-and-red howler *(Alouatta belzebul)* 174
Black-and-red tamarin *(Saguinus nigricollis)* 193, 198
Black-bearded saki *(Pithecia hirsuta)* 136-138, 172
Black-cheeked white-nosed monkey *(Cercopithecus ascanius)* 260, 268
Black colobus *(Colobus polykomos)* 322
Black-handed spider monkey *(Ateles geoffroyi)* 160, 163, 166, 168, 176
Black howler *(Alouatta caraya)* 148, 174
Black lemur *(Lemur macaco)* 52-57, 72
Black mangabey *(Cercocebus aterrinus)* 258, 262
Black-penciled marmoset *(Callithrix penicillata)* 188, 192, 198
Black saki *(Chiropotes satanas)* 138, 172
Black spider monkey *(Ateles paniscus)* 161, 163, 176
Black-tailed bush baby *(Galago alleni)* 80, 94
Black uakari *(Cacajao melanocephalus)* 139-142, 174
Blue monkey *(Cercopithecus mitis)* 260, 268-270
Bonnet macaque *(Macaca radiata)* 220, 230, 234
Bonobo *(Pan paniscus)* 361, 486

Booted macaque *(Macaca ochreata)* 236
Brachyteles arachnoides (woolly spider monkey or muriqui) 170-171, 176
Bradypus torquatus (maned sloth) 606, 620
Bradypus tridactylus (pale-throated three-toed sloth) 605, 620
Bradypus variegatus (brown-throated three-toed sloth) 605, 606, 620
Brazilian lesser long-nosed armadillo *(Dasypus septemcinctus)* 620
Brazilian three-banded armadillo *(Tolypeutes tricinctus)* 622
Brelich's snub-nosed monkey *(Pygathrix [Rhinopithecus] brelichi)* 315, 322
Broad-nosed gentle lemur *(Hapalemur simus)* 50, 51, 72
Brown capuchin *(Cebus apella)* 151, 158, 174
Brown-headed tamarin *(Saguinus fuscicollis)* 193, 198
Brown-headed spider monkey *(Ateles fusciceps)* 163, 176
Brown howler *(Alouatta fusca)* 148, 174
Brown lemur *(Lemur fulvus)* 59, 60, 72
Brown-throated three-toed sloth *(Bradypus variegatus)* 605, 606, 620
Buff-headed marmoset *(Callithrix flaviceps)* 188, 192, 198
Buffy saki *(Pithecia albicans)* 136, 172
Bush babies (Galagidae) 77-96
Bushy-eared dwarf lemur *(Allocebus trichotis)* 48, 72

Cabassous centralis (northern naked-tailed armadillo) 622
Cabassous chacoensis (Chacoan naked-tailed armadillo) 622
Cabassous tatouay (greater naked-tailed armadillo) 622
Cabassous unicinctus (naked-tailed armadillo) 622
Cabuella pygmaea (pygmy marmoset 184, 185, 186-188, 198
Cacajao calvus (uakari) 139-142, 174
Cacajao calvus rubicundus (red uakari) 139, 142
Cacajao melanocephalus (black uakari) 139-142, 174
Callicebus moloch (red or dusky titi) 131, 172
Callicebus personatus (masked titi) 131, 134, 172
Callicebus torquatus (widow monkey) 131, 135, 172
Callimico goeldii (Goeldi's monkey) 178-182
Callithrix argentata (silvery marmoset) 188, 190, 198
Callithrix aurita (white-eared marmoset) 188, 191, 198
Callithrix flaviceps (buff-headed marmoset) 188, 192, 198
Callithrix geoffroyi (white-fronted marmoset) 188, 191, 192, 198
Callithrix humeralifer (white-shouldered marmoset) 188, 190, 191, 198
Callithrix jacchus (common marmoset or marmoset) 188-193, 198
Callithrix penicillata (black-penciled marmoset) 188, 192, 198
Campbell's monkey *(Cercopithecus campbelli)* 260, 265, 268
Cape pangolin *(Manis [Smutsia] temminicki)* 631, 638
Capped gibbon *(Hylobates pileatus)* 350-351, 354
Capped langur *(Presbytis pileata)* 310, 320
Cebus albifrons (white-fronted capuchin) 174
Cebus apella (hooded or brown capuchin) 151, 158, 174
Cebus capucinus (white-faced capuchin) 151, 152, 174
Cebus nigrivittatus (weeper or wedge-capped capuchin) 160, 174
Celebes black ape *(Macaca nigra)* 231-232, 234
Celebes crested macaque *(Macaca nigra)* 231-232, 234
Celebes tarsier *(Tarsius spectrum)* 97, 102
Cercocebus (mangabeys) 257-262

Cercocebus albigena (gray-cheeked mangabey) 258, 262, 263
Cercocebus aterrimus (black mangabey) 258, 262
Cercocebus galeritus (agile or crested mangabey) 258, 262
Cercocebus torquatus (collared or sooty mangabey) 258, 262-263
Cercopithecus aethiops (grivet or savanna monkey) 275-279, 284
Cercopithecus aethiops aethiops (gray-green grivet) 275
Cercopithecus aethiops tantalus (tantalus monkey) 275
Cercopithecus albogularis (white-throated guenon or Sykes monkey) 260, 268-269
Cercopithecus ascanius (black-cheeked white-nosed monkey or redtail monkey) 260, 268
Cercopithecus campbelli (Campbell's monkey) 260, 265, 268
Cercopithecus cephus (moustached monkey) 260, 265-268, 272
Cercopithecus denti (Dent's guenon) 260, 268
Cercopithecus diana (diana monkey) 274, 284
Cercopithecus dryas (dryas monkey) 284
Cercopithecus erythrogaster (red-bellied monkey) 260, 270
Cercopithecus erythrotis (red-eared monkey) 260
Cercopithecus hamlyni (owl-faced monkey) 271, 274, 284
Cercopithecus l'hoesti (L'Hoest's monkey) 270, 274, 284
Cercopithecus mitis (blue monkey) 260, 268-270
Cercopithecus mona (mona monkey) 260, 264-265, 268
Cercopithecus neglectus (De Brazza's monkey) 271-274, 284
Cercopithecus nictitans (spot-nosed or greater white-nosed monkey) 260, 268-269, 271
Cercopithecus petaurista (lesser white-nosed monkey) 260, 270
Cercopithecus pogonias (crowned guenon) 260, 265, 268
Cercopithecus preussi (Preuss's monkey) 270-271, 274, 284
Cercopithecus pygerythrus (vervet or vervet monkey) 275, 284
Cercopithecus sabaeus (green monkey) 275, 284
Cercopithecus wolfi (Wolf's monkey) 260
Ceylon toque macaque *(Macaca sinica)* 218, 220-221, 234
Chacma baboon *(Papio ursinus)* 237-243, 258
Chacoan naked-tailed armadillo *(Cabassous chacoensis)* 622
Chaetophractus nationi (Andean hairy armadillo) 622
Chaetophractus vellerosus (screaming hairy armadillo) 622
Chaetophractus villosus (larger hairy armadillo) 622, 624-625
Cheirogaleidae (dwarf lemurs) 43-48
Cheirogaleus major (greater dwarf lemur) 47-48, 72
Cheirogaleus medius (fat-tailed dwarf lemur) 47, 72
Chimpanzee *(Pan troglodytes)* 360-400, 463-485
Chinese pangolin *(Manis pentadactyla)* 636, 638
Chiropotes albinasus (white-nosed saki) 139, 172
Chiropotes satanas (beared saki or black saki) 138, 172
Chlamyphorini (fairy armadillos) 625-626
Chlamyphorus retusus (greater fairy armadillo) 622, 625
Chlamyphorus truncatus (lesser fairy armadillo or pink fairy armadillo) 622, 625
Choloepidae (two-toed sloths) 606-612
Choloepus didactylus (two-toed sloth or unau) 606, 612, 620
Choloepus hoffmanni (Hoffmann's two-toed sloth) 606, 612, 620
Collared mangabey *(Cercocebus torquatus)* 258, 262-263
Colobus guereza (guereza or nothern black and white colobus) 319-324

APPENDIX

Colobus polykomos (black colobus or southern black and white colobus) 322
Common langur *see Presbytis entellus, Presbytis johnii*
Common long-nosed armadillo *(Dasypus novemcinctus)* 614–615, 620
Common marmoset *(Callithrix jacchus)* 188–193, 198
Concolor gibbon *(Hylobates concolor)* 338, 350–351, 354
Coquerel's dwarf (mouse) lemur *(Microcebus conquereli)* 46–47, 72
Cotton-top tamarin *(Saguinus oedipus)* 193, 196–200
Crab-eating monkey *(Macaca fascicularis)* 209, 210–213, 234
Crested gibbon *(Hylobates concolor)* 338, 350–351, 354
Crested mangabey *(Cercocebus galeritus)* 258, 262, 263
Crowned guenon *(Cercopithecus pogonias)* 260, 265, 268
Crowned sifaka *(Propithecus diadema)* 63, 74
Cyclopes didactylus (silky anteater or pygmy anteater) 596–597, 620

Dark-handed gibbon *(Hylobates lar agilis)* 354
Dasypodidae (armadillos) 612–626
Dasypodini (soft armadillos) 614–615
Dasypus hybridus (southern lesser long-nosed armadillo) 620
Dasypus kappleri (Kappler's armadillo or greater long-nosed armadillo) 620
Dasypus novemcinctus (common long-nosed armadillo or nine-banded armadillo) 614–615, 620
Dasypus pilosus (hairy long-nosed armadillo) 620
Dasypus sabanicola (northern lesser long-nosed armadillo) 620
Dasypus septemcinctus (Brazilian lesser long-nosed armadillo) 620
Daubentonia madagascariensis (aye-aye) 71, 74
De Brazza's monkey *(Cercopithecus neglectus)* 271–274, 284
Demidoff's bush baby *(Galago demidovii)* 80, 94
Dendrogale melanura (smooth-tailed tree shrew) 10
Dendrogale murina (smooth-tailed tree shrew) 10
Dent's guenon *(Cercopithecus denti)* 260, 268
Diana monkey *(Cercopithecus diana)* 274, 284
Douc (monkey) langur *(Pygathrix nemaeus)* 315, 322
Drill *(Mandrillus leucophaeus)* 249–252, 258
Dryas monkey *(Cercopithecus dryas)* 284
Dryopithecus 493
Dusky (langur) leaf monkey *(Presbytis obscura)* 308, 320
Dusky titi *(Callicebus moloch)* 131, 172
Dwarf bush baby *(Galago demidovii)* 80, 94
Dwarf lemurs (Cheirogaleidae) 43–48

Eastern needle-clawed bush baby *(Galago inustus)* 80, 94
Emperor tamarin *(Saguinus imperator)* 183, 193, 198
Erythrocebus patas (patas monkey or red guenon) 281–283, 284
Euphractini (hairy armadillos) 619–625
Euphractus sexcinctus (yellow armadillo or six-banded armadillo) 622, 624

Fairy armadillos (Chlamyphorini) 625–626
Fat-tailed dwarf lemur *(Cheirogaleus medius)* 47, 72
Fork-crowned dwarf lemur *(Phaner furcifer)* 48, 72
Formosan rock macaque *(Macaca cyclopis)* 216–220, 234
François' black leaf monkey *(Presbytis francoisi)* 312, 320

Galagidae (bush babies) 77–96
Galago alleni (Allen's or black-tailed bush baby) 80, 94
Galago crassicaudatus (thick-tailed bush baby or greater or giant galago) 78, 94
Galago demidovii (dwarf bush baby or Demidoff's bush baby) 80, 94

Galago elegantulus (western needle-clawed bush baby) 80, 94
Galago inustus (eastern needle-clawed bush baby) 80, 94
Galago senegalensis (Senegal or lesser bush baby) 78–80, 94
Gelada *(Theropithecus gelada)* 252–257, 258
Geoffroy's tamarin *(Saguinus geoffroyi)* 193, 200
Giant anteater *(Myrmecophaga tridactyla)* 583–594, 620
Giant armadillo *(Priodontes maximus)* 622
Giant galago *(Galago crassicaudatus)* 78, 94
Giant (ground) pangolin *(Manis [Smutsia] gigantea)* 631, 638
Giant lemur *(Megaladapis edwardsi)* 35
Goeldi's monkey *(Callimico goeldii)* 178–182
Golden langur *(Presbytis geei)* 310, 320
Golden potto *(Arctocebus calabarensis)* 78, 94
Golden snub-nosed monkey *(Pygathrix [Rhinopithecus] roxellanae)* 315, 322
Gorilla *(Gorilla gorilla)* 424–462
Gorilla gorilla beringei (mountain gorilla) 364, 424, 432, 434–437, 441–462
Gorilla gorilla gorilla (lowland gorilla) 426, 429
Gray-cheeked mangabey *(Cercocebus albigena)* 258, 262, 263
Gray gentle lemur *(Hapalemur griseus)* 48–49, 72
Gray gibbon *(Hylobates lar muelleri)* 336, 350–351, 353, 354
Gray-green grivet *(Cercopithecus aethiops aethiops)* 275
Gray langur *(Presbytis entellus)* 296–306, 311, 320
Great gibbon *(Hylobates syndactylus)* 325, 350, 354
Greater dwarf lemur *(Cheirogaleus major)* 47–48, 72
Greater fairy armadillo *(Chlamyphorus truncatus)* 622, 625
Greater galago *(Galago crassicaudatus)* 78, 94
Greater long-nosed armadillo *(Dasypus kappleri)* 620
Greater naked-tailed armadillo *(Cabassous tatouay)* 622
Greater white-nosed monkey *(Cercopithecus nictitans)* 260, 268–269, 271
Green baboon *(Papio anubis)* 237–245, 258
Green colobus *(Procolobus verus)* 317–318, 322
Green monkey *(Cercopithecus sabaeus)* 275, 284
Grivet *(Cercopithecus aethiops)* 275–279, 284
Guatemalan howler *(Alouatta villosa)* 148, 174
Guereza *(Colobus guereza)* 319–324
Guinea baboon *(Papio papio)* 237, 258

Hairy anteaters (Myrmecophagidae) 583–597
Hairy armadillos (Euphractini) 619–625
Hairy long-nosed armadillo *(Dasypus pilosus)* 620
Hairy-eared dwarf lemur *(Allocebus trichotis)* 48, 72
Hamadryas baboon *(Papio hamadryas)* 239, 258
Hanuman langur *(Presbytis entellus)* 296–306, 311, 320
Hapalemur griseus (gray gentle lemur) 48–49, 72
Hapalemur simus (broad-nosed gentle lemur) 50, 51, 72
Hill monkey *(Macaca assamensis)* 221, 234
Himalayan macaque *(Macaca assamensis)* 221, 234
Hoffmann's two-toed sloth *(Choloepus hoffmanni)* 606, 612, 614
Hominidae 483–573
Hominoids *(Proconsul africanus)* 492
Homo erectus 507–509
Homo habilis 491, 505–507
Homo sapiens 509–520
Hooded capuchin *(Cebus apella)* 151, 158, 174
Hoolock gibbon *(Hylobates hoolock)* 350, 354
Howler monkeys *(Alouatta)* 144–150
Humans 488–573
Hylobates concolor (crested, concolor, or white-cheeked gibbon) 338, 350–351, 354
Hylobates hoolock (white-browed gibbon or hoolock gibbon) 350, 354

Hylobates klossi (kloss gibbon or beeloh) 344, 350, 354
Hylobates lar agilis (agile gibbon or dark-handed gibbon) 354
Hylobates lar carpenteri (lar or white-handed gibbon) 351, 354
Hylobates lar entelloides (lar or white-handed gibbon) 351, 354
Hylobates lar moloch (moloch or silvery gibbon) 350–351, 354
Hylobates lar muelleri (gray or Müller's gibbon) 350–351, 353, 354
Hylobates lar pileatus (capped gibbon or pileated gibbon) 350–351, 354
Hylobates syndactylus (siamang or great gibbon) 325, 350, 354

Indian bonnet macaque *(Macaca radiata)* 220, 230, 234
Indian pangolin *(Manis [Phatages] crassicaudata)* 636, 638
Indian tupaya *(Anathana ellioti)* 6, 10
Indri *(Indri indri)* 69–70, 74

Japanese macaque *(Macaca fuscata)* 208, 216, 234, 286–295
Java monkey *(Macaca fascicularis)* 209, 210–213, 234
John's langur *(Presbytis johnii)* 306–307, 320

Kappler's armadillo *(Dasypus kappleri)* 620
Kenyapithecus 493
Kloss gibbon *(Hylobates klossi)* 344, 350, 354

L'Hoest's monkey *(Cercopithecus l'hoesti)* 270, 274, 284
Lagothrix flavicauda (yellow-tailed woolly monkey) 170, 176
Lagothrix lagotricha (woolly monkey) 162–163, 176
Langurs *(Presbytis)* 297–315
Lar *(Hylobates lar carpenteri* and *H. l. entelloides)* 351, 354
Larger hairy armadillo *(Chaetophractus villosus)* 622, 624–625
Leaf monkey *(Presbytis potenziani)* 310–312, 320
Lemur catta (ring-tailed lemur) 50–52, 72
Lemur fulvus (brown lemur) 59, 60, 72
Lemur leucomystax (white-bearded lemur) 52–56
Lemur macaco (black lemur) 52–57, 72
Lemur mongoz (mongoose lemur) 72
Lemur rubriventer (red-bellied lemur) 72
Leontopithecus rosalia (lion tamarin) 184, 200–204
Lepilemur mustelinus (sportive or weasel lemur) 58–62, 74
Lepilemur ruficaudatus (lesser sportive lemur) 74
Lepilemuridae (sportive lemurs) 58–62
Lesser bush baby *(Galago senegalensis)* 78–80, 94
Lesser fairy armadillo *(Chlamyphorus truncatus)* 622, 625
Lesser mouse lemur *(Microcebus murinus)* 43–46, 72
Lesser sportive lemur *(Lepilemur ruficaudatus)* 74
Lesser white-nosed monkey *(Cercopithecus petaurista)* 260, 270
Limnopithecus 492
Lion-tailed macaque *(Macaca silenus)* 229–230, 232, 233, 234
Lion tamarin *(Leontopithecus rosalia)* 184, 200–204
Long-haired spider monkey *(Ateles belzebuth)* 160, 163, 176
Long-tailed macaque *(Macaca fascicularis)* 209, 210–213, 234
Long-tailed pangolin *(Manis [Uromanis] tetradactyla [longicaudata])* 636, 638
Loris tardigradus (slender loris) 77–78, 94
Lowland gorilla *(Gorilla gorilla gorilla)* 426, 429
Lyonogale dorsalis (Malayan tree shrew) 6, 10
Lyonogale tana (Malayan tree shrew) 10

Macaca (macaques) 210-232
Macaca arctoides (bear macaque or stump-tailed macaque) 230-231, 236
Macaca assamensis (Assamese or Himalayan macaque or hill monkey) 221, 234
Macaca cyclopis (Formosan rock macaque) 216-220, 234
Macaca fascicularis (crab-eating or Java monkey or long-tailed macaque) 209, 210-213, 234
Macaca fuscata (Japanese or red-faced macaque) 208, 216, 234, 286-295
Macaca maura (moor macaque) 231, 232, 234
Macaca mulatta (Rhesus monkey) 213-216, 234
Macaca nemestrina (pig-tailed macaque) 224-229, 234
Macaca nigra (Celebese black ape or Celebes crested macaque) 231-232, 234
Macaca ochreata (booted macaque) 236
Macaca radiata (bonnet macaque) 220, 230, 234
Macaca silenus (lion-tailed macaque) 229-230, 232, 233, 234
Macaca sinica (toque macaque) 218, 220-221, 234
Macaca sylvanus (Barbary ape) 205, 222-224, 234
Macaca thibetana (Tibetan stump-tailed macaque) 221-222, 234
Macaca tonkeana (Tonkean black ape or Tonkean macaque) 236
Macaques *(Macaca)* 210-232
Malayan pangolin *(Manis [Paramanis] javanica)* 636, 638
Malayan tree shrew *(Lyonogale tana* and *Lyonogale dorsalis)* 6, 10
Mandrill *(Mandrillus sphinx)* 249-252
Mandrillus leucophaeus (drill) 249-252, 258
Mandrillus sphinx (mandrill) 249-252
Maned sloth *(Bradypus torquatus)* 606, 620
Mangabeys *(Cercocebus)* 257-262
Manis pentadactyla (Chinese pangolin) 636, 638
Manis (Paramanis) javanica (Malayan pangolin) 636, 638
Manis (Phatages) crassicaudata (Indian pangolin) 636, 638
Manis (Phataginus) tricuspis (African tree pangolin, three-pointed pangolin) 634, 638
Manis (Smutsia) gigantea (giant [ground] pangolin) 631, 638
Manis (Smutsia) temmincki (cape pangolin, Temminck's pangolin, or Temminck's ground pangolin) 631, 638
Manis (Uromanis) tetradactyla (longicaudata) (long-tailed pangolin) 636, 638
Mantled howler *(Alouatta palliata)* 144, 148, 174
Marmoset (Callithrix jacchus) 188-193, 198
Maroon leaf monkey *(Presbytis rubicunda)* 312-313, 320
Masked titi *(Callicebus personatus)* 131, 134, 172
Megaladapis edwardsi (giant lemur) 35
Mentawai langur *(Presbytis potenziani)* 310-312, 320
Microcebus coquereli (Coquerel's mouse lemur) 46-47, 72
Microcebus murinus (lesser mouse lemur) 43-46, 72
Micropithecus 492
Miopithecus talapoin (talapoin or talapoin monkey) 280, 284
Moloch gibbon *(Hylobates lar moloch)* 350-351, 354
Mona monkey *(Cercopithecus mona)* 260, 264-265, 268
Mongoose lemur *(Lemur mongoz)* 72
Moor Macaque *(Macaca maura)* 231, 232, 234
Mottled-faced tamarin *(Saguinus inustus)* 193, 200
Mountain gorilla *(Gorilla gorilla beringei)* 364, 424, 432, 434-437, 441-462
Mouse lemurs *(Cheirogaleidae)* 43-48
Moustached monkey *(Cercopithecus cephus)* 260, 265-268, 272
Moustached tamarin *(Saguinus mystax)* 193, 198
Müller's gibbon *(Hylobates lar muelleri)* 336, 350-351, 353, 354
Muriqui *(Brachyteles arachnoides)* 170-171, 176

Myrmecophagidae (anteaters) 583-597
Myrmecophaga tridactyla (giant anteater) 583-594, 620

Naked-tailed armadillos *(Cabassous)* 618-619
Nasalis concolor (pig-tailed snub-nosed monkey) 313, 314-315, 322
Nasalis larvatus (proboscis monkey) 313-314, 322
Neandertal 510-511
Night (owl) monkey *(Aotus trivirgatus)* 123-126, 172
Nilgiri langur *(Presbytis johnii)* 306-307, 320
Nine-banded armadillo *(Dasypus novemcinctus)* 614-615, 620
Northern black and white colobus *(Colobus guereza)* 319-324
Northern lesser long-nosed armadillo *(Dasypus sabanicola)* 620
Northern naked-tailed armadillo *(Cabassous centralis)* 622
Northern tamandua *(Tamandua mexicana)* 594, 620
Nycticebus coucang (slow loris) 77, 94

Olive baboon *(Papio anubis)* 237-245, 258
Olive colobus *(Procolobus verus)* 317-318, 322
Orangutan *(Pongo pygmaeus)* 398, 401-423
Owl-faced monkey *(Cercopithecus hamlyni)* 271, 274, 284
Owl monkey *(Aotus trivirgatus)* 123-126, 172

Pale-throated three-toed sloth *(Bradypus tridactylus)* 605, 620
Pan paniscus (pygmy chimpanzee or bonobo) 361, 486
Pan troglodytes (chimpanzee) 360-400, 463-485
Papio anubis (anubis or olive baboon) 237-245, 258
Papio cynocephalus (yellow baboon) 236, 258
Papio hamadryas (hamadryas or sacred baboon) 239, 258
Papio papio (Guinea baboon) 237, 258
Papio ursinus (chacma baboon) 237-243, 258
Patas monkey *(Erythrocebus patas)* 281-283, 284
Perodicticus potto (potto) 78, 94
Phaner furcifer (fork-crowned dwarf lemur) 48, 72
Phayre's leaf monkey *(Presbytis phayrei)* 309, 320
Philippine tarsier *(Tarsius syrichta)* 97, 102
Philippine tree shrew *(Urogale everetti)* 6, 10
Pichi *(Zaedyus pichiy)* 622
Pied tamarin *(Saguinus bicolor)* 192, 200
Pig-tailed macaque *(Macaca nemestrina)* 224-229, 234
Pig-tailed snub-nosed monkey *(Nasalis concolor)* 313, 314-315, 322
Pileated gibbon *(Hylobates lar pileatus)* 350-351, 354
Pilosa (sloths) 597-612
Pink fairy armadillo *(Chlamyphorus truncatus)* 622, 625
Pithecantropus erectus 490
Pithecia albicans (buffy saki) 136, 172
Pithecia hirsuta (black-bearded saki) 136-138, 172
Pithecia monachus (red-bearded saki) 136-138, 172
Pithecia pithecia (white-faced saki) 136-138, 172
Pongo pygmaeus (orangutan) 398, 401-423
Potto *(Perodicticus potto)* 78, 94
Presbytis (langurs) 297-315
Presbytis aygula (Sunda Island leaf monkey) 313, 320
Presbytis cristata (silvered leaf monkey) 309-310, 320
Presbytis entellus (common, gray, or hanuman langur) 296-306, 311, 320
Presbytis francoisi (François' black leaf monkey) 312, 320
Presbytis frontata (white-fronted leaf monkey) 313, 320
Presbytis geei (golden langur) 310, 320
Presbytis johnii (Nilgiri langur, common langur or John's langur) 306-307, 320
Presbytis melalophos (banded leaf monkey) 312, 320

Presbytis obscura (dusky langur or dusky leaf monkey) 308, 320
Presbytis phayrei (Phayre's leaf monkey) 309, 320
Presbytis pileata (capped langur) 310, 320
Presbytis potenziani (Mentawai langur or leaf monkey) 310-312, 320
Presbytis rubicunda (maroon leaf monkey) 312-313, 320
Presbytis senex (purple-faced langur) 307-308, 311, 320
Preuss's monkey *(Cercopithecus preussi)* 270-271, 274, 284
Priodontes maximus (giant armadillo) 622
Proboscis monkey *(Nasalis larvatus)* 313-314, 322
Procolobus badius (red colobus) 317-318, 320
Procolobus verus (green or olive colobus) 317-318, 322
Proconsul africanus 492
Propithecus diadema (crowned sifaka) 63, 74
Propithecus verreauxi (Verreaux's sifaka) 63-69, 74
Propliopithecus 492
Ptilocercus lowii (pentail tree shrew) 10
Purple-faced langur *(Lemur senex)* 307-308, 311, 320
Pygathrix (Rhinopithecus) avunculus (Tonkin snub-nosed monkey) 315, 322
Pygathrix (Rhinopithecus) brelichi (Brelich's snub-nosed monkey) 315, 322
Pygathrix (Rhinopithecus) nemaeus (douc monkey or douc langur) 315, 322
Pygathrix (Rhinopithecus) roxellanae (golden snub-nosed monkey) 315, 322
Pygmy anteater *(Cyclopes didactylus)* 596-597, 620
Pygmy chimpanzee *(Pan paniscus)* 361, 486
Pygmy marmoset *(Cebuella pygmaea)* 184, 185, 186-188, 198

Ramapithecus 493
Red-backed squirrel monkey *(Saimiri oerstedii)* 126-130, 172
Red-bearded saki *(Pithecia monachus)* 136-138, 172
Red-bellied lemur *(Lemur rubriventer)* 72
Red-bellied monkey *(Cercopithecus erythrogaster)* 260, 270
Red-bellied tamarin *(Saguinus labiatus)* 193, 198
Red colobus *(Procolobus badius)* 317-318, 322
Red-eared monkey *(Cercopithecus erythrotis)* 260
Red-faced macaque *(Macaca fuscata)* 208, 216, 234, 286-295
Red guenon *(Erythrocebus patas)* 281-283, 284
Red-handed tamarin *(Saguinus midas)* 193, 200
Red howler *(Alouatta seniculus)* 144, 148, 174
Red titi *(Callicebus moloch)* 131, 172
Red uakari *(Cacajao calvus rubicundus)* 139, 142
Redtail monkey *(Cercopithecus ascanius)* 260, 268
Rhesus monkey *(Macaca mulatta)* 213-216, 234
Rhinopithecus See *Pygathrix*
Ring-tailed lemur *(Lemur catta)* 50-52, 72
Ruffed lemur *(Varecia variegata)* 57-58, 72

Sacred baboon *(Papio hamadryas)* 239, 258
Saddle-backed tamarin *(Saguinus fuscicollis)* 193, 198
Saguinus bicolor (pied tamarin) 192, 200
Saguinus fuscicollis (brown-headed or saddleback tamarin) 193, 198
Saguinus geoffroyi (Geoffroy's tamarin) 193, 200
Saguinus imperator (emperor tamarin) 183, 193, 198
Saguinus inustus (mottle-faced tamarin) 193, 200
Saguinus labiatus (red-bellied tamarin) 193, 198
Saguinus leucopus (white-footed tamarin) 193, 200
Saguinus midas (red-handed tamarin) 193, 200
Saguinus mystax (moustached tamarin) 193, 198
Saguinus nigricollis (black-and-red tamarin) 193, 198
Saguinus oedipus (cotton-top tamarin) 193, 196-200
Saimiri oerstedii (red-backed squirrel monkey) 126-130, 172

APPENDIX

Saimiri sciureus (common squirrel monkey or squirrel monkey) 126–130, 172
Savanna monkey *(Cercopithecus aethiops)* 275–279, 284
Screaming hairy armadillo *(Chaetophractus vellerosus)* 622
Senegal bush baby *(Galago senegalensis)* 78–80, 94
Siamang *(Hylobates syndactylus)* 325, 350, 354
Silky anteater *(Cyclopes didactylus)* 596–597, 620
Silvered leaf monkey *(Presbytis cristata)* 309–310, 320
Silvery gibbon *(Hylobates lar moloch)* 350–351, 354
Silvery marmoset *(Callithrix argentata)* 188, 190, 198
Sivapithecus 493
Six-banded armadillo *(Euphractus sexcinctus)* 622, 624
Slender loris *(Loris tardigradus)* 77–78, 94
Sloths (Pilosa) 597–612
Slow loris *(Nycticebus coucang)* 77, 94
Smooth-tailed tree shrews *(Dendrogale melanura, D. murina)* 10
Soft armadillos (Dasypodini) 614–615
Sooty mangabey *(Cercocebus torquatus)* 258, 262–263
Southern black and white colobus *(Colobus polykomos)* 322
Southern lesser long-nosed armadillo *(Dasypus hybridus)* 620
Southern tamandua *(Tamandua tetradactyla)* 594, 620
Southern three-banded armadillo *(Tolypeutes matacus)* 618, 622
Spectral tarsier *(Tarsius spectrum)* 97, 102
Sportive lemur *(Lepilemur mustelinus)* 58–62, 74
Sportive lemurs (Lepilemuridae) 58–62
Spot-nosed monkey *(Cercopithecus nictitans)* 260, 268–269, 271
Squirrel monkey (Saimiriinae) 126–130
Stump-tailed macaque *(Macaca arctoides)* 230–231, 236
Sulawesi tarsier *(Tarsius spectrum)* 97, 102
Sunda Island leaf monkey *(Presbytis aygula)* 313, 320
Swamp guenon *(Allenopithecus nigroviridis)* 280–281, 284
Sykes monkey *(Cercopithecus albogularis)* 260, 268–269
Talapoin monkey *(Miopithecus talapoin)* 280, 284

Tamandua mexicana (northern tamandua) 594, 620
Tamandua tetradactyla (southern tamandua or yellow tamandua) 594, 620
Tantalus monkey *(Cercopithecus aethiops tantalus)* 275
Tarsius bancanus (western tarsier) 97, 102
Tarsius spectrum (spectral celebes or Sulawesi tarsier) 97, 102
Tarsius syrichta (Philippine tarsier) 97, 102
Temminck's (ground) pangolin *(Manis [Smutsia] temmincki)* 631, 638
Theropithecus gelade (gelada) 252–257, 258
Thick-tailed bush baby *(Galago crassicaudatus)* 78, 94
Three-pointed pangolin *(Manis [Phataginus] tricuspis)* 634, 638
Tibetan stump-tailed macaque *(Macaca thibetana)* 221–222, 234
Tolypeutes matacus (southern three-banded armadillo) 618, 622
Tolypeutes tricinctus (Brazilian three-banded armadillo) 622
Tonkean (black ape) macaque *(Macaca tonkeana)* 236
Tonkin snub-nosed monkey *(Pygathrix [Rhinopithecus] avunculus)* 315, 322
Toque macaque *(Macaca sinica)* 218, 220–221, 234
Tree shrews (Scandentia) 2–12
Tupaia (tree shrews) 6, 10
Two-foed sloths (Choloepidae) 606–612

Uakari *(Cacajao calvus)* 139–142, 174
Unau *(Choloepus didactylus)* 606, 612, 620
Urogale everetti (Philippine tree shrew) 6, 10

Varecia veriegata (ruffed lemur) 57–58, 72
Verreaux's sifaka *(Propithecus verreauxi)* 63–69, 74
Vervet monkey *(Cercopithecus pygerythrus)* 275, 284

Weasel lemur *(Lepilemur mustelinus)* 58–62, 74
Wedge-capped capuchin *(Cebus nigrivittatus)* 160, 174
Weeper capuchin *(Cebus nigrivittatus)* 160, 174

Western needle-clawed bush baby *(Galago elegantulus)* 80, 94
Western tarsier *(Tarsius bancanus)* 97, 102
White-bearded lemur *(Lemur leucomystax)* 52–56
White-browed gibbon *(Hylobates hoolock)* 350, 354
White-cheeked gibbon *(Hylobates concolor)* 338, 350–351, 354
White-eared marmoset *(Callithrix aurita)* 188, 191, 198
White-faced capuchin *(Cebus capucinus)* 151, 152, 174
White-faced saki *(Pithecia pithecia)* 136–138, 172
White-footed tamarin *(Saguinus leucopus)* 193, 200
White-fronted capuchin *(Cebus albifrons)* 174
White-fronted leaf monkey *(Presbytis frontata)* 313, 320
White-fronted marmoset *(Callithrix geoffroyi)* 188, 191, 192, 198
White-handed gibbon *(Hylobates lar carpenteri* and *H. l. entelloides)* 351, 354
White-nosed saki *(Chiropotes albinasus)* 139, 172
White-shouldered marmoset *(Callithrix humeralifer)* 188, 190, 191, 198
White-throated guenon *(Cercopithecus albogularis)* 260, 268–269
Widow monkey *(Callicebus torquatus)* 131, 135, 172
Wolf's monkey *(Cercopithecus wolfi)* 260
Woolly lemur *(Avahi laniger)* 69, 74
Woolly spider monkey *(Brachyteles arachnoides)* 170–171, 176
Woolly monkey *(Lagothrix lagotricha)* 162–163, 176

Xenarthra 576–626

Yellow baboon *(Papio cynocephalus)* 236, 258
Yellow armadillo *(Euphractus sexcinctus)* 622, 624
Yellow tamandua *(Tamandua tetradactyla)* 594, 620
Yellow-tailed woolly monkey *(Lagothrix flavicauda)* 170, 176

Zaedypus pichiy (pichi) 622
Zinjanthropus boisei 490, 491

599 GRZ
Grzimek's encyclopedia of mammals.

DEMCO